THE PHYSIOLOGY OF DOLPHINS

THE PHYSIOLOGY OF DOLPHINS

Edited by

ANDREAS FAHLMAN
Global Diving Research SL., Valencia, Spain
Fundación Oceanogràfic de la Comunitat Valenciana, Valencia, Spain
Kolmården Wildlife Park, Kolmården, Sweden

SASCHA K. HOOKER
Sea Mammal Research Unit, Scottish Oceans Institute, University of St Andrews, St Andrews, United Kingdom

ELSEVIER

ACADEMIC PRESS
An imprint of Elsevier

Academic Press is an imprint of Elsevier
125 London Wall, London EC2Y 5AS, United Kingdom
525 B Street, Suite 1650, San Diego, CA 92101, United States
50 Hampshire Street, 5th Floor, Cambridge, MA 02139, United States
The Boulevard, Langford Lane, Kidlington, Oxford OX5 1GB, United Kingdom

ISBN 978-0-323-90516-9

For information on all Academic Press publications
visit our website at https://www.elsevier.com/books-and-journals

Cover image:

Upper image: Bottlenose dolphin "Sylvia" upon release following a brief health assessment over the West Florida Shelf in September 2022. She was given a suction-cup-attached DTAG on her back and an earring-style satellite-linked transmitter on her dorsal fin for investigation of her diving behavior over periods of 24 hours and about two months, respectively, before the tags were released. Photo by the Chicago Zoological Society's Sarasota Dolphin Research Program, taken under NMFS Scientific Research Permit No. 20455.

Lower images: Common dolphin (left) and Atlantic white-sided dolphin (middle), North Atlantic; Commerson's dolphins (right), Falkland Islands. Photos by Sascha K. Hooker.

Publisher: Nikki P. Levy
Acquisitions Editor: Simonetta Harrison
Editorial Project Manager: Himani Dwivedi
Production Project Manager: Kumar Anbazhagan
Cover Designer: Matthew Limbert

Typeset by STRAIVE, India

Dedication

We would like to dedicate this book to Dr. Sam H. Ridgway (1936–2022), one of the great leaders in the field of bottlenose dolphin physiology. Sam was a veterinarian and scientist, and a founder of the US Navy's Marine Mammal Program. Affectionately known as "the Dolphin Doctor," he produced a huge volume of work on cetacean neuroscience, anatomy, and physiology. Sam's legacy can be seen throughout the chapters of this book.

Photo credit: National Marine Mammal Foundation/Shawn Houllis.

Contents

9. Neurophysiology

PETER F. COOK, STEFAN HUGGENBERGER, AND BRUNO COZZI

10. Sensory physiology in delphinids

FREDERIKE D. HANKE, T. ARAN MOONEY, AND VINCENT M. JANIK

11. Kidneys and osmoregulation

RUDY M. ORTIZ AND MIWA SUZUKI

12. Reproductive physiology of dolphins

SHANNON ATKINSON

13. Immunology

SYLVAIN DE GUISE, MILTON LEVIN, TRACY A. ROMANO, AND LAURA A. THOMPSON

14. Human impacts on dolphins: Physiological effects and conservation

RANDALL S. WELLS AND ANDREAS FAHLMAN

This book has a companion website hosting complementary materials. Visit this URL to access it: https://www.elsevier.com/books-and-journals/book-companion/9780323905169.

Contributors

Stephanie K. Adamczak Department of Ecology & Evolutionary Biology, Institute of Marine Sciences, University of California, Santa Cruz, CA, United States

Danielle S. Adams Department of Biology, Clemson University, Clemson, SC, United States

Shannon Atkinson Department of Fisheries, College of Fisheries and Ocean Sciences, University of Alaska Fairbanks, Juneau, AK, United States

Ashley M. Blawas Division of Marine Science and Conservation, Nicholas School of the Environment, Duke University Marine Laboratory, Beaufort, NC, United States

J. Chris McKnight Sea Mammal Research Unit, Scottish Oceans Institute, University of St Andrews, St Andrews, Scotland, United Kingdom

Peter F. Cook Department of Psychology, New College of Florida, Sarasota, FL, United States

Daniel P. Costa Department of Ecology & Evolutionary Biology, Institute of Marine Sciences, University of California, Santa Cruz, CA, United States

Bruno Cozzi Department of Comparative Biomedicine and Food Science, University of Padova, Padova, Italy

Randall W. Davis Department of Marine Biology, Texas A&M University, Galveston, TX, United States

Sylvain De Guise Department of Pathobiology and Veterinary Science, University of Connecticut, Storrs, CT, United States

Zeljko Dujic Department of Integrative Physiology, University of Split School of Medicine, Split, Croatia

Andreas Fahlman Fundación Oceanogràfic de la Comunitat Valenciana; Global Diving Research SL., Valencia, Spain; Kolmården Wildlife Park, Kolmården, Sweden

Arina B. Favilla Department of Ecology & Evolutionary Biology, Institute of Marine Sciences, University of California, Santa Cruz, CA, United States

Frank E. Fish Department of Biology, West Chester University, West Chester, PA, United States

William T. Gough Hopkins Marine Station of Stanford University, Pacific Grove, CA, United States

Frederike D. Hanke Neuroethology, Institute for Biosciences, University of Rostock, Rostock, Germany

Sascha K. Hooker Sea Mammal Research Unit, Scottish Oceans Institute, University of St Andrews, St Andrews, Scotland, United Kingdom

Stefan Huggenberger Institute of Anatomy and Clinical Morphology, Witten/Herdecke University, Witten, Germany

Vincent M. Janik Scottish Oceans Institute, School of Biology, University of St. Andrews, St. Andrews, United Kingdom

Gerald L. Kooyman Scripps Institution of Oceanography, La Jolla, CA, United States

Milton Levin Department of Pathobiology and Veterinary Science, University of Connecticut, Storrs, CT, United States

Birgitte I. McDonald Moss Landing Marine Laboratories at San Jose State University, Moss Landing, CA, United States

T. Aran Mooney The Sensory Ecology and Bioacoustics Lab, Biology Department, Woods Hole Oceanographic Institution, Woods Hole, MA, United States

Rudy M. Ortiz Department of Molecular & Cell Biology, School of Natural Sciences, University of California, Merced, CA, United States

Alexander Patrician Centre for Heart, Lung and Vascular Health, University of British Columbia, Okanagan campus, Kelowna, BC, Canada

Marina Piscitelli-Doshkov Department of Zoology, The University of British Columbia, Vancouver, BC, Canada; Department of Husbandry, North Carolina Aquarium on Roanoke Island, Manteo, NC, United States

Paul J. Ponganis Scripps Institution of Oceanography, La Jolla, CA, United States

Tracy A. Romano Research Department, Mystic Aquarium, Mystic; Department of Marine Sciences, University of Connecticut, Groton, CT, United States

Jason A. Somarelli Department of Medicine, Duke University Medical Center, Durham, NC, United States

Miwa Suzuki Department of Marine Science and Resources, College of Bioresource Sciences, Nihon University, Fujisawa, Kanagawa, Japan

Laura A. Thompson Research Department, Mystic Aquarium, Mystic; Department of Marine Sciences, University of Connecticut, Groton, CT, United States

Randall S. Wells Chicago Zoological Society's Sarasota Dolphin Research Program, c/o Mote Marine Laboratory, Sarasota, FL, United States

Terrie M. Williams Department of Ecology and Evolutionary Biology, University of California—Santa Cruz, Santa Cruz, CA, United States

Acknowledgments

A great many thanks are due to all our students and research collaborators who have been instrumental to our research efforts, and particularly the veterinarians and animal care specialists whose dedication and expertise have been vital for much of the work presented. We also thank our various institutional support structures that have enabled this work to take place, and the many funding agencies that have supported these efforts. In particular, the Office of Naval Research has supported the development of instrumentation and the physiological, functional, and energetic studies discussed in several chapters.

This book has taken a while to come to fruition, partly because of the pressures on us all with global pandemics, lockdowns, and the sudden need to home-school and home-work. We would like to thank all our chapters' incredible authors for their ongoing endeavors in the field of cetacean physiology and particularly for their time and diligence in preparing these chapters. In addition, chapter authors helped review other chapters, for which we are very grateful.

Support for illustrations has come in large part from Uko Gorter. However, several other illustrations have been taken from the book *Anatomy of Dolphins. Insights Into Body Structure and Function*, by Bruno Cozzi, Stefan Huggenberger, and Helmut Oelschläger. Helmut Oelschläger was an outstanding pioneer of comparative research on the nervous system of whales and dolphins, and his wife Jutta Oelschläger provided inspirational drawings that remain among the best examples of functional representation of the nervous system.

We also acknowledge the multidecadal collaborative research efforts that are often involved in the generation of this research, supported by a multitude of funding agencies. For example, since its establishment in the 1970s by Blair Irvine, Randall Wells, and Michael Scott, the Sarasota Dolphin Research Program has benefitted from the efforts of hundreds of staff members, research colleagues, graduate students, postdoctoral fellows, student interns, veterinarians, volunteers, and citizen scientists. Primary funding over the decades has come from Earthwatch Institute, Disney Conservation Fund, NOAA Fisheries, Dolphin Quest, Inc., Mote Scientific Foundation, and the Charles & Margery Barancik Foundation.

We have dedicated this book to Sam Ridgway, but there are many other "greats" in the world of dolphin physiology, too numerous to mention them all. However, we would like to specifically acknowledge Richard Harrison, who was strongly influential in Sam's and many others' early careers, and David Denison, who collaborated with Jerry Kooyman on the importance of strengthening the terminal airway, work that laid the foundation for understanding the explosive expirations of dolphins. Per Scholander laid the foundation of our current knowledge of diving physiology with his seminal publication in 1940, which summarized the physiological responses associated with diving. His work on alveolar collapse and his work in collaboration with Laurence Irving on the dive response continue to inspire current research in this area.

CHAPTER

1

Studying dolphin physiology

Sascha K. Hooker[a] *and Andreas Fahlman*[b,c,d]

[a]Sea Mammal Research Unit, Scottish Oceans Institute, University of St Andrews, St Andrews, Scotland, United Kingdom [b]Fundación Oceanogràfic de la Comunitat Valenciana, Valencia, Spain [c]Kolmården Wildlife Park, Kolmården, Sweden [d]Global Diving Research SL., Valencia, Spain

Abstract

How do mammals manage to live underwater? To us their environment presents many challenges: it is often cold, conductive, viscous, murky, saline, high pressured, and most critically—devoid of air. Aquatic mammals therefore need various physiological adaptations to cope with these environmental conditions. This book examines the state of knowledge of dolphin and whale organs and organ system functions. Of these species, the bottlenose dolphin features prominently largely due to its cosmopolitan and nearshore distribution which has enabled long-term field studies around the world. However, much of our knowledge, particularly of their physiology, has required in-depth analyses that could only be achieved from hands-on research. Here, we present a brief overview of how data are acquired—from studies of beach-cast cadavers, studies of animals in managed care, field studies catching and releasing wild animals, to studies applying instrumentation to monitor behavior of animals.

Dolphins are the iconic ambassadors of the oceans, with bodies, organs, and cells superbly adapted for their aquatic life. Humans have been fascinated by them for centuries, likely in part because of their similarities to us as fellow mammals but also because of their differences from us in terms of their ability to thrive in the marine environment where we cannot. The underwater world they inhabit can be cold, as water conducts heat 25 times as well as air; it is viscous, salty, and has no air to breathe. Excursions to depth can be dark and expose inhabitants to pressure many times that at the water's surface. This book explores the unique physiology that enables this marine existence. It is presented as a collection of past and current studies on different aspects of the ecophysiology of dolphins, together with insights from other marine mammal species.

The bottlenose dolphin

The bottlenose dolphin (*Tursiops* sp.) is the most familiar and probably the best-studied cetacean, and many physiological studies have focussed on this species. It is one of 78 species of odontocetes (toothed whales) and is a member of the family Delphinidae (the group of 38 oceanic dolphin species, Fig. 1). Originally classified as a single species, there are likely several species or subspecies of *Tursiops*, which show a variety of morphological and color variations. The genus is currently separated into three species: the common bottlenose dolphin (*Tursiops truncatus*), the Indo-Pacific bottlenose dolphin (*Tursiops aduncus*), and Tamanend's bottlenose dolphin (*Tursiops erebennus*), although until 1998 all were treated as the same species (*T. truncatus*). There have been suggestions of other species such as the Burrunan dolphin (*T. australis*), although taxonomic difference between this and *T. truncatus* is unclear (Jedensjo et al., 2020). Four subspecies of *T. truncatus* are currently recognized: the common bottlenose dolphin (*T. truncatus truncatus*), the Black Sea bottlenose dolphin (*T. t. ponticus*), Lahille's bottlenose dolphin (*T. t. gephyreus*), and the Eastern Tropical Pacific bottlenose dolphin (*T. t. nuuanu*) (Committee on Taxonomy, 2022—see https://marinemammalscience.org/science-and-publications/list-marine-mammal-species-subspecies/). Some also propose the subspecies Pacific bottlenose dolphin (*T. t. gilli*), although this has not been formally recognized yet. Coastal and offshore ecotypes have also been noted in several locations (Hoelzel et al., 1998). Indeed, the Tamanend's bottlenose dolphin was very recently proposed for the nearshore population of the Southwestern Atlantic, differentiating this from the offshore common bottlenose dolphin (Costa et al., 2021).

The Physiology of Dolphins
https://doi.org/10.1016/B978-0-323-90516-9.00014-2

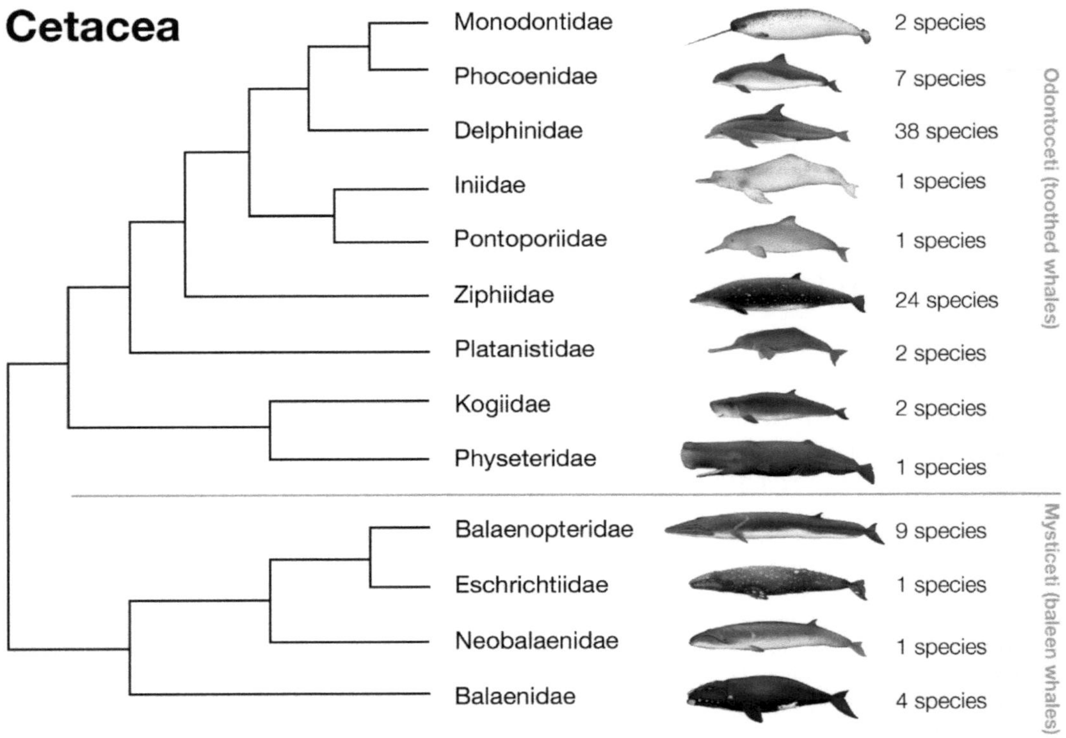

FIG. 1 Cetaceans are separated into the baleen whales (Mysticeti) containing four families and the toothed whales (Odontoceti) containing nine families. The bottlenose dolphin is part of the family Delphinidae, the most speciose of the toothed whale families. We refer the interested reader to the Society of Marine Mammalogy list of species updates: https://marinemammalscience.org/science-and-publications/list-marine-mammal-species-subspecies/ as taxonomic changes are fairly often revised. *Illustration by Uko Gorter.*

Bottlenose dolphins are one of the most cosmopolitan marine mammals, inhabiting warm and temperate seas worldwide. They are common in pelagic as well as coastal waters and are found in all but the polar oceans. They are often found nearshore, and this level of accessibility to humans is at least partially responsible for their dominance as research subjects.

Some populations of this species show long-term residency and site fidelity and usually have specific population home ranges (Fruet et al., 2014; Louis et al., 2014; Wells and Scott, 2018). The resulting geographical populations often show a great deal of variation in terms of their average length, weight, coloration, diet, and behavior. Since the 1970s, studies of individually recognizable bottlenose dolphins have been instigated for populations around the world (e.g., Argentina, Australia, Bahamas, Costa Rica, Croatia, Ecuador, Mexico, New Zealand, Portugal, Scotland, South Africa, and United States; Connor et al., 2000).

Bottlenose dolphins (and some other dolphin species) have a long history of association with humans in coastal waters (Lockyer, 1990) and appear in folklore, myths, and legends wherever people have sailed the oceans. Solitary-sociable dolphins, individuals which live apart from their own species and socialize with humans, have been observed all over the world (Nunny and Simmonds, 2019). A high proportion of these instances have involved bottlenose dolphins. It is possible that their social structure consisting of a fission-fusion society with frequently changing group membership as groups split and rejoin, together with varying food availability and loss of habitat, lead to this solitary behavior. Association with humans is then thought to have become more pronounced, progressing through stages of increasing interaction (Nunny and Simmonds, 2019). In some places, this has led to mutualism, such as in Laguna, Brazil, where bottlenose dolphins cooperate with local fishermen by driving fish toward their nets and eating the fish that escape (Pryor et al., 1990; Daura-Jorge et al., 2012).

Studies of stranded animals

Investigation of the physiology that underpins form and function generally requires hands-on access to animals. Much physiological research has therefore benefited from opportunistic access to live or deceased beach-cast animals

FIG. 2 Studies of dolphin physiology have included beach-cast cadavers, animals in managed care, captures of wild dolphins, or use of recording devices (tags) placed on wild dolphins. (A) Beach-cast bottlenose dolphin (*T. truncatus*) stranded 20 Jan 2019, Oak Island, North Carolina. (B) Bottlenose dolphin (*T. truncatus*) in managed care positioning its blow hole for flow meter respiratory measurement. (C) Wild capture of bottlenose dolphins (*T. truncatus*) for health assessment in Sarasota, Florida. (D) Atlantic spotted dolphin (*Stenella frontalis*) carrying suction cup attached D-tag, dorsal fin attached satellite transmitting tag, and dorsal fin tip attached ID tag. *(A) Photo taken under NOAA SER Stranding Agreement to the University of North Carolina Wilmington and UNCW IACUC #A1112–013; (B) Photo by Dolphin Quest Oahu; (C) Photo by the Sarasota Dolphin Research Program, taken under NMFS Scientific Research Permit No. 20455; (D) Photo taken 1 June 2022 by the Sarasota Dolphin Research Program, taken under NMFS Scientific Research Permit No. 20455.*

(Fig. 2). Post-mortem work can reveal information about morphology, disease, contaminant levels, reproductive patterns, diet, and more. Studies of live-stranded animals can be particularly informative, for example in detection of circulating bubbles in beach-cast dolphins (Dennison et al., 2012).

Many countries have long-term records and archived samples (e.g., bone, teeth, blubber) collected from dead stranded cetaceans. These can prove invaluable for comparative work. For example, anatomical analyses comparing body composition of long and deep divers to those of short and shallow diving species demonstrated less investment in metabolically expensive tissues and different muscle fiber types for long and deep divers (Pabst et al., 2016).

Studies of animals under human care

Many of the studies referred to in this book have also benefited (and continue to benefit) from the availability of animals housed in professional care. Bottlenose dolphins have been the most common cetacean species to be kept in aquaria (Defran and Pryor, 1980). They were first held in professional care in 1883 at the Brighton Aquarium in the United Kingdom and in 1914 at the New York Aquarium in the United States. The first commercial dolphinarium was opened in 1938 by Marine Studios in Florida, where it was discovered that dolphins could be trained. This effectively opened the window into these species' lives:

> "…Virtually nothing was known about the… behaviors of dolphins. Their underwater activities were effectively hidden from view, and since scientists had little comprehension of the behavioral attributes of these small toothed whales, there was neither incentive nor guidance for undertaking field studies… This situation changed rapidly when the first oceanarium, Marine Studios… opened in 1938. Here, for the first

time, scientists, along with the public could observe bottlenose dolphins at close range and for extended periods from below as well as above the surface." ***Forrest G. Wood (1986:331).***

In 1963, the Flipper movie and subsequent television series increased public appetite to see dolphins held in managed care for entertainment. However, early proliferation of facilities was tempered by new legislation (e.g., Marine Mammal Protection Act in the United States) and increasingly critical views about animal welfare, with many facilities (particularly in the United Kingdom) closing in the early 1970s. In both the United States and Russia, dolphins were also held for military purposes. These were highly trained and used to find mines or find and mark enemy divers, although they are now being phased out and replaced by less costly robotic mine hunters. Improvements in welfare, veterinary and husbandry care (high-quality food, exercise, and behavioral enrichment), and improved infrastructure of remaining facilities have led to decreased stress levels and increased lifespan, with birth and survivorship rates which can equal or exceed those of wild animals (Demaster and Drevenak, 1988; Ruiz et al., 2009; Brando, 2010; Jaakkola and Willis, 2019).

The keeping of dolphins in professional care has been vital for studies of physiology. Much of the basic information on the dolphin sensory system, the immune system, reproductive physiology, muscles and movements, and diving physiology has been collected from animals in managed care facilities. The ability to perform controlled experiments on animals that are trained and given a choice to participate in research studies is key to the study of normal physiology. These facilities allow examination of animal response to conservation and welfare issues such as reaction to noise, impact of contaminants, or the hydrodynamic impact of entanglement (e.g., Reddy et al., 1998; Branstetter et al., 2018; van der Hoop et al., 2018). They also allow development, testing, and validation of equipment before its use with wild animals, vastly reducing the trial phase and minimizing the risks to wild dolphins during field research. Advances that have come from research in marine mammal facilities could not have been achieved from wild studies and are likely crucial components for conserving these species in the wild. As will be seen throughout this book, a comprehensive understanding of these animals benefits from both in-situ and ex-situ studies.

Field studies of wild animals

Early field reports of wild dolphins were limited to descriptions of groups (e.g., Norris and Prescott, 1961), sightings of a single identifiable dolphin (e.g., Caldwell, 1955), or reports of unusual behaviors (e.g., Hubbs, 1953). Beginning in the 1970s, longitudinal studies of individual dolphins have been greatly facilitated by the development of photo-identification techniques allowing resightings of individual animals identified by scars and dorsal fin profile (Wursig and Wursig, 1977; Wells et al., 1980). For the most part, these studies have provided insights into the behavior and social structure of dolphin populations (e.g., Connor et al., 2000), and these have not been the primary source of data on dolphin physiology.

However, incorporating physiological measurements alongside these long-term longitudinal studies can prove extremely valuable—allowing the compilation of physiological data and life-history structures. One particular study, the Sarasota Dolphin Research Program (Chapter 14: Human impacts on dolphins: Physiological effects and conservation) engages in periodic catch-and-release efforts for health assessment and so has provided much information on health, survival, and reproductive rate changes over more than 50 years (Wells, 2020). This has not only improved understanding of how human impacts have affected the health and welfare of this wild population but also provided a vital control group to assess the impact of the Deep Water Horizon oil spill disaster (Schwacke et al., 2014; Venn-Watson et al., 2015; Smith et al., 2017).

Animal-attached instrumentation

In recent years, technological development of animal-attached instrumentation (Hooker et al., 2007; Wilmers et al., 2015; Fahlman et al., 2021) has begun to allow data collection from wild free-swimming animals, with tags deployed either remotely (by pole or crossbow) or attached by hand during temporary capture and restraint. Interpretation and development of tags is often assisted via calibration and validation studies using animals held in professional care.

One of the earliest attempts to attach an instrument to a whale was made by Scholander (1940) who attached a capillary manometer to a fin whale (*Balaenoptera physalus*) via harpoon, thus recording maximal dive depth before capture. Electronic time-depth recorders were then introduced but were initially used to study diving behavior of seals (Kooyman, 1965). Early studies applying these tags to dolphins mounted tags on molded saddles, or bolted tags to animal's fins (Norris et al., 1974), with trials on dusky dolphins (*Lagenorhynchus obscurus*) in Peninsula Valdes, Argentina, and bottlenose dolphins in Sarasota, United States. Radio tags provided insights into daily movements, showing that Sarasota bottlenose dolphin movements were approximately 2–5 km/h within a 40-km long home range (Irvine et al., 1981). Satellite-linked tag data from a dolphin in Tampa Bay, United States, showed movements of at least 23 km per day (Mate et al., 1995), while satellite-linked tags attached by harness to 14 bottlenose dolphins in Japan showed extensive movements at speeds of 30–40 m/min (Tanaka, 1987).

Attempts were made to record time-series of diving behavior using pole-deployed suction cup attached time-depth dataloggers. However, although successful with other dolphin species, bottlenose dolphins in New Zealand showed adverse reactions, with high-speed movements and breaching to dislodge the tags (Schneider et al., 1998). Diving studies from bottlenose dolphins have therefore come primarily from satellite-linked dive recorders, with consequent restrictions on data resolution (Klatsky et al., 2007; Fahlman et al., 2023).

High-resolution multi-sensor tags are now available which can provide data on depth, acceleration, orientation, and received sounds (Johnson and Tyack, 2003). Recent work using these tags is allowing great insights into odontocete sensory ecology (e.g., Vance et al., 2021). These suction-cup attached tags have been placed by hand on bottlenose dolphins in Sarasota Bay, and used to investigate links between animal vocalizations and agonistic movements (Casoli et al., 2022).

Animal-attached instrumentation used with trained animals in professional care allows the design of experimental tasks to probe specific research questions that examine aspects of behavior, physiology, welfare, and cognition. For instance, multi-sensor tags have been used with bottlenose dolphins in managed care to provide quantitative information on their welfare, examining different enrichment types and management schedules and the effect on subsequent behavior and activity level (Lauderdale et al., 2021a, 2021b). Thus, these new research tools provide information that can improve health and welfare of animals in professional care in addition to improving understanding of the physiology of wild populations (Fahlman et al., 2021).

Anatomy and physiology

It may be obvious to most that an animal's anatomy limits its physiological function. The bottlenose dolphin and other cetaceans possess a number of relatively unique anatomical structures (Fig. 3). There are several examples. During diving, their hinged ribs allow compression of the lung, the transitional epithelia of their airways may help to

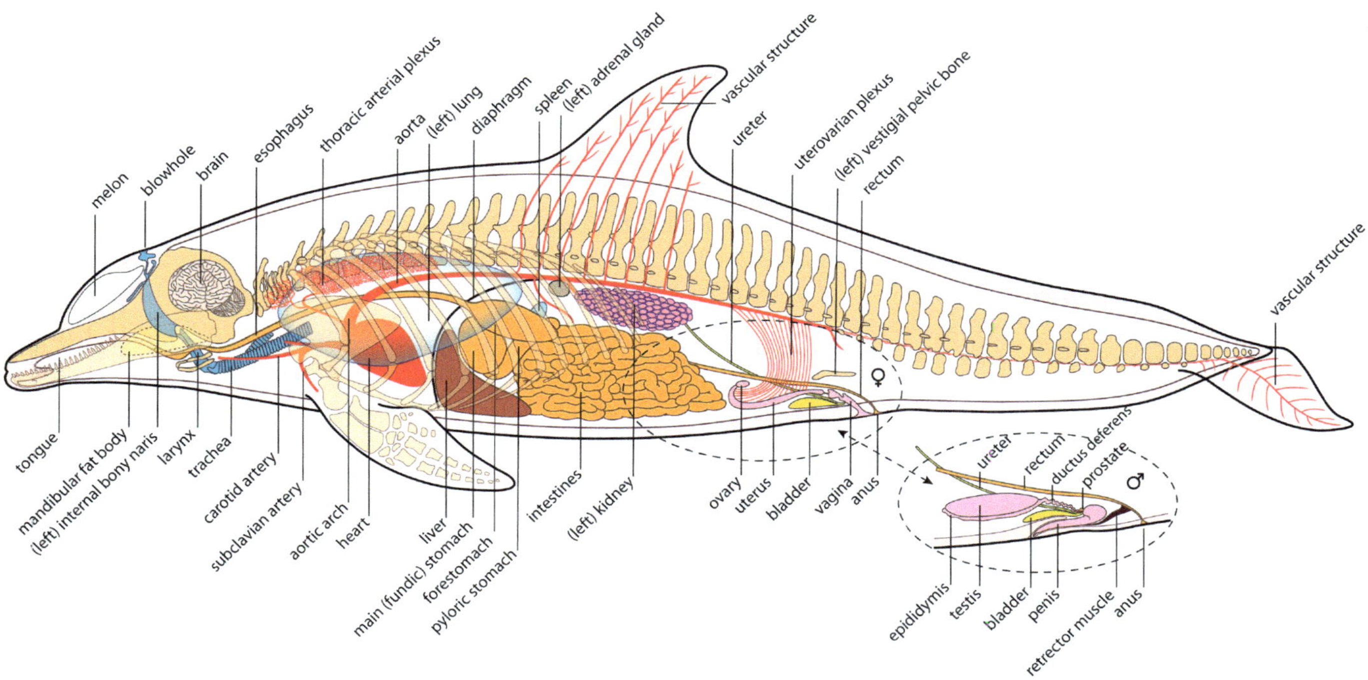

FIG. 3 Generalized anatomy of the dolphin, showing several features related to their unique physiology, for example, their circulatory system (red) including the thoracic arterial plexus (or rete) and countercurrent heat exchange systems in fluke and dorsal fin, and their respiratory system (blue) including their blowhole, air sacs above the bony naris, and their lungs beneath hinged ribs (facilitating lung collapse). *Illustration by Uko Gorter.*

prevent tracheal collapse and trauma from the pressure at depth, while their dorsal and ventral arterial and venous retes may help dampen blood pressure or prevent gas emboli from reaching the brain. As they surface, their stiff airway allows the highest known respiratory flow measured in the animal kingdom. To cope with the cold, their arterial and venous countercurrent systems in the dorsal fin and tail fluke provide thermal windows for heat exchange, and to cope with the dark, their sensory system allows dolphins to "see" at depth using echolocation. Thus, to better understand the physiology of the dolphin, it is also important to understand their anatomy. Several chapters in this book refer to anatomy in discussing physiology but we also take this opportunity to refer the interested reader to several excellent articles (Cozzi et al., 2005; Cotten et al., 2008; Cozzi et al., 2009; Costidis and Rommel, 2012; Mallette et al., 2016), books (Cozzi et al., 2017; Huggenberger et al., 2019), and book chapters (Pabst et al., 1999; Rommel et al., 2018) specifically on anatomy and function in cetaceans.

This book

The bottlenose dolphin has been, and continues to be, a model species to study cetacean physiology, providing crucial information to inform conservation actions for this and other more endangered cetaceans. A review of the IUCN Red List of Endangered Species shows how critical the current situation is for many dolphins and whales. Given that it is unlikely that we will reverse climate change anytime soon, halt ocean transport activities or control the impacts of pollution, understanding how the cetacean body functions is the essential first step to understanding and mitigating the multitude of stressors they currently face.

This book is structured as a collection of review chapters from researchers in the field of dolphin ecophysiology. Each chapter is targeted toward a specific physiological system and describes the physiological and anatomical adaptations for this system.

It begins with three chapters relating to energy expenditure. Chapter 2 looks at *the energetic costs of rest and locomotion*. The energetic costs of dolphins range from their resting metabolic rate, helping to maintain a stable temperature in water, up to sevenfold higher during swimming and diving depending on their speed, hydrodynamics, gliding and stroking mechanics. Chapter 3 looks at *thermoregulation*, examining the metabolic cost of living in the ocean and the importance of morphological, physiological, and behavioral adaptations to maintain thermal balance. Chapter 4 on *muscles and movement* considers how the cetacean body design enables maximization of thrust, minimization of drag, yet allows maneuverability required for feeding on small elusive prey.

Chapter 5 on *cardiovascular physiology* covers the heart and circulatory systems, describing some of the unusual anatomical features found in cetaceans and looking at how the heart responds to demands from diving and exercise. Chapter 6 on *respiratory physiology* looks at the adaptations found in the lungs, how gas exchange is organized for breath-holds and brief surface ventilation, and the effect of pressure on gas uptake. Chapter 7 on *diving physiology*, compares dolphin and human diving abilities, looking at the dive response, hypoxia, and the effect of pressure on blood gases. Chapter 8 on *genetic and molecular adaptations* explores how gene gain/loss, mutations, and alterations in regulatory networks underpin physiological adaptations allowing extreme tolerance of breath-hold and low oxygen levels.

Chapter 9 is dedicated to *neurophysiology*, examining the large brains of dolphins, noted for their unusual shape and convoluted cortex. This examines what we know about the central nervous system from cell type and brain structure to control of movement and sensory perception. Chapter 10 on *sensory physiology* further examines vision, hearing and touch, electroreception, and smell and taste.

Chapter 11 on *the kidneys and osmoregulation* takes a look at the multi-lobed kidneys of dolphins which help to cope with living in salt water and maintenance of electrolyte balance. Chapter 12 on *reproductive physiology* looks at sexual maturation, reproductive cycles, gestation, and lactation, including assisted reproductive and contraceptive technologies. Chapter 13 on *immunology* looks at blood health, hormones, and the immune response, and how these relate to stress, diving, and response to toxicants.

Finally, Chapter 14 on *human impacts and dolphin conservation* looks at the 50-year study of bottlenose dolphins in Sarasota, Florida. It examines how anthropogenic disturbance, such as entanglement in fishing gear, boat strikes and disturbance, pollution, and climate disruption may affect the health, survival, and reproductive success of dolphins.

References

Brando, S.I., 2010. Advances in husbandry training in marine mammal care programs. Int. J. Comp. Psychol. 23, 777–791.

Branstetter, B.K., Bowman, V.F., Houser, D.S., Tormey, M., Banks, P., Finneran, J.J., Jenkins, K., 2018. Effects of vibratory pile driver noise on echolocation and vigilance in bottlenose dolphins (*Tursiops truncatus*). J. Acoust. Soc. Am. 143, 429–439.

Caldwell, D.K., 1955. Evidence of home range of an Atlantic bottlenose dolphin. J. Mammal. 36, 304–305.

Casoli, M., Johnson, M., McHugh, K.A., Wells, R.S., Tyack, P.L., 2022. Parameterizing animal sounds and motion with animal-attached tags to study acoustic communication. Behav. Ecol. Sociobiol. 76, 59.

Connor, R.C., Wells, R.S., Mann, J., Read, A.J., 2000. The bottlenose dolphin, *Tursiops* spp.: social relationships in a fission-fusion society. In: Mann, J., Connor, R.C., Tyack, P., Whitehead, H. (Eds.), Cetacean Societies: Field Studies of Dolphins and Whales. University of Chicago Press, Chicago, pp. 199–218.

Costa, A.P.B., Fruet, P.F., Secchi, E.R., Daura-Jorge, F.G., Simoes-Lopes, P.C., Di Tullio, J.C., Rosel, P.E., 2021. Ecological divergence and speciation in common bottlenose dolphins in the western South Atlantic. J. Evol. Biol. 34, 16–32.

Costidis, A., Rommel, S.A., 2012. Vascularization of air sinuses and fat bodies in the head of the Bottlenose dolphin (*Tursiops truncatus*): morphological implications on physiology. Front. Physiol. 3, 243.

Cotten, P.B., Piscitelli, M.A., McLellan, W.A., Rommel, S.A., Dearolf, J.L., Pabst, D.A., 2008. The gross morphology and histochemistry of respiratory muscles in bottlenose dolphins, *Tursiops truncatus*. J. Morphol. 269, 1520–1538.

Cozzi, B., Bagnoli, P., Acocella, F., Costantino, M.L., 2005. Structure and biomechanical properties of the trachea of the striped dolphin *Stenella coeruleoalba*: evidence for evolutionary adaptations to diving. Anat. Rec. A Discov. Mol. Cell. Evol. Biol. 284A, 500–510.

Cozzi, B., Huggenberger, S., Oelschlager, H., 2017. Anatomy of Dolphins: Insights into Body Structure and Function. Academic Press/Elsevier Inc. ISBN: 978-0-12-407229-9.

Cozzi, B., Mazzariol, S., Podestà, M., Zotti, A., 2009. Diving adaptations of the cetacean skeleton. Open Zool. J. 2, 24–32.

Daura-Jorge, F.G., Cantor, M., Ingram, S.N., Lusseau, D., Simoes-Lopes, P.C., 2012. The structure of a bottlenose dolphin society is coupled to a unique foraging cooperation with artisanal fishermen. Biol. Lett. 8, 702–705.

Defran, R.H., Pryor, K., 1980. The behavior and training of cetaceans in captivity. In: Herman, L.M. (Ed.), Cetacean Behavior: Mechanisms and Functions. Wiley and Sons, New York, NY, pp. 247–305.

Demaster, D.P., Drevenak, J.K., 1988. Survivorship patterns in three species of captive cetaceans. Mar. Mamm. Sci. 4, 297–311.

Dennison, S., Moore, M.J., Fahlman, A., Moore, K., Sharp, S., Harry, C.T., Hoppe, J., Niemeyer, M., Lentell, B., Wells, R.S., 2012. Bubbles in live-stranded dolphins. Proc. R. Soc. Lond. B 279, 1396–1404.

Fahlman, A., Aoki, K., Bale, G., Brijs, J., Chon, K.H., Drummond, C.K., Føre, M., Manteca, X., McDonald, B.I., McKnight, J.C., Sakamoto, K.Q., Suzuki, I., Rivero, M.J., Ropert-Coudert, Y., Wisniewska, D.M., 2021. The new era of physio-logging and their grand challenges. Front. Physiol. 12, 669158.

Fahlman, A., Moore, R.T.B., Stone, R., Sweeney, J., Trainor, R.F., Barleycorn, A.A., McHugh, K., Allen, J.B., Wells, R.S., 2023. Deep diving by offshore bottlenose dolphins (*Tursiops spp.*). Mar. Mamm. Sci. 39, 1251–1266.

Fruet, P.F., Secchi, E.R., Daura-Jorge, F., Vermeulen, E., Flores, P.A.C., Simoes-Lopes, P.C., Genoves, R.C., Laporta, P., Di Tullio, J.C., Freitas, T.R.O., Dalla Rosa, L., Valiati, V.H., Beheregaray, L.B., Moller, L.M., 2014. Remarkably low genetic diversity and strong population structure in common bottlenose dolphins (*Tursiops truncatus*) from coastal waters of the Southwestern Atlantic Ocean. Conserv. Genet. 15, 879–895.

Hoelzel, A.R., Potter, C.W., Best, P.B., 1998. Genetic differentiation between parapatric 'nearshore' and 'offshore' populations of the bottlenose dolphin. Proc. R. Soc. Lond. B 265, 1177–1183.

Hooker, S.K., Biuw, M., McConnell, B.J., Miller, P.J., Sparling, C.E., 2007. Bio-logging science: logging and relaying physical and biological data using animal-attached tags. Deep-Sea Res. II Top. Stud. Oceanogr. 54, 177–182.

van der Hoop, J.M., Fahlman, A., Shorter, K.A., Gabaldon, J., Rocho-Levine, J., Petrov, V., Moore, M.J., 2018. Swimming energy economy in bottlenose dolphins under variable drag loading. Front. Mar. Sci. 5, 465.

Hubbs, C.L., 1953. Dolphin protecting dead young. J. Mammal. 34, 498.

Huggenberger, S., Oelschläger, H., Cozzi, B., 2019. Atlas of the Anatomy of Dolphins and Whales. Academic Press, Amsterdam.

Irvine, A.B., Scott, M.D., Wells, R.S., Kaufmann, J.H., 1981. Movements and activities of the Atlantic bottlenose dolphin, *Tursiops truncatus*, near Sarasota, Florida. Fish. Bull. 79, 671–688.

Jaakkola, K., Willis, K., 2019. How long do dolphins live? Survival rates and life expectancies for bottlenose dolphins in zoological facilities vs. wild populations. Mar. Mamm. Sci. 35, 1418–1437.

Jedensjo, M., Kemper, C.M., Milella, M., Willems, E.P., Krutzen, M., 2020. Taxonomy and distribution of bottlenose dolphins (genus Tursiops) in Australian waters: an osteological clarification. Can. J. Zool. 98, 461–479.

Johnson, M., Tyack, P.L., 2003. A digital acoustic recording tag for measuring the response of wild marine mammals to sound. IEEE J. Ocean. Eng. 28, 3–12.

Klatsky, L.J., Wells, R.S., Sweeney, J.C., 2007. Offshore bottlenose dolphins (*Tursiops truncatus*): movement and dive behavior near the Bermuda Pedestal. J. Mammal. 88, 59–66.

Kooyman, G.L., 1965. Techniques used in measuring diving capacities of Weddell seals. Polar Rec. 12, 391–394.

Lauderdale, L.K., Shorter, K.A., Zhang, D., Gabaldon, J., Mellen, J.D., Walsh, M.T., Granger, D.A., Miller, L.J., 2021a. Bottlenose dolphin habitat and management factors related to activity and distance traveled in zoos and aquariums. PLoS One 16, e0250687.

Lauderdale, L.K., Shorter, K.A., Zhang, D., Gabaldon, J., Mellen, J.D., Walsh, M.T., Granger, D.A., Miller, L.J., 2021b. Habitat characteristics and animal management factors associated with habitat use by bottlenose dolphins in zoological environments. PLoS One 16, e0252010.

Lockyer, C., 1990. Review of incidents of wild, sociable dolphins, worldwide. In: Leatherwood, S., Reeves, R.R. (Eds.), The Bottlenose Dolphin. Academic Press, San Diego, CA, pp. 337–353.

Louis, M., Fontaine, M.C., Spitz, J., Schlund, E., Dabin, W., Deaville, R., Caurant, F., Cherel, Y., Guinet, C., Simon-Bouhet, B., 2014. Ecological opportunities and specializations shaped genetic divergence in a highly mobile marine top predator. Proc. R. Soc. Lond. B 281, 20141558.

Mallette, S.D., McLellan, W.A., Scharf, F.S., Koopman, H.N., Barco, S.G., Wells, R.S., Pabst, D.A., 2016. Ontogenetic allometry and body composition of the common bottlenose dolphin (*Tursiops truncatus*) from the US mid-Atlantic. Mar. Mamm. Sci. 32, 86–121.

Mate, B.R., Rossbach, K.A., Nieukirk, S.L., Wells, R.S., Irvine, A.B., Scott, M.D., Read, A.J., 1995. Satellite-monitored movements and dive behavior of a bottlenose dolphin (*Tursiops truncatus*) in Tampa Bay, Florida. Mar. Mamm. Sci. 11, 452–463.

Norris, K.S., Evans, W.E., Ray, G.C., 1974. New tagging and tracking methods for the study of marine mammal biology and migration. In: Schevill, W.E. (Ed.), The Whale Problem. Harvard University Press, Cambridge, MA, pp. 395–408.

Norris, K.S., Prescott, J.H., 1961. Observations on Pacific cetaceans of Californian and Mexican waters. Univ. Calif. Publ. Zool. 63, 291–402.

Nunny, L., Simmonds, M.P., 2019. A global reassessment of solitary-sociable dolphins. Front. Vet. Sci. 5, 331.

Pabst, D.A., McLellan, W.A., Rommel, S., 2016. How to build a deep-diver: the extreme morphology of mesoplodonts. Integr. Comp. Biol. 56, 1337–1348.

Pabst, D.A., Rommel, S.A., McLellan, W.A., 1999. The functional morphology of marine mammals. In: Reynolds, J.E., Rommel, S.A. (Eds.), Biology of Marine Mammals. Smithsonian Institution Press, Washington, pp. 15–72.

Pryor, K., Lindbergh, J., Lindbergh, S., Milano, R., 1990. A dolphin-human fishing cooperative in Brazil. Mar. Mamm. Sci. 6, 77–82.

Reddy, M., Echols, S., Finklea, B., Busbee, D., Reif, J., Ridgway, S., 1998. PCBs and chlorinated pesticides in clinically healthy *Tursiops truncatus*: relationships between levels in blubber and blood. Mar. Pollut. Bull. 36, 892–903.

Rommel, S., Costidis, A.M., Lowenstine, L.J., 2018. Gross and microscopic anatomy. In: Gulland, F., Dierauf, L.A., Whitman, K.L. (Eds.), CRC Handbook of Marine Mammal Medicine. CRC Press, Boca Raton, FL, 48 pp.

Ruiz, C.L., Nollens, H.H., Venn-Watson, S., Green, L.G., Wells, R.S., Walsh, M.T., Nolan, E.C., McBain, J.F., Jacobson, E.R., 2009. Baseline circulating immunoglobulin G levels in managed collection and free-ranging bottlenose dolphins (*Tursiops truncatus*). Dev. Comp. Immunol. 33, 449–455.

Schneider, K., Baird, R.W., Dawson, S., Visser, I., Childerhouse, S., 1998. Reactions of bottlenose dolphins to tagging attempts using a remotely-deployed suction-cup tag. Mar. Mamm. Sci. 14, 316–324.

Scholander, P.F., 1940. Experimental investigations on the respiratory function in diving mammals and birds. Hvalrådets skr. 22, 1–131.

Schwacke, L.H., Smith, C.R., Townsend, F.I., Wells, R.S., Hart, L.B., Balmer, B.C., Collier, T.K., De Guise, S., Fry, M.M., Guillette, L.J., Lamb, S.V., Lane, S.M., McFee, W.E., Place, N.J., Turnlin, M.C., Ylitalo, G.M., Zolman, E.S., Rowles, T.K., 2014. Health of common bottlenose dolphins (*Tursiops truncatus*) in Barataria Bay, Louisiana, following the deepwater horizon oil spill. Environ. Sci. Technol. 48, 93–103.

Smith, C.R., Rowles, T.K., Hart, L.B., Townsend, F.I., Wells, R.S., Zolman, E.S., Balmer, B.C., Quigley, B., Ivancic, M., McKercher, W., Tumlin, M.C., Mullin, K.D., Adams, J.D., Wu, Q.Z., McFee, W., Collier, T.K., Schwacke, L.H., 2017. Slow recovery of Barataria Bay dolphin health following the Deepwater horizon oil spill (2013-2014), with evidence of persistent lung disease and impaired stress response. Endanger. Species Res. 33, 127–142.

Tanaka, S., 1987. Satellite radio tracking of bottle-nosed dolphins *Tursiops truncatus*. Nippon Suisan Gakkaishi 53, 1327–1338.

Taxonomy Co, 2022. List of Marine Mammal Species and Subspecies. Society for Marine Mammalogy. www.marinemammalscience.org. consulted on 17 Aug 2022.

Vance, H., Madsen, P.T., Aguilar de Soto, N., Wisniewska, D.M., Ladegaard, M., Hooker, S., Johnson, M., 2021. Echolocating toothed whales use ultra-fast echo-kinetic responses to track evasive prey. Elife 10, e68825.

Venn-Watson, S., Colegrove, K.M., Litz, J., Kinsel, M., Terio, K., Saliki, J., Fire, S., Carmichael, R., Chevis, C., Hatchett, W., Pitchford, J., Tumlin, M., Field, C., Smith, S., Ewing, R., Fauquier, D., Lovewell, G., Whitehead, H., Rotstein, D., McFee, W., Fougeres, E., Rowles, T., 2015. Adrenal gland and lung lesions in Gulf of Mexico common bottlenose dolphins (*Tursiops truncatus*) found dead following the Deepwater horizon oil spill. PLoS One 10, e0126538.

Wells, R.S., 2020. The Sarasota dolphin research program in 2020: celebrating 50 years of research, conservation, and education. Aquat. Mamm. 46, 502–503.

Wells, R.S., Irvine, A.B., Scott, M.D., 1980. The social ecology of inshore odontocetes. In: Herman, L.M. (Ed.), Cetacean Behavior: Mechanisms and Functions. John Wiley & Sons, New York, pp. 263–317.

Wells, R.S., Scott, M.D., 2018. Bottlenose dolphin, *Tursiops truncatus*, common bottlenose dolphin. In: Wursig, B., Thewissen, J.G.M., Kovacs, K.M. (Eds.), Encyclopedia of Marine Mammals. Academic Press/Elsevier, San Diego, CA, pp. 118–125.

Wilmers, C.C., Nickel, B., Bryce, C.M., Smith, J.A., Wheat, R.E., Yovovich, V., 2015. The golden age of bio-logging: how animal-borne sensors are advancing the frontiers of ecology. Ecology 96, 1741–1753.

Wood, F.G., 1986. Social behavior and foraging strategies of dolphins. In: Schusterman, R.J., Thomas, J.A., Wood, F.G. (Eds.), Dolphin Cognition and Behavior: A Comparative Approach. Lawrence Erlbaum Associates, Hillsdale, NJ, pp. 331–334.

Wursig, B., Wursig, M., 1977. Photographic determination of group-size, composition, and stability of coastal porpoises (*Tursiops truncatus*). Science 198, 755–756.

CHAPTER

2

Energetic costs of rest and locomotion in dolphins

Terrie M. Williams[a] *and Randall W. Davis*[b]

[a]Department of Ecology and Evolutionary Biology, University of California—Santa Cruz, Santa Cruz, CA, United States [b]Department of Marine Biology, Texas A&M University, Galveston, TX, United States

Abstract

Cetaceans, like other mammals, maintain energy balance for growth and reproduction through the regulation of metabolic processes. Energy obtained from the catabolism of food supports these processes, which in dolphins and whales can range sevenfold from rest to maximum exertion. In this chapter, we focus on the energetic costs of dolphins, discussing their metabolic versatility. On the low end of their metabolic range, dolphins maintain an elevated resting metabolic rate (RMR) characteristic of most marine mammals. At the opposite end are the high metabolic costs of swimming and diving that are dictated by speed, hydrodynamics, gliding, and stroking mechanics. Here, we provide allometric comparisons for RMR, swimming metabolism, cost of transport, and field metabolic rate (FMR) for cetaceans ranging in size from 30 kg harbor porpoises to 5000 kg killer whales. Importantly, these mass-specific relationships for dolphins and whales can be used to estimate the potential effects of environmental change and anthropogenic disturbances on energetics, performance, and ultimately the fitness of individuals and populations.

Introduction

Like all animals, dolphins and whales require energy for physiological functions, activity, and reproduction. Usable energy comes from underlying cellular processes and occurs in the form of adenosine triphosphate (ATP), an organic molecule produced primarily from the aerobic catabolism of dietary lipids, proteins, and carbohydrates. The amount of ATP produced depends on metabolic demand, which ranges from resting requirements essential for supporting basic physiological functions to maximum levels associated with high-performance locomotion. This range in metabolic capacity differs between marine and terrestrial mammals as a consequence of the physical characteristics of the environments in which they live (Williams, 2022).

To accommodate the energetic demands of marine living while enhancing locomotory performance and efficiency, Cetacea rely on a suite of morphologic, physiologic, and behavioral adaptations. This begins with adaptive thermogenesis resulting in an elevated resting metabolism for maintaining a stable core body temperature in water. Dolphins and whales are endothermic and homeothermic, which means that they maintain an elevated and constant core body temperature (~37°C) through regulated heat production and physiological thermoregulation (see Chapter 3: Thermoregulation). Superimposed on this energetic baseline are the metabolic costs of activity. Despite energetic challenges caused by elevated hydrodynamic drag when swimming, especially at high speeds, Cetacea can move efficiently through water. This cost-efficient use of energy relies on behavioral and morphological adaptations, which include the integration of different swimming gaits, specialized morphologies for enhanced body streamlining, and buoyancy control. Together, these benchmarks of metabolism from rest to cost-efficient activity provide the basic energetic metrics needed to estimate the total energetic cost of daily living, referred to as the field metabolic rate (FMR), which is vital for understanding the resource needs and survival of different cetacean species.

Resting metabolic rate: An energetic baseline for aquatic living

The resting metabolic rate (RMR) of an animal is the energy expended for (1) cellular metabolism and replacement, (2) physiological processes (e.g., respiration, blood circulation, hepatic function, osmoregulation), and (3) nervous

The Physiology of Dolphins
https://doi.org/10.1016/B978-0-323-90516-9.00001-4

system function. Adult mammals are in a resting metabolic state when they are sedentary, post-absorptive, thermally neutral, and not reproductively active (in estrous, pregnant, or lactating). For terrestrial mammals, this is often termed the basal metabolic rate (BMR; Kleiber, 1975). Although the measurement conditions are similar for marine mammals, there is an additional consideration as resting can occur when the animal is on the water's surface breathing or submerged and breath-holding. To allow for comparative analyses with terrestrial mammals, we define RMR as the level of metabolism measured when the marine mammal is sedentary and quietly breathing on the water's surface. This definition avoids complications resulting from localized metabolic adjustments associated with reduced heart rate (i.e., bradycardia) and altered perfusion of the liver and kidneys during prolonged apnea (breath-holding) when submerged.

The standard method for determining the RMR of dolphins and other small Cetacea is by measuring the rate of oxygen consumption ($\dot{V}O_2$; mL O_2 min^{-1} kg^{-1}) (Fig. 1). This method requires trained animals to rest quietly on the water's surface while breathing inside a metabolic dome. Air is pumped through the dome at a known rate, and

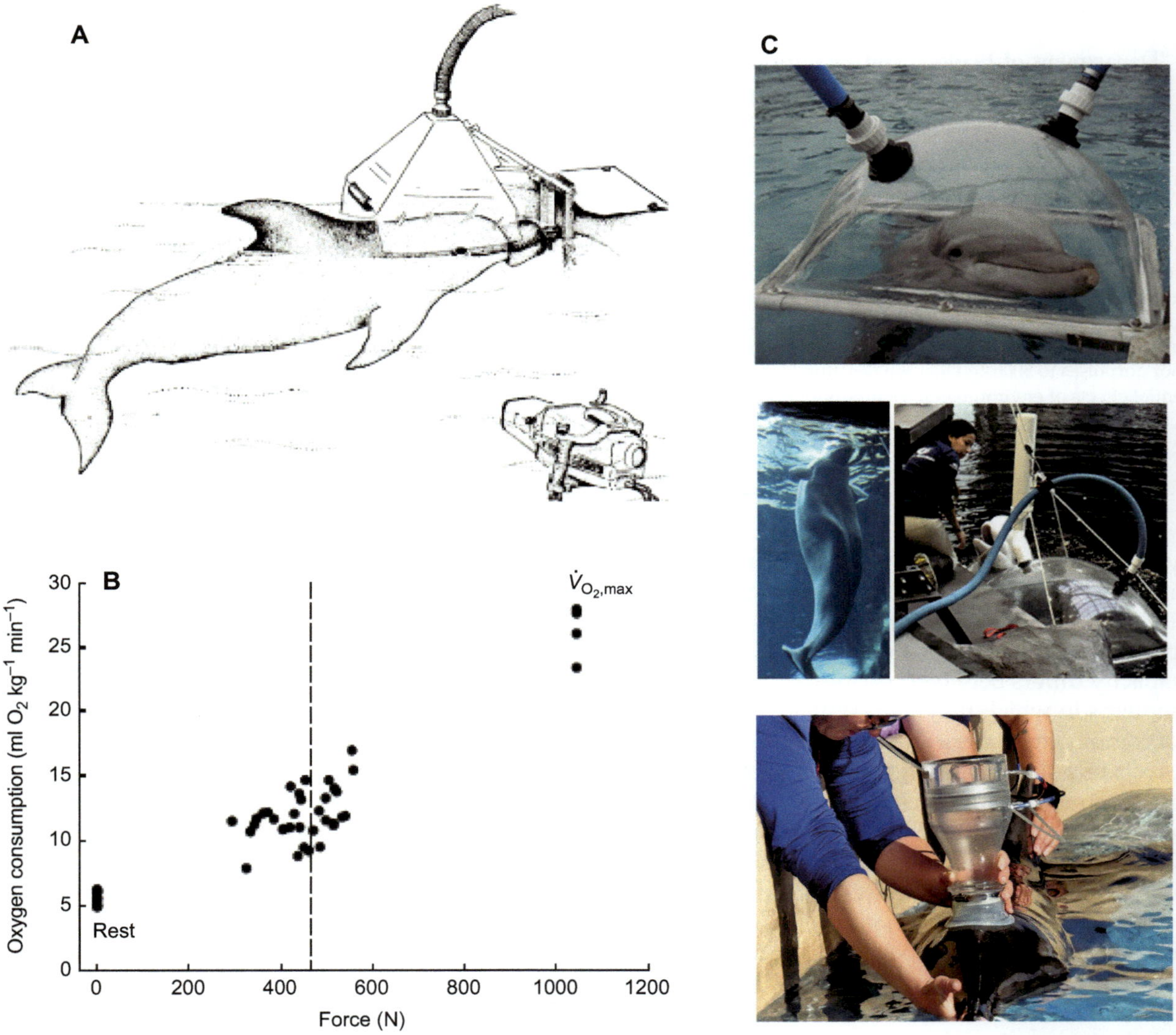

FIG. 1 Methods for measuring resting and active oxygen consumption in dolphins and whales. (A) Experimental design to simultaneously record stroking mechanics, force production, and oxygen consumption using open-flow respirometry with a trained bottlenose dolphin. The exercising dolphins push against a load cell as expired air is collected in a metabolic hood mounted overhead. Stroking movements can be recorded with submersible cameras (shown) or by the animal wearing an accelerometer tag. A load cell mounted on the wall monitors the force production of each stroke. (B) Results for oxygen consumption as a function of force production. The vertical dashed line indicates the preferred stroking forces of adult dolphins. (C) Images of a bottlenose dolphin (upper) and beluga whale (middle) breathing in a metabolic dome for open-flow respirometry tests, and a rough-toothed dolphin breathing into a pneumotachometer for breath-by-breath analyses (lower). *Redrawn from Williams et al. (2017b).*

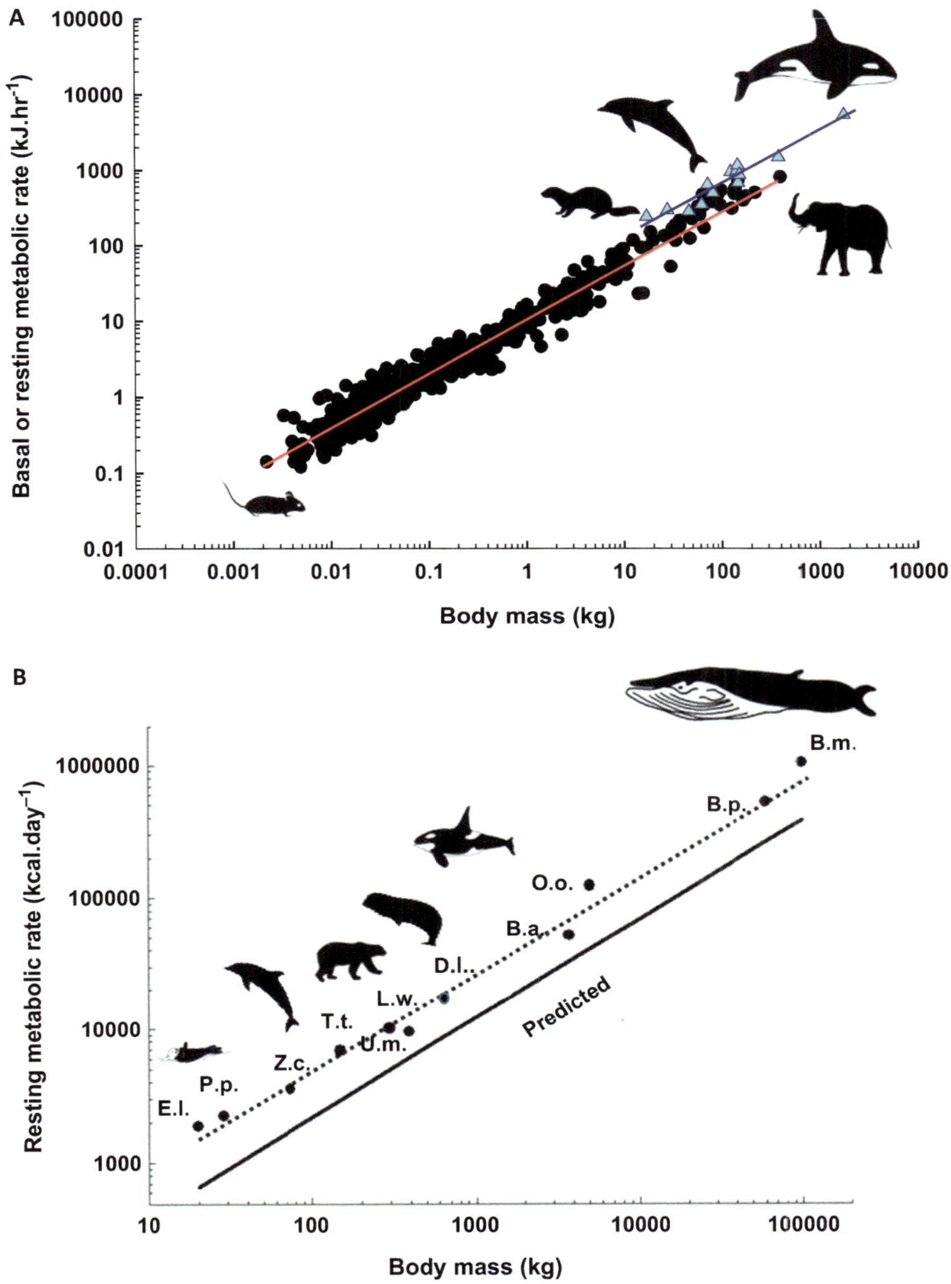

FIG. 2 Basal and resting metabolic rates of mammals as a function of body mass. (A) 538 species of eutherian terrestrial mammals (circles and red line described by Eq. 1) are compared to 15 species of marine mammals (triangles and blue line). (B) Comparison of BMR for 48 species of terrestrial carnivorans (Family Carnivora, "predicted" solid line as described by Eq. 2) and RMR measured for 15 species of marine mammals (circles, dashed line as described by RMR = 26.9Mass$^{0.69}$ where RMR is in kJ h^{-1} and body mass is in kg). Measured RMR for representative marine mammals graphed here in order of smallest to largest: sea otter (*Enhydra lutris*, 18 kg), harbor porpoise (*Phocoena phocoena*, 28 kg), California sea lion (*Zalophus californianus*, 73 kg), bottlenose dolphin (*T. truncatus*, 149 kg), polar bear (*Ursus maritimus*, 235 kg), Weddell seal (*Leptonychotes weddellii*, 389 kg), beluga whale (*Delphinapterus leucas*, 758 kg), and killer whale (*Orcinus orca*, 1800 kg). Estimates for the largest marine mammals, the baleen whales (B.a., minke whale; B.p., fin whale; B. m., blue whale), from Lockyer (1981) are provided for comparison and not included in the regression statistics. *Data adapted from Fedak and Anderson (1982), Irvine (1983), McNab (2008), Reed et al. (2000), Williams et al. (2001). Reprinted and adapted with permission from Davis (2019) and Williams and Davis (2021). The conversion factor for energetics is 1 kJ h^{-1} = 5.74 kcal day^{-1}.*

the exhaust air is analyzed for oxygen content. The decreased concentration of oxygen is then used to calculate the amount of oxygen removed from the ambient air entering the dome and to estimate the cetacean's $\dot{V}O_2$. The lowest, stable value for $\dot{V}O_2$ is the RMR (i.e., $\dot{V}O_{2rest}$), which can then be used for comparison with other marine mammals as well as the BMR of terrestrial mammals (Fig. 2; see Williams et al., 1993, 2017b for details and calibration methods). An alternative technique uses breath-by-breath spirometry with a pneumotachometer placed over the blowhole of

Cetacea (Fahlman et al., 2018; van der Hoop et al., 2018; Allen et al., 2022). Both methods have been used to estimate RMR as well as the swimming and diving metabolic rates of dolphins in zoological facilities. For active metabolic rates, the animals are trained to swim submerged before returning to the metabolic dome or pneumotach station. The post-performance $\dot{V}O_2$ initially exceeds the normal resting level but gradually returns to RMR during recovery. The additional oxygen consumed during recovery replenishes oxygen stores in the blood and muscles that were used during the previous submerged swim. The quotient of the additional oxygen consumed and the duration of submergence provides an estimate of the average metabolic rate for the dive. Both open-flow respirometry and breath-by-breath spirometry are considered indirect measurements of metabolism, as they are based on the calculated energy of oxygen consumption resulting in the production of ATP and heat.

The process of producing energy in the form of ATP from the catabolism of fuels also releases heat (Fig. 3). Here, oxygen consumption serves as a proxy for direct measurements of heat production (i.e., direct calorimetry). The latter would require monitoring core body temperature and measuring whole-body thermal conductance, a technique rarely used with large animals and impossible for marine mammals in water. The overall thermodynamic efficiency of ATP synthesis during oxidative metabolism in the mitochondria is ~40% for mammals, with the remainder of the energy producing heat. Such internal heat production (endothermy) warms the animal's body, a distinct advantage for marine species living in water with its high thermal conductivity compared with air.

As endotherms, marine mammals augment resting metabolism with *regulated* heat production through exothermic futile cycles within the cell. This occurs through the dissipation of the proton-motive force generated across the inner mitochondrial membrane to produce heat rather than support the synthesis of ATP (Wright et al., 2021; Fig. 3). As a result of this process, the overall RMR of Cetacea and other mammals is four- to eightfold greater than in ectotherms such as reptiles, amphibians, and fish. Consequently, mammals can maintain a much higher and stable core body temperature within their thermal neutral zone (the range of ambient temperatures over which resting metabolism is in balance with heat loss). Such regulated heat production is especially important for maintaining a stable core body temperature (homeothermy) for marine mammals, which spend all or much of their lives in water (see below and Chapter 3).

Predicting resting metabolic rate of dolphins and other mammals

Many factors (e.g., body mass and composition, taxonomy, habitat, diet, climate, circadian rhythm, reproductive status, age, sex, and season) may influence the BMR of terrestrial mammals and the RMR of dolphins and other marine mammals. One of the most important influencing factors is body mass. Both BMR and RMR (Y) scale allometrically with mass (M) in the general form $Y = a\,M^b$ where a is the proportionality constant and b is the mass exponent. An

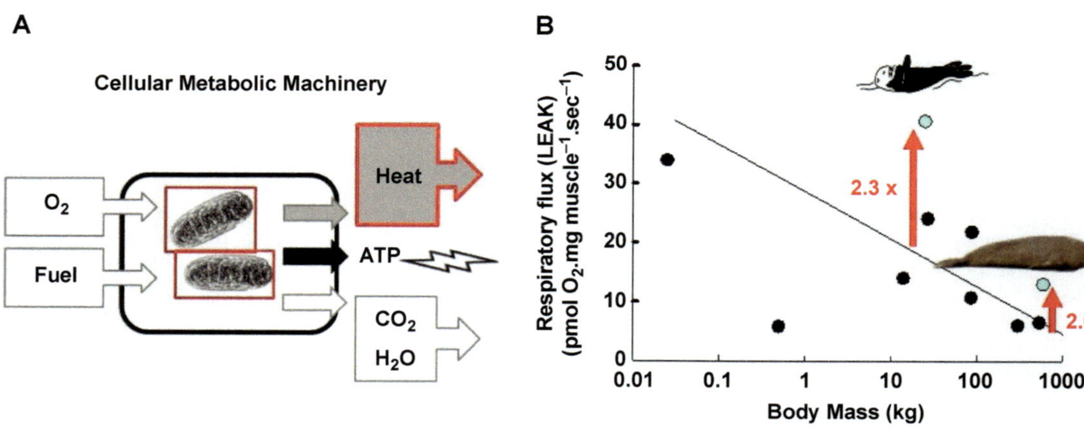

FIG. 3 Mitochondrial metabolic processes in marine mammals. (A) Pathway for ATP and heat production in mitochondria. Heat results from the oxidation of fuels (i.e., lipid, protein, carbohydrate) with the ensuing production of ATP, carbon dioxide, and water. (B) Respiratory flux (leak) as a function of body mass in the skeletal muscles of terrestrial mammals ranging from mice to quarter horses (black circles) compared to sea otters and elephant seals (cyan circles), which are 2.3- and 2.0-fold higher than the allometric prediction, respectively. Red arrows indicate the proportional difference between measured values for marine mammals and predicted values for terrestrial mammals (black line). *Reprinted and adapted from Williams (2022) with data from Wright et al. (2021).*

analysis of 538 species of eutherian mammals (excluding marine mammals) over a size range of 10^5 g gives the generalized equation

$$\text{BMR}\left(\text{kJ h}^{-1}\right) = 0.071\, M^{0.724} \quad (1)$$

as illustrated in (Table 1; Fig. 2A with data from McNab, 2008).

When compared to this broad group, the mean RMR of marine mammals including sea otters (*Enhydra lutris*), Pinnipedia, polar bears (*Ursus maritimus*), and Cetacea is 2.3-fold greater than the predicted BMR for terrestrial species (Davis, 2019). A further analysis of 48 species of mammals in the taxonomic order of Carnivora (excluding marine mammals) over a size range of 10^5 g gives the following generalized equation (from Tables 1 and 2; Fig. 2B with data McNab, 2008).

$$\text{BMR}\left(\text{kJ h}^{-1}\right) = 0.060\, M^{0.752} \quad (2)$$

where mass is in grams. Here the mean RMR of marine mammals is 2.0-fold greater than the predicted BMR of terrestrial carnivorans (Table 2; Fig. 2B; data from McNab, 2008).

One challenge when comparing the RMR and BMR of different species, particularly marine mammals, is the influence of body composition. Most marine mammals and all Cetacea have a blubber layer, the mass of which varies seasonally (Aarseth et al., 1999). Because most of the RMR results from metabolic processes in lean tissues, the ratio of lean-to-fat mass will alter the mass-specific RMR, which is the amount of energy needed per kilogram of body mass (reported as $\text{kJ h}^{-1}\,\text{kg}^{-1}$). Hence, body composition must be kept in mind when making comparisons of RMR among diverse species (Rea and Costa, 1992). This is particularly relevant when evaluating the metabolic demands of large Cetacea, which undergo significant seasonal changes in fat deposition and loss.

Resting metabolic rate of dolphins and other marine mammals

All aquatic mammals and birds face the same thermoregulatory challenge: the 23-fold increase in thermal conductivity of water relative to air at the same temperature. Rapid heat loss in water is especially demanding for smaller

TABLE 1 Resting metabolic rate (RMR; kJ h^{-1}) for 15 species of marine mammals relative to the allometric prediction for 538 species of eutherian terrestrial mammals as described by Eq. (1).

Species	Body mass (g)	Measured RMR (kJ h^{-1})	Predicted RMR (kJ h^{-1})	Ratio of measured to predicted RMR
Sea otter[1]	17,300	243	83	2.9
Harbor porpoise[2]	28,100	296	118	2.5
Ringed seal[3]	46,600	282	170	1.7
Harbor seal[4]	63,000	348	212	1.6
California sea lion[5]	73,000	641	236	2.7
Hawaiian monk seal[6]	82,000	499	256	1.9
Southern sea lion[7]	125,400	948	349	2.7
Steller sea lion[8]	146,000	992	389	2.5
Bottlenose dolphin[9]	148,600	1164	394	3.0
Harp seal[10]	149,500	685	396	1.7
Gray seal[11]	155,000	871	407	2.1
Polar Bear[12]	235,000	1086	549	2.0
Weddell seal[13]	388,500	1492	791	1.9
Killer whale[14]	1,800,000	5378	2990	2.1
Mean				2.3
Florida manatee[15]	428,000	239	848	0.3

Data for eutherian terrestrial mammals from McNab (2008). Data sources for marine mammal from: (1) Costa and Kooyman (1984), (2) Reed et al. (2000), (3) Ochoa-Acuña et al. (2009), (4) Davis et al. (1985), (5) Liao (1990), (6) Williams et al. (2011), (7) Dassis et al. (2012), (8) Rosen and Trites (2002), (9) Williams et al. (2001), (10) Ochoa-Acuña et al. (2009), (11) Fedak and Anderson (1982), (12) Pagano et al. (2018a), (13) Williams et al. (2001), (14) Williams et al. (2017b), and (15) Irvine (1983).
Reprinted and adapted with permission from Davis (2019).

TABLE 2 Resting metabolic rate (RMR; $kJ\,h^{-1}$) for 15 species of marine mammals relative to the allometric prediction for 49 species of terrestrial mammals in the taxonomic order Carnivora as described by Eq. (2).

Species	Body mass (g)	Measured RMR ($kJ\,h^{-1}$)	Predicted RMR ($kJ\,h^{-1}$)	Ratio measured to predicted RMR
Sea otter[1]	17,300	243	93	2.6
Harbor porpoise[2]	28,100	296	133	2.2
Ringed seal[3]	46,600	282	194	1.5
Harbor seal[4]	63,000	348	244	1.4
California sea lion[5]	73,000	641	273	2.4
Hawaiian monk seal[6]	82,000	499	297	1.7
Southern sea lions[7]	125,400	948	409	2.3
Steller sea lion[8]	146,000	992	459	2.2
Bottlenose dolphin[9]	148,600	1164	465	2.5
Harp seal[10]	149,500	685	467	1.5
Gray seal[11]	155,000	871	480	1.8
Polar bear[12]	235,000	1086	549	2.0
Weddell seal[13]	388,500	1498	958	1.6
Killer whale[14]	1,800,000	5378	3035	1.8
Mean				*2.0*
Florida manatee[15]	428,000	239	848	0.2

Data for terrestrial carnivores from McNab (2008). Data sources for marine mammals from (1) Costa and Kooyman (1984), (2) Reed et al. (2000), (3) Ochoa-Acuña et al. (2009), (4) Davis et al. (1985), (5) Liao (1990), (6) Williams et al. (2011), (7) Dassis et al. (2012), (8) Rosen and Trites (2002), (9) Williams et al. (2001), (10) Ochoa-Acuña et al. (2009), (11) Fedak and Anderson (1982), (12) Pagano et al. (2018a), (13) Williams et al. (2001), (14) Williams et al. (2017b), and (15) Irvine (1983).

species, which have a higher surface-to-volume ratio for heat transfer, and for species that live in polar regions where seawater temperatures are near the freezing point (~−1.8°C). Deep-diving species that travel to depths below 500–1000 m may also experience water temperatures near ~4°C even in temperate and tropical latitudes. To offset excessive heat loss, most marine mammals (Cetacea, Sirenia, and Pinnipedia) use blubber for thermal insulation (Davis, 2019). Indeed, there is evidence from fossilized soft tissues that ichthyosaurs had insulating blubber, indicating convergent evolution for homeothermic regulation in water (Lindgren et al., 2018). Although the terrestrial ancestors of Cetacea used fur for insulation, it was replaced by blubber as they became fully aquatic. Among marine mammals, only sea otters rely exclusively on the air trapped in fur for thermal insulation in the marine environment, whereas adult fur seals use both blubber and fur (see Chapter 3).

Neither blubber nor fur is sufficient to completely offset heat loss in submerged marine mammals that are smaller than a female killer whale. Thus, supplemental heat production is necessary to maintain a normal core body temperature. For many species of marine mammals, this is accomplished with an elevation in RMR, which is higher than predicted resting levels for terrestrial mammals of similar size (Fig. 2A and B). Among smaller Cetacea, the RMR of bottlenose dolphins (*Tursiops truncatus*, 1164 $kJ\,h^{-1}$ or 2.18 $W\,kg^{-1}$), harbor porpoises (*Phocoena phocoena*, 296 $kJ\,h^{-1}$ or 2.93 $W\,kg^{-1}$), and killer whales (*Orcinus orca*, 5378 $kJ\,h^{-1}$ or 0.83 $W\,kg^{-1}$) is 2.5-fold greater than the allometric prediction from Eq. (1) for terrestrial eutherian mammals and 2.1-fold greater than the allometric prediction from Eq. (2) for terrestrial carnivorans (Fig. 2; Tables 1 and 2). Although similar variability has been reported for resting metabolism in other eutherian and carnivorous mammalian groups (McNab, 2008), the consistency in elevated RMR of Cetacea, Pinnipedia, and sea otters is noteworthy (Davis, 2019). Among marine mammals, only herbivorous manatees (*Trichechus manatus*) (and probably dugongs, *Dugong dugon*) have an RMR that is considerably lower than predicted for terrestrial mammals (Table 1).

These comparative differences in metabolism begin at the subcellular level (Fig. 3A). Enhanced mitochondrial leak associated with uncoupling protein-1 (UCP1, thermogenin) in skeletal muscle serves as a thermogenic adaptation for living in water. This specialized biochemical pathway produces enough heat to adequately explain the elevated RMR of the smallest marine mammal, the sea otter, and is also evident in elephant seals, *Mirounga angustirostris* (Fig. 3B; Wright et al., 2021). Although further research is needed to determine the significance of such mitochondrial leak and regulated heat production in Cetacea, it likely explains the metabolic basis for an elevated RMR in most smaller species.

Overall, it is apparent that the RMR of mammals is a physiological variable that is adaptive across evolutionary time (Hart, 1957; Speakman, 1997; Lovegrove, 2005; Speakman and Król, 2010). It is influenced by factors mentioned previously (e.g., body mass and composition, taxonomy, habitat, diet, climate, circadian rhythm, and season), and is meditated in a variety of tissues. Adjusting the rate of mitochondrial leak through genetic mutation or phenotypic plasticity at the subcellular level provides an additional mechanism for heat production. Instead of the strict allometric relationship between RMR and body mass historically ascribed to mammals (as well as birds), metabolic plasticity may be fundamental for the heightened efficiency, performance, and fitness needed by mammals to adapt to varied thermal environments. In view of this, enhanced thermogenesis may have been a critical adaptation enabling the terrestrial ancestors of marine mammals to overcome the thermoregulatory challenges of aquatic life, and to eventually inhabit the world's oceans (Davis, 2019; Wright et al., 2021; Williams, 2022).

Special consideration of RMR in large whales

A question remains whether the same elevated RMR observed for smaller marine mammals occurs in the largest Cetacea. Reduced heat loss because of a lower surface-to-volume ratio (SVR) and potential for hyperthermia in a 100–150 ton blue whale may set upper metabolic limits for ocean giants (Speakman and Król, 2010). There have been no direct measurements of RMR in adult Cetacea larger than a killer whale. Hence, we do not know if the RMR of very large whales (e.g., blue whales, *Balaenoptera musculus*) is elevated as it is in smaller Cetacea. Davis (2019) examined this in a theoretical model comparing the thermal profiles of a 1.4 m long harbor porpoise with a 30 m long blue whale. Using a double cone model (i.e., two cones joined at their base to represent a fusiform body shape without flukes or flippers) and assuming a fineness ratio (body length/maximum body diameter) of 4.5 for the harbor porpoise and 7.0 for the blue whale to calculate the diameter of the base of the cones (Ahlborn et al., 2009), Davis estimated the surface area and volume of these two cetaceans of greatly different size. Assuming a body density of 0.99 kg L^{-1} allowed an estimate of body mass. The harbor porpoise in this model had a calculated surface area of 0.7 m^2, a volume of 0.035 m^3, and a body mass of 35 kg. The blue whale model had a surface area of 204 m^2, a volume of 144 m^3, and a body mass of 142,815 kg. These are realistic estimates for both animals based on their morphology. The resulting SVR is 19.8 for the harbor porpoise but only 1.4 for the blue whale, a 93% reduction. The measured RMR in a 35 kg harbor porpoise is 9.6 kJ h^{-1} kg^{-1}, which is 2.2-fold greater than the allometric prediction for carnivorans (McNab, 2008). If we use the same assumptions for a 142,815 kg blue whale, the estimated mass-specific RMR would be 1.2 kJ h^{-1} kg^{-1}, a value that is 87% less than for the harbor porpoise. Based on this calculation, the SVR decreases to a greater extent than the mass-specific RMR for the blue whale (93% vs 87%, respectively) compared to the harbor porpoise, which could pose a problem with overheating in such a large animal. This results from the difference in the mass exponent scaling factor (b) for body surface-to-volume geometry (0.66) and the scaling factor for RMR in carnivorous mammals (0.75 as in Eq. (2)), which becomes pronounced in very large Cetacea (Speakman and Król, 2010; Ballesteros et al., 2018; Davis, 2019). Alternatively, if we assume that the mass-specific RMR of a blue whale is *not* elevated over the allometric prediction, then the reduction in both SVR and RMR relative to the harbor porpoise is ~94% (i.e., 93% and 94% lower in the blue whale, respectively). Now the heat generated by the RMR closely matches the SVR necessary to lose excess heat to the environment and maintain homeothermy in the blue whale. This simple calculation indicates that the RMR of very large whales may converge with the allometric prediction based on the BMR for terrestrial carnivores. In contrast, the very high SVR of the harbor porpoise may explain the elevated RMR necessary for smaller dolphins and porpoises to achieve homeothermy.

Energetic costs of swimming and diving

Relationships between hydrodynamics, speed, and energetics

One of the defining features of dolphins is their athleticism. Fast swimming speeds, acrobatics (Fish et al., 2006), gliding, wave-riding (Williams et al., 1992, 2000), and leaping (Au and Weihs, 1980; Blake, 1983) are common for many species of Odontoceti (toothed whales). While some of these activities appear to be playful, others reduce the physical effort of moving through the water, especially at high speeds. Energetic consequences are associated with all aquatic performance, which result from hydrodynamics, the energetic cost of stroking, and physiological limits in aerobic metabolism.

The details of cetacean hydrodynamics and speed are reviewed in Chapter 4: Muscles and movement. Here, we discuss the energetic implications of the physical forces associated with hydrodynamic drag during aquatic activity.

As with other swimming vertebrates, including fish (Lee et al., 2003; Clark and Seymour, 2006; Blank et al., 2007), aquatic birds (Culik et al., 1994), Pinnipedia (seals and sea lions; Feldkamp, 1987; Williams et al., 1991), and even humans (Toussaint and Truijens, 2005), the energetic cost of swimming in Odontoceti increases exponentially with speed (Fig. 4; Davis et al., 1985; Williams and Kooyman, 1985; Williams et al., 1992; Kriete, 1995; Yazdi et al., 1999; Otani et al., 2001; Allen et al., 2022). These elevated swimming costs mirror the exponential increase in hydrodynamic drag with speed (Fish, 1998), resulting from the high (820-fold greater) density of water compared with air (Denny, 1993).

Ultimately, the exponential increase in the energy expended for swimming limits routine speeds to a relatively narrow and surprisingly similar range for most Cetacea measured to date, regardless of body size. This is evident from the changes in power, the cost per stroke (CPS), and cost of transport (COT) that occur with changes in swimming speed (Figs. 4–7). For example, the quotient of $\dot{V}O_2$ and the corresponding stroke frequency is the total energetic cost per stroke (CPS_{total}; $mL\,O_2\,stroke^{-1}\,kg^{-1}$), which includes the cost of stroking and the underlying RMR. When calculated for a range of swimming speeds, the result is a U-shaped curve, with the lowest CPS_{total} associated with routine or preferred swimming speeds ranging from $0.7\,m\,s^{-1}$ to $2.5\,m\,s^{-1}$ (Williams et al., 2017b). Even the most active dolphins and largest whales (Goldbogen et al., 2006; Gough et al., 2019) generally cruise within this energetically economical range of speeds when transiting or diving.

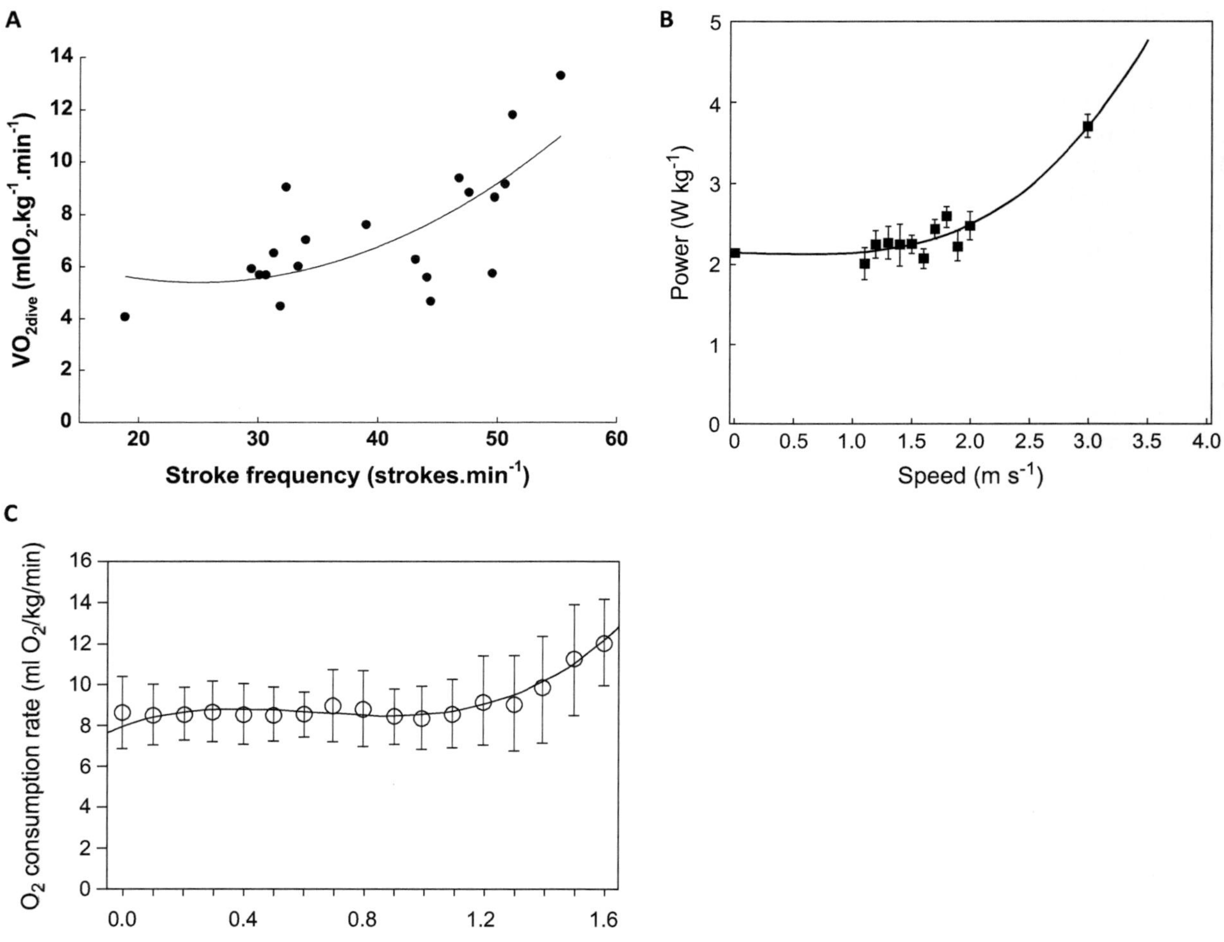

FIG. 4 Nonlinear relationships between oxygen consumption and power requirements measured for swimming dolphins and porpoises. In all three studies, the energetic demand increases exponentially with the level of exertion, and reflects the exponential increase in hydrodynamic drag with speed (see Chapter 4), hence, the challenge of moving through water at high speed. *Data from Williams et al. (2017b). (A), Yazdi et al. (1999). (B) for bottlenose dolphins, and Otani et al. (2001). (C) for harbor porpoises.*

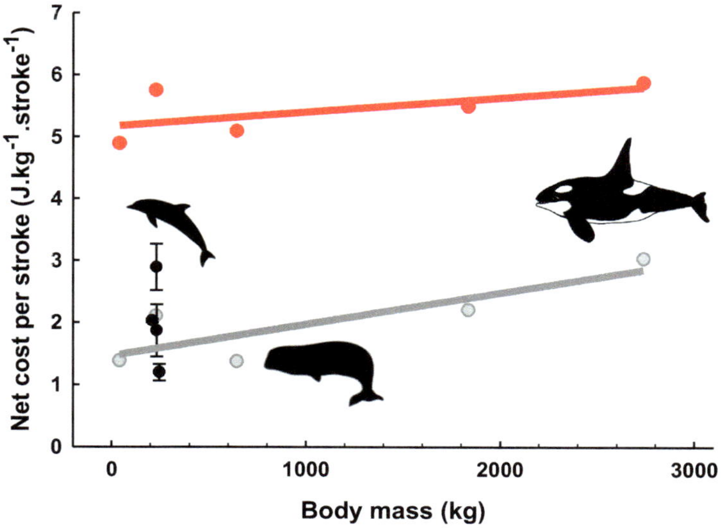

FIG. 5 The net cost per stroke (CPS_{net}) as a function of body mass for four species of Odontoceti. Measured values for bottlenose dolphins (black symbols for individual dolphins with the mean denoted by the gray symbol) are compared to calculated values for other odontocetes differing in body mass. Data for preferred or routine swimming speed (gray symbols and line) are compared to values estimated for maximum aerobic performance (red symbols and line). The solid lines are the least-squares linear regressions as described in Eqs. (4) and (5) in the text. Data points denote values for harbor porpoises (42 kg), bottlenose dolphins (229 kg), beluga whales (645 kg), and female killer whales (1837 and 2738 kg). *Reprinted from Williams et al. (2017b).*

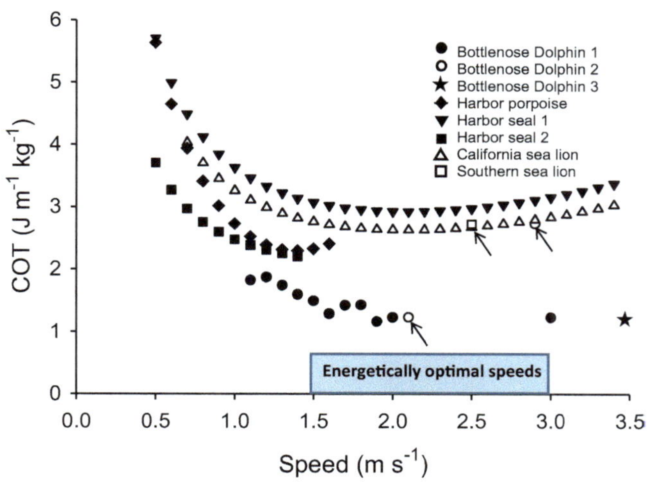

FIG. 6 Cost of transport (COT) in relation to swimming speed for five species of marine mammals. Symbols denote different studies with data adapted from bottlenose dolphin 1 (162 kg; Yazdi et al., 1999); bottlenose dolphin 2 (145 kg; Williams et al., 1992); bottlenose dolphin 3 (191 kg; van der Hoop et al., 2014); harbor porpoise (43 kg; Otani et al., 2001); harbor seal 1 (63 kg; Davis et al., 1985); harbor seal 2 (32 kg; Williams et al., 1991); California sea lion (21 kg; Williams et al., 1991); southern sea lion (113 kg; Dassis et al., 2013). The values for the harbor seals and California sea lions are based on regression equations at 0.1 m s^{-1} speed intervals (Williams et al., 1991). There was only one value for the southern sea lion, two values for dolphin 2 (denoted by arrows), and one value for dolphin 3 (star). The blue box denotes the energetically optimal swimming speed range to achieve maximum distance at the lowest energetic cost. *Reprinted and adapted with permission from Davis (2019).*

The high cost of speed

In the wild, there are many occasions when dolphins and whales move beyond their range of routine swimming speeds. Whether pursuing prey (Williams et al., 2004a; Guinet et al., 2007; Soto et al., 2008), avoiding predators (Ford et al., 2005), or evading anthropogenic threats (Todd et al., 1996; DeRuiter et al., 2013; Goldbogen et al., 2013; Williams et al., 2022), all Odontoceti are capable of short bursts of high-speed swimming (see Chapter 4). During foraging dives, 1500 kg short-finned pilot whales *(Globicephala macrorhynchus)* referred to as "cheetahs of the seas," can briefly exceed 9 m s^{-1}, a fourfold increase in speed compared to their routine efforts (Soto et al., 2008). Because the oxygen consumption (i.e., power) for such extraordinary swimming performance increases exponentially with speed (Fig. 4A–C), these efforts cannot be sustained for long periods and likely rely in part on anaerobiosis.

The aerobic power resulting from thrust production by the flukes is represented by the net cost per stroke (CPS_{net}), which excludes the underlying RMR ($\dot{V}O_{2RMR}$) (Fig. 5). It is calculated as:

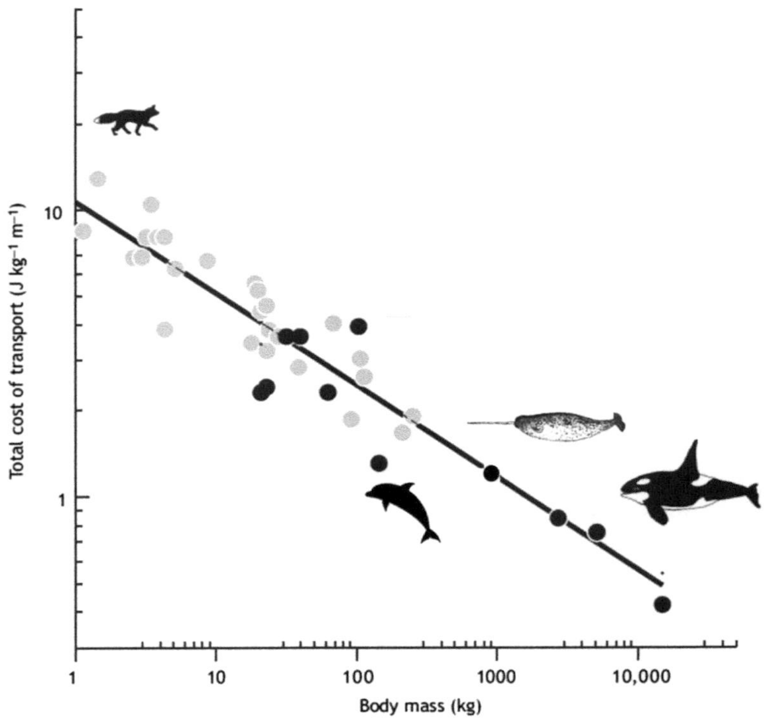

FIG. 7 Cost of transport (COT) in relation to body mass for swimming marine (black circles) and running terrestrial (gray circles) mammals. The black line is the least-squares regression for all mammals from Williams (1999) and differs little from the regression for swimming marine mammals predicted from Eq. (6) in the text. *Data are from Williams (1999) and Pagano and Williams (2021) with marine mammals represented by pinnipeds, harbor porpoises, bottlenose dolphins, polar bears, narwhals, beluga whales, and killer whales.*

$$CPS_{net} = (\dot{V}O_{2swim} - \dot{V}O_{2RMR}) \div \text{strokes min}^{-1} \tag{3}$$

where $\dot{V}O_2$ is in mL O_2 min^{-1} kg^{-1}, with CPS_{net} converted to J kg^{-1} $stroke^{-1}$ using the conversion factor 20.1 J/mL O_2^{-1} (Williams et al., 2004a).

The CPS_{net} depends on body mass (M) and the level of performance. In general, Odontoceti demonstrate a two-gear system of energetic performance corresponding to either routine swimming speed or high-stroke frequency, high-speed swimming, which is comparatively rare. At routine speeds and stroke frequencies typical of transiting and migrating, the CPS_{net} increases with M from harbor porpoises (42 kg) to killer whales (2738 kg; Fig. 5) according to the equation:

$$\text{Routine speed } CPS_{net} = 1.46 + 0.0005M \tag{4}$$

where CPS_{net} is in J kg^{-1} $stroke^{-1}$ and M is body mass in kg. In comparison, at maximum sustained speeds and aerobic performance, the CPS_{net} is described by the equation:

$$\text{Max speed } CPS_{net} = 5.17 + 0.0002M \tag{5}$$

At maximum aerobic effort, the CPS_{net} is ~3.5-fold greater than at routine speeds and may be associated with a greater stroke amplitude. Because of the non-linear relationship between oxygen consumption and speed, power, and stoke frequency (Fig. 4), small changes in stroking mechanics have little effect on energetic costs except at extreme levels of exertion. As a result, the two-gear system (Fig. 5) provides a reasonable first estimate of stroking costs in Cetacea. As we shall see in the Section below on "Accelerometry and energetics in the wild," these equations are useful for predicting the locomotory costs of free-ranging Cetacea by monitoring stroke frequency using animal-borne three-axis accelerometers in bio-logging tags. The accelerometer data, in turn, can be converted to stroking costs and then be used to predict activity costs in normal and disturbed environments.

Cost of transport

The relationship between swimming metabolic rate ($\dot{V}O_{2swim}$) and speed is referred to as the cost of transport (COT in $J\,kg^{-1}\,m^{-1}$), that is, the energetic cost (J) of moving 1 kg of body mass 1 m. The COT of individual animals is described by a U-shaped curve for a wide variety of swimmers from fish and humans to small and large marine mammals (Fig. 6). Calculated as the quotient of $\dot{V}O_{2swim}$ and swimming speed, the COT decreases to a minimum value, which scales allometrically in marine mammals as

$$\mathrm{COT} = 7.87M^{-0.29} \tag{6}$$

where COT is in $J\,kg^{-1}\,m^{-1}$ and M is mass in kg (Fig. 7). This minimum value corresponds to the speed that enables swimmers to cover the greatest distance for the least amount of energy (Schmidt-Nielsen, 1972) including bottlenose dolphins (Williams et al., 1992; Allen et al., 2022). When comparing different species of marine mammals, the decrease in minimum COT with increasing body mass results from a lower mass-specific RMR and higher optimum speeds for larger animals. For example, the mean minimum submerged COT for harbor seals (*Phoca vitulina*), California sea lions (*Zalophus californianus*), and South American sea lions (*Otaria flavescens*) with an average mass of 57 kg is $2.6\,J\,kg^{-1}\,m^{-1}$. This is similar to the minimum COT of $2.4\,J\,kg^{-1}\,m^{-1}$ for a 43 kg harbor porpoise but twofold greater than for bottlenose dolphins ($1.2\,J\,kg^{-1}\,m^{-1}$) with an average mass of 154 kg traveling at about the same swim speed ($\sim 1.8\,m\,s^{-1}$) (Figs. 6 and 7; Williams et al., 1991, 1992; Otani et al., 2001; Allen et al., 2022). Even at a swim speed of $3.6\,m\,s^{-1}$, the mean minimum submerged COT for bottlenose dolphins with an average mass of 191 kg is $1.2\,J\,kg^{-1}\,m^{-1}$ (van der Hoop et al., 2014) reflecting the energetic advantage associated with body streamlining (see Chapter 4). For the harbor seals and California sea lions, minimum COT increases a modest 16% as speed increases to $3.4\,m\,s^{-1}$, but there is little or no increase for bottlenose dolphins.

Although some variability in COT results from differences in body mass (Fig. 7), even using the body mass scaling factor of 0.75 does not eliminate all variability, which may result from swimming mode or experimental methodology. Nevertheless, the mean minimum COT for the five species of Pinnipedia and Cetacea discussed above ($2.2 \pm 0.70\,J\,kg^{-1}\,m^{-1}$) is similar to the predicted mean minimum COT of quadrupedal running mammals ($2.0 \pm 0.64\,J\,kg^{-1}\,m^{-1}$) with the same body masses (Taylor et al., 1970; Williams, 1999). Interestingly, the mean value for marine mammals is only 2.8-fold greater than the predicted minimum COT ($0.8 \pm 0.15\,J\,kg^{-1}\,m^{-1}$) for a salmon (an ectotherm), which does not devote a large part of its total locomotor budget to the maintenance of endothermy (Schmidt-Nielsen, 1972). Furthermore, at similar body masses and body temperatures scaled to a common mammalian level of 38°C (Dewar and Graham, 1994; Dewar et al., 1994), the COT_{net} of marine mammals (i.e., the energy used solely for propulsion) approaches the theoretical value predicted for a swimming tuna (Williams, 1999).

In contrast to subsurface swimming described above, the predicted minimum COT for surface swimming by marine mammals is significantly higher (see Chapter 4). This is largely attributed to increased drag due to surface wave generation and, in the case of the sea otter, the use of comparatively inefficient paddling or rowing forms of propulsion (Williams et al., 1992).

The total aerobic power expended for both surface and submerged swimming by marine mammals increases exponentially with speed, creating a steep wall preventing *sustained* swimming at speeds $>3\,m\,s^{-1}$ for most species (Fig. 4). At the other end of the range, the minimum COT also rises exponentially as speed approaches zero because of the influence of the underlying RMR (Fig. 6). Marine mammals tend to preferentially swim at the slowest speed (and the lowest metabolic rate) to save energy until the COT begins to rise exponentially at speeds $<1\,m\,s^{-1}$. The range of energetically optimum speeds will vary among species, and there are undoubtedly other factors that constrain swim speed (e.g., the use of hydrodynamic lift to counter negative buoyancy at depth). Nevertheless, marine mammals appear to be trapped in an energetic valley between exponentially rising COT at low ($<1\,m\,s^{-1}$) swim speeds because of the underlying RMR and an exponentially rising metabolic rate and COT at higher speeds ($>3\,m\,s^{-1}$). This results in a narrow range of energetically optimal swimming speeds from approximately $1–3\,m\,s^{-1}$ for marine mammals regardless of species or size. Consequently, extreme burst speeds for most marine mammals are comparably rare, brief, and rely on anaerobic metabolism.

The fiber type profile of the major locomotory muscles of Odontoceti reflects the capacity for metabolic support of high-speed performance, and is highly variable within this group (Fig. 8). In general, small and medium-sized Odontoceti show a mixed fiber type profile ranging from approximately 35%–50% slow twitch fibers in the epaxial muscles (*longissimus dorsi*). At the extremes are Cetacea with an exceptionally high proportion of fast twitch fibers, which include deep-diving beaked whales capable of high-speed sprints when foraging (Velten et al., 2013). Skeletal muscle fiber profiles of these whales are similar to those of the locomotory muscles of pouncing felids and human sprinters. This contrasts with comparatively slower swimming species including the pygmy sperm whale (*Kogia breviceps*, Sierra et al., 2015) and narwhal (*Monodon monoceros*, Williams et al., 2011a, b), which show the highest proportion of slow twitch fibers in the same locomotory muscle and is only exceeded by human marathoners (Fig. 8).

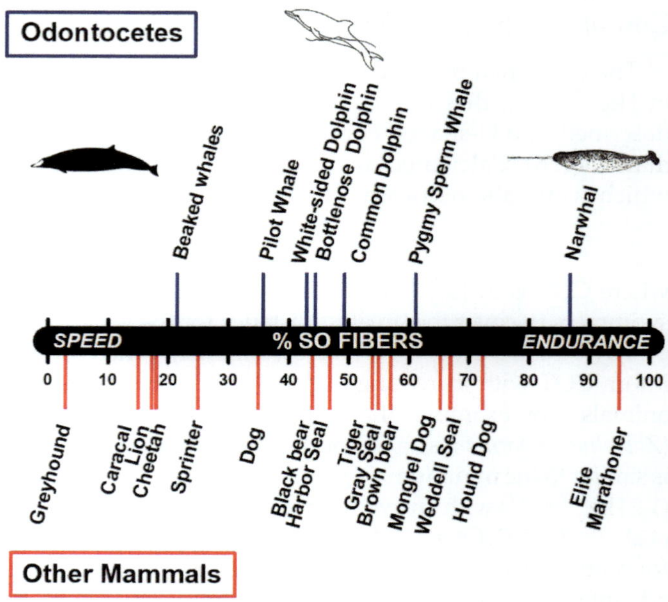

FIG. 8 Muscle fiber type profiles for marine and terrestrial mammals. The black horizontal bar indicates the percentage of slow oxidative (SO) fibers in the major locomotory muscles powering the flukes of odontocetes (upper blue bars), and supporting swimming and running performance in pinnipeds and terrestrial mammals (lower red bars). *Data are redrawn from Pagano and Williams (2021) with additional data from Velten et al. (2013) and Sierra et al. (2015).*

Conservation of energy through drafting, gliding, and leaping

The athletic performance of Odontoceti is made possible by a suite of behaviors, which minimize drag and reduce the metabolic cost of surface and subsurface swimming. These energy-saving behaviors begin early in life when a swimming bottlenose dolphin calf positions itself along the flank of its mother to draft. This position takes advantage of the reduced water pressure next to the female. As a result, overall pressure drag on the calf is reduced, conferring an energetic advantage by lowering stroke frequency compared with independent swimming (Noren et al., 2008). Similarly, schooling dolphins likely obtain an energetic advantage when swimming in echelon formation. During high-speed surface swimming, dolphins may leap clear of the water (i.e., porpoise: Au and Weihs, 1980; Blake, 1983) or ride on the bow or stern waves of boats, tidal bore, and even other whales to avoid wave drag, thereby saving energy and increasing speed (Williams et al., 1992).

As mentioned above, submerged swimming is less costly than surface swimming because surface wave drag is either low or absent. When diving, dolphins and whales use different modes of stroking (i.e., continuous stroking, stroke-and-glide, gliding) to save energy and conserve body oxygen stores. These gait changes act to decrease energetic costs and prolong breath-holding duration. For example, gliding during descent becomes possible for many Cetacea as the lungs compress with increased hydrostatic pressure at depth (see Chapter 6: Respiratory physiology in the dolphin and other whales). As the animal transitions from positive to negative buoyancy during descent, gliding becomes possible, which reduces the total number of strokes taken during the dive. Among Pinnipedia and Cetacea, gliding can reduce the energetic cost of diving by 10%–60% depending on the species (Williams et al., 2000). For example, bottlenose dolphins diving to 100 m realize a 10%–20% energetic savings by gliding during descent (Skrovan et al., 1999). Interestingly, Balaenidae, most notably, positively buoyant North Atlantic right whales, show the opposite behavior by stroking during the descent and gliding during ascent (Nowacek et al., 2001). Cetacea, including high-speed swimming dolphins (Skrovan et al., 1999) and deep-diving beaked whales (Martin Lopez et al., 2015) and narwhals (Williams et al., 2017a), routinely switch stroking modes to achieve the most economical locomotion. Importantly, these behavioral strategies which maximize the use of cost-efficient modes of stroking can extend dive duration and ultimately influence escape responses when exposed to threats (see Section "Concluding remarks").

Energetic limits: Aerobic scope and $\dot{V}O_{2max}$

As with other highly athletic mammals, there is a limit to the aerobic capabilities of dolphins and whales despite energy-saving adaptations. This limit can be expressed as the factorial aerobic scope, which is defined as the ratio of maximum aerobic capacity ($\dot{V}O_{2max}$) and the RMR ($\dot{V}O_{2RMR}$):

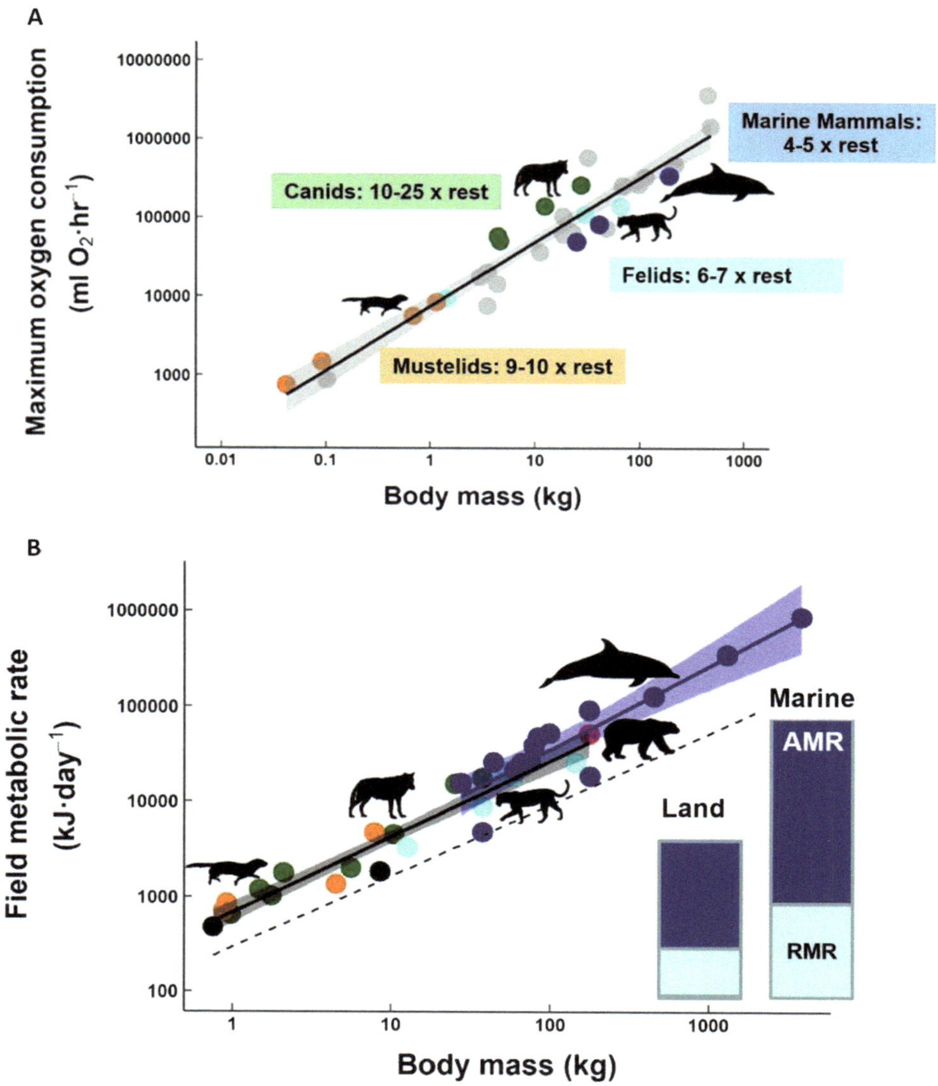

FIG. 9 Maximum oxygen consumption ($\dot{V}O_{2max}$, A) and field metabolic rate (FMR, B) of carnivorans as a function of body mass. Data for marine mammals (dark blue circles) are compared to Canidae (green circles), Felidae (cyan), and Mustelidae (orange). Noncarnivorans are denoted by the gray symbols. FMR values for Ursidae (red circle) and carnivorans not included in the other major groups (black) are also presented in B. Solid lines are the least-squares regressions for the data as described in Williams et al. (2020) and in the text. The dashed line in B is the allometric regression describing resting metabolic rate in relation to body mass for domestic mammals (Kleiber, 1975). Shaded boxes in (A) indicate the general range for factorial aerobic scope for each carnivore group. The inset bars in (B) indicate the relative contribution of resting (RMR, light blue) and active (AMR, dark blue) metabolic rate relative to the total FMR for terrestrial and marine mammals. *Modified from Williams et al. (2020).*

$$\text{Factorial Aerobic Scope} = \dot{V}O_{2max} \div \dot{V}O_{2RMR} \tag{7}$$

Typical values range from 10 to 25 in highly aerobic Canidae (i.e., dogs), and are more than double the scope of less athletic mammalian species (Fig. 9A; Williams et al., 2015). Although few measurements of $\dot{V}O_{2max}$ are available for Cetacea, it has been estimated for trained bottlenose dolphins pushing against a load cell (Fig. 1; Williams et al., 1993, 2017b). In these studies, $\dot{V}O_{2max}$ was six- to sevenfold higher than the RMR (i.e., a factorial aerobic scope of 6–7), considerably less than predicted for dogs (Fig. 9A). As reported for harbor seals (Elsner, 1986) and California sea lions (Butler et al., 1992), dolphins generally have lower aerobic scopes (~52%) compared with the allometric prediction for terrestrial carnivorans. Elevated post-exercise blood lactate concentrations confirmed that the trained dolphins had exceeded their aerobic capacity and relied on anaerobic metabolism when exceeding $\dot{V}O_{2max}$.

In part, the lower aerobic scopes of dolphins compared with terrestrial mammals results from their elevated RMR. Because the RMR of bottlenose dolphins (and other Odontoceti) is ~2.5-fold higher than that of terrestrial carnivorans of similar size (Fig. 2B, Table 2), the factorial aerobic scope is by default lower. The actual $\dot{V}O_{2max}$ of dolphins (i.e., 26 mL O_2 min^{-1} kg^{-1}; Fig. 1) is comparable with other athletic mammals, although the RMR makes up a larger proportion of overall aerobic costs (Fig. 9B; Williams, 2022).

Quantifying the energetic capabilities of marine mammals is challenging because of the unique respiratory-cardiovascular changes associated with the dive response and rapid postexercise recovery of oxygen stores when the animal surfaces. This is further complicated for Cetacea because of the limited availability of metabolic and physiological data for most species, especially the largest whales. Regardless, recent studies have indicated the extraordinary power requirements for locomotion in large whales. Based on the threefold increase in postdive respiratory rates for lunge-feeding humpback whales (*Megaptera novaeangliae*), feeding behaviors appear to result in high energetic costs in this and other Balaenopteridae (Goldbogen et al., 2008). Breaching by large whales may be the most expensive burst maneuver in any mammal, requiring even more energy than lunge-feeding (Segre et al., 2020). The calculated mass-specific energetic cost of breaching in humpback whales increases with body mass and ranges from 130 to 220 $J\,kg^{-1}$ for 7000–46,000 kg whales, respectively. The speed of the maneuver also contributes to these high costs. Ultimately, the capacity of mammalian skeletal muscle to generate power will limit the maximum size of whales that can perform such behaviors. How this will impact the aerobic scope of large whales relative to terrestrial mammals awaits further investigation with a wider range of cetacean species and additional comparative metrics such as relative heart mass and post-exercise recovery times.

Predicting free-ranging locomotory costs

Quantifying the energetic requirements of dolphins and whales in the wild is important for understanding their ecology. In the past, such studies for Cetacea were limited because of their large size and pelagic lifestyle. Common methods for measuring the metabolism of captive animals with respirometry or isotopic techniques (i.e., doubly-labeled water) are rarely possible for Cetacea (Costa, 2002; Rimbach et al., 2021). However, indirect methods that match free-ranging locomotory behavior (e.g., overall dynamic body acceleration, swim speed, fluke stroke frequency) or physiologic variables (i.e., respiratory rate, heart rate) with corresponding metabolic measurements on trained animals have proven to be a useful approach (Butler and Jones, 1997; Costa, 2002). Respiratory rate (Sumich, 1983; Kriete, 1995), heart rate (Williams et al., 1992; Boyd et al., 1995; Butler and Jones, 1997), and swimming speed (Kshatriya and Blake, 1988; Hind and Gurney, 1997) have been used to estimate the energetics of free-ranging marine mammals, including some Cetacea. In comparison, several limitations have been reported for isotopic techniques and animal-borne heart rate monitors (see Butler et al., 2004), especially with regard to invasiveness and variability resulting from the marine mammal dive response. Specifically, the effect of diving bradycardia (Butler, 1993) and prolonged gliding on estimated swimming metabolism based solely on measurements of speed can invalidate predicted energetic costs. Isotopic techniques require the administration of labeled water followed by blood samples collected 3–8 days apart, which is not possible for most Cetacea. Furthermore, the cost of isotopes is prohibitive for Cetacea larger than a dolphin. Future improvements in instrumentation and methodologies may overcome these limitations.

Accelerometry and energetics in the wild

Recent advances in animal-borne instruments, particularly time-depth recorders paired with tri-axial accelerometers, have produced reasonable energetic estimates for free-ranging dolphins and whales (see Ladds et al., 2017). In general, the recorders are calibrated in the laboratory or on animals in zoological settings where standard measurements of oxygen consumption can be paired with stroking rates and acceleration (i.e., John, 2020; Allen et al., 2022). The calibrated recorders are then deployed on free-ranging Cetacea, which enables the post-deployment calculation of instantaneous and sustained energetics. The following sections describe two common methods (stroke counting and dynamic body acceleration), which have been used to estimate the energetics of free-ranging marine mammals. For a comparison of these two methods, we recommend reading Halsey et al. (2009, 2011) and Qasem et al. (2012). When using these methods, care must be taken to account for the duration of specific activities (e.g., create an activity budget) rather than simply summing acceleration over long periods. This will avoid the confounding effects of time on energetic estimates (Jeanniard-du-Dot et al., 2017; Ladds et al., 2017).

The first method for estimating the energetic cost of diving and surface swimming (E_{dive} and E_{swim}, respectively; $J\,kg^{-1}$) for dolphins and whales is to use acceleration recordings to determine the number of swimming strokes taken during specific activities. The energetic cost ($J\,kg^{-1}$) of an activity is based on total strokes and the mass-specific net cost per stroke (CPS_{net}), which has been determined for some Pinnipedia (Williams et al., 2004a, b) and Cetacea (Williams et al., 2017b). The equation for calculating E_{dive} and E_{swim} including the RMR (Williams, 1999; Davis, 2019) with the CPS_{net} (Eq. 4) is described by:

$$E_{dive} \text{ or } E_{swim} = (\text{RMR} \times D) + (\text{CPS}_{net} \times S) \quad (8)$$

where RMR is in $J\,kg^{-1}min^{-1}$, D is the duration of the swim or dive in minutes, and S is the number of strokes during the recording period. The metabolic rate ($W\,kg^{-1}$) for diving and surface swimming (MR_{dive} and MR_{swim}, respectively) is the quotient of E_{dive} or E_{swim} and the duration of the activity in seconds. Note that during routine dives, the submerged RMR may be lower (e.g., 2%–35%) than resting at the surface depending on the species and dive duration (Hurley and Costa, 2001; Williams et al., 2004a; Fahlman et al., 2013), although additional data are needed to confirm this for Cetacea. For these calculations, we also assume that the animal is in a fasted state with no appreciable contribution from the heat increment of feeding (HIF) and that energy metabolism is primarily aerobic.

For maximum aerobic performance and maximum speeds, Eq. (5) should be used for the CPS_{net}. Because free-ranging activities may include periods of routine and high-performance stroking, the total cost of a dive may be calculated by parsing the dive into periods of routine and maximum performance (see Williams et al., 2017b). Overall, this method enables energetic costs to be assessed for free-ranging marine mammals in which direct energetic measurements may be impossible. The energetic cost of discrete behaviors such as searching, prey pursuit, or transiting can also be estimated by counting the number of strokes during each activity. Because few direct energetics data are available, extrapolation beyond Odontoceti is currently not recommended. In particular, there are size limitations with the use of these stroking methods for estimating the energetics of larger whales.

In cases where it is not possible to instrument animals, the availability of unmanned aerial vehicles (UAVs) with high-resolution video provides a cost-effective way to study free-ranging dolphin behavior and energetics (Fig. 10). Individual and group follows are possible using a UAV launched and retrieved from a skiff. The cost of locomotion can be estimated by counting fluke strokes along a trajectory of known distance and time near the surface. For example, in a theoretical study on mother-calf energetics, the net cost of a stroke can be calculated from Eq. (4) for routine swim

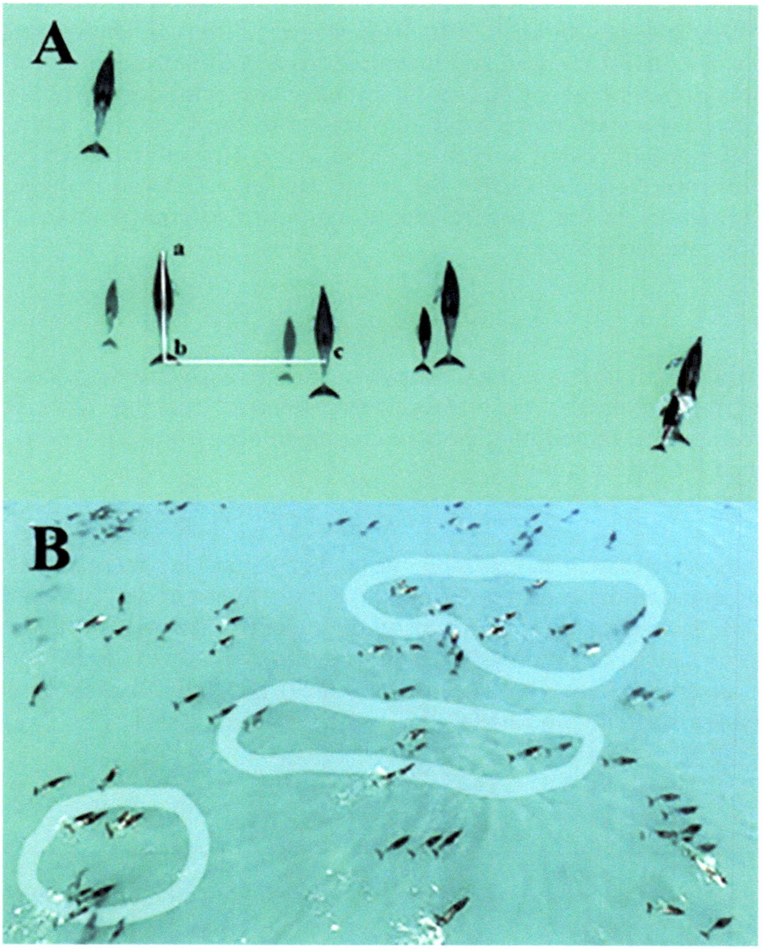

FIG. 10 Remote assessment of dolphin energetics using an unmanned aerial vehicle (UAV). Aerial video recordings of mother-calf pairs and a fifth adult without a calf enable relative stroke rates to be monitored and then used to calculate the energetic cost of traveling of each (A). Expanding the view (B) demonstrates how mothers and calves moved relative to the larger group (mother-calf pairs are circled for ease of viewing).

speeds, resulting in a predicted CPS_{net} for a 148 kg dolphin of $1.53\,J\,kg^{-1}\,stroke^{-1}$. For a dolphin swimming at modest stroke frequency of 1 Hz, the estimated mass-specific cost of swimming would then be:

$$1.53\,J\,kg^{-1}\,stroke^{-1} \times 1\ stroke\ s^{-1} = 1.53\ W\ kg^{-1} \tag{9}$$

where W is watts; the total energetic cost of swimming for this animal would be 226 W (viz. $1.53\,W\,kg^{-1} \times 148\,kg$). Swim speed at this stroke frequency would be $\sim 2\,m\,s^{-1}$ and classified as transit. At a stroke frequency of $2\,strokes^{-1}$ associated with higher swimming speeds ($>3\,m\,s^{-1}$), the energetic cost of stroking would be 453 W, possibly associated with prey pursuit or predator avoidance.

These calculated stroking costs do not include the underlying RMR, which must be added to the net cost of locomotion to calculate the total metabolic cost of the dolphin while swimming (MR_{swim}). Assuming a mass-specific RMR of $2.2\,W\,kg^{-1}$ or 323 W for a 148 kg dolphin, the total cost of swimming at stroke frequency at 1 Hz is the sum of the stroking cost and resting metabolic rate (viz. 226 W + 323 W = 549 W or $3.7\,W\,kg^{-1}$). This value is only 1.7-fold higher than the RMR of the animal. At a higher stroking rate of $2\,strokes^{-1}$, the total cost of swimming would be 776 W ($5.2\,W\,kg^{-1}$), which is 2.4-fold higher than the RMR, highlighting the effect of an elevated RMR on the dolphin's energy budget and the advantage of a cost-efficient form of swimming (Davis, 2019).

A second method for calculating the energetic cost of locomotor movements in wild animals also uses data from all three axes of accelerometer tags, referred to as overall body dynamic acceleration (ODBA). Here three-axis accelerometer outputs are summed to allow both stroking and non-stroking (i.e., rolling, spinning) maneuvers to be monitored. As with the stroke count method, these sums are paired with oxygen consumption of trained, instrumented animals, and the energy expended for various movements calculated. The use of the ODBA method was pioneered by Wilson et al. (2006, 2019), and has been used for estimating the energetic costs in a wide variety of free-ranging terrestrial, aerial, and aquatic animals, including Cetacea (John, 2020; Allen et al., 2022; Williams et al., 2022) and other marine mammals (Fahlman et al., 2008, 2013; Hindle et al., 2010; Pagano et al., 2018b). By monitoring both locomotor and non-locomotor body movements, the ODBA method and its variations account for turning, rolling, and other potentially energetically-costly behaviors in free-ranging Cetacea. Preliminary tests with captive bottlenose dolphins and beluga whales (*Delphinapterus leucas*) have shown that the stroke counting method and the ODBA method are equally effective for estimating the overall energetic cost of swimming (John, 2020). However, further research comparing the relationship between oxygen consumption and ODBA for small and large Cetacea is needed. Because accelerometer placement is critical for these measurements, issues related to anatomical location and tag slippage during deployment need to be addressed for accurate recordings.

Field metabolic rates

When the daily activity budgets of free-ranging Cetacea are known, E_{dive} and E_{swim} can be used to estimate a total daily energy budget, as well as the average daily metabolic rate (referred to as the FMR). The FMR has been reported for several marine mammal species, including Cetacea ranging in size from bottlenose dolphins to killer whales (Fig. 9B), which are described by the equation:

$$FMR = 651.2 \times M^{0.87} \tag{10}$$

where FMR is in $kJ\,day^{-1}$ and body mass is in kg (Williams et al., 2020). Using these estimates of FMR, we can assess intra- and inter-species differences in the energetic demands of daily living in dolphins and whales. We find that the combination of elevated RMR and energetic costs for locomotion contributes to a 1.3-fold higher FMR in Odontoceti compared to similarly-sized terrestrial carnivores. In this regard, the toothed whales follow the pattern for highly aerobic terrestrial mammalian carnivores (Fig. 9B; Williams et al., 2020), but differ from the low FMR previously reported for some Phocidae (i.e., true seals) (Reilly and Fedak, 1991; Maresh et al., 2014).

Concluding remarks

Energy balance in animals is essential for individual fitness and population stability, thus forming the foundation for conservation modeling and species management. Dolphins and whales, which are some of the largest and most mobile mammals on earth, have an important top-down effect on prey populations, and consequently on the structure of marine and coastal ecosystems (Estes et al., 1998; Springer et al., 2003). Because of this ecological role, understanding the metabolic rates and energy requirements of both large and small Cetacea is essential for determining trophic dynamics (Williams et al., 2004b). By providing the range and variability of metabolic rates and energetic costs of free-ranging

Cetacea, this chapter lays the foundation for conservation modeling and the assessment of the energetic impacts of anthropogenic disturbance. Streamlining and efficient modes of locomotion used by dolphins and whales typically reduce the energetic cost of swimming and diving, which in turn prolongs aerobic diving time. This increases the available time for energy acquisition while foraging at depth (see Chapter 5: Cardiovascular physiology in dolphins and other cetaceans). Conversely, current research suggests that anthropogenic disturbances may come at a high energetic cost for Cetacea by quickly depleting onboard oxygen stores during escape responses. This is evident in the elevated stroke frequencies reported for whales exposed to anthropogenic noise and other activities (i.e., DeRuiter et al., 2013; Pagano and Williams, 2021; Williams et al., 2017a, b, 2022). Future research concerning the energetics of large and small Cetacea will be needed to determine how both acute and chronic environmental changes and anthropogenic disturbances, will influence the cost of daily living, and ultimately, survival of dolphin and whale populations.

References

Aarseth, J.J., Nordøy, E.S., Blix, A.S., 1999. The effect of body fat on basal metabolic rate in adult harp seals (*Phoca groenlandica*). Comp. Biochem. Physiol. A: Physiol. 124, 69–72.

Ahlborn, B.K., Blake, R.W., Chan, K.H., 2009. Optimal fineness ratio for minimum drag in large whales. Can. J. Zool. 87 (2), 124–131.

Allen, A.S., Read, A.J., Shorter, K.A., Gabaldon, J., Blawas, A.M., Rocho-Levine, J., Fahlman, A., 2022. Dynamic body acceleration as a proxy to predict the cost of locomotion in bottlenose dolphins. J. Exp. Biol. 225 (4), jeb243121.

Au, D., Weihs, D., 1980. At high-speed dolphins save energy by leaping. Nature 284, 548–550.

Ballesteros, F.J., Martinez, V.J., Luque, B., Lacasa, L., Valor, E., Moya, A., 2018. On the thermodynamic origin of metabolic scaling. Sci. Rep. 8.

Blake, R., 1983. Energetics of leaping in dolphins and other aquatic animals. J. Mar. Biol. Assoc. U. K. 63 (1), 61–70.

Blank, J.M., Farwell, C.J., Morrissette, J.M., Schallert, R.J., Block, B.A., 2007. Influence of swimming speed on metabolic rates of juvenile Pacific bluefin tuna and yellowfin tuna. Physiol. Biochem. Zool. 80, 167–177.

Boyd, I.L., Woakes, A.J., Butler, P.J., Davis, R.W., Williams, T.M., 1995. Validation of heart rate and doubly labelled water as measures of metabolic rate during swimming in California sea lions. Funct. Ecol., 151–160.

Butler, P.J., 1993. To what extent can heart rate be used as an indicator of metabolic rate in free-living marine mammals. In: Boyd, L.L. (Ed.), Marine Mammals: Advances in Behavioural and Population Biology. Oxford University Press, UK, pp. 317–332. Zoological Society of Long Symposia 66.

Butler, P.J., Green, J.A., Boyd, I.L., Speakman, J.R., 2004. Measuring metabolic rate in the field: the pros and cons of the doubly labelled water and heart rate methods. Funct. Ecol. 18 (2), 168–183.

Butler, P.J., Jones, D.R., 1997. Physiology of diving of birds and mammals. Physiol. Rev. 77 (3), 837–899.

Butler, P.J., Woakes, A.J., Boyd, I.L., Kanatous, S., 1992. Relationship between heart rate and oxygen consumption during steady-state swimming in California sea lions. J. Exp. Biol. 170 (1), 35–42.

Clark, T.D., Seymour, R.S., 2006. Cardiorespiratory physiology and swimming energetics of a high-energy-demand teleost, the yellowtail kingfish (*Seriola lalandi*). J. Exp. Biol. 209, 3940–3951.

Costa, D.P., 2002. Energetics. In: Perrin, W.F., Wursig, B., Thewissen, J.G.M. (Eds.), Encyclopedia of Marine Mammals. Academic Press, San Diego, pp. 387–394.

Costa, D.P., Kooyman, G.L., 1984. Contribution of specific dynamic action to heat balance and thermoregulation in the sea otter, *Enhydra lutris*. Physiol. Zool. 57, 199–203.

Culik, B.M., Wilson, R.P., Bannasch, R., 1994. Underwater swimming at low energetic cost by pygoscelid penguins. J. Exp. Biol. 197, 65–78.

Dassis, M., Rodriguez, D.H., Ieno, E.N., Davis, R.W., 2012. Submerged swimming and resting metabolic rates in Southern sea lions. J. Exp. Mar. Biol. Ecol. 432, 106–112.

Davis, R.W., 2019. Marine Mammals: Adaptations for an Aquatic Life. Springer Nature Switzerland, Cham (Switzerland), 314 pp.

Davis, R.W., Williams, T.M., Kooyman, G.L., 1985. Swimming metabolism of yearling and adult harbor seals. Physiol. Zool. 58 (5), 590–596.

Denny, M.W., 1993. Air and Water: The Biology and Physics of life's Media. Princeton University Press, Princeton (NJ), p. 342.

DeRuiter, S.L., Southall, B.L., Calambokidis, J., Zimmer, W.M.X., Sadykova, D., Falcone, E.A., Tyack, P.L., 2013. First direct measurements of behavioural responses by Cuvier's beaked whales to mid-frequency active (MFA) sonar. Biol. Lett. 9, 20130223.

Dewar, H., Graham, J.B., 1994. Studies of tropical tuna swimming performance in a large water tunnel. I. Energetics. J. Exp. Biol. 192, 13–31.

Dewar, H., Graham, J.B., Brill, R.W., 1994. Studies of tropical tuna swimming performance in a large water tunnel. II. Thermoregulation. J. Exp. Biol. 192, 33–44.

Elsner, R.E., 1986. Limits to exercise performance: some ideas from comparative studies. Acta Physiol. Scand. Suppl., 45–51.

Estes, J.A., Tinker, M.T., Williams, T.M., Doak, D.F., 1998. Killer whale predation on sea otters linking oceanic and nearshore ecosystems. Science 282, 473–476.

Fahlman, A., Brodsky, M., Wells, R., McHugh, K., Allen, J., Barleycorn, A., Sweeney, J.C., Fauquier, D., Moore, M., 2018. Field energetics and lung function in wild bottlenose dolphins, *Tursiops truncatus*, in Sarasota Bay Florida. Royal Soc. Open Sci. 5 (1), 171280.

Fahlman, A., Svärd, C., Rosen, D.A.S., Wilson, R.S., Trites, A.W., 2013. Activity as a proxy to estimate metabolic rate and to partition the metabolic cost of diving vs. breathing in pre- and post-fasted Steller Sea lions. Aquat. Biol. 18, 175–184.

Fahlman, A., Wilson, R., Svärd, C., Rosen, D.A.S., Trites, A.W., 2008. Activity and diving metabolism correlate in Steller sea lion *Eumetopias jubatus*. Aquat. Biol. 2, 75–84.

Fedak, M.A., Anderson, S.S., 1982. The energetics of lactation: accurate measurements from a large wild mammal, the grey seal (*Halichoerus grypus*). J. Zool. 198 (4), 473–479.

Feldkamp, S., 1987. Swimming in the California Sea lion: morphometrics, drag, and energetics. J. Exp. Biol. 131, 117–135.

Fish, F.E., 1998. Comparative kinematics and hydrodynamics of odontocete cetaceans: morphological and ecological correlates with swimming performance. J. Exp. Biol. 201, 2867–2877.

Fish, F.E., Nicastro, A.J., Weihs, D., 2006. Dynamics of aerial maneuvers of spinner dolphins. J. Exp. Biol. 209, 590–598.

Ford, J.K., Ellis, G.M., Matkin, D.R., Balcomb, K.C., Briggs, D., Morton, A.B., 2005. Killer whale attacks on minke whales: prey capture and antipredator tactics. Mar. Mamm. Sci. 21 (4), 603–618.

Goldbogen, J.A., Calambokidis, J., Croll, D.A., Harvey, J.T., Newton, K.M., Oleson, E.M., Schorr, G., Shadwick, R.E., 2008. Foraging behavior of humpback whales: kinematic and respiratory patterns suggest a high cost for a lunge. J. Exp. Biol. 211 (23), 3712–3719.

Goldbogen, J.A., Calambokidis, J., Shadwick, R.E., Oleson, E.M., McDonald, M.A., Hildebrand, J.A., 2006. Kinematics of foraging dives and lunge-feeding in fin whales. J. Exp. Biol. 209 (7), 1231–1244.

Goldbogen, J.A., Southall, B.L., DeRuiter, S.L., Calambokidis, J., Friedlaender, A.S., Hazen, E.L., Falcone, E.A., Schorr, G.S., Douglas, A., Moretti, D.J., Kyburg, C., 2013. Blue whales respond to simulated mid-frequency military sonar. Proc. R. Soc. Lond. B 280 (1765), 20130657.

Gough, W.T., Segre, P.S., Bierlich, K.C., Cade, D.E., Potvin, J., Fish, F.E., Dale, J., di Clemente, J., Friedlaender, A.S., Johnston, D.W., Kahane-Rapport, S.R., Kennedy, J., Long, J.H., Oudejans, M., Penry, G., Savoca, M.S., Simon, M., Videsen, S.K.A., Visser, F., Wiley, D.N., Goldbogen, J.A., 2019. Scaling of swimming performance in baleen whales. J. Exp. Biol. 222, jeb204172.

Guinet, C., Domenici, P., de Stephanis, R., Barrett-Lennard, L., Ford, J.K.B., Verborgh, P., 2007. Killer whale predation on bluefin tuna: exploring the hypothesis of the endurance-exhaustion technique. Mar. Ecol. Prog. Ser. 347, 111–119.

Halsey, L.G., Shepard, E.L.C., Quintana, F., Gomez Laich, A., Green, J.A., Wilson, R.P., 2009. The relationship between oxygen consumption and body acceleration in a range of species. Comp. Biochem. Physiol. A: Physiol. 152, 197–202.

Halsey, L.G., Shepard, E.L.C., Wilson, R.P., 2011. Assessing the development and application of the accelerometry technique for estimating energy expenditure. Comp. Biochem. Physiol. A: Physiol. 158, 305–314.

Hart, J.S., 1957. Climatic and temperature induced changes in the energetics of homeotherms. Rev. Can. Biol. 16, 133–174.

Hind, A.T., Gurney, W.S., 1997. The metabolic cost of swimming in marine homeotherms. J. Exp. Biol. 200 (3), 531–542.

Hindle, A.G., Rosen, D.A., Trites, A.W., 2010. Swimming depth and ocean currents affect transit costs in Steller sea lions *Eumetopias jubatus*. Aquat. Biol. 10 (2), 139–148.

Hurley, J.A., Costa, D.P., 2001. Standard metabolic rate at the surface and during trained submersions in adult California sea lions (*Zalophus californianus*). J. Exp. Biol. 204 (19), 3273–3281.

Irvine, A.B., 1983. Manatee metabolism and its influence on distribution in Florida. Biol. Conserv. 25 (4), 315–334.

Jeanniard-du-Dot, T., Guinet, C., Arnould, J.P., Speakman, J.R., Trites, A.W., 2017. Accelerometers can measure total and activity-specific energy expenditures in free-ranging marine mammals only if linked to time-activity budgets. Funct. Ecol. 31 (2), 377–386.

John, J.S., 2020. Energetics of Rest and Locomotion in Diving Marine Mammals: Novel Metrics for Predicting the Vulnerability of Threatened Cetacean, Pinniped and Sirenian Species. PhD Thesis, University of California, Santa Cruz, CA.

Kleiber, M., 1975. The Fire of Life: An Introduction to Animal Energetics. RE Kreiger Publishing Co, Huntington (NY), p. 453.

Kriete, B., 1995. Bioenergetics of the Killer Whale, *Orcinus orca*. Ph.D. thesis, University of British Columbia, Canada.

Kshatriya, M., Blake, R.W., 1988. Theoretical model of migration energetics in the blue whale, Balaenoptera musculus. J. Theor. Biol. 133 (4), 479–498.

Ladds, M.A., Rosen, D.A., Slip, D.J., Harcourt, R.G., 2017. Proxies of energy expenditure for marine mammals: an experimental test of "the time trap". Sci. Rep. 7 (1), 1–10.

Lee, C.G., Farrell, A.P., Lotto, A., MacNutt, M.J., Hinch, S.G., Healey, M.C., 2003. The effect of temperature on swimming performance and oxygen consumption in adult sockeye (Oncorhynchus nerka) and coho (O. kisutch) salmon stocks. J. Exp. Biol. 206, 3239–3251.

Liao, J.A., 1990. An Investigation of the Effect of Water Temperature on the Metabolic Rate of the California Sea Lion *Zalophus californianus*. CA Master thesis, University of California, Santa Cruz, p. 55.

Lindgren, J., Sjövall, P., Thiel, V., et al., 2018. Soft-tissue evidence for homeothermy and crypsis in a Jurassic ichthyosaur. Nature 564, 359–368.

Lockyer, C., 1981. Growth and energy budgets of large baleen whales from the southern hemisphere. Food Agric. Organ. 3, 379–487.

Lovegrove, B.G., 2005. Seasonal thermoregulatory responses in mammals. J. Comp. Physiol. B 175, 231–247.

Maresh, J.L., Simmons, S.E., Crocker, D.E., McDonald, B.I., Williams, T.M., Costa, D.P., 2014. Free-swimming northern elephant seals have low field metabolic rates that are sensitive to an increased cost of transport. J. Exp. Biol. 217 (9), 1485–1495.

Martin Lopez, L.M., Miller, P.J., Aguilar de Soto, N., Johnson, M., 2015. Gait switches in deep-diving beaked whales: biomechanical strategies for long-duration dives. J. Exp. Biol. 218 (9), 1325–1338.

McNab, B.K., 2008. An analysis of the factors that influence the level and scaling of mammalian BMR. Comp. Biochem. Physiol. A: Physiol. 151, 5–28.

Noren, S.R., Biedenbach, G., Redfern, J.V., Edwards, E.F., 2008. Hitching a ride: the formation locomotion strategy of dolphin calves. Funct. Ecol. 22, 278–283.

Nowacek, D.P., Johnson, M.P., Tyack, P.L., Shorter, K.A., McLellan, W.A., 2001. Buoyant balaenids: the ups and downs of buoyancy in right whales. Proc. R. Soc. Lond. B 268 (1478), 1811–1816.

Ochoa-Acuña, H.G., McNab, B.K., Miller, E.H., 2009. Seasonal energetics of northern phocid seals. Comp. Biochem. Physiol. A: Physiol. 152, 341–350.

Otani, S., Naito, Y., Kato, A., Kawamura, A., 2001. Oxygen consumption and swim speed of the harbor porpoise Phocoena phocoena. Fish. Sci. 67 (5), 894–898.

Pagano, A.M., Carnahan, A.M., Robbins, C.T., Owen, M.A., Batson, T., Wagner, N., Cutting, A., Nicassio-Hiskey, N., Hash, A., Williams, T.M., 2018a. Energetic costs of locomotion in bears: is plantigrade locomotion energetically economical? J. Exp. Biol. 221, jeb175372.

Pagano, A.M., Durner, G.M., Rode, K.D., Atwood, T.C., Atkinson, S.N., Peacock, E., Costa, D.P., Owen, M.A., Williams, T.M., 2018b. High-energy, high-fat lifestyle challenges an Arctic apex predator, the polar bear. Science 359, 568–572.

Pagano, A., Williams, T.M., 2021. Physiological consequences of Arctic sea ice loss on large marine carnivores: unique responses by polar bears and narwhals. J. Exp. Biol. 224, jeb228049.

Qasem, L., Cardew, A., Wilson, A., Griffiths, I., Halsey, L.G., Shepard, E.L.C., Gleiss, A.C., Wilson, R., 2012. Tri-axial dynamic acceleration as a proxy for animal energy expenditure; should we be summing values or calculating the vector?. PLoS One 7, e31187.

Rea, L.D., Costa, D.P., 1992. Changes in standard metabolism during long-term fasting in northern elephant seal pups (*Mirounga-Angustirostris*). Physiol. Zool. 65 (1), 97–111.

Reed, J.Z., Chambers, C., Hunter, C.J., Lockyer, C., Kastelein, R., Fedak, M.A., Boutilier, R.G., 2000. Gas exchange and heart rate in the harbour porpoise, *Phocoena phocoena*. J. Comp. Physiol. B 170, 1–10.

Reilly, J.J., Fedak, M.A., 1991. Rates of water turnover and energy expenditure of free-living male common seals (*Phoca vitulina*). J. Zool. 223 (3), 461–468.

Rimbach, R., Amireh, A., Allen, A., Hare, B., Guarino, E., Kaufman, C., Salomons, H., Pontzer, H., 2021. Total energy expenditure of bottlenose dolphins (*Tursiops truncatus*) of different ages. J. Exp. Biol. 224 (15), jeb242218.

Rosen, D.A., Trites, A.W., 2002. Changes in metabolism in response to fasting and food restriction in the Steller Sea lion (Eumetopias jubatus). Comp. Biochem. Physiol. B: Biochem. 132 (2), 389–399.

Schmidt-Nielsen, K., 1972. Locomotion: energy cost of swimming, flying, and running. Science 177, 222–228.

Segre, P.S., Potvin, J., Cade, D.E., Calambokidis, J., Di Clemente, J., Fish, F.E., Friedlaender, A.S., Gough, W.T., Kahane-Rapport, S.R., Oliveira, C., Parks, S.E., 2020. Energetic and physical limitations on the breaching performance of large whales. Elife 9, e51760.

Sierra, E., Fernández, A., de Los, E., Monteros, A., Díaz-Delgado, J., Bernaldo de Quirós, Y., García-Álvarez, N., Arbelo, M., Herráez, P., 2015. Comparative histology of muscle in free ranging cetaceans: shallow versus deep-diving species. Sci. Rep. 5 (1), 1–13.

Skrovan, R.C., Williams, T.M., Berry, P.S., Moore, P.W., Davis, R.W., 1999. The diving physiology of bottlenose dolphins (*Tursiops truncatus*). II. Biomechanics and changes in buoyancy at depth. J. Exp. Biol. 202 (20), 2749–2761.

Soto, N.A., Johnson, M.P., Madsen, P.T., Díaz, F., Domínguez, I., Brito, A., Tyack, P., 2008. Cheetahs of the deep sea: deep foraging sprints in short-finned pilot whales off Tenerife (Canary Islands). J. Anim. Ecol., 936–947.

Speakman, J., 1997. Factors influencing the daily energy expenditure of small mammals. Proc. Nutr. Soc. 56, 1119–1136.

Speakman, J.R., Król, E., 2010. Maximal heat dissipation capacity and hyperthermia risk: neglected key factors in the ecology of endotherms. J. Anim. Ecol. 79, 726–746.

Springer, A.M., Estes, J.A., van Vliet, G.B., Williams, T.M., Doak, D.F., Danner, E.M., Forney, K.A., Pfister, B., 2003. Sequential megafaunal collapse in the North Pacific Ocean: an ongoing legacy of industrial whaling? Proc. Natl. Acad. Sci. U. S. A. 100 (21), 12223–12228.

Sumich, J.L., 1983. Swimming velocities, breathing patterns and estimated costs of locomotion in migrating gray whales, *Eschrichtius robustus*. Can. J. Zool. 61, 647–652.

Taylor, C.R., Schmidt-Nielsen, K., Raab, J.L., 1970. Scaling of energetic cost of running to body size in mammals. Am. J. Physiol. 219 (4), 1104–1107.

Todd, S., Lien, J., Marques, F., Stevick, P., Ketten, D., 1996. Behavioural effects of exposure to underwater explosions in humpback whales (*Megaptera novaeangliae*). Can. J. Zool. 74 (9), 1661–1672.

Toussaint, H., Truijens, M., 2005. Biomechanical aspects of peak performance in human swimming. Anim. Biol. 55, 17–40.

van der Hoop, J.M., Fahlman, A., Hurst, T., Rocho-Levine, J., Shorter, K.A., Petrov, V., Moore, M.J., 2014. Bottlenose dolphins modify behavior to reduce metabolic effect of tag attachment. J. Exp. Biol. 217 (23), 4229–4236.

van der Hoop, J.M., Fahlman, A., Shorter, K.A., Gabaldon, J., Rocho-Levine, J., Petrov, V., Moore, M.J., 2018. Swimming energy economy in bottlenose dolphins under variable drag loading. Front. Mar. Sci., 5. https://doi.org/10.3389/fmars.2018.00465.

Velten, B.P., Dillaman, R.M., Kinsey, S.T., McLellan, W.A., Pabst, D.A., 2013. Novel locomotor muscle design in extreme deep-diving whales. J. Exp. Biol. 216 (10), 1862–1871.

Williams, T.M., 1999. The evolution of cost-efficient swimming in marine mammals: limits to energetic optimization. Philos. Trans. R. Soc. Lond. B. 354, 193–201.

Williams, T.M., 2022. Racing time: how physiological rates dictate the scaling of metabolic locomotory costs in marine mammals. Integr. Comp. Biol., icac054. https://doi.org/10.1093/icb/icac054. 2022.

Williams, T.M., Davis, R.W., 2021. Physiological resiliency in diving mammals: insights on hypoxia protection using the Krogh principle to understand COVID-19 symptoms. Comp. Biochem. Physiol. A: Physiol. 253, 110849.

Williams, T.M., Bengtson, P., Steller, D.L., Croll, D.A., Davis, R.W., 2015. The healthy heart: lessons from nature's elite athletes. (Reviews). Physiology 30 (5), 349–357.

Williams, T.M., Blackwell, S.B., Richter, B., Sinding, M.S., Heide-Jørgensen, M.P., 2017a. Paradoxical escape responses by narwhals (*Monodon monoceros*). Science 358, 1328–1331.

Williams, T.M., Blackwell, S.B., Tervo, O., Garde, E., Sinding, M.S., Richter, B., Heide-Jørgensen, M.P., 2022. Physiological responses of narwhals to anthropogenic noise: a case study with seismic airguns and vessel traffic in the Arctic. Funct. Ecol. 36 (9), 2135–2428.

Williams, T.M., Davis, R.W., Fuiman, L.A., Francis, J., LeBoeuf, B.J., Horning, M., Calambokidis, J., Croll, D.A., 2000. Sink or swim: strategies for cost-efficient diving by marine mammals. Science 288, 133–136.

Williams, T.M., Estes, J.A., Doak, D.F., Springer, A.M., 2004b. Killer appetites: assessing the role of predators in ecological communities. Ecology 85 (12), 3373–3384.

Williams, T.M., Friedl, W.A., Fong, M.L., Yamada, R.M., Sedivy, P., Haun, J.E., 1992. Travel at low energetic cost by swimming and wave-riding bottlenose dolphins. Nature 355, 821–823.

Williams, T.M., Friedl, W.A., Haun, J.E., 1993. The physiology of bottlenose dolphins (*Tursiops truncatus*): heart rate, metabolic rate and plasma lactate concentration during exercise. J. Exp. Biol. 179, 31–46.

Williams, T.M., Fuiman, L.A., Horning, M., Davis, R., 2004a. The cost of foraging by a marine predator, the Weddell seal *Leptonychotes weddellii*: pricing by the stroke. J. Exp. Biol. 207 (6), 973–982.

Williams, T.M., Haun, J., Davis, R.W., Fuiman, L., Kohin, S., 2001. A killer appetite: metabolic consequences of carnivory in marine mammals. G.L. Kooyman Symposium, Comp. Biochem. Physiol. A: Physiol. 129, 785–796.

Williams, T.M., Heide-Jørgensen, M.P., Pagano, A.M., Bryce, C., 2020. Hunters versus hunted: new perspectives on the physiological costs of survival at the top of the food chain. Funct. Ecol. https://doi.org/10.1111/1365-2435.13649.

Williams, T.M., Kendall, T., Richter, P., Ribeiro-French, C., John, J., Odell, K., Losch, B., Feuerbach, D., Stamper, M.A., 2017b. Swimming and diving energetics in dolphins: a stroke-by-stroke analysis for predicting the cost of flight responses in wild odontocetes. J. Exp. Biol. 220, 1135–1145.

Williams, T.M., Kooyman, G.L., 1985. Swimming performance and hydrodynamic characteristics of harbor seals (*Phoca vitulina*). Physiol. Zool. 58 (5), 576–589.

Williams, T.M., Kooyman, G.L., Croll, D., 1991. The effects of submergence on heart rate and oxygen consumption of swimming seals and sea lions. J. Comp. Physiol. 160, 637–644.

Williams, T.M., Noren, S.R., Glenn, M., 2011b. Extreme physiological adaptations as predictors of climate-change sensitivity in the narwhal, *Monodon monoceros*. Mar. Mamm. Sci. 27 (2), 334–349.

Williams, T.M., Richter, B., Kendall, T., Dunkin, R., 2011a. Metabolic demands of a tropical marine carnivore, the Hawaiian monk seal (*Monachus schauinslandi*): implications for fisheries competition. Aquat. Mamm. 37 (3), 372–376.

Wilson, R.P., Börger, L., Holton, M.D., Scantlebury, D.M., Gómez-Laich, A., Quintana, F., Rosell, F., Graf, P.M., Williams, H., Gunner, R., et al., 2019. Estimates for energy expenditure in free-living animals using acceleration proxies: a reappraisal. J. Anim. Ecol. 89, 161–172.

Wilson, R.P., White, C.R., Quintana, F., Halsey, L.G., Liebsch, N., Martin, G.R., Butler, P.J., 2006. Moving towards acceleration for estimates of activity-specific metabolic rate in free-living animals: the case of the cormorant. J. Anim. Ecol. 75, 1081–1090.

Wright, T., Davis, R.W., Pearson, H.C., Murray, M., Sheffield-Moore, M., 2021. Skeletal muscle thermogenesis enables aquatic life in the smallest marine mammal. Science 373, 223–225.

Yazdi, P., Kilian, A., Culik, B.M., 1999. Energy expenditure of swimming bottlenose dolphins (*Tursiops truncatus*). Mar. Biol. 134 (4), 601–607.

CHAPTER

3

Thermoregulation

Arina B. Favilla, Stephanie K. Adamczak, and Daniel P. Costa

Department of Ecology & Evolutionary Biology, Institute of Marine Sciences, University of California, Santa Cruz, CA, United States

Abstract

The aquatic environment poses unique thermoregulatory challenges for marine mammals due to its high thermal conductivity. Despite these challenges, dolphins have adapted to a multitude of thermal regimes across their cosmopolitan distribution. This chapter addresses the thermoregulatory adaptations of dolphins, and other whale species, that allow them to inhabit such a wide range of habitats. We discuss the role of blubber, body size, and morphology in maintaining thermal balance, as well as associated physiological adaptations (e.g., countercurrent heat exchangers, peripheral vasoconstriction). Additionally, we discuss the importance of behavior for thermoregulation, such as the role of exercise and diving. We demonstrate the importance of morphological, physiological, and behavioral adaptations that enable dolphins to survive in their respective thermal habitats. We show that thermoregulatory demands often conflict with other physiological processes such as reproduction, integument repair, and energy storage. This chapter provides a holistic view of the thermoregulatory adaptations of a widely distributed species, the bottlenose dolphin.

Introduction

Water is roughly 25 times more conductive than air (Dejours, 1987). Thus, heat is quickly drawn away from the body in the aquatic environment, even across relatively small temperature gradients. Still, marine mammals maintain a constant and relatively high core body temperature (~37°C) across a range of ambient temperatures defined by their thermoneutral zone (Fig. 1). This gives marine endotherms an energetic advantage over their ectothermic equivalents at colder temperatures (Cairns et al., 2008; Grady et al., 2019). However, maintaining a large temperature gradient between the core and surrounding water requires significant energetic expenditure.

Marine mammals have a basal metabolism that ranges from 1–3 times that of terrestrial eutherian mammals. The thermoregulatory requirements associated with resting metabolism account for 70% of their daily energy expenditure (Williams et al., 2001; Davis, 2019). Within their thermoneutral zone, the metabolic rate is constant over a range of ambient temperatures and is bounded by the lower and upper critical temperature thresholds (Fig. 1). Over this range, compensation to heat flux variations is managed by insulation adjustments (i.e., peripheral blood flow) and blood flow to the appendages. At ambient temperatures outside of this range, animals employ mechanisms to produce or dissipate heat that requires increased energy expenditure and consequently an increase in metabolism (Fig. 1). For most marine mammals, heat conservation is the primary concern. As such, marine mammals have various adaptations for minimizing heat loss, such as increasing their insulation. Still, mechanisms for heat dissipation are needed, especially during periods of activity or when they are in warmer habitats that exceed their upper critical temperature (UCT).

Yet, it is important to remember that thermoregulation is a balancing act—not only in terms of heat balance but also for the coordination of competing physiological demands associated with diving, exercise, digestion, molting, and skin growth/repair (Favilla and Costa, 2020). For example, daily activities, such as foraging and transiting, elevate metabolism above basal levels and thus increase heat production, impacting calculations of the thermoneutral zone. Additionally, cardiovascular adjustments are integral to these responses, and those required for thermoregulation may, at times, conflict with other concurrent demands. For example, blood flow patterns for dissipating excess heat generated from intense foraging activity would conflict with oxygen conservation during a dive. Thus, thermoregulation is not an isolated response, but rather it depends on the coordination of several responses based on the animal's current homeostatic conditions.

The Physiology of Dolphins
https://doi.org/10.1016/B978-0-323-90516-9.00009-9

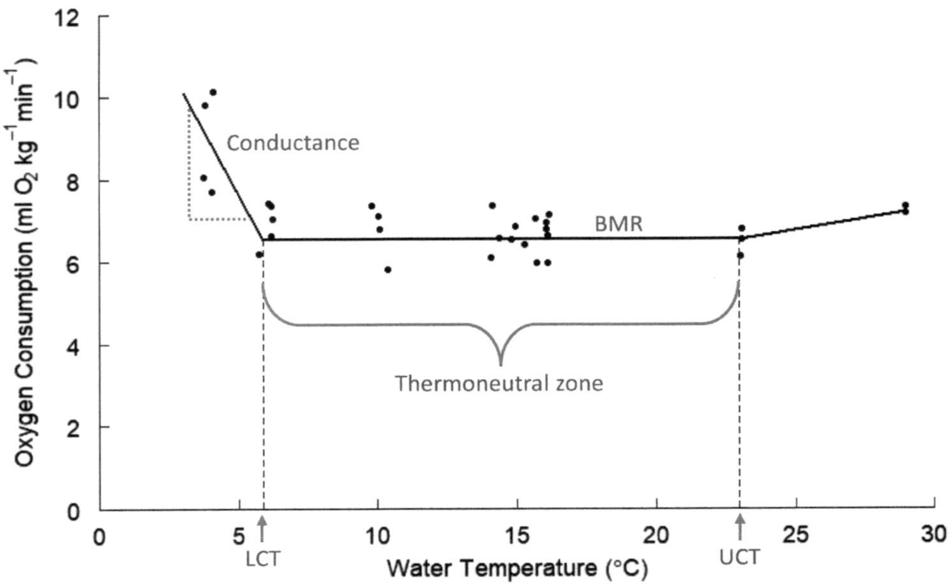

FIG. 1 The relationship between resting metabolic rate and water temperature for three adult male, post-absorptive bottlenose dolphins (mean body mass = 148.6 kg). Dolphins were acclimatized to a water temperature of 15.4°C. Each point represents a single measurement. The thermoneutral zone is the range of temperatures where the metabolic rate is at a minimum, i.e., BMR = 6.53 0.16 mL O_2 kg^{-1} min^{-1}, mean for all three dolphins. Above the UCT (23.0°C) and below the LCT (5.9°C), metabolic rate increases due to the physiological mechanisms required to reestablish thermal balance. *Adapted with permission from Williams et al. (2001).*

To take our understanding of thermoregulation to the next level, it is best to use a study species that inhabits a range of thermal regimes and can be accessed both in laboratory and wild settings. The best-studied dolphin is the bottlenose dolphin (*Tursiops* spp.), a cosmopolitan species inhabiting coastal and offshore waters (Wells and Scott, 2018) and kept in managed care in several research facilities and aquaria worldwide. Coastal nonmigratory populations in Sarasota Bay, FL, experience considerable seasonal changes in water temperature, ranging from 11 to 33°C (Barbieri et al., 2010), whereas other populations off Scotland withstand temperatures as low as ~5°C in the winter (Wilson et al., 1997; Fougères, 2008). In addition to seasonal temperature variation, populations that perform deep dives offshore or periodically transition to inshore habitats will experience abrupt temperature changes over short timescales.

This chapter describes the morphological, physiological, and behavioral adaptations that allow homeothermy in bottlenose dolphins, and some other cetaceans, across a broad range of thermal habitats. In addition to region-specific thermal challenges, the tradeoffs between thermoregulation, reproduction, integument repair, and energy storage are considered. We examine how form meets function by comparing the thermoregulatory adaptations and costs in bottlenose dolphins across different thermal habitats, different ages, and reproductive states.

Avenues of heat transfer

Dolphins are fully aquatic and must maintain thermal balance in a highly conductive environment with limited avenues of heat transfer. Typically, animals exchange heat with their environment via radiation, convection, conduction, and evaporation. However, for aquatic animals, such as the bottlenose dolphin, heat exchange predominantly occurs via conduction and convection. Conductive heat transfer occurs through direct contact with a solid or nonmoving fluid material. In contrast, convective heat loss is associated with heat transfer as a fluid moves across the animal's body surface. Both mechanisms depend on the temperature difference between the animal's body surface and environment. Animals minimize heat loss by reducing the temperature differential between their skin and the environment. Marine mammals employ various adaptations to maintain their skin temperature within a few degrees of water temperature (Irving and Hart, 1957; Barbieri et al., 2010).

In contrast, radiation and evaporation are less important mechanisms of heat exchange in bottlenose dolphins. Radiation is the emission of infrared energy by all objectives due to its temperature. This form of heat transfer is relatively insignificant compared to the other forms of heat flux because radiation is quickly absorbed by water. Similarly, respiratory evaporative heat loss only occurs while breathing at the water's surface. Dolphins, like other cetaceans, lack nasal turbinates that reduce heat and water loss while breathing (Hillenius, 1992). In seals, respiratory heat loss contributes to roughly 10% of the total heat dissipation [gray seals; (Folkow and Schytte Blix, 1987)]. As dolphins lack nasal turbinates, their respiratory evaporative heat loss is likely to be higher, however, differences in respiratory patterns between pinnipeds and cetaceans may also contribute to different levels of respiratory evaporative heat loss. Due to the increased permeability of their skin, some cetaceans have high cutaneous water loss. However, this water loss is primarily associated with osmoregulation (Hui, 1981; Ortiz, 2001; Costa, 2018). As dolphins are entirely aquatic, they cannot rely on evaporative heat loss. They must therefore rely on other mechanisms besides sweating to dump heat.

Thermal adaptations span morphology, physiology, and behavior

Heat loss (Q; $\mathrm{W\,m^{-2}}$) can be estimated using Newton's law of cooling: $Q = C(T_b - T_a)$ where C is thermal conductance ($\mathrm{W\,m^{-2}\cdot°C^{-1}}$), T_b is core body temperature (°C), and T_a is ambient temperature (°C). However, this requires that skin temperature is at or near water temperature, core body temperature is constant (no heat storage or net heat loss), and there is no evaporative heat loss. As water temperature changes, marine mammals can maintain their core body temperature at ~37°C by adjusting their thermal conductance or heat production. For example, metabolism increases when ambient temperatures drop below the lower critical temperature (LCT), whereas thermal conductance decreases to keep the body temperature constant (Fig. 1). Thermal conductance is a measure of how much heat transfers through a material and is dependent on the material's thermal properties and thickness. Thermal conductance can be broken down into two components: whole-body conductance and blubber conductance. Whole-body conductance is represented by the slope of the line relating changes in metabolism to ambient temperatures below the LCT (Ryg et al., 1993). Long-term changes in whole-body conductance (days to months) vary with changing blubber thickness and composition. Blubber generally has low thermal conductance and thus high insulating capability because insulation is the inverse of conductance. However, blubber conductance can change within seconds as peripheral blood flow through the blubber layer changes. Both mechanisms effectively modify the thermal conductance of the periphery, albeit on different timescales (i.e., days vs seconds, respectively).

As changes in insulation are constrained by morphology and physiology, marine mammals can also increase their metabolic heat production to maintain thermal balance in cold water. Marine mammals have elevated metabolisms compared to terrestrial animals as a mechanism to offset the greater thermal demands of the aquatic environment (Williams et al., 2001; Costa and Maresh, 2017; Davis, 2019). Some authors dispute this and argue that there is no difference in the metabolism between terrestrial and marine mammals (Lavigne et al., 1986a, b; Innes and Lavigne, 1991). However, many factors determine mammals' resting or basal metabolic rate (BMR), including body mass, body composition, diet, reproductive status, age, sex, and phylogeny (McNab, 2001). More importantly, marine mammals are composed of at least five different evolutionary lineages and may exhibit distinctly different metabolic adaptations associated with their phylogeny.

The BMR of bottlenose dolphins has been measured (Williams et al., 1999a; Noren et al., 2013; van der Hoop et al., 2014; Noren et al., 2017; Pedersen et al., 2020), in addition to measurements from the spinner dolphin (*Stenella longirostris*; Hampton and Whittow, 1976), Pacific white-sided dolphin (*Lagenorhynchus obliquidens*; Rechsteiner et al., 2013), harbor porpoise (*Phocoena phocoena*; Kanwisher and Sundnes, 1965), killer whale (*Orcinus orca*; Worthy et al., 2014), beluga whale (*Delphinapterus leucas*), and a gray whale calf (*Eschrictius robustus*; Wahrenbrock et al., 1974). Although results are equivocal among cetacean species (e.g., Worthy et al., 2014), bottlenose dolphins have a BMR 2.5–3 times the predicted BMR of terrestrial carnivores or eutherian mammals, respectively (Williams et al., 2001; Davis, 2019). They are also among the few cetacean species for which their lower and upper critical temperatures have been empirically determined (Fig. 1). However, these thermal limits that define the thermoneutral zone curve will differ when daily activities are considered since field metabolic rate is up to 3–6 times the predicted basal metabolic level in bottlenose dolphins (Costa and Williams, 2000; Rimbach et al., 2021).

Activity is essential in maintaining thermal balance for some delphinids, a family of small to medium-sized odontocetes. Like the Hawaiian spinner dolphin, smaller dolphin species remain active in cold water to keep warm (Hampton et al., 1971; Hampton and Whittow, 1976). This allows them to use the heat generated by working muscles to offset thermoregulatory costs. In contrast, the larger Pacific bottlenose dolphins may need to dissipate excess heat during activity to prevent overheating (McGinnis et al., 1972). As discussed further below, size plays an important role in mediating heat flux and thus thermal balance.

The role of size and morphology

Morphology, or an animal's size and shape, plays an important role in thermoregulation. The surface area-to-volume helps determine the heat lost to the environment relative to metabolic heat production. Heat loss occurs across the body surface, whereas metabolic heat production occurs throughout the body's volume. The surface area-to-volume ratio is predominantly modulated by size due to the scaling of each component relative to body length. Surface area scales with size squared (length^2), whereas volume scales with size cubed (length^3). Therefore, larger animals have a greater volume for a given surface area, resulting in lower ratios. The smaller exposed surface area for a given amount of heat produced also means larger animals will have a greater ability to retain heat. Conversely, smaller animals have larger surface area-to-volume ratios, yielding higher heat loss.

Facilitated by the reduced gravitational field and fluid dynamics of the marine environment, marine mammals have converged on a hydrodynamic body shape and relatively large body size. Cetaceans in particular contain many of the largest species (e.g., the blue whale, *Balaenoptera musculus*) that have a much lower surface area-to-volume ratio than their terrestrial counterparts. In addition to thermoregulation, body size and shape also play an important role in hydrodynamics. The streamlined body shape of marine mammals aids in reducing drag and locomotion costs (Fish, 1994; Fish et al., 2008). Additionally, an animal's shape can inform its foraging strategy (Woodward et al., 2006). Therefore, morphology is ultimately a result of optimizing body size and shape for all these functions. Interestingly, the thermoregulatory implications of body size are thought to be a fundamental constraint that has determined the minimum body size of marine mammals (Gearty et al., 2018).

Bergmann's and Allen's rules

Thermoregulatory costs are not only influenced by body size (and thus surface area-to-volume ratios) but also appendage size. Latitudinal changes in these morphological features are associated with the thermal constraints of living in a cold environment (Worthy and Edwards, 1990; Adamczak et al., 2020). Larger animals with lower surface area-to-volume ratios typically inhabit colder climates at higher latitudes, which is known as Bergmann's rule. Allen's rule posits that terrestrial animals inhabiting colder climates typically have smaller appendages to decrease surface area. Cetaceans and other marine endotherms follow Bergmann's rule across taxa (Torres-Romero et al., 2016) and within taxa (e.g., Adamczak et al., 2020). However, there is conflicting evidence for Allen's rule in cetaceans (e.g., Fougères, 2008; Adamczak et al., 2020).

The broad distribution of bottlenose dolphins provides the opportunity to further explore Bergmann's rule within a species (Fougères, 2008). When examining dolphin populations spanning 48 degrees latitude in the northern hemisphere, the largest dolphins in total length inhabit the coldest waters off Scotland. In contrast, the smallest dolphins in total length and surface area were found where the mean summer temperature was warmest (i.e., coastal FL in Fougères, 2008). Similarly, when looking at populations spanning both the northern and southern hemispheres (Fig. 2), dolphins in Shark Bay, Australia, are smaller in total length than bottlenose dolphins in FL, as well as greater in surface area-to-volume ratio than bottlenose dolphins in FL and Scotland. Thus, Shark Bay dolphins have a much larger surface area relative to size across which to dissipate heat. Their small size may be in part due to the thermoregulatory challenges of living in warm waters. However, optimal body size is also influenced by other physiological and ecological processes, such as the increased diving ability associated with larger body size and the environmental advantages or disadvantages of body size related to prey distribution (Hückstädt et al., 2016).

Allometric morphological changes also influence thermal dynamics with ontogenetic growth. As dolphins mature, their surface area-to-volume relationship changes. Their surface area increases slower than the corresponding increase in volume. These allometric changes have thermal consequences due to the scaling between body volume and metabolic heat production. Younger (i.e., smaller) animals have higher total energy expenditure than those of conspecific adults, yielding more internal heat production relative to their size (Rimbach et al., 2021). The large size of Scottish dolphins should correspond with lower mass-specific metabolic rates than their warmer water counterparts, but this would conflict with the need to maintain thermal balance in their cold habitat. It remains to be determined if there are differences in metabolism along with the factors that drive these ontogenetic and population differences in surface area-to-volume relationships.

Appendage shape and size play an important role in thermoregulation as appendages can increase an animal's surface area without significantly increasing internal heat production, thereby increasing heat loss. Allen's rule was formulated based on terrestrial animals, with those inhabiting colder climates having smaller appendages to reduce

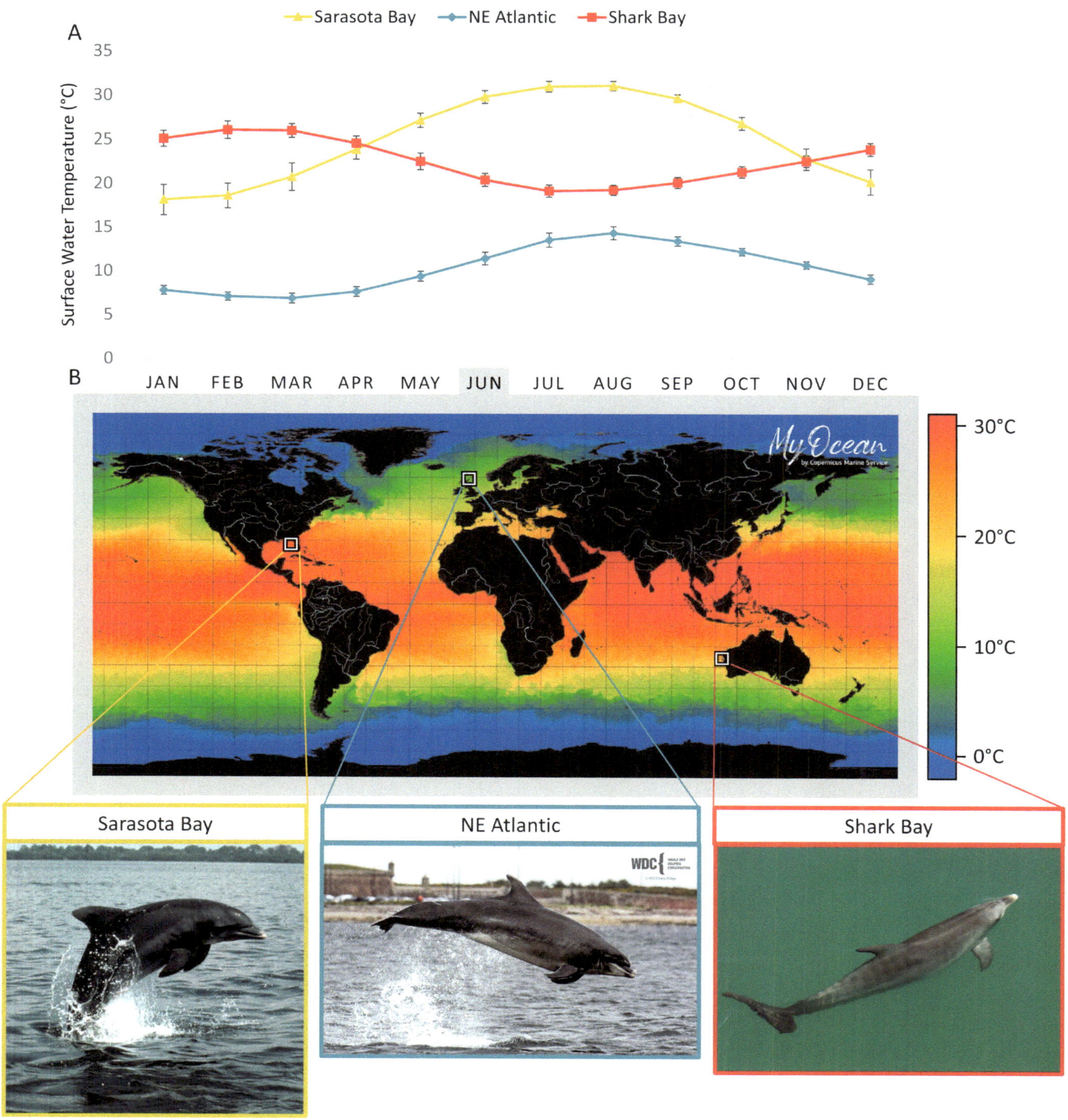

FIG. 2 Annual trend of average monthly climatology (1993–2019) of surface water temperatures off northern UK in Moray Firth (58.0°N, 3.0°W), Sarasota Bay, FL (27.36°N, 82.59°W), and Shark Bay, Australia (25.65°S, 113.65°E), with error bars showing standard deviation. The locations of the three bottlenose dolphin populations are shown on the map, with average monthly surface water temperatures in June 2019 depicted by colour. *Data and maps obtained from the E.U. Copernicus Marine Service Information, Product GLOBAL_MULTIYEAR_ PHY_001_030. Sarasota Bay dolphin photo by the Chicago Zoological Society's Sarasota Dolphin Research Program, taken under NMFS permit no. 638. NE Atlantic dolphin photo by Charlie Phillips WDC. Shark Bay dolphin photo by Jana Watson-Capps, Shark Bay Dolphin Research Project.*

heat loss. Interestingly, some cetaceans exhibit a pattern of appendage size opposite that of Allen's rule. For example, the northern species (higher latitude) of pilot whale has a larger overall appendage surface area than their more southern (lower latitude) counterparts (Adamczak et al., 2020). Given the large body size and thick blubber layer of the northern species of pilot whale, larger appendages likely provide excellent thermal windows that can dissipate heat during activity and conserve heat while resting. However, the potential reversal of Allen's rule is less clear for bottlenose dolphins as appendage size appears to be more related to coastal or offshore habitats than a function of water temperature or latitude. Coastal bottlenose dolphin populations have a greater appendage surface area than offshore populations, which could improve maneuverability in a more complex coastal habitat (Fougères, 2008). If chasing prey in the coastal environment requires greater bursts of activity, larger appendages could also facilitate heat loss as thermal windows.

Appendages as thermal windows: Where morphology and physiology meet

Large appendages serve as dynamic thermoregulatory control surfaces because they are mostly uninsulated and contain countercurrent heat exchangers (CCHEs; Elsner et al., 1974; Pabst et al., 1999). CCHEs function by juxtaposing outgoing, warm arterial blood with cool venous blood returning from the periphery (Fig. 3). The vascular structure of CCHEs in appendages allows for efficient heat exchange control between the arterial and venous blood supply. Major arteries are located centrally within a ring of thin-walled veins to conserve heat in the body's core. The rate of blood flow is key for controlling heat regulation through CCHEs. High blood flow into the appendages expands the circumscribed artery, reducing venous return within the CCHE and providing little time for heat exchange (Scholander and Schevill, 1955). This effectively dumps heat into the environment. When heat conservation is a priority, a reduction in peripheral blood flow reduces the flow of warm blood to the appendage, facilitating heat transfer to the returning venous blood. Thus, by altering blood flow in the appendages, the tissue conductance is rapidly changed to conserve or dissipate heat, respectively (Elsner et al., 1974).

CCHEs are found in the dorsal fin, pectoral flippers, and flukes of cetaceans (Fig. 4). Many studies have investigated their thermoregulatory function relative to water temperature, activity level, and diving behavior, focusing on the dorsal fin of dolphins. The underlying vasculature in the dorsal fin results in a spatially heterogeneous thermal surface (Fig. 5, Meagher et al., 2002; Plön et al., 2018). The regulation of blood flow results in temporal changes in skin temperature and heat flux. Skin temperature of the dorsal fin responds to changes in behavioral state (McGinnis et al., 1972; Westgate et al., 2007) and water temperature and is usually within 1°C of water temperature (Noren et al., 1999; Williams, 1999; Meagher et al., 2002, 2008; Barbieri et al., 2010). The temperature differential between the dorsal fin and ambient water is constant across seasons, thus resulting in a large change in the temperature gradient between

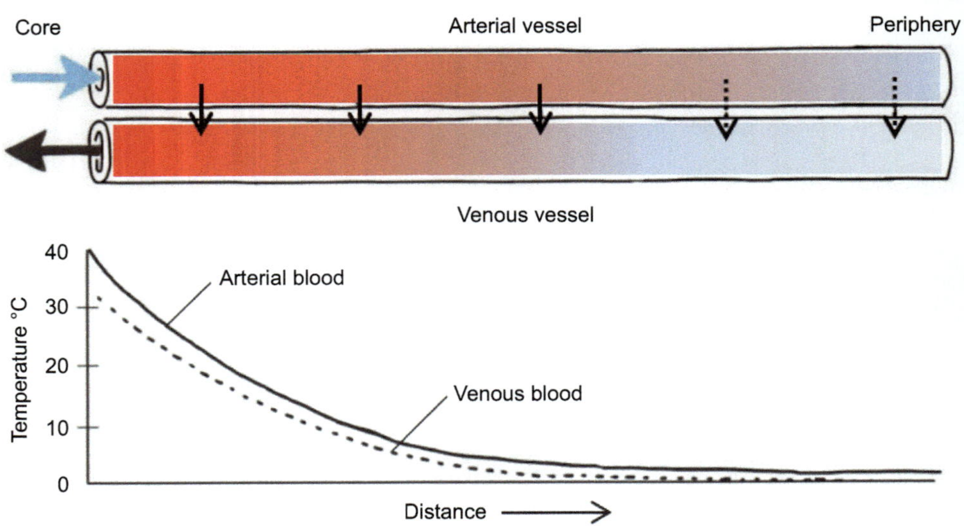

FIG. 3 Schematic of countercurrent heat exchanger showing how heat transfers between juxtaposed arterial and venous vessels. Warm arterial blood traveling from the core transfers heat to cool venous blood returning from the periphery, resulting in temperature gradients along the length of the vessel such that blood in the core body is kept warm. *Reprinted with permission from Davis (2019).*

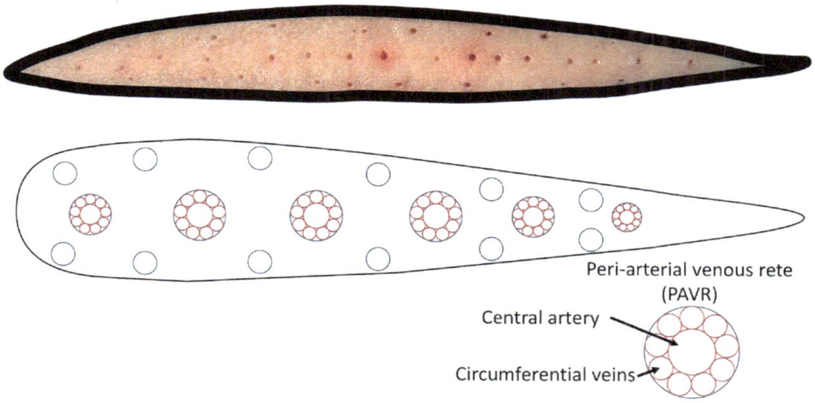

FIG. 4 Dorsal fin cross section from a harbor porpoise (photo courtesy of The Long Marine Lab Marine Mammal Stranding Network) and schematic showing the arrangement of countercurrent heat exchangers in the dorsal fin. Countercurrent heat exchangers contain central arteries surrounded by veins that form peri-arterial venous retia (PAVR). *Redrawn from Meagher et al. (2002).*

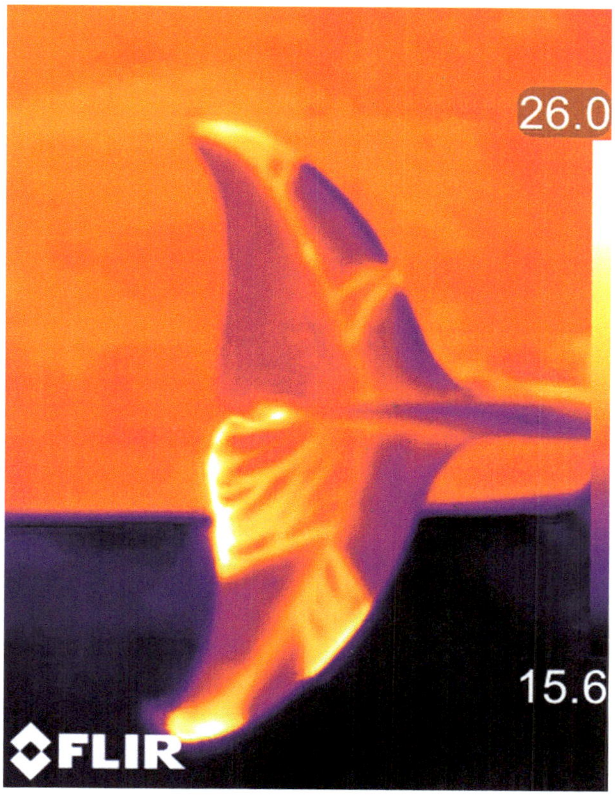

FIG. 5 Infrared thermal image of a bottlenose dolphin fluke showing warmer areas in white and yellow that correspond to vasculature in the appendage. *Image used with permission from NMFS, permit #24054.*

the body core and surface when water temperatures are lower (Fig. 6; Barbieri et al., 2010). A temperature gradient between body core and surface as large as 23°C in the winter (compared to 4–7°C in the summer) is possible due to changes in blubber thickness and lipid content (Barbieri et al., 2010). Seasonal patterns in heat flux suggest that appendages consistently dump heat in the summer but are more variable thermoregulatory surfaces in the winter. Greater variability within seasons in heat flux in the appendages also supports their flexible role as heat dissipators (Meagher et al., 2008).

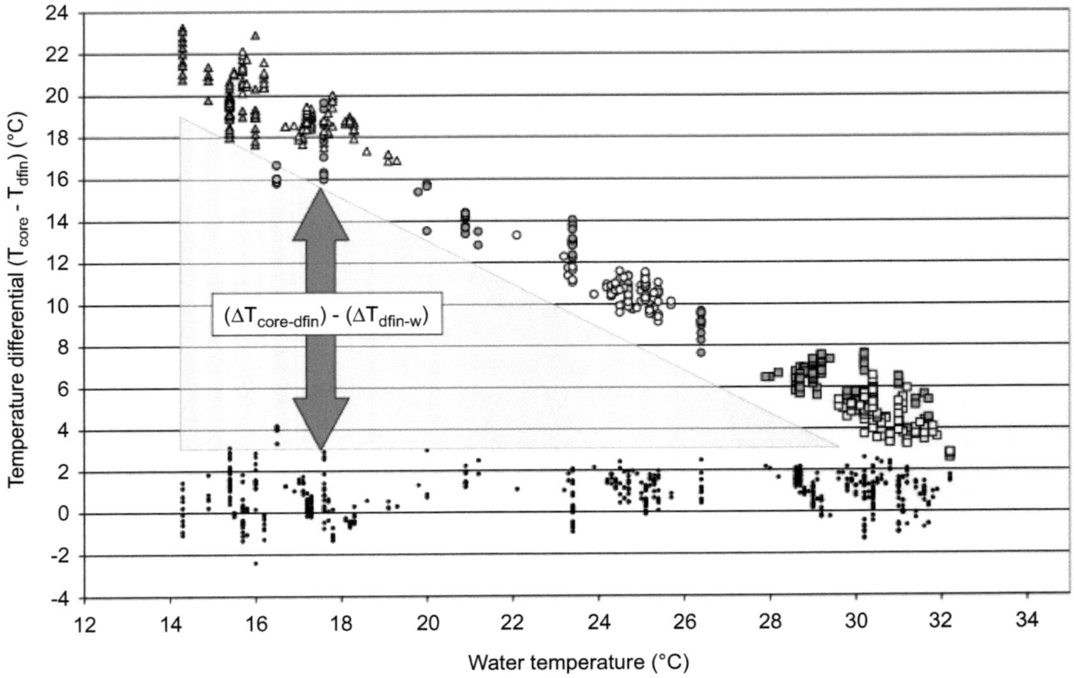

FIG. 6 Difference between temperature differentials of bottlenose dolphins. Temperature differential decreases with increasing water temperatures from 14 to 33°C as depicted by the shaded region. Small dots represent temperature differential between the dorsal fin and water ($\Delta T_{dfin\text{-}w}$). Larger shapes represent temperature differential between core body temperature (assumed to be 37°C) and dorsal fin ($\Delta T_{core\text{-}dfin}$) in different field seasons (filled circle, November 2002; open circle, November 2003; filled square, June 2003; open square, June 2004; filled triangle, February 2003; open triangle, February 2004). *Reprinted with permission from Barbieri et al. (2010).*

Thermal effects of exercise and diving

The ability to induce rapid, short-term changes in heat flux using CCHEs in the appendages is beneficial for adjusting thermal balance after exercise (Kanwisher and Sundnes, 1966; Hampton et al., 1971; Meagher et al., 2008). Another mechanism to avoid activity-induced thermal stress is to use the separate thick-walled superficial veins in the appendages as conduits to bypass the CCHE (Fig. 4). When cold blood preferentially returns in these vessels, the body is cooled (Scholander and Schevill, 1955). A separate CCHE near the reproductive organs facilitates regional cooling with the body core (Fig. 7). Mammalian testes need to be cooler than the core body temperature, or they become infertile. However, dolphins have internal testes. They use cold blood from the dorsal fin and fluke to keep the intraabdominal testes cooler than 37°C (Rommel et al., 1992). This mechanism is especially critical during exercise since the reproductive organs of dolphins lie in between two major locomotor muscles, which generate a lot of heat. Colonic temperatures taken at rest and immediately after exercise in male dolphins showed regional cooling (up to 1.3 and 0.5°C, respectively) and provided evidence for the function of the reproductive CCHE (Rommel et al., 1994; Pabst et al., 1995).

Similarly, for pregnant females, a CCHE brings cool blood to the uterus. This dissipates excess heat and avoids increases in fetal temperature that could result in detrimental developmental effects (Fig. 7; Rommel et al., 1993, 1998). For most mammals, some fetal heat (~15%) is generally dissipated through the mother's abdomen. However, in dolphins, the locomotory muscles and the insulating blubber layer make this unlikely. The CCHE near the uterus allows the temperature gradient between the fetus and surrounding maternal tissue to be favorable for dissipating fetal heat even during exercise.

As previously described, changes in peripheral blood flow are associated with the thermal demands of the animal. However, peripheral blood flow is also modulated by the dive response. To conserve oxygen during a dive, blood flow is reduced to the more hypoxic tolerant organs and tissues, including the appendages. While diving in cold water, the dive response and the thermal response are synchronized, and decreased blood flow to the appendages attenuates heat dissipation at depth (Noren et al., 1999; Williams et al., 1999b). Studies using trained dolphins showed that when the animal returns to the surface at the end of the dive, heat flux increases at the dorsal fin—likely due to enhanced blood flow associated with the anticipatory tachycardia—and remains elevated during the surface interval. Dissipating heat at depth would require temporarily overriding the dive response to perfuse peripheral thermal windows. However,

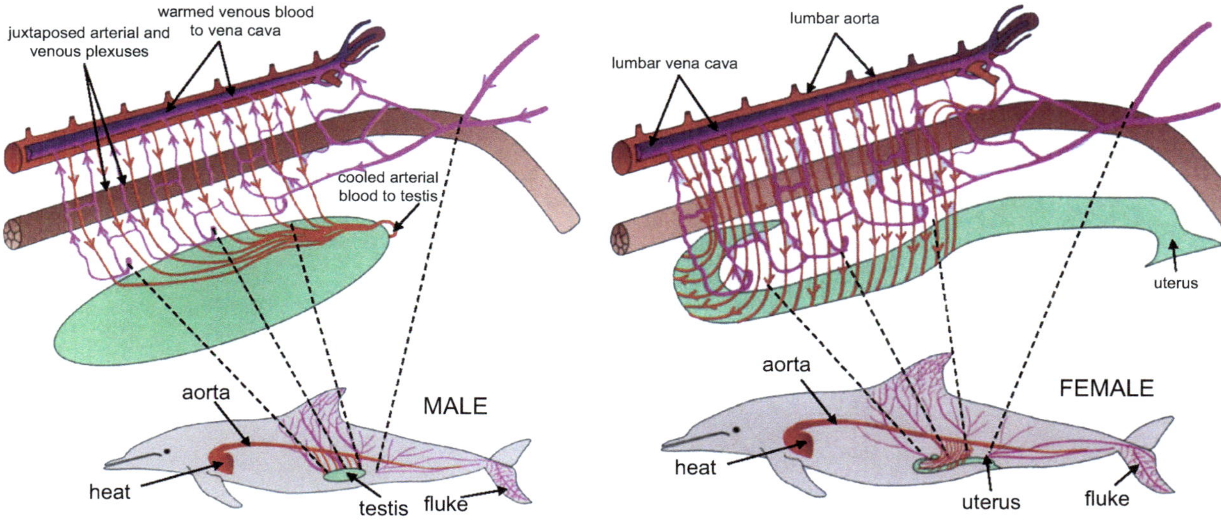

FIG. 7 Countercurrent heat exchanger associated with the intra-abdominal testis of males (left) and uterus of females (right). Cool blood returning from the dorsal fin and fluke forms a venous plexus that is juxtaposed with arteries. Blood from the aorta is cooled before reaching the reproductive organ to prevent overheating of the testis or developing fetus in the uterus. *Adapted with permission from Rommel et al. (1998).*

this does occur in certain situations. Attenuation of heat dissipation at depth was recorded in a bottlenose dolphin that had performed several jumps out of the water before diving (Noren et al., 1999). Because cardiovascular adjustments are integral to the dive response, exercise response, and thermoregulation, a balancing act is required to meet these potentially conflicting physiological demands to maintain homeostasis.

While appendages, particularly the dorsal fin, are the principal body surfaces to regulate thermal balance, the body trunk provides a large surface area for heat transfer. The vascularization of the blubber layer also facilitates changes in the body conductance in response to changing water temperature or activity level. For example, dolphins chased in the Eastern Tropical Pacific tuna fishery had elevated body surface temperatures, consistent with greater peripheral blood flow to the integument for dissipation of excess heat (Pabst et al., 2002). The fact that blubber is internal and vascularized allows for the easy bypass of this thick insulation when necessary. The following section details how blubber's structure and composition affects its thermal properties and how that changes with ontogeny, reproductive state, season, and diet. All these factors influence the thermoregulatory role of blubber, which is only one of its many roles, resulting in many tradeoffs.

Blubber structure and its thermoregulatory function

Blubber is the vascularized adipose tissue that serves important physiological roles supporting energy balance, maintaining water balance while fasting, streamlining, and aiding in thermoregulation for dolphins and other cetaceans (Iverson, 2009). Dolphin blubber is a lipid-rich tissue filled with adipocytes, or fat cells, typically composed of triacylglycerols (TAGs). While deeper-diving cetacean species have blubber that is predominantly composed of waxy esters (WEs), most dolphin species examined to date have little to no WEs in their blubber (Pabst et al., 1999; Koopman, 2007, 2018; Bagge et al., 2012). Here, we focus on how blubber quality (i.e., the insulative properties) and quantity (i.e., the amount of insulation) relate to dolphin thermoregulation, while noting the competing functions that it serves.

The insulative properties of blubber are heavily dictated by blubber lipid content. Blubber lipid content is driven by the size and density of adipocytes. Adipocytes form during fetal development and proliferate throughout the gestational period (Struntz et al., 2004). However, it is posited that postnatal changes in blubber lipid content are driven by changes in adipocyte size rather than the addition or subtraction of adipocytes (Struntz et al., 2004). Mean adipocyte size typically increases until maturity (Fig. 8).

In contrast, bottlenose dolphin blubber lipid content increases until the animal reaches a juvenile stage and declines into adulthood (Fig. 9; Struntz et al., 2004; Dunkin et al., 2005). Blubber thickness, relative to total body size, begins to decline at the juvenile stage as the rate of adipocyte growth is slower than total body growth. In emaciated animals, the

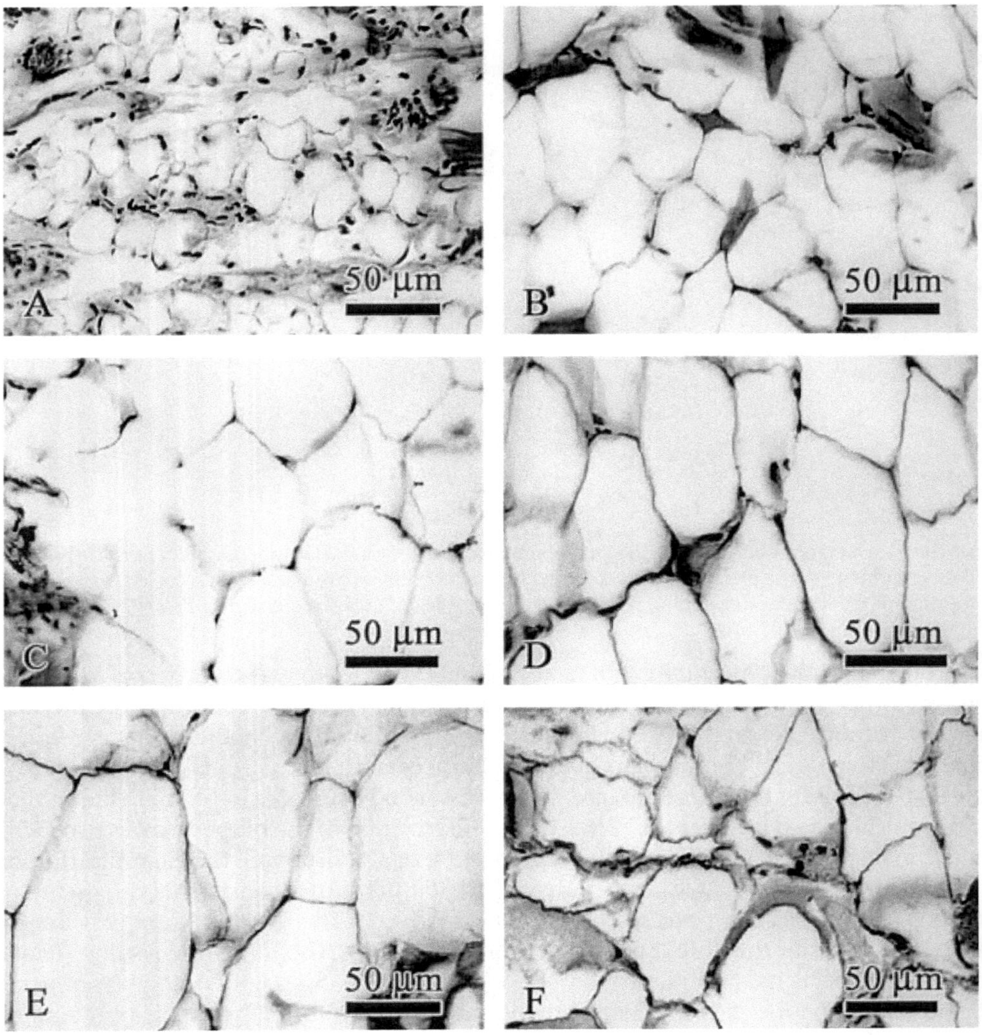

FIG. 8 Light micrographs of blubber from the mid-thoracic region of bottlenose dolphins showing the difference in adipocyte size in a fetus (A), neonate (B), juvenile (C), subadult (D), adult (E), and emaciated adult (F). *Reprinted with permission from Struntz et al. (2004).*

adipocytes shrink, resulting in a blubber lipid content that mimics a neonate's (Fig. 8; Struntz et al., 2004; Dunkin et al., 2005). Similarly, emaciated animals and neonates have a high blubber water content due to the inverse relationship between blubber lipid and water content (Dunkin et al., 2005). Sub-adult and pregnant dolphins have the most insulative blubber due to the complex relationship between body size, blubber lipid content, and blubber insulative properties. In contrast, fetuses and emaciated adults have blubber with higher conductivity and thus are more poorly insulated (Fig. 9; Dunkin et al., 2005).

Although the entire blubber layer serves some insulative purpose for a dolphin, blubber is stratified into superficial, middle, and deep layers that serve slightly different functions. The superficial blubber layer has a high density of structural fibers and a low adipocyte cell count, suggesting its importance in streamlining rather than insulating (Montie et al., 2008). The middle blubber layer has the highest blubber lipid content resulting in a greater insulative capacity (Montie et al., 2008; McClelland et al., 2012). Middle and deep blubber layers typically have more vasculature than superficial layers (McClelland et al., 2012). Blood flow to these layers is critical for accessing and depositing fat stores, but peripheral perfusion also increases blubber conductivity resulting in a tradeoff between blubber as insulation and energy storage. However, the initial thermoregulatory costs of depositing fat stores may be offset by the thicker blubber layer that is often a response to seasonal changes in temperature or food availability, which is discussed further in the next section. This tradeoff is less of a concern in the superficial blubber layer that has reduced vasculature. With fewer veins carrying warm blood reaching the animal's surface, heat dissipation is reduced in this superficial layer. At times when heat conservation is a priority, reducing blood flow to the entire blubber layer will decrease blubber conductivity and minimize heat loss (McGinnis et al., 1972).

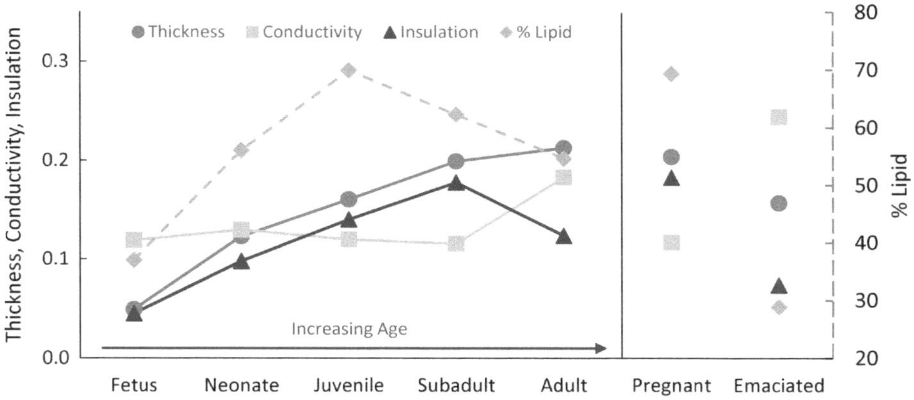

FIG. 9 Blubber thickness (dm, circles), thermal conductivity ($W m^{-1} deg.^{-1}$, squares), insulation ($m^2 deg. W^{-1}$, triangles), and lipid content (%, diamonds) shown for bottlenose dolphins of different age classes. For comparison, values for pregnant and emaciated adults are shown. *Redrawn from Dunkin et al. (2005).*

Intrinsic and extrinsic factors affecting blubber

The thickness of the blubber layer, often driven by the size of adipocyte cells, can change as a function of intrinsic factors such as sex, age, reproductive state, and nutritional state (Fig. 9) (Struntz et al., 2004; Dunkin et al., 2010; Ball et al., 2015). However, blubber thickness alone is typically a poor indicator of the insulative properties of blubber across different ontogenetic stages (Dunkin et al., 2005). Blubber thickness steadily increases from birth to maturation but can be modulated by energetically costly reproductive events such as pregnancy (Struntz et al., 2004; Dunkin et al., 2005).

Because of the multipurpose nature of blubber, dolphins must balance their thermoregulatory needs with the other necessary functions that blubber serves, including buoyancy and energy stores. During nutritional stress, adipocyte cells typically shrink as the lipid is mobilized, decreasing the thickness and insulating quality of the blubber layer, which may become too thin to provide sufficient insulation. A thick blubber layer is essential for thermoregulation and provides energy stores when food is scarce. However, a thick blubber layer also increases buoyancy and may increase energy expenditure to overcome buoyancy when diving (e.g., Adachi et al., 2014; Nousek-McGregor et al., 2014). The animal may also be at greater risk of heat stress due to the increased insulative quality of the blubber as adipocyte cells grow.

Blubber thickness and its associated tradeoffs between thermoregulation and energy stores vary seasonally with changes in water temperature. Typically, blubber thickness and mass decrease in warmer water (Meagher et al., 2008; Noren and Wells, 2009; Adamczak et al., 2021). Decreases in blubber thickness typically yield higher thermoregulatory costs for both male and female dolphins. We examined this relationship by modeling the average adult male and female dolphin (mass = 225 and 184 kg, respectively) using morphometric data provided by the Sarasota Bay Dolphin Research Project. For individuals with an average blubber depth of approximately 2 cm, thermoregulatory costs are relatively low, with temperatures above ~17°C resulting in no thermoregulatory costs for both sexes (Fig. 10). In colder water, a thicker blubber layer is required (~2.5 cm at 5°C) to reach a thermoregulatory cost of zero. However, in warmer water, the primary challenge may be dumping heat. In contrast to adults, younger dolphins are less capable of adjusting their blubber thickness in response to water temperatures despite being more vulnerable to the cold due to their small size (Noren and Wells, 2009).

Blubber's dynamic role in thermoregulation

Interestingly, blubber may aid in both heat generation and dissipation for dolphins. Dolphins have brown adipocytes within the deep blubber layer. These form a layer of brown adipose tissue along the entire body (Fig. 11; Hashimoto et al., 2015). Brown adipose tissue (BAT) is found in most mammals. It plays a vital role in producing heat through non-shivering thermogenesis. Through the futile cycling of protons across a protein (UCP1) in the mitochondrial membrane, BAT tissue generates heat without producing usable energy in the form of ATP. Thus, BAT upgrades the function of the blubber layer from merely insulating as a "conventional blanket" to producing heat as an "electric blanket."

Four species—bottlenose dolphin, Pacific white-sided dolphin, Dall's porpoise (*Phocoenoides dalli*), and harbor porpoise—that are known to have BAT all occur in the North Pacific Ocean where water temperatures are relatively cool. This enhanced heat generation capacity allows these species to survive in waters colder than their insulation alone

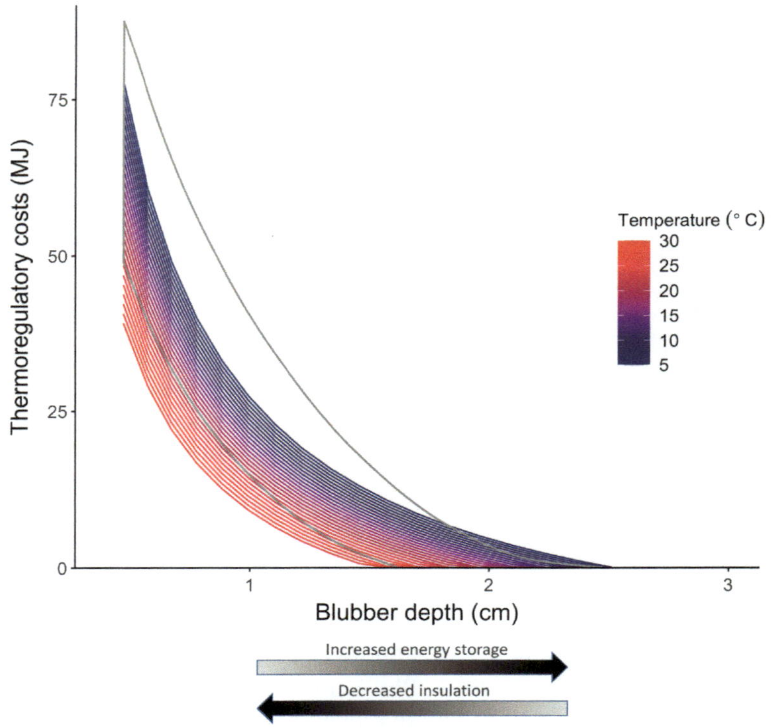

FIG. 10 Thermoregulatory costs were calculated as a function of blubber depth for an adult female bottlenose dolphin for water temperatures ranging from 5 to 30°C. Thermoregulatory costs between 5 and 30°C for an adult male bottlenose dolphin are represented by the gray polygon. Arrows depict the conflicting energy demands of blubber relative to its thickness. Costs are calculated using equations put forth by Hind and Gurney (1997).

FIG. 11 CT scan of a Pacific white-sided dolphin (*Lagenorhynchus obliquidens*) showing the distribution of brown adipose tissue (denoted by the white arrows) along the entire body. *Reprinted from Hashimoto et al. (2015).*

would allow. It may also help maintain thermal homeostasis during periods of inactivity (Hashimoto et al., 2015). Interestingly, the body sites where BAT was not found included the rostrum, fins, and flukes, of which the latter two are important thermal windows.

In contrast to the heat-generating properties of BAT, blubber may also play a role in dissipating heat for dolphins in warmer waters. Dolphins typically swim in warm coastal waters without overheating. This may be attributed to the redistribution of heat from the internal body to the blubber in times of thermal stress, which simultaneously decreases the rate of heat transfer from the warm water to the dolphin and cools the inner body to maintain normal function (Heath and Ridgway, 1999). Further, the deep layer of blubber has been described as a phase change material, meaning that it may act to store and release heat when necessary (Dunkin et al., 2005). When heat is shunted to the periphery via blood flow, blubber may temporarily store this heat and serve as a buffer zone when experiencing large temperature gradients (Dunkin et al., 2005; Bagge et al., 2012).

Thermoregulatory considerations for epidermal growth and repair

Besides the hypodermal blubber layer, the integument is composed of the dermal and epidermal layers, which provide some additional thermal resistance. These layers' quality and proper functioning are critical for ensuring a protective yet dynamic barrier.

As bare-skinned mammals, dolphins typically maintain their skin at or near ambient temperature. In cold water, this creates a problem as most mammalian tissues, including epidermal cells, have a limited ability to survive prolonged exposure at low temperatures. Further, most are not capable of cell division at low temperatures (Feltz and Fay, 1966). Wound repair and epidermal regeneration are typically associated with increased peripheral blood flow to the skin. However, in a cold environment, thermoregulatory requirements necessitate a reduction in peripheral blood flow. To circumvent this potential conflict, sea lions and seals haul out of the cold water to molt in a more favorable thermal environment (Paterson et al., 2022). While on land, they can increase the blood flow to the periphery to regenerate new hair and epidermal tissue without compromising their heat balance. How dolphins and other fully aquatic marine mammals solve this dilemma is unclear but is likely related to the complex structure of their dermal papillae (see below).

Unlike the annual molt of amphibious marine mammals, dolphins experience continuous epidermal growth facilitated by the structure of their dermal layer. Dolphins have a thick epidermis (roughly 15–20 times thicker than that of humans) with highly innervated dermal papillae protruding roughly half its thickness (Sokolov, 1960; Palmer and Weddell, 1964; Brown et al., 1983). The convoluted nature of the papillae provides a high surface area for epidermal cell proliferation (Fig. 12). This, alongside an increased rate of cell replication, leads to enhanced skin renewal and wound healing capacities (Bruce-Allen and Geraci, 1985). The outermost layer of skin is renewed roughly 12 times per day. While this high sloughing rate is energetically expensive, it is critical for maintaining a clean, smooth outer surface. Besides improving hydrodynamics (Pavlov, 2006), it allows cells to be mitotically active and replicate for wound healing (Hicks et al., 1985).

Observations of skin lesions in wild dolphin populations in regions with different environmental conditions have shown that the prevalence and severity of skin lesions are correlated with water temperature and salinity (Fig. 13) (Wilson et al., 1999; Croft et al., 2020; McClain et al., 2020). Dolphins near Scotland experience the lowest water temperatures and salinity and have the most severe lesions (Wilson et al., 1999). Although epidermal cells can proliferate at temperatures as cold as 17°C, the optimal temperature for proliferation is close to core body temperature (pinnipeds; Feltz and Fay, 1966). The grooves of the dermal papillae that sit deeper in the epidermis will be more shielded from the cold water. They may allow for faster proliferation compared to more superficial cells (Feltz and Fay, 1966). While warmer waters enhance cellular mitotic rate, a higher salinity helps create a degenerative layer of cells to serve as a buffer zone for the underlying wound since dolphins do not form scabs in the healing process (Bruce-Allen and Geraci, 1985; Corkeron et al., 1987). Therefore, the effectiveness of the skin as a physical barrier is dependent on environmental factors; water temperature is a significant determinant of skin health and blubber thickness.

The major effects of water temperature, specifically on skin health and regeneration, are clearly seen in another cosmopolitan delphinid species, the killer whale. Although much larger than bottlenose dolphins, they are the smallest cetacean to regularly inhabit Antarctic waters, where subzero water surface temperatures are typical (Durban and

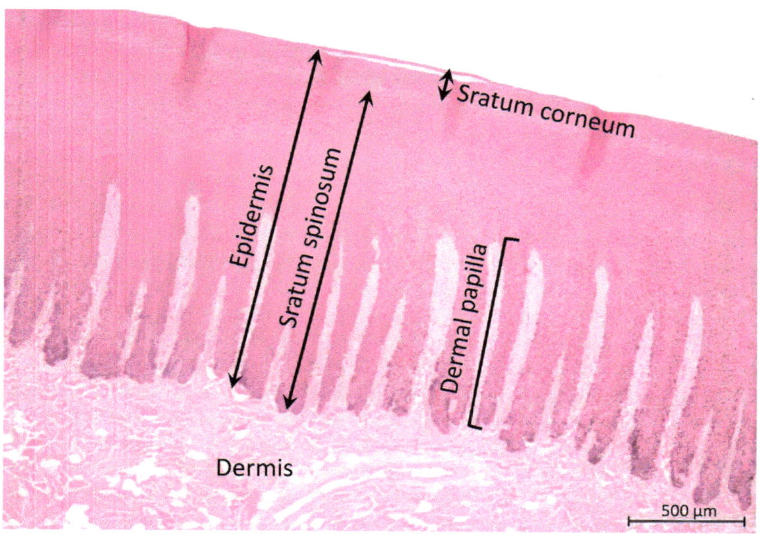

FIG. 12 Hematoxylin and eosin (H&E) stained photomicrograph of the dorsal skin of a bottlenose dolphin depicting the dermis, epidermis, stratum spinosum, dermal papilla, and stratum corneum. *Adapted with permission from Garten and Fish (2020).*

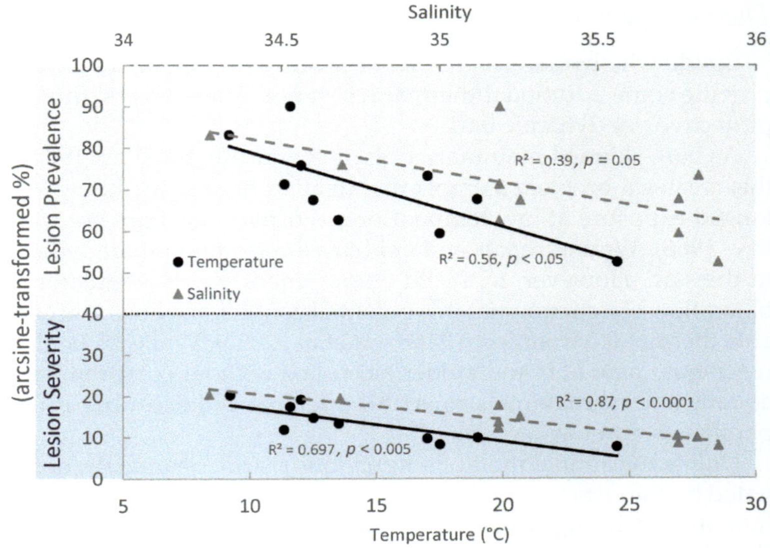

FIG. 13 Prevalence (data > 40%) and severity (data < 40%) of epidermal lesions in bottlenose dolphins from 10 coastal populations (Moray Fifth, Scotland; Shannon Estuary, Ireland; Cardigan Bay, Wales; Cornwall, England; Brittany, France; Doubtful Sound, New Zealand; Kvarnevic, Croatia; Sado Estuary, Portugal; Beaufort, North Carolina USA; Sarasota Bay, FL) in relation to mean annual water temperature (circles, solid lines) and salinity (triangles, dotted lines). *Redrawn from Durban and Pitman (2012).*

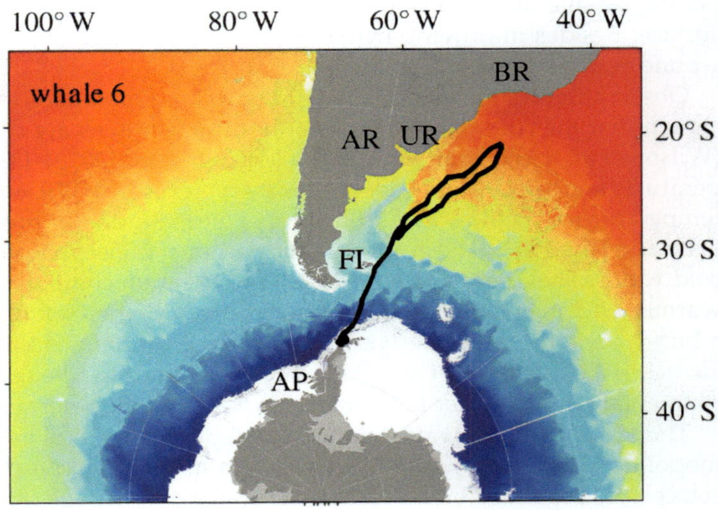

FIG. 14 Track (black line) of an adult female killer whale (ecotype B) tagged in 2010 (Feb 13–May 21) showing movement away from the Antarctic Peninsula (AP) northward past the Falkland Islands (FI) and reaching latitudes off Uruguay (UR) and Brazil (BR). *Adapted from Durban and Pitman (2012).*

Pitman, 2012). Tracking data revealed directed long-distance migrations from their Antarctic foraging grounds to subtropical waters off Uruguay and Brazil, where warmer waters would allow epidermal replacement (Fig. 14). These migrations were short (6–8 weeks) and occurred on a facultative, rather than seasonal, time frame (Pitman et al., 2019). Additionally, a yellowish film of diatoms on the whales' skin was observed following several weeks of foraging in Antarctic waters (Fig. 15). These observations support the hypothesis that trips to warmer waters serve a thermoregulatory role. Here they can molt in more favorable conditions to maintain healthy skin. Another strategy that other cetacean species foraging in the Antarctic may employ is timing their molt to coincide with their migration to warmer breeding grounds (Pitman et al., 2019).

Energetic tradeoffs with thermoregulation

The cost of thermoregulation can be represented by the energy needed to maintain constant body temperature minus internal heat generation from other metabolic processes (Hind and Gurney, 1997). Thermoregulatory costs vary with size, activity level, and reproductive state because internal heat generation depends on these factors. To explore

FIG. 15 A killer whale (ecotype B2) photographed near Gerlache Strait, Western Antarctic Peninsula showing a skin condition possibly related to an inability to molt properly while in frigid waters. *Reprinted from Pitman et al. (2019).*

how different metabolic demands influence thermoregulatory costs, we used equations put forth by Hind and Gurney (1997) for an average female bottlenose dolphin (184 kg and 249 cm; data provided by Sarasota Dolphin Research Program). As basal metabolism increases, which often occurs due to changes in mass resulting from growth, thermoregulatory costs steadily decline. This is because increases in basal metabolism result in greater heat generation, which will offset thermoregulatory costs in a cold environment. In contrast, the impact of increased activity levels resulting from locomotion on thermoregulatory costs is not quite as straightforward. At low velocities, internal heat generation is high enough to counteract heat loss from the animal's surface. However, at higher velocities (around 6 m/s for an average adult female dolphin) heat loss from the skin rapidly increases as cool water passes over the skin, resulting in convective heat loss. This results in an increase in thermoregulatory costs at high swim speeds (Hind and Gurney, 1997; Gallagher et al., 2018).

Increased costs from reproduction may similarly influence thermoregulation. A pregnant female has increased internal heat production due to the heat of gestation and energy input into fetal tissue development. Although gestation results in a slight increase in internal heat production from the growing fetus, it does not counteract heat loss from the skin surface (Fig. 16). However, lactation, the most energy-intensive part of reproduction, results in no thermoregulatory costs due to excess internal heat generation. Interestingly, the swimming speed at which heat production

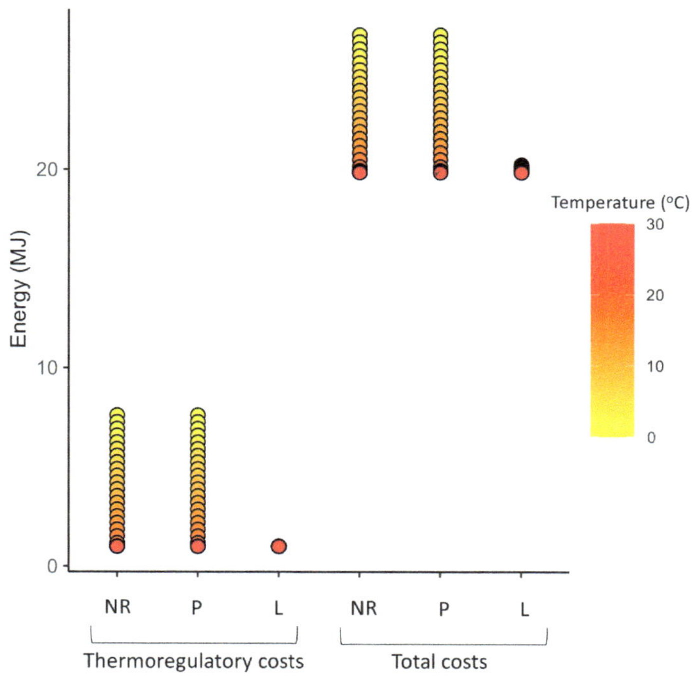

FIG. 16 Thermoregulatory and total energetic costs of a mature female in different reproductive states (NR, nonreproductive; P, pregnant; L, lactating) for water temperatures ranging from 5 to 30°C. Thermoregulatory costs are calculated using equations put forth by Hind and Gurney (1997). Pregnancy costs are calculated for a female with a near-term fetus and include the cost of fetal tissue investment (Boult et al., 2018) and the heat of gestation (Brody, 1968). Lactation cost derived from milk energy output (Cockcroft, 1989).

exceeds thermoregulatory needs for pregnant bottlenose dolphins was found to be only 1.5 m/s, while the optimal speed for dolphins is higher between 1.9 m/s and 3.2 m/s (Yazdi et al., 1999). This indicates that offloading heat may be more important to reproductively-active dolphins than generating heat due to their already heightened metabolic heat production.

There are two contrasting costs associated with foraging. The first is the cost of warming cold prey. The second is the energy liberated as a function of that prey's digestion and biochemical processing. The heat generated from the digestion of prey is known as the heat increment of feeding (HIF), also known as the specific dynamic action (Kleiber, 1975). HIF is characterized by an increase in metabolic rate following food consumption. For dolphins, it has been found that metabolic rate increases by a maximum of 57.1% following consumption of fish (Yeates and Houser, 2008). In other marine mammal species, such as the sea otter, HIF is an essential mechanism for heat production to offset heat loss while resting (Costa and Kooyman, 1984). It is unknown whether dolphins incorporate HIF as part of their thermal budget.

Due to these numerous energetic tradeoffs, the cost of thermoregulation fluctuates throughout an animal's lifetime (as a result of age, mass, and reproductive state) and even throughout the day (as a result of activity level and prey intake). Although the information presented here focuses on the costs associated with heat generation, there are likely costs associated with heat dissipation. The energetic consequences of dissipating heat in times of thermal stress have not yet been quantified. Still, they are likely an important part of a dolphin's thermal budget. As environments continue to change due to human influence, we can expect changes in temperature and activity budgets to influence the energy allotted to thermoregulation.

Concluding remarks

The wide range of thermal regimes that bottlenose dolphins inhabit due to their cosmopolitan distribution makes them an ideal species for studying thermoregulation. From warm, inshore habitats to colder, pelagic waters, bottlenose dolphins persist in temperatures ranging from 5 to 30°C, providing unique insight into the physiological and behavioral mechanisms that maintain homeostasis. Additionally, this broad habitat has made bottlenose dolphins one of the most accessible and widely studied marine mammal species globally. There is a breadth of knowledge about dolphin biology and physiology that is unparalleled in other cetaceans. Long-term monitoring programs, such as those in Sarasota Bay (Wells et al., 2004; https://sarasotadolphin.org/) and Shark Bay (Connor and Smolker, 1985; https://www.monkeymiadolphins.org/), have taken advantage of coastal populations to study bottlenose dolphins in their natural environmental regimes and collect longitudinal data on wild populations. Additionally, studies on captive dolphins have allowed us to collect crucial physiological data on important thermoregulatory features such as internal heat generation through basal metabolism and lower critical temperatures. Similar studies would not be possible for larger cetaceans. Technological advances in field studies have enhanced our knowledge of thermoregulation by measuring physiological variables relevant to thermoregulation (e.g., skin temperature and heat flux) in wild bottlenose dolphins that are freely swimming or diving (e.g., Westgate et al., 2007; Barbieri et al., 2010).

Although bottlenose dolphins have become the model species for thermoregulation in cetaceans, questions remain regarding how these animals withstand life in an aquatic environment. Offshore bottlenose dolphin populations living in colder, more pelagic waters are less accessible, thus limiting our knowledge of bottlenose dolphin thermoregulation in more extreme environments. Further, while the costs of heat generation through metabolism and activity are well-known for bottlenose dolphins and other cetacean species, the costs associated with heat dissipation and rapid offloading of heat have not been documented in any cetacean. Preventing hyperthermia plays an important role in the maintenance of homeostasis for cetaceans to prevent protein denaturation. Several studies have identified mechanisms for offloading heat in bottlenose dolphins (e.g., using their dorsal fin as a thermal window); however, the energetic costs associated with these mechanisms are unknown. Further research into these topics will complete our understanding of thermoregulation in a harsh, aquatic environment.

References

Adachi, T., Maresh, J.L., Robinson, P.W., Peterson, S.H., Costa, D.P., Naito, Y., Watanabe, Y.Y., Takahashi, A., 2014. The foraging benefits of being fat in a highly migratory marine mammal. Proc. R. Soc. Lond. B 281, 1–9.

Adamczak, S.K., Pabst, D.A., McLellan, W.A., Thorne, L.H., 2020. Do bigger bodies require bigger radiators? Insights into thermal ecology from closely related marine mammal species and implications for ecogeographic rules. J. Biogeogr. 47, 1193–1206.

Adamczak, S.K., Holser, R.R., Costa, D.P., Berens McCabe, E.J., Wells, R.S., 2021. Body composition of common bottlenose dolphins in Sarasota Bay, Florida. Front. Mar. Sci. 8 (June), 1–12.

Bagge, L.E., Koopman, H.N., Rommel, S.A., McLellan, W.A., Pabst, D.A., 2012. Lipid class and depth-specific thermal properties in the blubber of the short-finned pilot whale and the pygmy sperm whale. J. Exp. Biol. 215, 4330–4339.

Ball, H.C., Stavarz, M., Oldaker, J., Usip, S., Londraville, R.L., George, J.C., Thewissen, J.G.M., Duff, R.J., 2015. Seasonal and ontogenetic variation in subcutaneous adipose of the bowhead whale (*Balaena mysticetus*). Anat. Rec. 298 (8), 1416–1423.

Barbieri, M.M., McLellan, W.A., Wells, R.S., Blum, J.E., Hofmann, S., Gannon, J., Pabst, D.A., 2010. Using infrared thermography to assess seasonal trends in dorsal fin surface temperatures of free-swimming bottlenose dolphins (*Tursiops truncatus*) in Sarasota Bay, Florida. Mar. Mamm. Sci. 26 (1), 53–66.

Boult, V.L., Quaife, T., Fishlock, V., Moss, C.J., Lee, P.C., Sibly, R.M., 2018. Individual-based modelling of elephant population dynamics using remote sensing to estimate food availability. Ecol. Model. 387, 187–195.

Brody, S., 1968. Bioenergetics and Growth. Hafner, New York.

Brown, W.R., Geraci, J.R., Hicks, B.D., St. Aubin, D.J., 1983. Epidermal cell proliferation in the bottlenose dolphin (*Tursiops truncatus*). Can. J. Zool. 61 (7), 1587–1590.

Bruce-Allen, L.J., Geraci, J.R., 1985. Wound healing in the bottlenose dolphin (*Tursiops truncatus*). Can. J. Fish. Aquat. Sci. 42, 216–228.

Cairns, D.K., Gaston, A.J., Huettmann, F., 2008. Endothermy, ectothermy and the global structure of marine vertebrate communities. Mar. Ecol. Prog. Ser. 356 (1977), 239–250.

Cockcroft, V.G., 1989. Biological Studies of Bottlenose Dolphins from Natal Coastal Waters (PhD thesis). University of Natal.

Connor, R.C., Smolker, R.S., 1985. Habituated dolphins (*Tursiops sp.*) in Western Australia. Am. Soc. Mammal. 66 (2), 398–400.

Corkeron, P.J., Morris, R.J., Bryden, M.M., 1987. A note on healing of large wounds in bottlenose dolphins, *Tursiops truncatus*. Aquat. Mamm. 13 (3), 96–98.

Costa, D.P., 2018. Osmoregulation. In: Würsig, B., Thewissen, J.G.M., Kovacs, K.M. (Eds.), Encyclopedia of Marine Mammals. Academic Press, pp. 659–664.

Costa, D.P., Kooyman, G.L., 1984. Contribution of specific dynamic action to heat balance and thermoregulation in the sea otter *Enhydra lutris*. Physiol. Zool. 57 (2), 199–203.

Costa, D.P., Maresh, J.L., 2017. Energetics. In: Encyclopedia of Marine Mammals, pp. 329–335.

Costa, D.P., Williams, T.M., 2000. Marine mammal energetics. In: Reynolds, J., Twiss, J. (Eds.), The Biology of Marine Mammals. Smithsonian Institution Press, Washington, D.C., USA, pp. 176–217.

Croft, L.A., Laughlin, R., Manley, M., Nollens, H.H., 2020. Water temperature fluctuations as a key driver of cetacean pox (tattoo) lesions in bottlenose dolphins *Tursiops truncatus*. Dis. Aquat. Org. 139, 69–79.

Davis, R.W., 2019. Metabolism and thermoregulation. In: Davis, R.W. (Ed.), Marine Mammals: Adaptations for an Aquatic Life. Springer Nature, Cham, Switzerland, pp. 57–88.

Dejours, P., 1987. Water and air physical characteristics and their physiological consequences. In: Dejours, P., et al. (Eds.), Comparative Physiology: Life in Water and on Land. Liviana Press, Padova, pp. 3–11.

Dunkin, R.C., McLellan, W.A., Blum, J.E., Pabst, D.A., 2005. The ontogenetic changes in the thermal properties of blubber from Atlantic bottlenose dolphin *Tursiops truncatus*. J. Exp. Biol. 208, 1469–1480.

Dunkin, R.C., McLellan, W.A., Blum, J.E., Pabst, D.A., 2010. The buoyancy of the integument of Atlantic bottlenose dolphins (*Tursiops truncatus*): effects of growth, reproduction, and nutritional state. Mar. Mamm. Sci. 26 (3), 573–587.

Durban, J.W., Pitman, R.L., 2012. Antarctic killer whales make rapid, round-trip movements to subtropical waters: evidence for physiological maintenance migrations? Biol. Lett. 8, 274–277.

Elsner, R., Pirie, D., Kenney, D.D., Schemmer, S., 1974. Functional circulatory anatomy of cetacean appendages. In: Harrison, R.J. (Ed.), Functional Anatomy of Marine Mammals. vol. 2. Academic Press, London, pp. 143–159.

Favilla, A.B., Costa, D.P., 2020. Thermoregulatory strategies of diving air-breathing marine vertebrates: a review. Front. Ecol. Evol. 11 (8) (34 pp).

Feltz, E.T., Fay, F.H., 1966. Thermal requirements in vitro of epidermal cells from seals. Cryobiology 3 (3), 261–264.

Fish, F.E., 1994. Influence of hydrodynamic design and propulsize mode on mammalian swimming energetics. Aust. J. Zool. 42, 79–101.

Fish, F.E., Howles, L.E., Murray, M.M., 2008. Hydrodynamic flow control in marine mammals. Integr. Comp. Biol. 48 (6), 788–800.

Folkow, L.P., Schytte Blix, A., 1987. Nasal heat and water exchange in gray seals. Am. J. Physiol. 253 (6), R883–R889.

Fougères, E.M., 2008. Thermoregulation in Bottlenose Dolphins (*Tursiops truncatus*) (PhD thesis). University of North Carolina Wilmington.

Gallagher, C.A., Stern, S.J., Hines, E., 2018. The metabolic cost of swimming and reproduction in harbor porpoises (*Phocoena phocoena*) as predicted by a bioenergetic model. Mar. Mamm. Sci. 34 (4), 875–900.

Garten, J.L., Fish, F.E., 2020. Comparative histological examination of the integument of odontocete flukes. Aquat. Mamm. 46, 367–381.

Gearty, W., McClain, C.R., Payne, J.L., 2018. Energetic tradeoffs control the size distribution of aquatic mammals. Proc. Natl. Acad. Sci. U. S. A. 115 (16), 4194–4199.

Grady, J.M., Maitner, B.S., Winter, A.S., Kaschner, K., Tittensor, D.P., Record, S., Smith, F.A., Wilson, A.M., Dell, A.I., Zarnetske, P.L., Wearing, H.J., Alfaro, B., Brown, J.H., 2019. Metabolic asymmetry and the global diversity of marine predators. Science 366 (January), aat4220.

Hampton, I.F.G., Whittow, G.C., 1976. Body temperature and heat exchange in the Hawaiian spinner dolphin, *Stenella longirostris*. Comp. Biochem. Physiol. A: Physiol. 55, 195–197.

Hampton, I.F.G., Whittow, G.C., Szekerczes, J., Rutherford, S., 1971. Heat transfer and body temperature in the Atlantic bottlenosed dolphin, *Tursiops truncatus*. Int. J. Biometeorol. 15 (2–4), 247–253.

Hashimoto, O., Ohtsuki, H., Kakizaki, T., Amou, K., Sato, R., Doi, S., Kobayahi, S., Matsuda, A., Sugiyama, M., Funaba, M., Matsuishi, T., Terasawa, F., Shindo, J., Endo, H., 2015. Brown adipose tissue in cetacean blubber. PLoS One 10 (2), e0116734.

Heath, M.E., Ridgway, S.H., 1999. How dolphins use their blubber to avoid heat stress during encounters with warm water. Am. J. Physiol. 276, R1188–R1194.

Hicks, B.D., St. Aubin, D.J., Geraci, J.R., Brown, W.R., 1985. Epidermal growth in the bottlenose dolphin, *Tursiops truncatus*. J. Invest. Dermatol. 85 (1), 60–63.

Hillenius, W.J., 1992. The evolution of nasal turbinates and mammalian endothermy. Paleobiology 18 (1), 17–29.

Hind, A.T., Gurney, W.S.C., 1997. The metabolic cost of swimming in a marine homeotherm. J. Exp. Biol. 200, 531–542.

Hicks, B.D., St. Aubin, D.J., Geraci, J.R., Brown, W.R., 2014. Bottlenose dolphins modify behavior to reduce metabolic effect of tag attachment. J. Exp. Biol. 217 (23), 4229–43336.

Hückstädt, L.A., Tift, M.S., Riet-Sapriza, F., Franco-Trecu, V., Baylis, A.M.M., Orben, R.A., Arnould, J.P.Y., Sepulveda, M., Santos, M., Burns, J.M., Costa, D.P., 2016. Regional variability in diving physiology and behavior in a widely distributed air-breathing marine predator, the South American sea lion (*Otaria byronia*). J. Exp. Biol. 219 (15), 2320–2330.

Hui, C.A., 1981. Seawater consumption and water flux in the common dolphin *Delphinus delphis*. Physiol. Zool. 54 (4), 430–440.

Innes, S., Lavigne, D.M., 1991. Do cetaceans really have elevated metabolic rates? Physiol. Zool. 64 (4), 1130–1134.

Irving, L., Hart, J.S., 1957. The metabolism and insulation of seals as bare-skinned mammals in cold water. Can. J. Zool. 35 (4), 497–511.

Iverson, S.J., 2009. Blubber. In: Perrin, W.F., Wursig, B., Thewissen, H.G. (Eds.), Encyclopedia of Marine Mammals. Academic Press, San Diego, pp. 107–112.

Kanwisher, J., Sundnes, G., 1965. Physiology of a small cetacean. Hvalrådets skr. 48, 45–53.

Kanwisher, J., Sundnes, G., 1966. Thermal regulation in cetaceans. In: Norris, K.S. (Ed.), Whales Dolphins and Porpoises. University of California Press, Los Angeles, pp. 397–407.

Kleiber, M., 1975. The metabolic rate per unit body weight. J. Theor. Biol. 53 (1), 199–204.

Koopman, H.N., 2007. Phylogenetic, ecological, and ontogenetic factors influencing the biochemical structure of the blubber of odontocetes. Mar. Biol. 151, 277–291.

Koopman, H.N., 2018. Function and evolution of specialized endogenous lipids in toothed whales. J. Exp. Biol. 121, jeb161471.

Lavigne, D.M., Innes, S., Worthy, G.A.J., Kovacs, K.M., 1986a. Metabolic rate-body size relations in marine mammals. J. Theor. Biol. 122 (1), 123–124.

Lavigne, D.M., Innes, S., Worthy, G.A.J., Kovacs, K.M., Schmitz, O.J., Hickie, J.P., 1986b. Metabolic rates of seals and whales. Can. J. Zool. 64 (2), 279–284.

McClain, A.M., Daniels, R., Gomez, F.M., Ridgway, S.H., Takeshita, R., Jensen, E.D., Smith, C.R., 2020. Physiological effects of low salinity exposure on bottlenose dolphins (*Tursiops truncatus*). J. Zool. Bot. Gard. 1 (1), 61–75.

McClelland, S., Gay, M., Pabst, D.A., Dillaman, R., Westgate, A.J., Koopman, H.N., 2012. Microvascular patterns in the blubber of shallow and deep diving Odontocetes. J. Morphol. 273, 932–942.

McGinnis, S.M., Whittow, G.C., Ohata, C.A., Huber, H., 1972. Body heat dissipation and conservation in two species of dolphins. Comp. Biochem. Physiol. A: Physiol. 43, 417–423.

McNab, B.K., 2001. Vertebrate metabolic variation. In: Encyclopedia of Life Sciences (online). John Wiley & Sons, Ltd.

Meagher, E.M., McLellan, W.A., Westgate, A.J., Wells, R.S., Frierson Jr., D., Pabst, D.A., 2002. The relationship between heat flow and vasculature in the dorsal fin of wild bottlenose dolphins *Tursiops truncatus*. J. Exp. Biol. 205, 3475–3486.

Meagher, E.M., McLellan, W.A., Westgate, A.J., Wells, R.S., Blum, J.E., Pabst, D.A., 2008. Seasonal patterns of heat loss in wild bottlenose dolphins (*Tursiops truncatus*). J. Comp. Physiol. B 178, 529–543.

Montie, E.W., Garvin, S.R., Fair, P.A., Bossart, G.D., Mitchum, G.B., McFee, W.E., Speakman, T., Starczak, V.R., Hahn, M.E., 2008. Blubber morphology in wild bottlenose dolphins (*Tursiops truncatus*) from the Southeastern United States: influence of geographic location, age class, and reproductive state. J. Morphol. 269 (4), 496–511.

Noren, S.R., Wells, R.S., 2009. Blubber deposition during ontogeny in free-ranging bottlenose dolphins: balancing disparate roles of insulation and locomotion. J. Mammal. 90 (3), 629–637.

Noren, D.P., Williams, T.M., Berry, P., Butler, E., 1999. Thermoregulation during swimming and diving in bottlenose dolphins (*Tursiops truncatus*). J. Comp. Physiol. B 169, 93–99.

Noren, D.P., Holt, M.M., Dunkin, R.C., Williams, T.M., 2013. The metabolic cost of communicative sound production in bottlenose dolphins (*Tursiops truncatus*). J. Exp. Biol. 216 (9), 1624–1629.

Noren, D.P., Holt, M.M., Dunkin, R.C., Williams, T.M., 2017. Echolocation is cheap for some mammals: dolphins conserve oxygen while producing high-intensity clicks. J. Exp. Mar. Biol. Ecol. 495 (February), 103–109.

Nousek-McGregor, A.E., Miller, C.A., Moore, M.J., Nowacek, D.P., 2014. Effects of body condition on buoyancy in endangered North Atlantic right whales. Physiol. Biochem. Zool. 87 (1), 160–171.

Ortiz, R.M., 2001. Osmoregulation in marine mammals. J. Exp. Biol. 204 (11), 1831–1844.

Pabst, D.A., Rommel, S.A., McLellan, W.A., 1999. Functional anatomy of marine mammals. In: Reynolds III, J.E., Rommel, S.A. (Eds.), Biology of Marine Mammals. Smithsonian Institution Press, Washington, D.C, pp. 15–72.

Pabst, D.A., Rommel, S.A., McLellan, W.A., Williams, T.M., Rowles, T.K., 1995. Thermoregulation of the intra-abdominal testes of the bottlenose dolphin (*Tursiops truncatus*) during exercise. J. Exp. Biol. 198, 221–226.

Pabst, D.A., McLellan, W.A., Meagher, E.M., Westgate, A.J., 2002. Measuring Temperatures and Heat Flux from Dolphins in the Eastern Tropical Pacific: Is Thermal Stress Associated with Chase and Capture in the ETP-Tuna Purse Seine Fishery? (Report No. LJ-02-34C). NOAA Southwest Fisheries Science Center. (La Jolla, CA).

Palmer, E., Weddell, G., 1964. The relationship between structure, innervation, and function of the skin of the bottlenose dolphin (*Tursiops truncatus*). Proc. Zool. Soc. London 143, 553–568.

Paterson, W.D., Chaise, L.L., McKnight, C., Curie, J.I., Thompson, D., Ancel, A., Gilbert, C., McCafferty, D.J., 2022. Metabolic heat loss in southern elephant seals (*Mirounga leonina*) differs with stage of moult and between habitats. J. Therm. Biol. 104 (2022).

Pavlov, V.V., 2006. Dolphin skin as a natural anisotropic compliant wall. Bioinspir. Biomim. 1 (2), 31.

Pedersen, M.B., Fahlman, A., Borque-Espinosa, A., Madsen, P.T., Jensen, F.H., 2020. Whistling is metabolically cheap for communicating bottlenose dolphins (*Tursiops truncatus*). J. Exp. Biol. 223, jeb.212498.

Pitman, R.L., Durban, J.W., Joyce, T., Fearnbach, H., Panigada, S., Lauriano, G., 2019. Skin in the game: epidermal molt as a driver of long-distance migration in whales. Mar. Mamm. Sci. 36, 565–594.

Plön, S., Frainer, G., Wedderburn-Maxwell, A., Cliff, G., Huggenberger, S., 2018. Dorsal fin and hump vascular anatomy in the indo–pacific humpback dolphin (*Sousa plumbea*) and the indo-pacific bottlenose dolphin (*Tursiops aduncus*). Mar. Mamm. Sci 35 (2), 684–695.

Rechsteiner, E.U., Rosen, D.A.S., Trites, A.W., 2013. Energy requirements of Pacific white-sided dolphins (*Lagenorhynchus obliquidens*) as predicted by a bioenergetic model. J. Mammal. 94 (4), 820–832.

Rimbach, R., Amireh, A., Allen, A., Hare, B., Guarino, E., Kaufman, C., Salomons, H., Pontzer, H., 2021. Total energy expenditure of bottlenose dolphins (*Tursiops truncatus*) of different ages. J. Exp. Biol. 224 (15).

Rommel, S.A., Pabst, D., McLellan, W.A., 1993. Functional morphology of the vascular plexuses associated with the cetacean uterus. Anat. Rec. 2 (6).

Rommel, S.A., Pabst, D.A., McLellan, W.A., Mead, J.G., Potter, C.W., 1992. Anatomical evidence for a countercurrent heat exchanger associated with dolphin testes. Anat. Rec. 232 (1), 150–156.

Rommel, S.A., Pabst, D.A., McLellan, W.A., Williams, T.M., Friedl, W.A., 1994. Temperature regulation of the testes of the bottlenose dolphin (*Tursiops truncatus*): evidence from colonic temperatures. J. Comp. Physiol. B 164 (2), 130–134.

Rommel, S.A., Pabst, D.A., McLellan, W.A., 1998. Reproductive thermoregulation in marine mammals: how do male cetaceans and seals keep their testes cool without a scrotum? It turns out to be the same mechanism that keeps the fetus cool in a pregnant female. Am. Sci. 86 (5), 440–448.

Ryg, M., Lydersen, C., Knutsen, L.Ø., Bjørge, A., Smith, T.G., Øritsland, N.A., 1993. Scaling of insulation in seals and whales. J. Zool. 230 (2), 193–206.

Scholander, P.F., Schevill, W.E., 1955. Counter-current vascular heat exchange in the fins of whales. J. Appl. Physiol. 8, 279–282.

Sokolov, W., 1960. The skin structure in Pinnipedia of the U.S.S.R. fauna. J. Morphol. 107, 285–296.

Struntz, D.J., McLellan, W.A., Dillaman, R.M., Blum, J.E., Kucklick, J.R., Pabst, D.A., 2004. Blubber development in bottlenose dolphins (*Tursiops truncatus*). J. Morphol. 259, 7–20.

Torres-Romero, E.J., Morales-Castilla, I., Olalla-Tárraga, M., 2016. Bergmann's rule in the oceans? Temperature strongly correlates with global interspecific patterns of body size in marine mammals. Glob. Ecol. Biogeogr. 25, 1206–1215.

Wahrenbrock, E.A., Maruschak, G.F., Elsner, R., Kenney, D.W., 1974. Respiration and metabolism in two baleen whale calves. Mar. Fish. Rev. 36 (4), 3–8.

Wells, R.S., Scott, M.D., 2018. Bottlenose dolphin, *Tursiops truncatus*, common bottlenose dolphin. In: Encyclopedia of Marine Mammals, pp. 118–125.

Wells, R., Rhinehart, H.L., Hansen, L.J., Sweeney, J.C., Townsend, F.I., Stone, R., Casper, D.R., Scott, M.D., Hohn, A.A., Rowles, T.K., 2004. Bottlenose dolphins as marine ecosystem sentinels: developing a health monitoring system. Ecohealth 1 (3), 246–254.

Westgate, A.J., McLellan, W.A., Wells, R.S., Scott, M.D., Meagher, E.M., Pabst, D.A., 2007. A new device to remotely measure heat flux and skin temperature from free-swimming dolphins. J. Exp. Mar. Biol. Ecol. 346, 45–59.

Williams, T.M., 1999. The evolution of cost efficient swimming in marine mammals: limits to energetic optimization. Philos. Trans. R. Soc. Lond. B. 354, 193–201.

Williams, T.M., Haun, J.E., Friedl, W.A., 1999a. The diving physiology of bottlenose dolphins (*Tursiops truncatus*): I. Balancing the demands of exercise for energy conservation at depth. J. Exp. Biol. 202 (20), 2739–2748.

Williams, T.M., Noren, D., Berry, P., Estes, J.A., Allison, C., Kirtland, J., 1999b. The diving physiology of bottlenose dolphins (*Tursiops truncatus*): III. Thermoregulation at depth. J. Exp. Biol. 202, 2763–2769.

Williams, T.M., Haun, J., Davis, R.W., Fuiman, L.A., Kohin, S., 2001. A killer appetite: metabolic consequences of carnivory in marine mammals. Comp. Biochem. Physiol. A: Physiol. 129 (4), 785–796.

Wilson, B., Thompson, P.M., Hammond, P.S., 1997. Habitat use by bottlenose dolphins: seasonal distribution and stratified movement patterns in the Moray Firth, Scotland. J. Appl. Ecol. 34 (6), 1365.

Wilson, B., Arnold, H., Bearzi, G., Fortuna, C.M., Gaspar, R., Ingram, S., Liret, C., Pribanić, S., Read, A.J., Ridoux, V., Schneider, K., Urian, K.W., Wells, R.S., Wood, C., Thompson, P.M., Hammond, P.S., 1999. Epidermal diseases in bottlenose dolphins: impacts of natural and anthropogenic factors. Proc. R. Soc. Lond. B 266 (1423), 1077–1083.

Woodward, B.L., Winn, J.P., Fish, F.E., 2006. Morphological specializations of baleen whales associated with hydrodynamic performance and ecological niche. J. Morphol. 267, 1284–1294.

Worthy, G.A.J., Edwards, E.F., 1990. Morphometric and biochemical factors affecting heat loss in a small temperate cetacean (*Phocoena phocoena*) and a small tropical cetacean (*Stenella attenuata*). Physiol. Zool. 63 (2), 432–442.

Worth, G.A.J., Worthy, T.A.M., Yochem, P.K., Dold, C., 2014. Basal metabolism of an adult male killer whale (*Orcinus orca*). Mar. Mamm. Sci. 30 (3), 1229–1237.

Yazdi, P., Kilian, A., Culik, B.M., 1999. Energy expenditure of swimming bottlenose dolphins (*Tursiops truncatus*). Mar. Biol. 134, 601–607.

Yeates, L.C., Houser, D.S., 2008. Thermal tolerance in bottlenose dolphins (*Tursiops truncatus*). J. Exp. Biol. 211 (20), 3249–3257.

CHAPTER

4

Muscles and movement

Frank E. Fish[a], *William T. Gough*[b], *and Danielle S. Adams*[c]

[a]Department of Biology, West Chester University, West Chester, PA, United States [b]Hopkins Marine Station of Stanford University, Pacific Grove, CA, United States [c]Department of Biology, Clemson University, Clemson, SC, United States

Abstract

Swimming by dolphins requires the efficient generation of thrust to overcome resistive drag on the body. The longitudinal muscles actuate dorsoventral motions of the propulsive flukes. The flukes act as an oscillating wing to generate thrust that is derived from a lift force. The pitch angle of the flukes that is controlled through tendinous insertions from the musculature increases thrust generation and propulsive efficiency. The propulsive movements and thrust generated by dolphins are on a continuum with larger whales. Drag is minimized by streamlining of the body and appendages. Special mechanisms that were suggested to reduce drag for dolphins have not been demonstrated. However, various behaviors (e.g., porpoising, drafting, and bow riding) can lower the energetic expenditure to swim. Maneuverability by dolphins is dependent on body size, body stiffness, and use of control surfaces. The ability to turn rapidly is necessary to forage on small elusive prey.

Introduction

Dolphins represent a group of marine mammals that are considered the epitome of morphological convergence for fast-swimming, aquatic predators. These servants of Neptune evolved from terrestrial quadrupeds to the highly streamlined biological torpedoes of the oceans. The locomotor capabilities of dolphins extend to high-speed movement through a highly dense and viscous aquatic medium, agile maneuverability, and spectacular aerial leaps. These capabilities utilize the large muscle mass from the trunk of the body in concert with morphological specializations associated with enhanced thrust production, high propulsive efficiency, and reduced drag (Weihs and Webb, 1983; Fish and Hui, 1991; Fish and Rohr, 1999). For dolphins, these morphological specializations include a streamlined body, tight skin incorporating stiff collagen fibers in the underlying blubber layer, strongly compressed caudal peduncle, and high aspect ratio flukes and flippers with sweepback that are actuated as an oscillatory lifting wing (Webb, 1975; Fish, 1993a, 1998a). The specialized morphology and swimming mode allow dolphins to achieve high-performance swimming movements with a low energetic cost (see Chapter 2: Energetic costs of rest and locomotion in dolphins). The high-performance locomotion and energy economy are tied tightly to the aquatic ecology of these animals to accomplish high accelerations to catch prey or evade predators, migrate and forage over long distances, and execute aerial behaviors (Weihs, 2002; Fish et al., 2006; Goldbogen et al., 2016).

The body morphology, musculature, and gross swimming motions of dolphins were described early on. Aristotle (*Historia Animalium*) considered dolphins the fastest of all animals with the ability to leap over the masts of tall sailing ships. Borelli (1689) noted the up and down movements of the tail of swimming cetaceans. This oscillatory motion of the flukes was considered superior in thrust production to the screw propeller (Pettigrew, 1893; Petersen, 1925). Cayley (circa 1800) considered the streamlined shape of the dolphin body as a solid of least resistance design (Gibbs-Smith, 1962). It was not until James Gray performed the first biomechanical analysis of the hydrodynamics of dolphins that these animals fostered a paradigm shift in animal mechanics (Gray, 1936). Although controversial, Gray's work on dolphins was the impetus for studies on swimming performance in animals and nautical engineering (Kramer, 1960; Webb, 1975; Fish and Rohr, 1999; Fish, 2006).

This chapter will review the mechanics of locomotion by dolphins. An examination of the muscular architecture and power output, transmission of muscular power to the propulsive tail, and hydrodynamics of the low-drag body and the lift-based generation of thrust by the caudal flukes will provide a comprehensive picture of the mechanical energy expenditure of swimming by dolphins. Furthermore, mechanisms of energy capture, scaling, and maneuvering

The Physiology of Dolphins
https://doi.org/10.1016/B978-0-323-90516-9.00002-6

performance will be explored. Particular attention will be paid to the 37 species of the family Delphinidae (Jefferson and LeDuc, 2018), which display a high level of swimming performance, although to put the performance of the delphinids in context, other cetaceans will also be examined.

Streamlining

Fusiform body design

Dolphins reside in an aqueous environment for their entire lives and must minimize the energetic effort of swimming through an aquatic medium that can be 890 times denser and at least 50 times more viscous than air (Webb, 1988; Denny, 1993).

Overall, body shape is the major determinant of resistance (i.e., drag) to movement through a fluid (Fish and Rohr, 1999). Drag is minimized primarily by streamlining the shape of the body and appendages (i.e., flukes, flippers, and dorsal fin) (Fish, 1993a). Streamlining minimizes drag by reducing the magnitude of the pressure gradient over the body and allows water to flow smoothly over the surface (Vogel, 1994). The streamlined profile is characterized by a fusiform shape emulating an elongate teardrop with a rounded leading edge extending to a maximum thickness and a slowly tapering tail. This fusiform shape is displayed by all dolphins (Lighthill, 1969; Fish and Hui, 1991).

An indicator of the degree of streamlining is the fineness ratio (FR = BL/D, where BL is the body length and D is the maximum diameter). The FR value of 4.5 produces the least drag and surface area for the maximum volume (von Mises, 1945; Webb, 1975; Fish, 1993a; Ahlborn et al., 2009). Based on blimp design, a streamlined body with FR = 4.5 will have a 75% reduction in drag from that for a sphere of equal volume (von Mises, 1945). The drag on axisymmetrical bodies indicated a higher optimal FR of 6–7 (Gertler, 1950; Joubert, 2004). Dolphins should have minimal drag in the FR range of 4.5–7 with only a slight increase in drag at higher values of FR. Most dolphins fall into this optimal range, although FR can be as high as 11 for the northern right whale dolphin (*Lissodelphis borealis*) (Fish and Rohr, 1999).

FR is a crude indicator of the streamlining of the body because it does not provide information on changes in body contour. As a better indicator of body shape, profiles of dolphins were compared to standard two-dimensional airfoils. The body form of a dolphin is similar in outline with the NACA 66-018 foil (Hertel, 1966), whereas the killer whale (*Orcinus orca*) is similar to NACA 66-026 (Pershin, 1983). The choice of foil design to emulate body shape is subjective and overly simplistic for not accounting for body contours that do not match the idealized foil shape (Weber et al., 2009). In addition, the dolphin's body is three-dimensional with attached appendages that are not accounted for in the two-dimensional foil design. The position of the dolphin's maximum thickness is at a position 0.34–0.45 BL from the rostrum (Fish and Hui, 1991). The position of the maximum thickness will maintain flow over a large portion of the body to minimize drag (Lang, 1963; Webb, 1975; Fish and Rohr, 1999; Fish et al., 2008).

Control surfaces

The control surfaces extending from the body of the dolphin act as hydrofoils to generate a lift force to stabilize the body and effect maneuverability (Fish, 2002; Fish et al., 2003; Fish and Lauder, 2017). These control surfaces include the pectoral flippers, dorsal fin, and caudal flukes, which also are used for propulsion (as discussed further). All have a streamlined, fusiform, cross-sectional design (Felts, 1966; Lang, 1966a; Pershin, 1988; Fish et al., 2007; Weber et al., 2009a, b). The addition of skeletal elements in the flippers makes these appendages relatively thick compared with the dorsal fin and flukes. For dolphins, the relative proportions of the control surfaces vary substantially between the different species. The percentage of total control surface area ranges from 14% to 42% for the flippers, 0% to 18% for the dorsal fin, 16% to 48% for the peduncle, and 22% to 37% for the flukes (Fish, 1997).

Flipper planforms (Fig. 1) vary from the large sweepback, highly tapered flipper of the long-finned pilot whale (*Globicephala melas*) to rounded paddle-like forms in killer whale (Fish and Rohr, 1999). For dolphins, flipper length directly increases with increasing BL. The interaction between the flipper span (S, i.e., linear distance from base to tip of the control surface) and its planform surface area (A) as related to effectiveness of the hydrofoil design is expressed as the aspect ratio (AR). AR is calculated as S^2/A (Vogel, 1994). High AR indicates long narrow flukes, whereas low AR indicates broad flukes with a short S. For dolphin flippers, AR values range from 1.9 for killer whaler to 7.7 for long-finned pilot whale (Fish and Rohr, 1999). In general, species with high *AR* control surfaces are capable of higher speeds than species with low AR control surfaces (Fish, 1998a, b).

When present, dorsal fins are shaped as falcate, triangular, or rounded (Minasian et al., 1984; Pavlov, 2003; Morteo et al., 2017). The largest dorsal fin (1.8 m high) is found in male killer whales and is a distinct sexually dimorphic characteristic in this species. No association between swimming speed and dorsal fin design is apparent. Finless dolphins,

FIG. 1 Images of flipper planforms for the striped dolphin (*Stenella coeruleoalba*, left), and killer whale (*Orcinus orca*, right).

such as northern right whale dolphin, are considered to be rapid swimmers (Winn and Olla, 1979). The largest dorsal fin AR of 2.4 is found for a male killer whale. For dolphins that possess a dorsal fin, the smallest AR is displayed for the triangular fin of the Amazon river dolphin (*Inia geoffrensis*, Fish and Rohr, 1999).

An additional control surface is the peduncle that is located just anterior of the caudal flukes. The peduncle is laterally compressed (narrow-necking) and highly streamlined in the direction of oscillation of the flukes to reduce drag (Felts, 1966; Lighthill, 1969; Hamilton et al., 2004; Etnier et al., 2008). The peduncle can be bent laterally and twisted to act as a rudder during turning maneuvers (Fish, 2002).

Propulsive fluke design

The flukes are lateral extensions of the distal tail of dolphins. The flukes act biomechanically like a pair of wings to generate a lift force (Vogel, 1994). As static wings, the flukes can act to stabilize the body of a dolphin when gliding (Fish, 2002). However, during active swimming, the flukes are vertically oscillated to generate a lift-derived thrust (Fish, 1993b, 1998a, b; Goldbogen et al., 2016).

The shape of the flukes influences the effective generation of high lift for thrust with low-drag performance. Fluke planforms of dolphins (Fig. 2) have a swept-back (i.e., angle between a perpendicular to RC and one-quarter chord position) tapered design, which varies between different species. *S* and *A* are directly related to increasing BL for

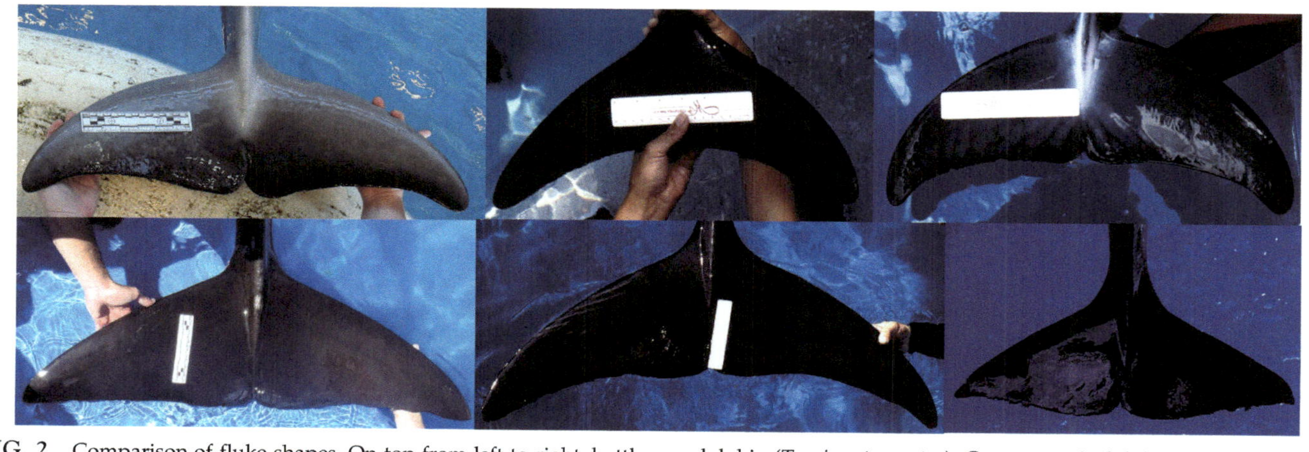

FIG. 2 Comparison of fluke shapes. On top from left to right, bottlenose dolphin (*Tursiops truncatus*), Commerson's dolphin (*Cephalorhynchus commersonii*), Atlantic spotted dolphin (*Stenella frontalis*), and on bottom from left to right, short-finned pilot whale (*Globicephala macrohynchus*), false killer whale (*Pseudorca crassidens*), and killer whale (*Orcinus orca*).

dolphins with a slight positive allometry (Fish and Rohr, 1999). *S* and *A* determine the mass of water that is affected for thrust generation. A large *A* generates large thrust (Bose et al., 1990). In neonates and prepubescent dolphins, the increase in *S* with respect to BL is not as rapid as observed for adults (Perrin, 1975; Amano and Miyazaki, 1993). Young animals, therefore, might be at a disadvantage when swimming, thus requiring the use of free-riding behaviors to maintain speed with the parent (Lang, 1966b; Weihs, 2004; Edwards, 2006; Noren and Edwards, 2007).

Sections of the fluke along the longitudinal axis display a conventional streamlined foil profile with a rounded leading edge and long tapering trailing edge. Contrary to the observation of asymmetry about the chord (i.e., straight line from leading to trailing edge) by Purves (1969), the flukes are symmetrical about the chord (Lang, 1966a; Pershin, 1975; Bose et al., 1990; Fish et al., 2006, 2007). The symmetrical design of the flukes indicates that thrust is generated on both up- and downstrokes. The NACA 634-021 and 0018 foils (Abbott and von Doenhoff, 1959; Kayan, 1979) provide reasonable facsimiles of the fluke cross sections.

Well-designed flukes maximize the ratio of lift (*L*) to drag (*D*) generated by their action (Webb, 1975; Weihs, 1989). AR varies from 2.0 for the Amazon river dolphin to high values of 6.1 and false killer whale (*Pseudorca crassidens*), respectively (Bose and Lien, 1989; Fish, 1998b). These values correspond to the swimming performance in these species. High AR hydrofoils have high lift-to-drag ratios, and thus, species with high AR are capable of higher speeds than species with flukes of low AR (Fish, 1998a, b). The longer trailing edge of a high AR fluke increases the mass of water deflected posteriorly augmenting thrust (Felts, 1966; Webb, 1975).

Fluke internal structure

Structurally, the flukes are composed of layers of collagen fibers, including an outer cutaneous layer, a subcutaneous blubber layer, a ligamentous layer, and a core of dense fibrous tissue, constituting the bulk of the flukes (Felts, 1966; Gough et al., 2018). Both the cutaneous and subcutaneous layers are continuations of their respective layers covering the rest of the body. Because the fibrous composition not only strengthens the flukes but also increases flexibility, extreme increase in span with increased AR, although potentially generating higher lift, would exaggerate the bending of the appendage in an oscillatory mode and reduce performance.

Flukes show some degree of both spanwise and chordwise flexibility. The center of the flukes is more rigid than the tips. During the upstroke, fluke tips are bent down slightly from the plane of the fluke and lag behind the center, whereas bending in the opposite direction occurs during the downstroke. Bose et al. (1990) suggested that the phase difference due to this spanwise flexibility would prevent the total loss of thrust at the end of the stroke. The flexing fluke tips could act as winglets to enhance lift and thrust generation (Fish and Lauder, 2017). Chordwise flexibility at the trailing edge of the flukes potentially could increase the efficiency of the flukes by up to 20% with only a moderate decrease in the overall thrust (Katz and Weihs, 1978).

Propulsive muscles and tendons

The propulsive muscles of the axial body are the biological motors that actuate the principal swimming motions of dolphins. The power required for swimming is determined by the speed and stress developed by contraction of the muscles (Pennycuick, 1992). This power is delivered from the muscles through the tendinous drivetrain to the oscillating flukes for thrust generation.

The propulsive musculature of cetaceans is composed of the longitudinal muscles associated with the vertebral column (Fig. 3). The muscles are divided into dorsally positioned epaxial muscles (*multifidus* and *longissimus*) and ventrally located hypaxial muscles (*hypaxialis lumborum* and *intertransverarius caudae ventralis*) (Howell, 1930; Parry, 1949b; Agarkov and Lukhanin, 1970; Pilleri, 1976; Smith et al., 1976; Strickler, 1980; Goforth, 1984; Bello et al., 1985; Pabst, 1990, 1993; Adams and Fish, 2019). Though the former has a larger proportion of the propulsive musculature, the epaxial and hypaxial muscles likely produce equivalent propulsive forces given the similar arrangement of the fasciculi, tendons, and muscle insertions (Arkowitz and Rommel, 1985), and equivalent numbers of slow-oxidative and fast-glycolytic muscle fibers (Mankovskaya, 1975; Ponganis and Pierce, 1978; Bello et al., 1983, 1985; Suzuki et al., 1983). The long spinous and transverse processes of the trunk vertebral column serve for tendinous attachment and act as lever arms that increase mechanical advantage and amplify muscle force compared with insertions closer to the central axis of the vertebral column (Slijper, 1961; Agarkov and Lukhanin, 1970; Smith et al., 1976; Pabst, 1993; Fish and Hui, 1991). In addition, Pabst (1988) has suggested that the subdermal connective tissue sheath under the blubber acts as a force transmission system. The distal position of the sheath far from the vertebral column would provide a large mechanical advantage for flexing the spine and the peduncle.

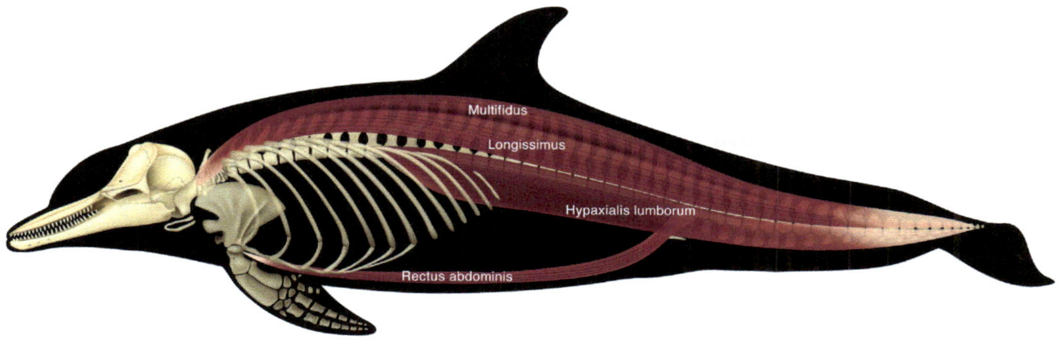

FIG. 3 Positions of the major propulsive epaxial and hypaxial muscles are shown. *Illustration by Uko Gorter.*

There are two epaxial muscles that are important for the extension of the peduncle and flukes (Pabst, 1990). These muscles are the *multifidus* and the *longissimus* (Fig. 3) with their caudal extensions, *extensor caudae medialis* and *extensor caudae lateralis.* The *multifidus* arises from the cervical vertebrae and inserts by two tendons: (1) a superficial set that runs caudo-medially to the dorsal midline and a (2) deep tendon that is attached to the first three caudal vertebrae. The latter is continuous with the *extensor caudae medialis*, which is the major elevator of the peduncle (Strickler, 1980; Pabst, 1990). The *longissimus* originates at the skull and runs posteriorly, fusing with the *semispinalis* in the thoracic region. The *extensor caudae lateralis* arises from this combined muscle mass posteriorly and inserts exclusively on caudal vertebrae associated with the flukes. It is considered to be the elevator of the flukes based on its insertion points (Smith et al., 1976; Strickler, 1980; Pabst, 1990).

The primary flexor of the trunk and flukes is the *hypaxialis lumborum* (Fig. 3). This large muscle originates from the transverse processes of the posterior thoracic and all lumbar vertebrae, as well as the ribs and chevron bones in the anterior portion of the caudal vertebrae (Strickler, 1980; Pabst, 1990). Some studies report a full or partial split of the *hypaxialis lumborum* at the level of the lumbar vertebrae into two distinct muscles (the *dorso-lateralis* and *ventro-medialis*) and their caudal tendon extensions (Strickler, 1980; Parry, 1949a), while others report no posterior separation in the *hypaxialis lumborum* muscle (Purves, 1963; Pabst, 1990). However, these studies do agree that the *hypaxialis lumborum* muscle mass terminates in two tendon bundles that insert into the caudal peduncle and caudal flukes, respectively. The *hypaxialis lumborum* muscle acts as an antagonist to the *extensor caudae lateralis*.

The movements of the peduncle and flukes are actuated by the epaxial and hypaxial muscles described previously, which insert onto the vertebrae through a series of tendons known collectively as the caudal tendons. There are two distinct epaxial tendons of the posterior peduncle: (1) extensor caudae medialis tendon (continuation of the caudal aspect of the *extensor caudae medialis*) and the (2) extensor caudae lateralis tendon (ECL tendon: continuation of the caudal aspect of the *extensor caudae lateralis*) (Fig. 3). Both tendons have their caudal most insertion point posterior to the level of the anterior fluke insertion (Adams and Fish, 2019). However, only the ECL tendon reaches the level of the terminal caudal vertebra (Parry, 1949a; Pabst, 1990; Adams and Fish, 2019). The different insertion points of the two epaxial tendons suggest different functions associated with flexion of the terminal caudal vertebra and chordwise bending of the flukes (Adams and Fish, 2019).

There is agreement in the tendinous continuations of the large trunk flexor muscle, the *hypaxialis lumborum* (Strickler, 1980; Pabst, 1990; Adams and Fish, 2019). There are two tendon bundles originating from the belly of the *hypaxialis lumborum* with two insertions: (1) the ventral aspects of the anterior caudal vertebrae (ventral tendon) and (2) the ventral aspects of vertebral bodies and chevron bones of the vertebrae within the fluke (medial tendon). The medial tendon of the *hypaxialis lumborum* (MHL tendon) is the only hypaxial caudal tendon, as the ventral tendon does not reach past the level of the anterior caudal vertebrae (Pabst, 1990).

Differing from the general perception that tendons insert on bone, the ECL and MHL tendons also have connections into the fibrous core that composes the bulk of the flukes (Felts, 1966; Gough et al., 2018). This physical connection to the internal core fibrous layer suggests an element of active control of the flexibility of the flukes via the longitudinal trunk muscles. The small branches from the ECL and MHL tendons do not penetrate far into the fibrous matrix. However, the highly organized nature of the fibrous core layer could transmit muscular contraction through the tendons to affect spanwise fluke flexibility. The ECL and MHL tendons not only affect the vertical displacement of the tailbase (Sun et al., 2010a, b) but also can decrease the spanwise flexibility of the fluke blade (Adams, 2018). Changing the stiffness of the fluke could modulate thrust and efficiency over an extended operational range of swimming speeds in dolphins.

Speed

The one locomotor parameter for dolphins that has the greatest significance for survival is swimming speed. Regulation of speed allows for pursuit of prey, escape from predators, and increased maneuverability. Swimming speed is a function of the power expended by the muscles that is translated into the propulsive movements of the flukes. Low swimming speeds are used for foraging and migrating to minimize locomotor costs (Würsig and Würsig, 1979; Lockyer and Morris, 1987; Fish and Hui, 1991; Hanson and Defran, 1993) and to maximize distance traveled (Williams, 1987). High swimming speeds require large energy expenditures because the power to generate the propulsive thrust is directly related to the cube of speed.

Research on swimming speed was motivated by accounts of extraordinary speeds by dolphins that were associated with unresolved mechanisms to reduce hydrodynamic drag. Reports on swimming speeds of dolphins and other cetaceans have in many instances been anecdotal and often unreliable (Fish and Hui, 1991; Fish and Rohr, 1999). Lane (1941) reported a dolphin swimming at 112.7 km/h based on an airplane observation, a speed that has never been confirmed. One record was for a killer whale swimming at 45.0–55.4 km/h for 20 min (Johannessen and Harder, 1960). This record was based on the killer whale "playing" around the ship in which the animal was probably drafting and swimming at a higher speed over a prolonged period than physiologically capable. A speed 54.0 km/h was reported for bursting by bottlenose dolphins (*Tursiops truncatus*) (Lockyer and Morris, 1987), although this record was made from a cliff-top observation and was probably an overestimate (Fein, 1998; Fish, 1998a).

Swimming speeds of dolphins performing routine behaviors have been collected from boat measurements, timing of captive animals, theodolite tracking, observations correlated with map locations, sonar tracking, and radio and archival tagged animals. These methods still have limitations regarding accurate speed measurements. Even with the use of satellite tracking systems, the resolution and sampling rate are too low to accurately measure short-duration swimming efforts as occurs for sprints. The diving pattern and high probability that animals do not swim in straight lines between consecutive locations measured by satellites tend to skew the data toward minimum transit speeds, which are well below the routine and maximum speeds (Martin et al., 1993; Davis et al., 1996).

Swimming speeds are categorized as (1) sustained and (2) sprint swimming (Webb, 1975). Sustained swimming is associated with steady low-level activities over long time durations. Sustained speeds are performed for routine swimming and cruising associated with migration, transit to foraging sites, and diving. These speeds are preferred and encompass the optimal speeds for the most efficient generation of thrust. Sprints are high-speed, maximum effort movements. These are short-duration activities. Sprints are used during chases and rapid accelerations. The range of reported maximum speeds varies considerably for odontocetes. For dolphins, the Platanistidae are the slowest swimmers, whereas the highest speeds are attained by the Delphinidae (Fish and Rohr, 1999).

Maximum speeds for trained animals over a 0.7–2.8 s swim were 27.9 km/h for Pacific white-sided dolphin (*Lagenorhynchus obliquidens*) and bottlenose dolphin, and 28.8 km/h for common dolphin (*Delphinus delphis*) and false killer whale (Lang and Daybell, 1963; Rohr et al., 1998; Fish and Rohr, 1999). In open ocean tests, maximum speeds of 39.8 km/h for pantropical spotted dolphin (*Stenella attenuata*) and 29.9 km/h for bottlenose dolphin were recorded (Lang and Norris, 1966; Lang and Pryor, 1966). Swimming speed is related to duration of activity. Bottlenose dolphin can swim at 11.1 km/h indefinitely, 21.9 km/h for 50 s, 25.2 km/h for 10 s, and 29.9 km/h for 7.5 s (Lang, 1975). Maximum swimming speed of a school of free-ranging common dolphin responding to multiple passes by an airplane was 24.1 km/h, although the average speed for the school was 17.3 km/h (Rohr et al., 2002). The lower average speed may have resulted from the presence of newborn dolphins within the school.

Although there is a considerable range of overlap for maximum swimming speeds regardless of taxonomic status, there is a marked dependence of specific speed on size (Webb, 1975). In ascending order of body mass, length-specific speeds for various dolphins were recorded at 6.0 BL/s (pantropical spotted dolphin, 52.7 kg), 4.4 BL/s (common dolphin, 104.8 kg), 4.3 BL/s (Pacific coast bottlenose dolphins [*Tursiops truncatus gilli*], 89 kg), 3.8 BL/s (bottlenose dolphin, 149.2 kg), 2.2 BL/s (false killer whale, 461.8 kg), and 1.5 BL/s (killer whale, 1995.8 kg) (Lang and Norris, 1966; Lang and Pryor, 1966; Fish, 1998a; Rohr et al., 1998). This trend is explained as a matter of scaling whereby animals maintain the same proportion of muscle mass in the body independent of size, but smaller animals have relatively greater muscular power outputs than larger animals (Hill, 1950; Pedley, 1977).

Hydrodynamics

Drag

For a streamlined body that is fully submerged, the total drag has two components: frictional or viscous drag and pressure or form drag (Webb, 1975; Vogel, 1994). The frictional drag is due to the viscosity of the fluid, giving rise to tangential forces from skin friction. Because the water immediately attached to the surface of a body does not move relative to the body (no slip condition), a thin layer of water (boundary layer) encompasses a velocity gradient from the body surface to the free stream velocity outside of the boundary layer (i.e., outer flow). This velocity gradient generates shear (frictional) forces that consume kinetic energy from the movement of the body and transfers it to the water. The frictional drag is proportional to the wetted surface area of the body.

The pressure drag component is due to the distortion of the flow outside the boundary layer. Pressure drag is dependent on pressure distribution as the body deflects the water around it. The pressure acts perpendicularly to the body surface. The pressure is highest at the rostrum or leading edge, where the flow stagnates. As the flow moves around the rostrum, it accelerates and reaches a high velocity at the maximum thickness of the body. By Bernoulli's principle, the pressure at the rostrum is high as the velocity of the flow is zero, but as the velocity increases downstream, the pressure is reduced. The region anterior to the maximum thickness has a favorable pressure gradient (i.e., water flow from high to low pressure). Downstream of the maximum thickness pressure increases as the flow decelerates, but is not equivalent to the pressure at the rostrum. This region has an adverse pressure gradient (i.e., water must move against an increasing pressure), and the boundary layer flow around the body must have enough energy to continue alongside the body before separating into the wake. Separation of the boundary layer occurring at the trailing edge of the body will produce a narrow wake. If there is insufficient energy and momentum in the boundary layer, the increased pressure in the aft of the body will decelerate the flow. This reduction in flow velocity can lead to instabilities and flow reversal. Reversals in the form of eddies and vortices (i.e., cyclonic flow) can interact with the boundary layer and cause it to prematurely detach from the body interacting with the outer flow. The premature separation further increases the kinetic energy losses and associated drag on the body, manifested as a broad wake.

The flow within both the boundary layer and outer flow can be laminar, turbulent, or transitional (Webb, 1975). The type of flow is dependent on the Reynolds number (Re), which is the ratio of inertial to viscous forces in the fluid. Re is calculated as the product of the flow velocity (U), characteristic length (L), and density (ρ) of the fluid divided by the viscosity (μ).

$$\mathrm{Re} = UL\rho/\mu$$

Typically, the flow will be laminar at $\mathrm{Re} < 5 \times 10^5$, turbulent at $\mathrm{Re} > 5 \times 10^6$, and transitional in between (Webb, 1975; Vogel, 1994). In a laminar flow, the flow is envisioned as parallel layers sliding smoothly over each other, whereas, in turbulent conditions, the flow is disordered and chaotic. The ordered structure of laminar flow has less friction consuming less energy and producing a lower drag than turbulent flow. However, laminar flow in the boundary layer is more susceptible to premature separation with a concomitant increase in the pressure drag. A turbulent flow will energize and maintain attachment of the boundary layer and minimize the pressure drag at the expense of a slight increase in the frictional drag (Webb, 1975).

Dolphins swimming at routine speeds will be in the turbulent regime with Re above 10^6, although the local Re in the forebody of the dolphin can be in the laminar regime. Experiments using pressure sensors in the boundary layer of an actively swimming dolphin indicated that although agitated the boundary layer did not become completely turbulent (Kozolv and Shakalo, 1973; Romanenko and Yanov, 1973; Romanenko, 1976, 1981). The inferred transition was anterior to the dorsal fin and corresponded with a local Re of about 3×10^6 (Aleyev, 1977).

Early studies of dolphin swimming energetics and hydrodynamics used estimates of drag based on rigid bodies models (Gray, 1936). These models assumed that the thrust generated by swimming dolphins was equal to estimates of drag from gliding dolphins (Lang and Daybell, 1963; Lang and Pryor, 1966; Kayan, 1974), towed reproductions of dolphins (Purves et al., 1975; Aleyev, 1977), or hydrodynamic formula based on flat plates (Gray, 1936; Kayan and Pyatetskiy, 1978; Hui, 1987).

Estimates of drag assuming equivalence to a flat plate use the equation:

$$\mathrm{Drag} = 1/2\,\rho\,C_D\,S_w\,U^2$$

where ρ is the density of the medium, C_D is the nondimensional drag coefficient, S_w is the wetted surface area of the body, and U is the velocity. The equivalence of drag on a flat plate with drag on the body of a dolphin assumes that the drag is predominantly frictional, arising from the boundary layer (Webb, 1975).

A dolphin, however, does not swim as a rigid body, but oscillates its tail and flukes in large-amplitude motions (Joh, 1925; Slijper, 1961; Lang and Daybell, 1963; Pyatetskiy and Kayan, 1975; Kayan, 1979; Videler and Kamermans, 1985; Fish, 1993b, 1998a). Hydromechanical models based on kinematics provide drag values for actively swimming animals, which exceed equivalent rigid bodies by three to seven times (Lighthill, 1971; Kayan, 1974; Webb, 1975; Kayan and Pyatetskiy, 1978; Fish, 1998a).

The addition of appendages increases the drag. The body of the harbor porpoise (*Phocoena phocoena*) encompasses 87.6% of the total surface area, but only 64.3% of the drag (Yasui, 1980). The remaining 35.7% of total drag comes from the dorsal fin (4.3%), pectoral flippers (18.0%), and flukes (13.4%) comprising only 2.6%, 4.2%, and 5.6% of the total surface area, respectively. The drag added by the appendages of bottlenose dolphin was estimated as 28% of the total drag (Lang and Pryor, 1966). The added increase in drag due to the appendages is caused by interference drag as flow over the body is distorted by the appendages and induced drag from lift generation by the appendages (Hoerner, 1965; Fish, 1998b; Vogel, 1994). The induced drag component is the energy lost due to tip vortices that are generated by pressure differences between the two sides of the flukes during the oscillating propulsive strokes. In addition, induced drag can be produced by the flippers and dorsal fin when canted at an angle to the oncoming flow during maneuvers (Webb, 1975; Fish, 1993a). The induced drag is highly dependent on AR, with high drag associated with low AR hydrofoils.

Higher C_D than the theoretical minimum for a rigid body is expected for a swimming dolphin because the oscillating motions of the flukes and body will produce "boundary layer thinning" (Lighthill, 1971). Lighthill (1971) estimated that skin friction could increase up to a factor of five. In addition, the pressure component of drag will increase because the propulsive motions of the body will produce a deviation from a streamlined body (Fish and Rohr, 1999). The frontal area of an actively swimming dolphin will be larger compared to one passively gliding as the rostrum oscillates in the vertical plane by 0.02–0.06 BL (Fish et al., 2003). Such gyrations will also increase drag on control surfaces. Unless the flippers can be manipulated in tune with the oscillations of the flukes to present a small frontal area, drag will increase. Increased drag would occur also as dolphins bend their bodies to maneuver (Fish, 1997). Dolphins swimming in small radius circles have higher calculated power outputs than for linear swimming (Hui, 1987).

Drag reduction mechanisms

Observations of dolphins swimming at high speeds in what was regarded as an effortless manner were the incentive for examination of possible drag reduction mechanisms (Fish and Rohr, 1999; Fish, 2006). The consideration of specialized drag reduction mechanisms was initiated by the drag calculations of Gray (1936), who concluded that a dolphin did not have sufficient muscular power to overcome the drag with a turbulent boundary layer during high-speed swimming (36 km/h). Gray's solution in what became known as "Gray's Paradox" was to consider the boundary layer to remain laminar, despite the dolphin operating at a high Re (1.8×10^7), indicating turbulent conditions. As pointed out by Fish (2006), Gray had a number of errors in his calculations associated with the initial speed observation of a dolphin swimming along the side a ship for a 7 s burst compared with sustained muscle power outputs of human oarsmen and dogs, and no consideration of the effect of deformation of the dolphin shape due to swimming movements.

The use of various visualization techniques such as water shed from leaping into the air, dye applied to the dolphin's melon, and swimming through air bubbles, particle curtains, and water with bioluminescent plankton were inconclusive regarding the flow within the boundary layer or demonstrated only the anterior 0.32 BL possessed laminar flow (Rosen, 1962, 1963; Latz et al., 1995; Rohr et al., 1998). The presence of bioluminescence clearly showed that the boundary layer remained attached over most of the dolphin (Rohr et al., 1998). The boundary layer remained attached up to the base of the flukes during rectilinear glides. No difference in drag was apparent when tripping rings were carried by a dolphin during gliding experiments (Lang and Daybell, 1963). The implication was that separation not turbulence was suppressed to minimize drag (Webb, 1975; Fish, 1998b). Despite the high Re indicating a turbulent boundary flow, the lure of Gray's Paradox encouraged investigators to seek special drag reduction mechanisms to laminarize flow over the dolphins (Fish, 2006).

As originally proposed by Gray (1936) but never tested, acceleration of the boundary layer due to propulsive fluke actions could re-laminarize the boundary layer to reduce drag. Oscillations of the flukes generate unsteady velocity and pressure gradients by accelerating water over the body (Gray, 1936; Lang, 1963; Romanenko, 1995). The pressure in the wake will be low as fluid is ejected by the flukes into the wake at a velocity greater than the free stream (Webb, 1975). This action reduces the pressure gradient over the posterior portion of the body (Jackson, 1992; Platzer et al., 1998). Acceleration of boundary flow along with fluid accelerated by the flukes into the propulsive wake would delay separation by reducing the pressure gradient.

The integument or skin is the one part of the body that is intimate contact with the environment and can function in enhancing locomotion (Garten and Fish, 2020). An active mechanism to dampen turbulence in the boundary was hypothesized by changing skin compliance with muscular control by microvibrations (Haider and Lindsley, 1964; Babenko et al., 1982; Haun et al., 1983, 1984; Koslov and Pershin, 1983; Ridgway and Carder, 1993). Microvibrations are small tremor-like vibrations (1–5 mm: 7–13 Hz) that occur over the body of warm-blooded animals (Haider and Lindsley, 1964; Ridgway and Carder, 1993). This mechanism relies upon a sensory input from the richly innervated skin.

Dolphin skin generates microvibrations with an amplitude 3–4 times higher than for humans. It was suggested that the dolphin skin could move or vibrate to improve hydrodynamic performance (Ridgway and Carder, 1993). In that the skin is sensitive enough to detect turbulence, drag could be decreased by the skin actively flexing away from areas of higher pressure toward areas of lower pressures and thus decrease the pressure gradient.

Active deformation of the skin is under the control of the *musculus cutaneous*. This muscle is found in the bottlenose dolphin as a sheet lying 2–4 cm below the skin surface. Bottlenose dolphins initiated microvibrations in the 25–40 Hz range, which was considered to lend support to active compliance as a drag reducer (Haun et al., 1984). However, Aleyev (1974) argued that the function of the *musculus cutaneous* was not only to generate deformation waves in the skin but also to maintain a smooth contour to the body.

The magnitude of the frictional drag component is a function of the texture of the skin. The skin of cetaceans is generally described as smooth (Shoemaker and Ridgway, 1991). However, cutaneous ridges are present at the surface of the skin in many dolphins. The ridges are formed from the dermal crests and papillae (Sokolov, 1960; Purves, 1963, 1969; Purves et al., 1975; Aleyev, 1977; Haun et al., 1983; Pershin, 1988). A longitudinal orientation of the ridges would be analogous to riblets, which can reduce drag by 7%–9%. Riblets are streamwise microgrooves that act as fences to break up spanwise vortices and reduce the surface shear stress and momentum loss from the boundary layer (Walsh, 1990). The orientation of the cutaneous ridges on dolphins is not oriented parallel with the flow (Purves, 1963; Pilleri, 1976; Shoemaker and Ridgway, 1991). The ridges in skin samples from seven species of cetaceans showed the ridges spaced 0.41–2.35 mm apart and 7–114 μm in height (Shoemaker and Ridgway, 1991). Wainwright et al. (2019) validated that the skin of dolphins was smooth, particularly when compared to other pelagic swimmers. It was shown that the size of the ridges on the body was small (2–180 μm) or absent on skin of the control surfaces for most species.

The addition of dilute solutions of long-chain polymers into flow is well established as a means of drag reduction (Rosen and Cornford, 1971; Hoyt, 1975; Daniel, 1981). The epidermal cells of dolphins contain lipid droplets that are similar to mucopolysaccharides (Harrison and Thurley, 1972). Epidermal cell production in bottlenose dolphins occurs at a rate 250–290 times that of humans (Palmer and Weddell, 1964). This high rate is associated with increased skin sloughing. Cells shed from the epidermis were investigated as a mechanism to reduce drag but were found to have a negligible effect on hydrodynamic drag (Sokolov et al., 1969). The surface chemical/physical properties of the skin in conjunction with its high rate of sloughing could maintain low-drag characteristics by preventing biofouling on the dolphin's surface (Gucinski and Baier, 1983; Fish and Hui, 1991).

Secretions from the dolphin eye are a highly viscous complex of proteins and polysaccharides (Uskova et al., 1983). Sokolov et al. (1969) found that these secretions have no effect of drag characteristics. However, the secretions may fill in any disruptions in skin contour around the eyes. The area covered by the secretions is generally too small to aid in any significant drag reduction for the body.

Increasing the temperature of water will change its viscosity (Webb, 1975), thereby making the boundary layer laminar flow more stable. The surface temperatures of spinner dolphin (*Stenella longirostris*) and Pacific coast bottlenose dolphin were reported to be higher than water temperature by as much as 9°C (McGinnis et al., 1972; Hampton and Whittow, 1976). The swimming speed of the animal precludes sufficient contact with the skin to effectively heat the water. The effectiveness of heating as a drag reduction mechanism is considered limited (Lang and Daybell, 1963; Lang, 1966b) or insignificant (Webb, 1975).

Mobile skin folds have been observed on accelerating dolphins (Essapian, 1955; Backhouse, 1960; Pershin, 1988). These folds are believed to result from vorticity along the body surface creating pressure differences that could deform the flexible skin (Backhouse, 1960; Aleyev, 1977) or from active control by muscles (Sokolov et al., 1969). The folds move in a wavelike manner perpendicular to direction of the dolphin's movement from anterior to posterior (Essapian, 1955). Large skin folds, however, increase drag (Aleyev, 1977). Skin folds along the dolphin's body during swimming probably are minimized by the crossed helically wound layer of collagen fibers within and underlying the blubber (Wainwright et al., 1985; Pabst, 1996a, b, 1988; Toedt et al., 1997; Hamilton et al., 1998).

The idea of viscous dampening from a compliant skin as a mechanism for drag reduction in dolphins was developed by Kramer (1960, 1965). Kramer (1960) argued that viscous dampening by passive compliance of dolphin skin would delay the transition from laminar to turbulent flow in the boundary layer. It was hypothesized that energy can be absorbed by the blubber to smooth the flow and reduce drag (Fitzgerald et al., 1995, 1998). Passive changes in compliance were envisioned by regulating blood pressure within the blood vessels of the dermal papillae (Babenko et al., 1982; Pershin, 1983). However, a review of compliant surfaces by Choi (2012) indicated no direct evidence for drag reduction by dolphin skin for this mechanism.

Swimming kinematics

The swimming kinematics of cetaceans are characteristic of the thunniform mode (Lighthill, 1969, 1970; Webb, 1975; Lindsey, 1978). The thunniform mode is typical of some of the fastest marine vertebrates (Lighthill, 1969). Dolphins generate thrust in the thunniform mode exclusively with high AR, caudal flukes (Fish and Hui, 1991; Fish, 1993a). The flukes are oscillated dorsoventrally (i.e., heaving) by bending of the posterior one-third of the body (Fig. 4; Parry 1949b; Slijper, 1961; Lang and Daybell, 1963; Videler and Kamermans, 1985; Yanov, 1990, 1991; Fish, 1993b; Long et al., 1997; Fish et al., 2014).

Superimposed on the heaving motion (i.e., linear vertical movements), the flukes are pitched at a joint at their base. Pitching (i.e., rotation within the vertical plane) at the base of the flukes occurs because of the double-hinge mechanism of the caudal vertebrae, including the "ball" vertebrae (Watson and Fordyce, 1993; Tsai, 1998; Fish et al., 2006). These heaving and pitching motions are responsible for changes of angle of attack and thrust generation throughout the stroke cycle (Lang and Daybell, 1963; Lighthill, 1969; Videler and Kamermans, 1985; Fish, 1993a, 1998a, b).

Up/downstroke

Dolphins use their axial muscles to propel themselves by vertical oscillations of flukes (Parry, 1949a; Strickler, 1980; Bello et al., 1985; Pabst, 1990, 1993). The movement of the flukes is exclusively up and down. Flukes follow a sinusoidal pathway (Fig. 4) that is symmetrical about the longitudinal axis of the body and in time (Pyatetskiy and Kayan, 1975; Goforth, 1990; Fish, 1993b, 1998a). Previously, it was assumed that cetaceans swam with an asymmetrical propulsive stroke. This assumption was predicated on differences in the epaxial and hypaxial muscle masses (Parry, 1949b; Purves, 1963). Smith et al. (1976) found that the hypaxial muscles were smaller than the epaxial muscles in the harbor porpoise, but the muscles were considered powerful enough to flex the tail during a downstroke. Parry (1949a) was able to confirm differences in duration between upstroke and downstroke from counts of film frames of a dolphin swimming away from a camera. Unfortunately, the film records of the swimming dolphin showed the animal giving birth. This would have changed the normal swimming pattern. Subsequent reproductions of the swimming sequence excluded the tail flukes of the neonate (Slijper, 1961; Hertel, 1966). For rapid accelerations, Tanaka et al. (2019) found greater acceleration from the downstroke than the upstroke.

Maximum heaving amplitude occurs at the tips of the flukes. At low swimming speeds (<7.9 km/h for Atlantic white-sided dolphin (*Lagenorhynchus acutus*), heave amplitude appears to increase with speed (Curren et al., 1994).

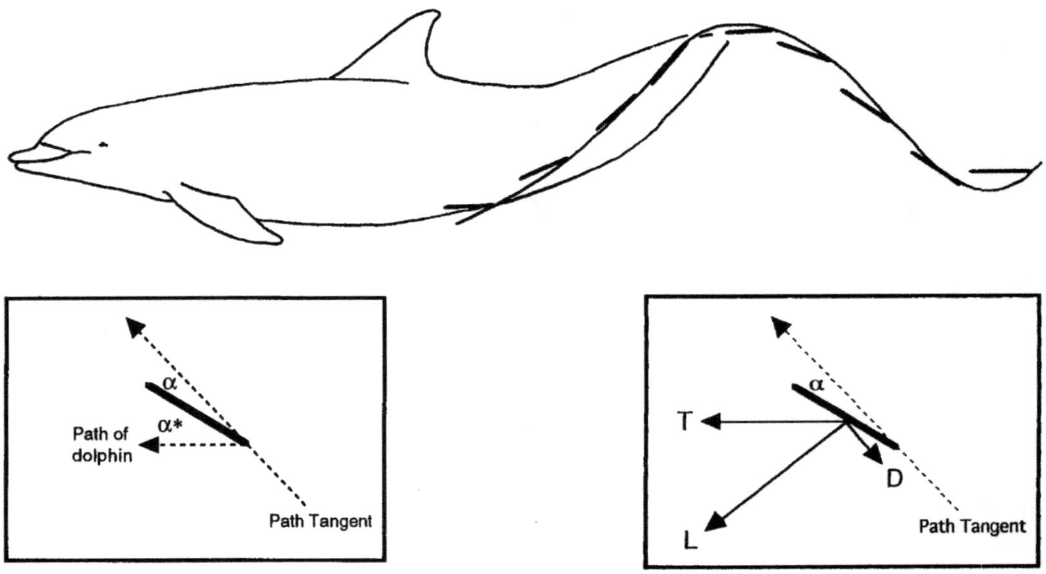

FIG. 4 Path of oscillating dolphin flukes through a stroke cycle. The flukes move along a sinusoidal path. Sequential fluke positions along the path are illustrated as straight lines. Box on the lower left shows the relationship between the tangent to the fluke path with the angle of attack, α, and the pitch angle α^*. The box on the lower right shows the relationship between the major forces produced by the fluke motion, where D is drag, L is lift, and T is thrust resolved from L. *From Fish (1993b).*

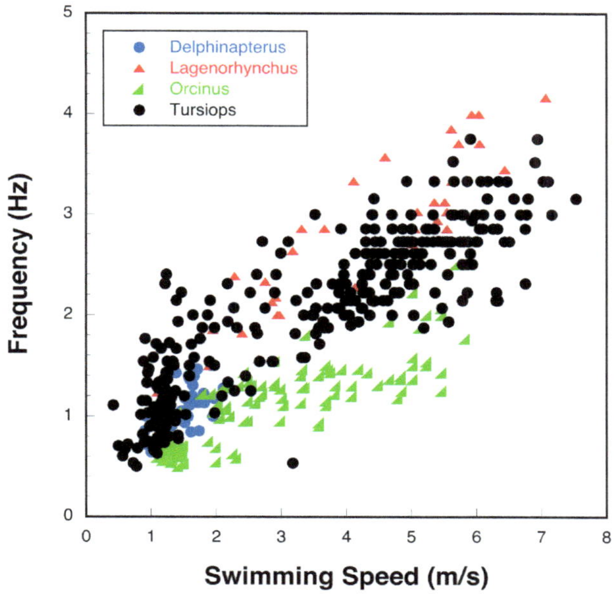

FIG. 5 Tailbeat frequency in Hz as a function of swimming velocity for four cetacean species (beluga, Pacific white-sided dolphin, killer whale and bottlenose dolphin).

Heave amplitude (h), however, is independent of swimming speed at routine and sprint speeds (Kayan and Pyatetskiy, 1977; Fish, 1993b, 1998a; Rohr et al., 1998). Maximum peak-to-peak amplitude ($2h$) measured over a complete tail beat cycle remains a constant of 0.2 BL (Fish, 1998a). Videler and Kamermans (1985) observed that despite equal time periods for up- and downstrokes, they determined that the downstroke generated more thrust than the upstroke. The low swimming speed ($\leq$11.4 km/h) may have resulted in the thrust asymmetry. In addition, asymmetrical motions of the flukes can be observed during accelerations (Skrovan et al., 1999).

The stroke frequency varies directly with swimming speed (Pyatetskiy and Kayan, 1975; Kayan and Pyatetskiy, 1977; Fish, 1993b, 1998a; Rohr et al., 1998). Maximum tailbeat frequencies of approximately 3 Hz were exhibited by the bottlenose dolphin and false killer whale (Goforth, 1990; Fish, 1998a; Rohr et al., 1998). Differences in slope of tailbeat frequency are dependent on body size (Fig. 5). The linear slope for tailbeat frequency regressed against length-specific swimming speed (BL/s) decreases with increasing body size (Fish, 1998a).

Thrust generation

Thrust is generated almost continuously throughout a stroke cycle using lift-based oscillating flukes (Lighthill, 1969; Fish and Lauder, 2006). Lift is produced from differential flow between the surfaces of the flukes (i.e., Bernoulli effect) as the flukes are canted at an angle of attack (α; i.e., incident angle to oncoming flow). The heaving and pitching motions of the flukes are responsible for changes of pitch angle (α^*) and α throughout the stroke cycle (Fig. 4). As the flukes move along a sinusoidal path through the water, the incident flow encountered by the flukes is a combination of the forward movement of the dolphin and the heaving motion of the tail, where the pitch of the flukes relative to its path is α (Fig. 4). Lift is directed perpendicular to the pathway traversed by the flukes and can be resolved into an anteriorly directed thrust vector (Lighthill, 1969). Thrust is derived from a combination of this horizontal component of the lift force and leading-edge suction (Ahmadi and Widnall, 1986).

The action of the flukes in generating lift induces a rotation of fluid in the wake, known as vorticity. The increase in vorticity transfers the kinetic energy of the dolphin to the water, which is eventually lost in the formation of eddies and frictional forces in the water. Vorticity shed into the wake reflects the transfer of momentum from the flukes to the fluid. Momentum shed into the wake of a swimming dolphin was visualized by bubble digital particle image velocimetry (DPIV; Fish et al., 2014, 2018). The dorsoventral oscillations of the flukes produce a thrust-type wake (Fig. 6), which takes the form of two alternating rows of cyclonic vortices and a posteriorly directed jet stream (Weihs, 1972; Fish et al., 2014; Smits, 2019). Each vortex is produced by the fluke trailing edge at the end of each half stroke.

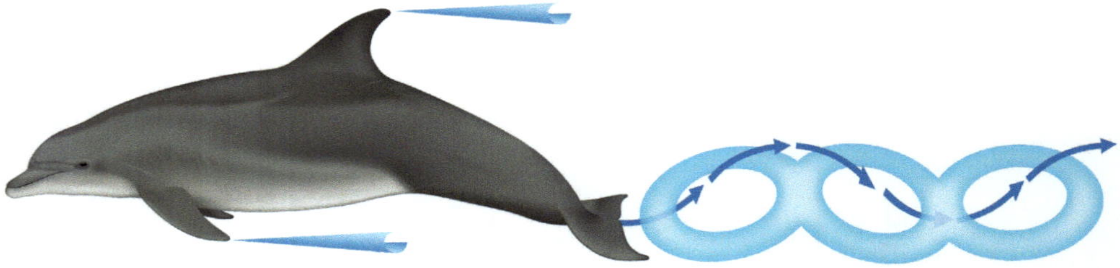

FIG. 6 Pattern of vorticity shed into the wake of a dolphin. The wake is shown as alternating vortices that are interconnected with a momentum jet running through their centers. *Illustration by Uko Gorter.*

Gray's (1936) original calculations of thrust power (1939 W) for a dolphin with turbulent conditions provided a mass-specific thrust power of 21.4 W/kg. At a maximum speed of 39.96 km/h for 1.5 s acceleration by pantropical spotted dolphin, thrust power was 4517.8 W with a mass-specific thrust power of 85.7 W/kg (Lang and Pryor, 1966). Pacific white-sided dolphin swimming at 30.24 km/h produced a thrust power of 9900 W with a mass-specific thrust power of 90 W/kg (Tanaka et al., 2019). Fish (1998a) found that thrust power based on unsteady lifting wing theory (Chopra and Kambe, 1977) was highly dependent on swimming velocity and size. The killer whale had a thrust power of 36.3 kW at U of 28.8 km/h, representing a mass-specific power output of 22.0 W/kg, whereas the false killer whale and bottlenose dolphin had equivalent maximum mass-specific power outputs of 22.5 and 23.7 W/kg, respectively. Bubble DPIV directly showed large thrust power production even at low speeds (Fish et al., 2014).

Control of pitch angle

Thrust from lift increases directly with increasing angle of attack. The heaving and pitching motions of the flukes result in pitch angles, α^*, between the flukes and the horizontal plane (Fig. 4). At their maximum vertical displacement, flukes have α^* of zero so that the axis of the fluke chord is parallel to the axis of progression. This orientation effectively minimizes the drag on the flukes but generates no thrust. As the flukes are swept downward, α^* increases by flexion at the peduncle-fluke junction. Through the middle of the downward excursion of the stroke, α^* is at a maximum (Fig. 4). The end of the stroke is accompanied by a decrease in α^* with the flukes again oriented parallel to the direction of forward progression. Maximum α^* ranges from 33 to 40 degrees (Fish, 1998a). The average α^* for the harbor porpoise and Atlantic white-sided dolphin was 34 and 33 degrees, respectively (Curren et al., 1994).

The ability to rotate the flukes about a pitching axis allows for control of α. Maintenance of a positive α ensures thrust generation throughout the majority of the stroke cycle (Lang and Daybell, 1963; Lighthill, 1969, 1970; Goforth, 1990).

The magnitude of α will affect the propulsive efficiency and the thrust generated in lift-based swimming (Webb, 1975). As α is increased for a hydrofoil, there is increase in both lift and drag. Lift will increase faster than the drag with increasing α up to a critical level. Furthermore, increase in α leads to an increase in drag and precipitous loss of lift in a condition called stall. Stall is due to separation of the flow from the foil surface, which is unavoidable above the critical α.

The α of oscillating dolphin flukes increases rapidly at the initiation of the stroke cycle, reaching a maximum within the first third of the stroke (Fish, 1993b). Maximum α is 22.5–24 degrees for Pacific white-sided dolphin (Lang and Daybell, 1963). Maximum α varies indirectly with swimming speed for bottlenose dolphin (Fish, 1993b). Mean values range from 4.6 to 17.5 degrees for bottlenose dolphin, although maximum values can approach 30 degrees at low velocities (Fish, 1993b).

The α observed for dolphins is higher than would be expected compared with nonoscillating airfoils. Conventional airfoils stall at low α in steady flow. An oscillating foil appears to perform more efficiently at higher α than nonoscillating hydrofoils (Fierstine and Walters, 1968; Maresca et al., 1979). At high frequencies, the lift of an oscillating foil will increase with increased α above the stall angle (Maresca et al., 1979). Between 15 and 25 degrees, α provided optimal thrust production for a two-dimensional oscillating foil (Anderson et al., 1998). Hydrofoils like the dolphin flukes can sustain lift without stalling at α of 30 degrees (Triantafyllou and Triantafyllou, 1995). This value may represent an upper limit to α for an oscillating system above which thrust and efficiency are reduced (Chopra, 1976). An oscillating foil can function at high α due to dynamic stall or delayed stall caused by unsteady effects (Maresca et al., 1979; Ellington et al., 1996; Anderson et al., 1998).

Propulsive efficiency

Propulsive efficiency (η_p) is defined as the ratio of the mean thrust power required to overcome the drag on the animal divided by the mean rate at which the animal is doing work against the surrounding water (Lighthill, 1970). Efficient thrust production requires high-lift production while minimizing energy loss into the wake (Blake, 1983b).

Wu (1971) estimated that the propulsive efficiency of a dolphin could be as high as 0.99. This η_p was assumed to be an overestimate because Wu used a two-dimensional analysis, which underestimated trailing vorticity and wake energy loss (Ahmadi and Widnall, 1986; Karpouzian et al., 1990). In general, η_p is expected to be lower than 0.99 but above 0.7 at routine swimming speeds (Chopra and Kambe, 1977; Yates, 1983; Karpouzian et al., 1990; Fish, 1993b, 1996, 1998a). For cetaceans, η_p is 0.75–0.9 (Fish and Rohr, 1999), where maximum η_p is achieved within the range of normal cruising speeds (0.8–1.5 BL/s). Efficiencies in this range are considered good because few engineered propellers achieve efficiencies higher than 0.7 (Larrabee, 1980; Liu and Bose, 1993).

Optimal η_p is associated with the rate that vortices from the oscillating flukes are shed into the wake. The principal wake parameter is the Strouhal number (St; Triantafyllou et al., 1991, 1993; Rohr and Fish, 2004; Saadat et al., 2017). St is a dimensionless number, representing the ratio of unsteady and steady motion. St is defined as $St = Af/U$, where A is the width of the wake, taken to be equal to the peak-to-peak maximum excursion of the trailing edge of the caudal propulsor; f is the frequency of oscillation; and U is the mean forward velocity (Triantafyllou et al., 1991). The optimum creation of thrust-producing vortices lies in a narrow range of St. The predicted St range for maximum spatial amplification occurs between 0.25 and 0.4, where η_p peaks (Triantafyllou et al., 1991; Triantafyllou and Triantafyllou, 1995).

St of 0.30 for a dolphin was calculated by Triantafyllou et al. (1993) from data obtained from Lang and Daybell (1963). Rohr et al. (1998) examined the distribution of St with respect to swimming speed for the false killer whale and bottlenose dolphin housed in managed care. St derived for the false killer whale ranged from 0.21 to 0.37 and St for the bottlenose dolphin ranged from 0.15 to 0.36 with a mean of 0.25. Comparison data sets for the bottlenose dolphin showed mean St of 0.25 (Kayan and Pyatetskiy, 1977) and 0.27 (Fish, 1993b). Although the mean St for each of the dolphin data sets was within the 0.25–0.4 range predicted by Triantafyllou et al. (1993), this range included only 56% of all the data compared with 89% of the data residing between St of 0.2 and 0.35 (Fig. 7; Rohr et al., 1998; Rohr and Fish, 2004). Indeed, η_p decreases only slightly at $St = 0.2$ from peak η_p and is approximately the same as η_p at $St = 0.35$ (Triantafyllou et al., 1991).

Analysis of data from Fish (1998a) demonstrated the relationship between St and η_p for the beluga, killer whale, false killer whale, and bottlenose dolphin (Fig. 8). Maximum efficiency at St of 0.26 was found for the killer whale, false killer whale, and bottlenose dolphin, but for the beluga, maximum efficiency occurred at 0.31. Within the optimal St range (0.25–0.4), η_p remained above 0.83.

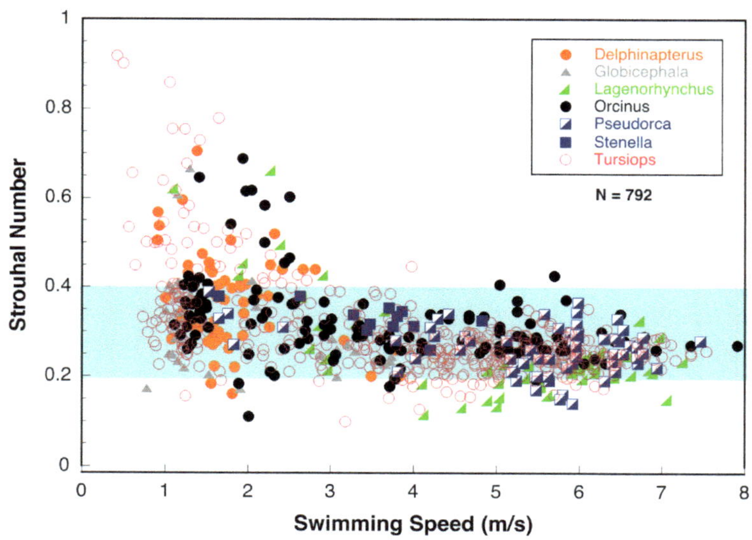

FIG. 7 Relationship between the Strouhal number and swimming speed for seven cetacean species (beluga, long-finned pilot whale, Pacific white-sided dolphin, killer whale, false killer whale, spotted dolphin and bottlenose dolphin). The light blue area indicates the optimal range of the Strouhal number, where cetaceans would been swimming with a high efficiency.

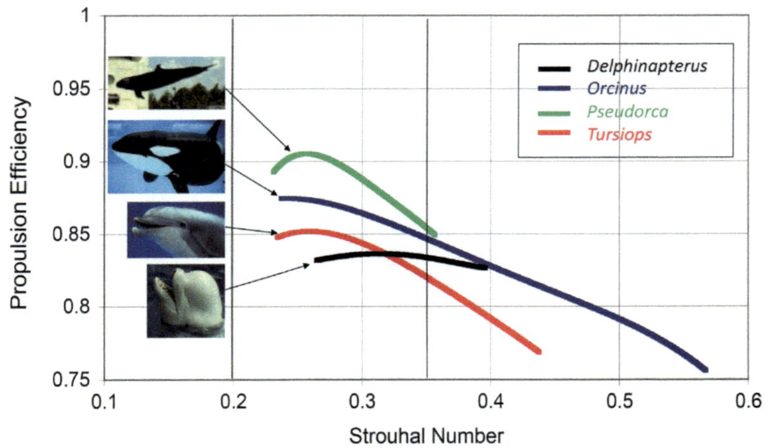

FIG. 8 Propulsive efficiency of cetaceans as a function of the Strouhal number. The efficiencies were calculated from the computational model by Chopra and Kambe (1977). The maximum propulsive efficiency for each species (beluga, killer whale, false killer whale and bottlenose dolphin) falls within the optimal range of the Strouhal number.

Flexibility and spring-like effects

Elastic tissues in the bodies of animals can store strain energy and later release it as kinetic energy to reduce the energetic cost of swimming (Alexander, 1988). The spring-like action of elastic structures (e.g., intervertebral joints, blubber, and tendons) can reduce the energy used by muscles to decelerate and reaccelerate propulsive appendages during locomotion (Wainwright et al., 1985; Pabst, 1998, 1996a, b; Pabst et al., 1995; Long et al., 1997; Toedt et al., 1997; Hamilton et al., 1998). These structures are composed of collagen fibers, which can store and transmit elastic energy. Collagenous tendons have a resilience of >90% at low strains (Bennett et al., 1986). It has been suggested that elastic structures are utilized in cetaceans to reduce the energy cost of swimming, although the idea remains controversial (Russell et al., 1983; Bello et al., 1985; Pabst, 1996a).

Elastic compliance of tendons that attach to the vertebral column for a harbor porpoise and Atlantic white-sided dolphin is above the optimal compliance to minimize the energy cost swimming (Bennett et al., 1987). However, using a mathematical model, Blickhan and Cheng (1994) indicated that the elastic elements in the body had the correct properties for near maximum energy savings of 55%. Under normal swimming conditions, the vertebral column would not act like a spring, but as a shock absorber. Higher energy return (~50%) might occur during rapid accelerations when the flukes move through a large vertical displacement (Long et al., 1997).

The properties of spring-like fibers may reside in the blubber. Blubber is an elastic structure (Orton and Brodie, 1985; Pabst et al., 1995). The blubber and subdermal connective sheath have fibers arranged in a cross-helical pattern (Pabst et al., 1995; Pabst, 1996a, b; Toedt et al., 1997; Hamilton et al., 1998). Angles between fibers greater than 60 degrees can store and release strain energy during swimming to straighten the body (Pabst, 1996a). The pattern of fiber orientation and the resilience of blubber in the tailstock of dolphins indicate that this structure might act as a biological spring (Hamilton et al., 2004).

Williams et al. (1993) demonstrated that dolphins swimming at speeds up to 7.6 km/h showed no significant increase in aerobic or anaerobic metabolism, despite increasing drag. Above 7.6 km/h, both oxygen consumption and lactate production increased. This trend is similar to the metabolic response observed in kangaroos, which store elastic energy in tendons to maintain a constant metabolic rate with increasing hopping speed (Baudinette, 1994). Pershin (1970) argued that swimming dolphins show resonance at an oscillatory frequency of 1.5–1.9 Hz. For bottlenose dolphin, this frequency range corresponds to 10.1–14.4 km/h (Fish, 1993b, 1998a). This range is at the swimming speeds at which dolphins increase metabolism (Williams et al., 1993). The use of resonating springs in dolphins may be limited, because there is substantial dampening by the highly dense and viscous medium, as opposed to air where such mechanisms are employed (Alexander, 1988; Clark and Fish, 1994).

Scaling

Cetaceans exist at the small end of a size continuum of cetaceans extending from the tiny 1.5-m-long Vaquita (*Phocoena sinus*) up to the enormous 30-m-long Blue Whale (*Balaenoptera musculus*). Many of these species have broadly similar morphologies, and all swim with the same oscillatory tail heaving motion. Hill (1950) first noted the effects of

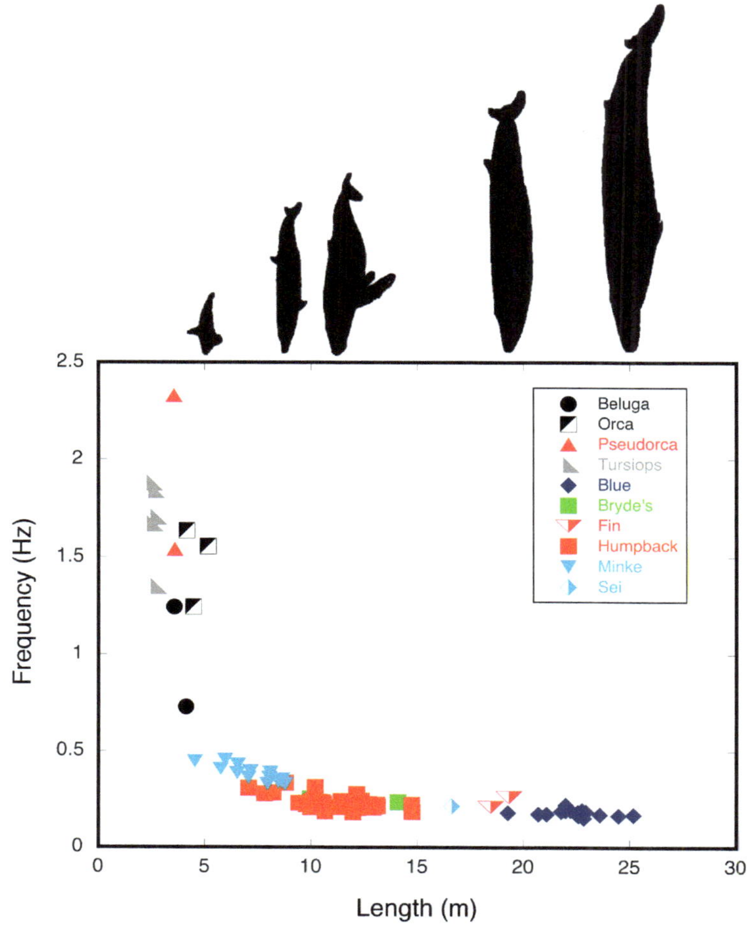

FIG. 9 Oscillatory frequency (Hz) of swimming in relation to total body length (m) over an extended body size range of cetaceans. Silhouettes above the graph represent (from left to right) killer whale, minke, humpback, fin, and blue whales.

body size on the oscillatory movements of propulsive structures. Using simple muscle models, he predicted that the frequency of oscillations should decrease in proportion to increasing body length for a given speed to balance the production of useful work and wasteful kinematic energy. This relationship between the oscillation of the tail and body size has been found for the full range of cetacean body sizes (Fig. 9).

Energy expended during swimming results from the interaction of oscillatory kinematics and body morphology. As body size increases, the drag power on that body increases exponentially, necessitating an exponential increase in energy expenditure in the form of thrust power to move that body through the water. Because of the consistent scaling of kinematics such as oscillatory frequency with body size, estimates of thrust power for all cetaceans fall along a smooth continuum associated with the Reynolds number (Fig. 10). The implications of these scaling relationships are that cetaceans, from the smallest dolphins to the largest whales, share similar hydrodynamic constraints and physical limitations, regardless of life history.

Behavioral strategies for energy economy

Submerged swimming

The need to breathe gaseous oxygen to fulfill metabolic demands incurs a potentially high drag and associated energy demand for cetaceans. Respiratory exchange can only take place at the water surface where the animal experiences increased drag from the production of surface waves (Scholander, 1959a; Hoerner, 1965; Hertel, 1966). Kinetic energy from the animal's motion is lost as it is changed into potential energy in the formation of waves (Denny, 1993; Vogel, 1994). This

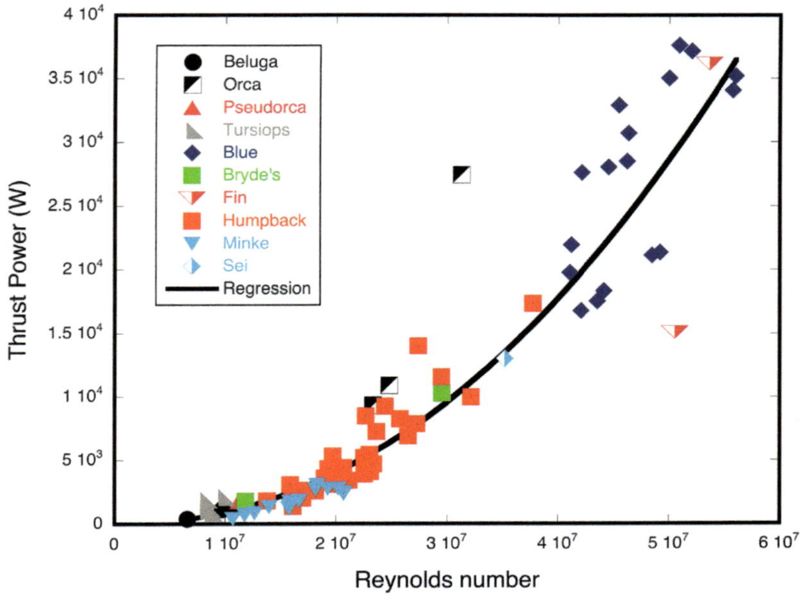

FIG. 10 Thrust power (W) output in relation to the Reynolds number over an extended body size range of cetaceans. The black line represents the regression for a power function with an exponent of 2.13 with a correlation coefficient of 0.94. All cetaceans fall along this continuum, from small dolphins to large whales.

energy exchange results in increased wave drag in addition to pressure and frictional drag components compared to the drag when the body is fully submerged (Lang and Daybell, 1963; Hoerner, 1965; Hertel, 1966; Fish, 1992, 1993a).

Estimates of wave drag on a Pacific white-sided dolphin were made by J. A. Poore and presented as an appendix in Lang and Daybell (1963). Wave drag estimates (based on towed bodies: Hertel, 1966) suggest that near the surface wave drag becomes important and exceeds frictional drag by five times. Wave drag will peak at a Froude number of 0.45 (Lang and Daybell, 1963; Hoerner, 1965). The Froude number (Fr) is the dimensionless ratio of inertial forces to gravitational forces experienced by a body moving at or close to a fluid/fluid interface. Fr is calculated as follows:

$$\mathrm{Fr} = U/(g\,L_{\mathrm{w}})^{1/2}$$

where L_{w} is the water line length along the longitudinal axis of the body protruding through the water surface. Because waves can be produced when the body does not pierce the surface, BL can substitute for L_{w}.

Au and Weihs (1980a) considered the maximum drag increase for a dolphin to be 4.5. The maximum wave drag occurs when the half the body is submerged so that the relative depth is 0.5 of maximum body diameter (Hertel, 1966). Wave drag decreases with increasing and decreasing submergence depths and becomes unimportant at depths greater than three times body diameter or when the animal becomes airborne (Hertel, 1966). Lang and Daybell (1963) found below a depth of two body diameters, the wave drag is essentially zero at U of 22.0–32.8 km/h for a dolphin.

Energy saved by swimming beneath the water surface offsets increased energy expended in surfacing to breathe. Dolphins could gain large energy benefits by swimming submerged for long distances between breaths (Hui, 1989). Furthermore, dolphins quickly ventilate the lungs before submerging and limit time at the surface. Dolphins can ventilate 90% of their lung volume in about 0.3–0.6 s (Irving et al., 1941; Ridgway et al., 1969; Kooyman and Cornell, 1981; Hui, 1989; Law and Blake, 1994; Norris and Johnson, 1994a). An alternate approach to prolong ventilation time while reducing drag is by porpoising (Au and Weihs, 1980a, b; Hui, 1987, 1989; Fish and Hui, 1991).

Porpoising

Porpoising is performed as rhythmic, serial leaping (Fish and Hui, 1991). Au and Weihs (1980a, b) and Blake (1983a) proposed models, which showed that the energy to swim a given distance increases with swimming speed faster than the energy to leap that distance. Therefore, above a critical swimming speed, U_c, where the energies converge, there is an energetic advantage to swimming by porpoising (Fig. 11). The estimated U_c computed by Au and Weihs (1980a) for a 2.69-m-long dolphin was 22.25 km/h and from the model by Blake (1983a) was 10.91–21.71 km/h.

Fish and Hui (1991) questioned the concept of U_c based on field data. Porpoising *Stenella* swim twice the distance they leap, whereas the assumption of U_c predicts that dolphins would spend more time leaping than swimming at $>U_c$

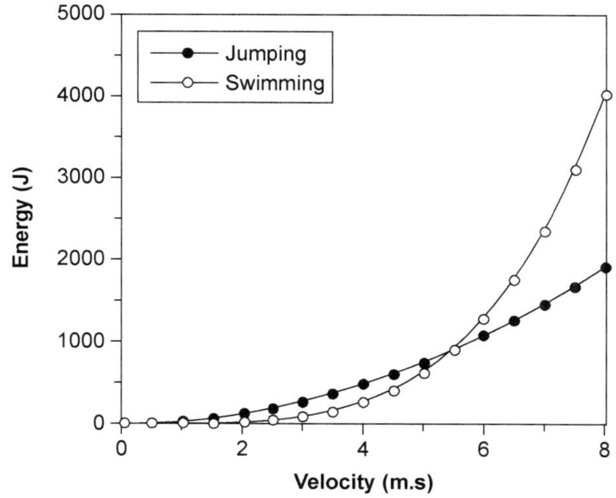

FIG. 11 Energies required by a dolphin swimming close to the water surface and for jumping according to the model by Au and Weihs (1980a). The dolphin was assumed to be neutrally buoyant with a volume of 0.1 m^3 and drag coefficient of 0.02 and 4.5 increase in drag due to wave drag at the surface. Crossover velocity (i.e., where it becomes more economical to jump a given distance than swim) was approximately 5.5 m/s (19.8 km/h).

(Au et al., 1988). Video data of free-ranging common dolphins and pantropical spotted dolphins indicate a graded transition from minimal blowhole exposure at the surface at the low swimming speeds, to quasi-leaps in which the dolphin is never completely out of the water at any instant at the medium swimming speeds, and to complete porpoising leaps at the highest swimming speeds (Hui, 1989). These observations are consistent with maintaining a minimum blowhole exposure time for respiratory inhalation as swimming speed increases. Porpoising may be an energy conservation behavior, which is directed more to economical breathing than as an energetically cheap method of swimming (Hui, 1989; Fish and Hui, 1991). The greater energetic demands of rapid swimming necessitate the ability to prolong ventilation without potential interference from waves.

Emergence angles of leaps by dolphins were widely distributed between a 30 degrees angle for maximum speed and 45 degrees angle predicted for maximum leap distance (Au and Weihs 1980a; Gordon, 1980; Hui, 1989). Gordon (1980) indicated that to maintain forward speed, an emergence angle of 30 degrees would be desirable as a compromise between maximum distance and maximum forward speed of a leap. In reality, the emergence angle of the leaps is approximately 39 degrees with wide variation (Hui, 1989; Weihs, 2002), which is between the optimal predicted angles.

Gliding

A substantial amount of time during swimming may be occupied by gliding to reduce drag (Lighthill, 1970; Weihs and Webb, 1983; Williams et al., 1996). Gliding is beneficial while diving (see Chapter 2). During deep dives, dolphins can reduce energy costs by approximately 20% when transiting to the bottom by using intermittent swimming behaviors (i.e., alternating swim and glide; Williams et al., 1996, 2000; Williams, 2001). Diving dolphins use gliding to reduce locomotor energy costs when they are negatively buoyant during descent and positively buoyant during ascent (Skrovan et al., 1999).

The density of the body increases with depth due to increased hydrostatic pressure. When diving deeply (>20 m), lung collapse reduces the net buoyant force causing the animal to sink (Ridgway and Howard, 1979; Kooyman and Ponganis, 1998; Moore et al., 2011). The dolphin can descend using its negative buoyancy to glide as gravity now supplies the motile force, thus saving energy over active swimming. During ascent, the reverse occurs, and the dolphin accelerates by actively swimming until its lung re-inflate sufficiently to provide positive buoyancy (Skrovan et al., 1999). By allowing the body to be neutrally or slightly negatively buoyant, a dolphin foraging on the bottom can conserve its oxygen reserves and increase its dive time.

Bow riding

Dolphins are well known to ride the bow waves of ships and large whales (Woodcock, 1948; Woodcock and McBride, 1951; Hayes, 1953, 1959; Caldwell and Fields, 1959; Scholander, 1959a, b; Fejer and Backus, 1960; Norris and Prescott, 1961; Lang, 1966b; Leatherwood, 1974; Herman and Tavolga, 1980; Fish and Hui, 1991; Johnson and Norris, 1994).

Dolphins make use of the pressure front created by the boat or whale (Scholander, 1959a; Fejer and Backus, 1960). Bow-riding behavior takes advantage of the thrust produced by another body by swimming in the hydrodynamically favorable area arising from the interaction of the pressure fields (Lang, 1966b). This behavior will occur from minutes to hours (Williams et al., 1992). Williams et al. (1992) found that bow wave-riding dolphins could swim at a higher speed while reducing or maintaining metabolic rate, heart rate, lactate production, and respiratory rate.

This behavior is complex with any energy savings to the dolphin related to bow design, swimming depth, body orientation, and distance from the ship (Fish and Hui, 1991). Scholander (1959a) experimentally tested a foil section in the bow wave and found that the optimal orientation to derive thrust from the ship pressure wave was with the foil pitched down at 28 degrees. Scholander's results were contrary to observations of the tail flukes of bow-riding dolphins with an upward pitch of 30 degrees (Norris and Prescott, 1961; Yuen, 1961). Alternately, Newman (1975) predicted that as the bow wave of a ship is directed laterally at an acute angle a wave riding dolphin positioned off the bow is expected to swim on its side with its flukes in the vertical plane (Newman and Wu, 1974; Newman, 1975). Fejer and Backus (1960) reported dolphins to bow ride on their sides. When shifting from between port and starboard sides of the front of a bow wave, dolphins will roll from side to side (Yuen, 1961; Hertel, 1969). However, this orientation requires that dolphins periodically roll 90 degrees to surface and breathe. It therefore seems that dolphins can take advantage of bow waves and are flexible in body and fluke orientation.

Wave riding

Wind-wave riding and surf-wave riding also can reduce the energy of surface swimming (Caldwell and Fields, 1959; Amundin, 1974). These wave-riding behaviors differ from bow wave riding because they use the interaction of the dolphin's weight and slope of the wave front to effect movement analogous to human surfers (Hayes, 1953; Fejer and Backus, 1960; Perry et al., 1961). When the slope of the wave front is large, dolphins have been observed to surf on in-shore waves and breaking wind waves (Norris and Prescott, 1961). Gravity is used to ride these waves (Hayes, 1953; Caldwell and Fields, 1959; Fejer and Backus, 1960; Perry et al., 1961). Dolphins ride waves with a forward slope of 10–18 degrees at velocities of 18.0–21.6 km/h (Hertel, 1969).

Drafting

Large groups of dolphins are observed in side-by-side and echelon formations (Norris and Prescott, 1961; Leatherwood and Walker, 1979; Johnson and Norris, 1994; Norris and Johnson, 1994b; Weihs, 2004; Shoele and Zhu, 2015). The occurrence of highly organized formations by cetaceans has been suggested as an adaptation for energy economy (Kelly, 1959; Lang, 1966b; Brodie, 1977; Norris and Johnson, 1994b). Animals swimming in formation influence water flow around adjacent individuals decreasing drag with a concomitant decrease in overall energy cost of locomotion (Weihs, 1973; Fish, 1999).

Depending on age, young dolphins will draft beside and slightly behind the maximum diameter of a larger animal (Tavolga and Essapian, 1957; Norris and Prescott, 1961; Dohl et al., 1974; Reid et al., 1995; Marino and Stowe, 1997; Noren and Edwards, 2011). In the neonate position, a calf will be positioned above the flank of the mother (Shoele and Zhu, 2015). Later, the calf moves to an echelon position within 10 cm of the mother's midbody close to the dorsal fin, and with increased size, the calf swims under the mother's tail (McBride and Kritzler, 1951; Cockcroft and Ross, 1990; Reid et al., 1995; Mann et al., 2000). Throughout the first year, up to 75% of observations showed calves to swim next to the midsection or near the genital region of the mother (Gubbins et al., 1999). These positions would afford increased energy saving due to drafting. While the larger dolphins will experience increased drag, the smaller gains an energetic benefit (Kelly, 1959; Lang, 1966b; Weihs, 2002; Noren and Edwards, 2011). This effect is beneficial particularly for young dolphins in order to maintain speed with their mothers. A neonatal dolphin could use this mechanism to gain up to 90% of the thrust needed to move alongside its mother.

In addition, as the two bodies are in close proximity, the water flow between them is accelerated resulting in an attractive force due to the Bernoulli effect (Kelly, 1959). This attractive force helps to keep the infant from drifting away from the mother to maintain high speed with the adult and reduce energy expenditure (Shoele and Zhu, 2015).

Maneuverability

Maneuverability by dolphins is dependent on body size, body stiffness, and use of control surfaces (e.g., flippers, fin, flukes, and peduncle). Control surfaces are structures that allow for adjustments and management of the attitude of a body in a fluid (Fish and Lauder, 2017). The ability to maneuver (i.e., turn) with speed becomes imperative in the

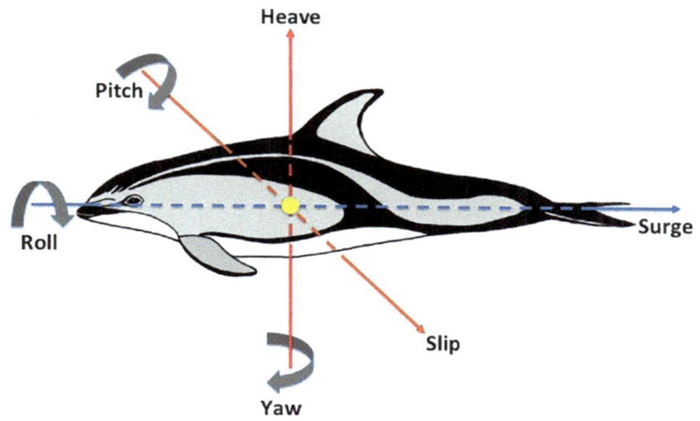

FIG. 12 Six degrees of freedom and three orthogonal axes that specify the position and orientation of the dolphin body. Rotational movements include roll (rotation around the longitudinal axis), pitch (rotation about the transverse axis), and yaw (rotation around the vertical axis). Translational movements along the three axes include surge (longitudinal axis), heave (vertical axis), and slip (transverse axis).

acquisition and capture of prey. Maneuverability is defined as the space required to execute a turn and agility is defined as the rapidity that direction can be changed and is measured as the rate of turn (Walker, 2000). Conversely, stability allows for self-correction opposing disturbances to maintain a desired postural attitude and trajectory (Fish, 2002).

Within the three-dimensional realm of water, maneuvering can move freely about three orthogonally arranged axes that intersect at the center of gravity (CG). CG for the bottlenose dolphin is located at a position of 0.41 BL (Fish, 2002). Movement about CG permits translational and rotational movements that give six degrees of freedom. The degrees of freedom of the translational movements are surge (anterior–posterior), heave (vertical displacement), and slip (lateral displacement), whereas rotation about the axes is termed roll for the longitudinal axis, pitch for the lateral axis, and yaw for the vertical axis (Fig. 12; Webb, 2004, 2006). Although the forward position of CG enhances stability, it also appears to be nearly coincident with the center of buoyancy (Slijper, 1979; Weihs, 1993; Fish, 2002). As a result, delphinids can be unstable with respect to roll and can side-swim, swim upside down, and barrel roll (Layne and Caldwell, 1964; Klima et al., 1987; Marino and Stowe, 1997).

Control of each degree of freedom for maneuverability and stability is dictated by various combinations of control surfaces (Fish and Lauder, 2017). The flippers of dolphins function as hydrofoils to control moments that produce pitch, yaw, and roll. Reduced motion of the flippers of fast-swimming dolphins is necessary to enhance stability (Fish, 1997). The mobility of the flippers of fast swimmers appears to be more constrained compared to the flippers of slow-swimming, highly maneuverable animals (Vasilevskaya, 1974; Pilleri et al., 1976; Klima et al., 1987; Fish, 2002). The shoulder muscular of Amazon river dolphin is highly differentiated in contrast to the faster-swimming white-beaked dolphin (*Lagenorhynchus albirostris*), harbor porpoise, and bottlenose dolphin (Klima et al., 1987). The Amazon river dolphin can perform a turn with radius that is 0.10 BL (Fish, 2002). This degree of maneuverability is necessary to operate in the complex environment of a shallow river habitat.

Dorsal fins are not mobile. The position of the dorsal fin is approximately over CG (Parry, 1949b; Maslov, 1970; Fish, 2002; Triantafyllou et al., 2020). This position limits the dorsal fin's effectiveness in developing a turning moment but allows the fin to prevent side-slip and oppose rolling (Fish, 2002).

Although the caudal flukes have a propulsive function, these lateral extensions of the tail can do double duty as control surfaces. Horizontal orientation of the dolphin flukes permits control of pitch, and the compressed peduncle can be lateral flexed to induce yawing motions (Fish, 2002). The dual function of the caudal appendages for both turning and propulsion presents a restriction to simultaneously maintain high speed during tight turns. To produce a small turn radius (0.11–0.17 BL), dolphins must use unpowered gliding maneuvers by uncoupling the flukes from thrust production (Fish, 2002). During unpowered turns, the peduncle and flukes are diverted from their propulsive orientation and used like a rudder. The tail of the dolphins can be twisted as much as 53 degrees during unpowered gliding turns creating an X-configuration of the flukes and peduncle (Fish, 2002). The combination of aft dive planes and rudders in a X-stern configuration provides submarines with reduced turn radii and increased maneuverability (Largess and Mandelblatt, 1999).

Flexibility in the body of dolphins is generally constrained (Long et al., 1997). The highly compressed cervical vertebrae and fusiform body restrict movement in the neck, although some species (e.g., Amazon river dolphin) have mobile necks. The generally stiff body and placement of control surfaces give dolphins a morphology that enhances stability, thereby constraining maneuvering performance (Fish, 2002). Dolphins with more flexible bodies and mobile flippers (e.g., Amazon river dolphin) sacrifice speed for maneuverability, whereas delphinids with more restricted morphologies (e.g., Pacific white-sided dolphin and Commerson's dolphin) produce faster but wider turns (Fish, 2002).

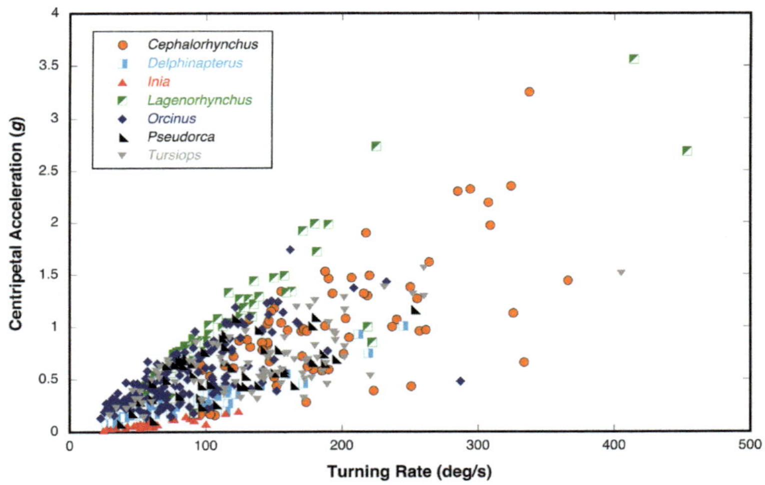

FIG. 13 Relationship between centripetal acceleration and turning rate for cetacean species (Commerson's dolphin, beluga, Amazon river dolphin, Pacific white-sided dolphin, killer whale, false killer whale and bottlenose dolphin). Centripetal is given as a multiple of gravitational acceleration ($g = 9.8\,m/s^2$).

The turning rate performance is illustrated in Fig. 13 by a plot of centripetal acceleration and turning rate. Most data for cetaceans are clustered at accelerations <1.5 g with turning rates <200 degrees/s. Individuals of Commerson's dolphin, Pacific white-sided dolphin, and bottlenose dolphin were able to exceed these lower values for cetaceans with white-sided dolphin displaying the maximum performance with an acceleration of 3.6 g and turning rate of 453 degrees/s.

The force necessary to maintain a curved trajectory of a given radius is directly related to the square of the velocity and the mass of the body so that turn radius increases directly with body mass (Howland, 1974; Weihs, 1981). In addition, the relatively stable design of delphinids limits their maneuverability. Small flexible fishes have an advantage with respect to maneuverability and agility compared with the larger dolphins (Howland, 1974; Fish, 1999). Escape by small prey fishes is possible as they turn in smaller radii with higher angular velocities than larger dolphins. The limitations of the cetacean design may be a causative reason for the use of cooperative foraging behaviors in pelagic habitats by dolphins that prey on smaller, more maneuverable organisms (Smith et al., 1981; Bel'kovich, 1991; Similä and Ugarte, 1993; Fish, 1997).

Despite the potential disadvantage in turning performance compared to smaller prey (Howland, 1974), dolphins hold an advantage in that their absolute swimming speed is substantially greater than the speed of the prey and they have developed a number of behavioral strategies to capture their elusive prey (Maresh et al., 2004). Groups of pelagic delphinids often use behaviors, involving encirclement of prey and division of labor during an attack (Similä and Ugarte, 1993; Similä, 1997; Berta et al., 2006; Vaughn et al., 2011). Dolphins will herd fish into bait balls and capture fish on the sides of the ball (Vaughn et al., 2011). Stunning behaviors are employed in some instances. Killer whales will herd herring into tight schools and slap the fish with the underside of the tail (Domenici et al., 2000). Female killer whales are mainly responsible for corralling prey (Barrett-Lennard, and Heise, 2006). Often females and subadults execute turning lunges on large prey species (Jefferson et al., 1991; Barrett-Lennard and Heise, 2006).

Mud ring feeding is a cooperative foraging behavior used to corral fish (Torres and Read, 2009). A single dolphin swims in a circle in close proximity to a muddy bottom. The downstroke of the flukes produces a mud plume. The ring of plumes encircles the fish, and the dolphin and any other conspecifics can lunge for the fish.

The pinwheel strategy is a foraging technique used by bottlenose dolphin in shallow water (Maresh et al., 2004). The maneuver is preceded by a sudden acceleration and 90° body roll with rapid flexing of the body (Fig. 14). The dolphin keeps its rostrum at a fixed point, rotating its body around that point. The dolphin spins around its prey, using its body to encircle and cut off the fish's escape. This strategy allows the dolphin to turn at a high rate (656.3 deg/s) within a small turning radius (0.10 BL) (Maresh et al., 2004).

A consideration of maneuverability is important in a full understanding of the ecology of dolphins. With respect to bioenergetic modeling, the additional movements of dolphins within a three-dimensional space increase the distance, and therefore, the energy expended to travel over a given linear distance. Deviation from straight line swimming also adds extra drag that requires increased energy to maintain momentum. Monitoring the maneuvering of dolphins in the wild with greater temporal fidelity will provide a more accurate energetic profile.

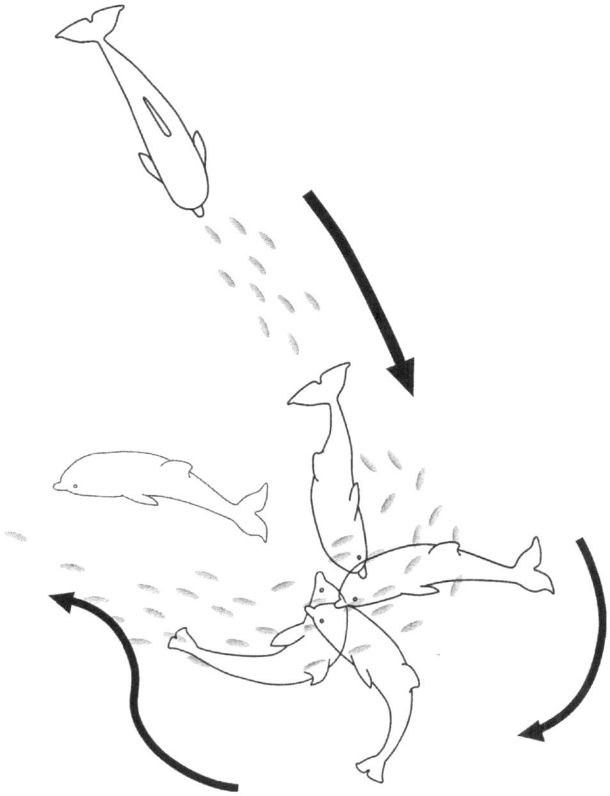

FIG. 14 Sequence of maneuvers involving pinwheeling for a single dolphin chasing a school of fish (gray ovals). The arrows show the direction of travel of dolphin and fish school. *From Maresh et al. (2004).*

References

Abbott, I.H., von Doenhoff, A.E., 1959. Theory of Wing Sections. Dover, New York.

Adams, D.S., 2018. Properties and Functions of Tendons in the Peduncle of Cetaceans. Master's Thesis, West Chester University.

Adams, D.S., Fish, F.E., 2019. Odontocete peduncle tendons for possible control of fluke orientation and flexibility. J. Morphol. 280, 1323–1331.

Agarkov, G.B., Lukhanin, V.Y., 1970. Propulsive tail musculature of the Atlantic common dolphin. Bionika 4, 61–64 (translated from Russian).

Ahlborn, B.K., Blake, R.W., Chan, K.H.S., 2009. Optimal fineness ratio for minimum drag in large whales. Can. J. Zool. 87, 124–131.

Ahmadi, A.R., Widnall, S.E., 1986. Energetics and optimum motion of oscillating lifting surfaces of finite span. J. Fluid Mech. 162, 261–282.

Alexander, R.M.N., 1988. Elastic Mechanisms in Animal Movement. Cambridge University Press, Cambridge.

Aleyev, Y.G., 1974. Hydrodynamic resistance and speed of movement of nekters. Zool. Zh. 53, 493–507 (translated from Russian).

Aleyev, Y.G., 1977. Nekton. Junk, The Hague.

Amano, M., Miyazaki, N., 1993. External morphology of Dall's porpoise (*Phocoenoides dalli*): growth and sexual dimorphism. Can. J. Zool. 71, 1124–1130.

Amundin, M., 1974. Functional analysis of the surfacing behavior in the harbour porpoise, *Phocoena phocoena*. Z. Saugetierkd 39, 313–318.

Anderson, J.M., Streitlien, K., Barrett, D.S., Triantafyllou, M.S., 1998. Oscillating foils of high propulsive efficiency. J. Fluid Mech. 360, 41–72.

Arkowitz, R., Rommel, S., 1985. Force and bending moment of the caudal muscles in the short fin pilot whale. Mar. Mamm. Sci. 1, 203–209.

Au, D., Scott, M.D., Perryman, W.L., 1988. Leap-swim behavior of "porpoising" dolphins. Cetus 8, 7–10.

Au, D., Weihs, D., 1980a. At high speeds dolphins save energy by leaping. Nature 284, 548–550.

Au, D., Weihs, D., 1980b. Leaping dolphins. Nature 287, 759.

Babenko, V.V., Koslov, L.F., Pershin, S.V., Sokolov, V.Y., Tomilin, A.G., 1982. Self-adjustment of skin dampening in cetaceans in active swimming. Bionika 16, 3–10 (translated from Russian).

Backhouse, K., 1960. Locomotion and direction-finding in whales and dolphins. New Sci. 7, 26–28.

Barrett-Lennard, L.G., Heise, K.A., 2006. The natural history and ecology of killer whales. In: Estes, J.A., Demaster, D.P., Doak, D.F., Williams, T.M., Brownell Jr., R.L. (Eds.), Whales, Whaling, and Ocean Ecosystems. University of California Press, Berkeley, CA, pp. 163–173.

Baudinette, R.V., 1994. Locomotion in macropodoid marsupials: gaits, energetics and heat balance. Aust. J. Zool. 42, 103–123.

Bel'kovich, V.M., 1991. Herd structure, hunting, and play: bottlenose dolphins in the Black Sea. In: Pryor, K., Norris, K. (Eds.), Dolphin Societies: Discoveries and Puzzles. University of California Press, Berkeley, pp. 17–77.

Bello, M.A., Roy, R.R., Martin, T.P., Goforth Jr., H.W., Edgerton, V.R., 1985. Axial musculature in the dolphin (*Tursiops truncatus*): some architectural and histochemical characteristics. Mar. Mamm. Sci. 1, 324–336.

Bello, M.A., Roy, R.R., Oxman, I., Martion, T., Edgerton, V.R., 1983. Fiber type and fiber size distributions in the axial musculature of the dolphin (*Tursiops truncatus*). Physiologist 26, A-26.

Bennett, M.B., Ker, R.F., Alexander, R.M.N., 1987. Elastic properties of structures in the tails of cetaceans (*Phocoena* and *Lagenorhynchus*) and their effect on the energy cost of swimming. J. Zool. (Lond.) 211, 177–192.

Bennett, M.B., Ker, R.F., Imery, N.J., Alexander, R.McN., 1986. Mechanical properties of various mammalian tendons. J. Zool. 209, 537–548.

Berta, A., Sumich, J.L., Kovacs, K.M., 2006. Marine Mammals: Evolutionary Biology. Academic Press, Amsterdam.

Blake, R.W., 1983a. Energetics of leaping in dolphins and other aquatic animals. J. Mar. Biol. Assoc. U. K. 63, 61–70.

Blake, R.W., 1983b. Fish Locomotion. Cambridge University Press, Cambridge.

Blickhan, R., Cheng, J.-Y., 1994. Energy storage by elastic mechanisms in the tail of large swimmers-a re-evaluation. J. Theor. Biol. 168, 315–321.

Borelli, G.A., 1689. On the Movement of Animals. Springer-Verlag, Berlin.

Bose, N., Lien, J., 1989. Propulsion of a fin whale (*Balaenoptera physalus*): why the fin whale is a fast swimmer. Proc. R. Soc. Lond. B 237, 175–200.

Bose, N., Lien, J., Ahia, J., 1990. Measurements of the bodies and flukes of several cetacean species. Proc. R. Soc. Lond. B 242, 163–173.

Brodie, P.F., 1977. Cetacean energetics, an overview of intraspecific size variation. Ecology 56, 152–161.

Caldwell, D.K., Fields, H.M., 1959. Surf-riding by Atlantic bottle-nosed dolphins. J. Mammal. 40, 454–455.

Choi, H., Park, H., Sagong, W., Lee, S.-i., 2012. Biomimetic flow control based on morphological features of living creatures. Phys. Fluids 24, 121302.

Chopra, M.G., 1976. Large amplitude lunate-tail theory of fish locomotion. J. Fluid Mech. 74, 161–182.

Chopra, M.G., Kambe, T., 1977. Hydrodynamics of lunate-tail swimming propulsion. Part 2. J. Fluid Mech. 79, 49–69.

Clark, B.D., Fish, F.E., 1994. Scaling of the locomotory apparatus and paddling rhythm in swimming mallard ducklings (*Anas platyrhynchos*): test of a resonance model. J. Exp. Zool. 270, 245–254.

Cockcroft, V.G., Ross, G.J.B., 1990. Observations on the early development of a captive bottlenose dolphin calf. In: Leatherwood, S., Reeves, R.R. (Eds.), The Bottlenose Dolphin. Academic Press, San Diego, pp. 461–478.

Curren, K.C., Bose, N., Lien, J., 1994. Swimming kinematics of a harbor porpoise (*Phocoena phocoena*) and an Atlantic white-sided dolphin (*Lagenorhynchus acutus*). Mar. Mamm. Sci. 10, 485–492.

Daniel, T.L., 1981. Fish mucus: in situ measurements of polymer drag reduction. Biol. Bull. 160, 376–382.

Davis, R.W., Worthy, G.A.J., Würsig, B., Lynn, S.K., 1996. Diving behavior and at-sea movements of an Atlantic spotted dolphin in the Gulf of Mexico. Mar. Mamm. Sci. 12, 569–581.

Denny, M.W., 1993. Air and Water. Princeton University Press, Princeton.

Dohl, T.P., Norris, K.S., Kang, I., 1974. A porpoise hybrid: *Tursiops* x *Steno*. J. Mammal. 55, 217–221.

Domenici, P., Batty, R.S., Similä, T., Ogam, E., 2000. Killer whales (*Orcinus orca*) feeding on schooling herring (*Clupea harengus*) using underwater tail-slaps: kinematic analyses of field observations. J. Exp. Biol. 203, 283–294.

Edwards, E.F., 2006. Duration of unassisted swimming activity for spotted dolphin (*Stenella attenuata*) calves: implications for mother-calf separation during tuna purse-seine sets. Fish. Bull. 104, 125–135.

Ellington, C.P., van den Berg, C., Willmott, A.P., Thomas, A.L.R., 1996. Leading-edge vortices in insect flight. Nature 384, 626–630.

Essapian, F.S., 1955. Speed-induced skin folds in the bottle-nosed porpoise, *Tursiops truncatus*. Breviora Mus. Comp. Zool. 43, 1–4.

Etnier, S.A., Mclellan, W.A., Blum, J., Pabst, D.A., 2008. Ontogenetic changes in the structural stiffness of the tailstock of bottlenose dolphins (*Tursiops truncatus*). J. Exp. Biol. 211, 3205–3213.

Fein, J.A., 1998. Dolphin drag reduction: myth or magic. In: Meng, J.C.S. (Ed.), Proceedings of the International Symposium on Seawater Drag Reduction. Newport, Rhode Island, pp. 429–433.

Fejer, A.A., Backus, R.H., 1960. Porpoises and the bow-riding of ships under way. Nature 188, 700–703.

Felts, W.J.L., 1966. Some functional and structural characteristics of cetaceans flippers and flukes. In: Norris, K.S. (Ed.), Whales, Dolphins and Porpoises. University of California Press, Berkeley, pp. 255–276.

Fierstine, H.L., Walters, V., 1968. Studies of locomotion and anatomy of scombrid fishes. Mem. South. Calif. Acad. Sci. 6, 1–31.

Fish, F.E., 1992. Aquatic locomotion. In: Tomasi, T.E., Horton, T.H. (Eds.), Mammalian Energetics: Interdisciplinary Views of Metabolism and Reproduction. Cornell University Press, Ithaca, New York, pp. 34–63.

Fish, F.E., 1993a. Influence of hydrodynamic design and propulsive mode on mammalian swimming energetics. Aust. J. Zool. 42, 79–101.

Fish, F.E., 1993b. Power output and propulsive efficiency of swimming bottlenose dolphins (*Tursiops truncatus*). J. Exp. Biol. 185, 179–193.

Fish, F.E., 1996. Transitions from drag-based to lift-based propulsion in mammalian swimming. Am. Zool. 36, 628–641.

Fish, F.E., 1997. Biological designs for enhanced maneuverability: analysis of marine mammal performance. In: Tenth International Symposium on Unmanned Untethered Submersible Technology: Proceedings of the Special Session on Bio-Engineering Research Related to Autonomous Underwater Vehicles. Autonomous Undersea Systems Institute, Lee, New Hampshire, pp. 109–117.

Fish, F.E., 1998a. Comparative kinematics and hydrodynamics of odontocete cetaceans: morphological and ecological correlates with swimming performance. J. Exp. Biol. 201, 2867–2877.

Fish, F.E., 1998b. Biomechanical perspective on the origin of cetacean flukes. In: Thewissen, J.G.M. (Ed.), The Emergence of Whales: Evolutionary Patterns in the Origin of Cetacea. Plenum Press, New York, pp. 303–324.

Fish, F.E., 1999. Energetics of swimming and flying in formation. Comments Theor. Biol. 5, 283–304.

Fish, F.E., 2002. Balancing requirements for stability and maneuverability in cetaceans. Integr. Comp. Biol. 42, 85–93.

Fish, F.E., 2006. The myth and reality of Gray's paradox: implication of dolphin drag reduction for technology. Bioinspir. Biomim. 1, R17–R25.

Fish, F.E., Beneski, J.T., Ketten, D.R., 2007. Examination of the three-dimensional geometry of cetacean flukes using CT-scans: hydrodynamic implications. Anat. Rec. 290, 614–623.

Fish, F.E., Howle, L.E., Murray, M.M., 2008. Hydrodynamic flow control in marine mammals. Integr. Comp. Biol. 211, 1859–1867.

Fish, F.E., Hui, C.A., 1991. Dolphin swimming—a review. Mammal Rev. 21, 181–195.

Fish, F.E., Lauder, G.V., 2006. Passive and active flow control by swimming fishes and mammals. Annu. Rev. Fluid Mech. 38, 193–224.

Fish, F.E., Lauder, G.V., 2017. Control surfaces of aquatic vertebrates in relation to swimming modes. J. Exp. Biol. 220, 4351–4363.

Fish, F.E., Legac, P., Williams, T.M., Wei, T., 2014. Measurement of hydrodynamic force generation by swimming dolphins using bubble DPIV. J. Exp. Biol. 217, 252–260.

Fish, F.E., Nusbaum, M.K., Beneski, J.T., Ketten, D.R., 2006. Passive cambering and flexible propulsors: cetacean flukes. Bioinspir. Biomim. 1, S42–S48.

Fish, F.E., Peacock, J.E., Rohr, J.J., 2003. Stabilization mechanism in swimming odontocete cetaceans by phased movements. Mar. Mamm. Sci. 19, 515–528.

Fish, F.E., Rohr, J., 1999. Review of Dolphin Hydrodynamics and Swimming Performance. SPAWARS System Center Technical Report 1801, San Diego, CA.

Fish, F.E., Williams, T.M., Sherman, E., Moon, Y.E., Wu, V., Wei, T., 2018. Experimental measurement of dolphin thrust generated during a tail stand using DPIV. Fluids 3, 33.

Fitzgerald, J.W., Fitzgerald, E.R., Carey, W.M., von Winkle, W.A., 1995. Blubber and compliant coatings for drag reduction in water. II. Matched shear impedance for compliant layer drag reduction. Mater. Sci. Eng. C2, 215–220.

Fitzgerald, J.W., Martin, J.E., Modert, E.F., 1998. Blubber and compliant coatings for drag reduction in fluids: VI. Rotating disc apparatus for drag measurement on compliant layers. In: Meng, J.C.S. (Ed.), Proceedings of the International Symposium on Seawater Drag Reduction. Newport, Rhode Island, pp. 215–218.

Garten, J.L., Fish, F.E., 2020. Comparative histological examination of the integument of cetacean flukes. Aquat. Mamm. 46, 367–381.

Gertler, M., 1950. Resistance Experiments on a Systematic Series of Streamlined Bodies of Revolution-for Application to the Design of High-Speed Submarines. Rep. C-297, David W. Taylor Model Basin, Washington, D.C.

Gibbs-Smith, C.H., 1962. Sir George Cayley's Aeronautics 1796–1855. Her Majesty's Stationery Office, London.

Goforth, H.W., 1990. Ergometry (exercise testing) of the bottlenose dolphin. In: Leatherwood, S., Reeves, R.R. (Eds.), The Bottlenose Dolphin. Academic Press, San Diego, pp. 559–574.

Goforth Jr., H.W., 1984. Biochemical and Anatomical Characteristics of Dolphin Muscles. NOSC Tech. Rpt. 940.

Goldbogen, J.A., Fish, F.E., Potvin, J., 2016. Hydrodynamics. In: Castellini, M.A., Mellish, J.-A. (Eds.), Marine Mammal Physiology: Requisites for Ocean Living. CRC Press, Boca Raton, FL, pp. 3–28.

Gordon, C.N., 1980. Leaping dolphins. Nature 287, 759.

Gough, W.T., Fish, F.E., Wainwright, D.K., Bart-Smith, H., 2018. Morphology of the core fibrous layer of the cetacean tail fluke. J. Morphol. 279, 757–765.

Gray, J., 1936. Studies in animal locomotion VI. The propulsive powers of the dolphin. J. Exp. Biol. 13, 192–199.

Gubbins, C., McCowan, B., Spencer, L.K., Hooper, S., Reiss, D., 1999. Mother-infant spatial relations in captive bottlenose dolphins, *Tursiops truncatus*. Mar. Mamm. Sci. 15, 751–765.

Gucinski, H., Baier, R.E., 1983. Surface properties of porpoise and killer whale skin *in vivo*. Am. Zool. 23, 959.

Haider, M., Lindsley, D.B., 1964. Microvibrations in man and dolphin. Science 146, 1181–1183.

Hamilton, J.L., Dillaman, R.M., McLellan, W.A., Pabst, D.A., 2004. Structural fiber reinforcement of keel blubber in harbor porpoise (*Phocoena phocoena*). J. Morphol. 261, 105–117.

Hamilton, J.L., McLellan, W.A., Pabst, D.A., 1998. Functional morphology of harbor porpoise (*Phocoena phocoena*) tailstock blubber. Am. Zool. 38, 203A.

Hampton, I.F.G., Whittow, G.C., 1976. Body temperature and heat exchange in the Hawaiian spinner dolphin, *Stenella longirostris*. Comp. Biochem. Physiol. A: Physiol. 55, 195–197.

Hanson, M.T., Defran, R.H., 1993. The behaviour and feeding ecology of the Pacific coast bottlenose dolphin, *Tursiops truncatus*. Aquat. Mamm. 19 (3), 127–142.

Harrison, R.J., Thurley, K.W., 1972. Fine structural features of delphinid epidermis. J. Anat. 111, 498–499.

Haun, J.E., Hendricks, E.W., Borkat, F.R., Kataoka, R.W., Carder, D.A., Chun, N.K., 1983. Dolphin Hydrodynamics: Annual Report FY 82. NOSC Tech. Rep. 935.

Haun, J.E., Hendricks, E.W., Borkat, F.R., Kataoka, R.W., Carder, D.A., Dooley, C.A., Lindner, E., Stromberg, M.W., 1984. Dolphin Hydrodynamics: FY83 and FY84 Report. NOSC Tech. Rep. 998.

Hayes, W.D., 1953. Wave riding of dolphins. Nature 172, 1060.

Hayes, W.D., 1959. Wave-riding dolphins. Science 130, 1657–1658.

Herman, L.M., Tavolga, W.N., 1980. The communication systems of cetaceans. In: Herman, L.M. (Ed.), Cetacean Behavior: Mechanisms and Functions. Robert E. Krieger Publishing, Malabar, Florida, pp. 149–209.

Hertel, H., 1966. Structure, Form and Movement. Rheinhold, New York.

Hertel, H., 1969. Hydrodynamics of swimming and wave-riding dolphins. In: Andersen, H.T. (Ed.), The Biology of Marine Mammals. Academic Press, New York, pp. 31–63.

Hill, A.V., 1950. The dimensions of animals and their muscular dynamics. Sci. Prog. 38, 209–230.

Hoerner, S.F., 1965. Fluid-Dynamic Drag. Published by author, Brick Town, New Jersey.

Howell, A.B., 1930. Aquatic Mammals. Charles C. Thomas, Springfield, Illinois.

Howland, H.C., 1974. Optimal strategies for predator avoidance: the relative importance of speed and manoeuvrability. J. Theor. Biol. 47, 333–350.

Hoyt, J.W., 1975. Hydrodynamic drag reduction due to fish slimes. In: Wu, T.Y., Brokaw, C.J., Brennen, C (Eds.), Swimming and Flying in Nature. Plenum Press, New York, pp. 653–672.

Hui, C., 1987. Power and speed of swimming dolphins. J. Mammal. 68, 126–132.

Hui, C., 1989. Surfacing behavior and ventilation in free-ranging dolphins. J. Mammal. 70, 833–835.

Irving, L., Scholander, P., Grinnell, S., 1941. The respiration of the porpoise *Tursiops truncatus*. J. Cell. Comp. Physiol. 17, 145–168.

Jackson, H.A., 1992. Fundamentals of submarine concept design. SNAME Trans. 100, 419–448.

Jefferson, T.A., LeDuc, R., 2018. Delphinids overview. In: Perrin, W.F., Würsig, B., Thewissen, J.G.M. (Eds.), Encyclopedia of Marine Mammals, third ed. Academic Press, San Diego, pp. 242–246.

Jefferson, T.A., Stacey, P.J., Baird, R.W., 1991. A review of killer whale interactions with other marine mammals: predation to co-existence. Mammal Rev. 21, 151–180.

Joh, C.G., 1925. The motion of whales during swimming. Nature 116, 327–329.

Johannessen, C.L., Harder, J.A., 1960. Sustained swimming speeds of dolphins. Science 132, 1550–1551.
Johnson, C.M., Norris, K.S., 1994. Social behavior. In: Norris, K.S., Würsig, B., Wells, R.S., Würsig, M. (Eds.), The Hawaiian Spinner Dolphin. University of California Press, Berkeley, pp. 243–286.
Joubert, P.N., 2004. Some Aspects of Submarine Design. Part 1. Hydrodynamics. Defence Science and Technology Organisation. DSTO-TR-1622.
Karpouzian, G., Spedding, G., Cheng, H.K., 1990. Lunate-tail swimming propulsion. Part 2. Performance analysis. J. Fluid Mech. 210, 329–351.
Katz, J., Weihs, D., 1978. Hydrodynamic propulsion by large amplitude oscillation of an airfoil with chordwise flexibility. J. Fluid Mech. 88, 485–497.
Kayan, V.P., 1974. Resistance coefficient of the dolphin. Bionika 8, 31–35 (translated from Russian).
Kayan, V.P., 1979. The hydrodynamic characteristics of the caudal fin of the dolphin. Bionika 13, 9–15 (translated from Russian).
Kayan, V.P., Pyatetskiy, V.Y., 1977. Kinematics of bottlenosed dolphins swimming as related to acceleration mode. Bionika 11, 36–41 (translated from Russian).
Kayan, V.P., Pyatetskiy, V.Y., 1978. The hydrodynamic characteristics of the Black Sea dolphin in different acceleration modes. Bionika 12, 48–55 (translated from Russian).
Kelly, H.R., 1959. A Two -Body Problem in the Echelon-Formation Swimming of Porpoise. U.S. Nav. Ord. Test Sta, China Lake, CA (Tech. Note 40606-1).
Klima, M., Oelschläger, H.A., Wünsch, D., 1987. Morphology of the pectoral girdle in the Amazon dolphin *Inia geoffrensis* with special reference to the shoulder joint and the movements of the flippers. Z. Saeugertierkd 45, 288–309.
Kooyman, G.L., Cornell, L.H., 1981. Flow properties of expiration and inspiration in a trained bottlenose porpoise. Physiol. Zool. 54, 55–61.
Kooyman, G.L., Ponganis, P.J., 1998. The physiological basis of diving to depth: birds and mammals. Annu. Rev. Physiol. 60, 19–32.
Koslov, L.F., Pershin, S.V., 1983. Comprehensive research on hydrodynamic drag reduction in dolphins by the active regulation of the skin. Bionika 17, 3–12 (translated from Russian).
Kozolv, L.F., Shakalo, V.M., 1973. Some results of measuring velocity pulsations in the boundary layer of dolphins. Bionika 7, 50–52 (translated from Russian).
Kramer, M.O., 1960. The dolphins' secret. New Sci. 7, 1118–1120.
Kramer, M.O., 1965. Hydrodynamics of the dolphin. In: Chow, V.T. (Ed.), Advances in Hydroscience. vol. 2. Academic Press, New York, pp. 111–130.
Lane, F.W., 1941. How fast do fish swim? Country Life 90, 534–535.
Lang, T.G., 1963. Porpoise, whale, and fish: comparison of predicted and observed speeds. Nav. Eng. J. 75, 437–441.
Lang, T.G., 1966a. Hydrodynamic analysis of dolphin fin profiles. Nature 209, 1110–1111.
Lang, T.G., 1966b. Hydrodynamic analysis of cetacean performance. In: Norris, K.S. (Ed.), Whales, Dolphins and Porpoises. Ed. University of California Press, Berkeley, California, pp. 410–432.
Lang, T.G., 1975. Speed, power, and drag measurements of dolphins and porpoises. In: Wu, T.Y., Brokaw, C.J., Brennen, C. (Eds.), Swimming and Flying in Nature. Plenum Press, New York, pp. 553–571.
Lang, T.G., Daybell, D.A., 1963. Porpoise Performance Tests in a Seawater Tank. Nav. Ord. Test Sta. Tech. Rep. 3063.
Lang, T.G., Norris, K.S., 1966. Swimming speed of a Pacific bottlenose porpoise. Science 151, 588–590.
Lang, T.G., Pryor, K., 1966. Hydrodynamic performance of porpoises (*Stenella attenuata*). Science 152, 531–533.
Largess, R.P., Mandelblatt, J.L., 1999. U.S.S. Albacore: Forerunner of the Future. Portsmouth Marine Society, Portmouth, NH.
Larrabee, E.E., 1980. The screw propeller. Sci. Am. 243, 134–148.
Latz, M.L., Rohr, J., Case, J.F., 1995. Description of a novel flow visualization technique using bioluminescent marine plankton. In: Crowder, J.P. (Ed.), Flow Visualization. vol. VII. Begell House, New York, pp. 28–33.
Law, T.C., Blake, R.W., 1994. Swimming behaviors and speeds of wild Dall's porpoises (*Phocoenoides dalli*). Mar. Mamm. Sci. 10, 208–213.
Layne, J.N., Caldwell, D.K., 1964. Behavior of the Amazon dolphin, *Inia geoffrensis* (Blainville), in captivity. Zoologica 49, 81–111.
Leatherwood, S., 1974. A note on gray whale behavioral interactions with other marine mammals. Mar. Fish. Rev. 26, 50–51.
Leatherwood, S., Walker, W.A., 1979. The northern right whale dolphin *Lissodelphis borealis* Peale in the eastern North Pacific. In: Winn, H.E., Olla, B.L. (Eds.), Behavior of Marine Animals. vol. 3. Plenum Press, New York, pp. 85–141.
Lighthill, J., 1969. Hydrodynamics of aquatic animal propulsion—a survey. Annu. Rev. Fluid Mech. 1, 413–446.
Lighthill, J., 1970. Aquatic animal propulsion of high hydromechanical efficiency. J. Fluid Mech. 44, 265–301.
Lighthill, J., 1971. Large-amplitude elongate-body theory of fish locomotion. Proc. R. Soc. Lond. B 179, 125–138.
Lindsey, C.C., 1978. Form, function, and locomotory habits in fish. In: Hoar, W.S., Randall, D.J. (Eds.), Fish Physiology: Locomotion. vol. 7. Academic Press, New York, pp. 1–100.
Liu, P., Bose, N., 1993. Propulsive performance of three naturally occurring oscillating propeller planforms. Ocean Eng. 20, 57–75.
Lockyer, C., Morris, R., 1987. Observations on diving behaviour and swimming speeds in wild juvenile *Tursiops truncatus*. Aquat. Mamm. 13 (1), 31–35.
Long Jr., J.H., Pabst, D.A., Shepherd, W.R., McLellan, W.A., 1997. Locomotor design of dolphin vertebral columns: bending mechanics and morphology of *Delphinus delphis*. J. Exp. Biol. 200, 65–81.
Mankovskaya, I.N., 1975. The content and distribution of myoglobin in muscle tissue of Black Sea dolphins. Zh. Evol. Biokhim. Fiziol. 11, 263–267.
Mann, J., Connor, R.C., Barre, L.M., Heithaus, M.R., 2000. Female reproductive success in bottlenose dolphins (*Tursiops* sp.): life history, habitat, provisioning, and group-size effects. Behav. Ecol. 11, 210–219.
Maresca, C., Favier, D., Rebont, J., 1979. Experiments on an aerofoil at high angle of incidence in longitudinal oscillations. J. Fluid Mech. 92, 671–690.
Maresh, J.L., Fish, F.E., Nowacek, D.P., Nowacek, S.M., Wells, R.S., 2004. High performance turning capabilities during foraging by bottlenose dolphins (*Tursiops truncatus*). Mar. Mamm. Sci. 20, 498–509.
Marino, L., Stowe, J., 1997. Lateralized behavior in two captive bottlenose dolphins (*Tursiops truncatus*). Zoo Biol. 16, 173–177.
Martin, A.R., Smith, T.G., Cox, O.P., 1993. Studying the behaviour and movements of high Arctic belugas with satellite telemetry. Symp. Zool. Soc. Lond. 66, 195–210.
Maslov, N.K., 1970. Maneuverability and controllability of dolphins. Bionika 4, 46–50 (translated from Russian).
McBride, A.F., Kritzler, H., 1951. Observations on pregnancy, parturition, and postnatal behavior in the bottlenose dolphin. J. Mammal. 32, 251–266.

McGinnis, S.M., Whittow, G.C., Ohata, C.A., Huber, H., 1972. Body heat dissipation and conservation in two species of dolphins. Comp. Biochem. Physiol. A: Physiol. 43, 417–423.

Minasian, S.M., Balcomb III, K.C., Foster, L., 1984. The World's Whales. Smithsonian Books, Washington, DC.

von Mises, R., 1945. Theory of Flight. Dover, New York.

Moore, M.J., Hammar, T., Arruda, J., Cramer, S., Dennison, S., Montie, E., Fahlman, A., 2011. Hyperbaric computed tomographic measurement of lung compression in seals and dolphins. J. Exp. Biol. 214, 2390–2397.

Morteo, E., Rocha-Olivares, A., Morteo, R., Weller, D.W., 2017. Phenotypic variation in dorsal fin morphology of coastal bottlenose dolphins (*Tursiops truncatus*) off Mexico. PeerJ 5, e3415.

Newman, J.N., 1975. Swimming of slender fish in a non-uniform velocity field. J. Aust. Math. Soc. B 19, 95–111.

Newman, J.N., Wu, T.Y., 1974. Hydromechanical aspects of fish swimming. In: Wu, T.Y., Brokaw, C.J., Brennen, C. (Eds.), Swimming and Flying in Nature. Plenum Press, New York, pp. 615–634.

Noren, S.R., Edwards, E.F., 2007. Physiological and behavioral development in delphinid calves: implications for calf separation and mortality due to tuna purse-seine sets. Mar. Mamm. Sci. 23, 15–29.

Noren, S.R., Edwards, E.F., 2011. Infant position in mother-calf dolphin pairs: formation locomotion with hydrodynamic benefits. Mar. Ecol. Prog. Ser. 424, 229–236.

Norris, K.S., Johnson, C.M., 1994a. Breathing at sea. In: Norris, K.S., Würsig, B., Wells, R.S., Würsig, M. (Eds.), The Hawaiian Spinner Dolphin. University of California Press, Berkeley, pp. 206–215.

Norris, K.S., Johnson, C.M., 1994b. Schools and schooling. In: Norris, K.S., Würsig, B., Wells, R.S., Würsig, M. (Eds.), The Hawaiian Spinner Dolphin. University of California Press, Berkeley, pp. 232–242.

Norris, K.S., Prescott, J.H., 1961. Observations on Pacific cetaceans of California and Mexican waters. Univ. Calif. Publ. Zool. 63, 291–402.

Orton, L.S., Brodie, P.F., 1985. Whale blubber elasticity. Am. Zool. 25, 13A.

Pabst, D.A., 1988. The subdermal connective tissue sheath in dolphin is not a typical crossed helical fiber array. Am. Zool. 28, 194A.

Pabst, D.A., 1990. Axial muscles and connective tissues of the bottlenose dolphin. In: Leatherwood, S., Reeves, R.R. (Eds.), The Bottlenose Dolphin. Academic Press, San Diego, pp. 51–67.

Pabst, D.A., 1993. Intramuscular morphology and tendon geometry of epaxial swimming muscles of dolphins. J. Zool. (Lond.) 230, 159–176.

Pabst, D.A., 1996a. Springs in swimming animals. Am. Zool. 36, 723–735.

Pabst, D.A., 1996b. Morphology of the subdermal connective tissue sheath of dolphins: a new fibre-wound, thin-walled, pressurized cylinder model for swimming vertebrates. J. Zool. (Lond.) 238, 35–52.

Pabst, D.A., McLellan, W.A., Gosline, J.G., Piermarini, P.M., 1995. Morphology and mechanics of dolphin blubber. Am. Zool. 35, 44A.

Palmer, E., Weddell, G., 1964. The relationship between structure, innervation and function of the skin of the bottle nose dolphin (*Tursiops truncatus*). Proc. Zool. Soc. London 143, 553–567.

Parry, D.A., 1949a. The swimming of whales and a discussion of Gray's paradox. J. Exp. Biol. 26, 24–34.

Parry, D.A., 1949b. Anatomical basis of swimming in whales. Proc. Zool. Soc. London 119, 49–60.

Pavlov, V.V., 2003. Wing design and morphology of the harbor porpoise dorsal fin. J. Morphol. 258, 284–295.

Pedley, T.J., 1977. Scale Effects in Animal Locomotion. Academic Press, London.

Pennycuick, C.J., 1992. Newton Rules Biology. Oxford University Press, Oxford.

Perrin, W.F., 1975. Variation of Spotted and Spinner Porpoise (Genus *Stenella*) in Eastern Tropical Pacific and Hawaii. Bull. Scripps Inst. Oceanogr. No. 21.

Perry, B., Acosta, A.J., Kiceniuk, T., 1961. Simulated wave-riding dolphins. Nature 192, 148–150.

Pershin, S.V., 1970. Resonance conditions in the swimming of dolphins. Bionika 4, 31–36 (translated from Russian).

Pershin, S.V., 1975. Hydrodynamic analysis of dolphin and whale fin profiles. Bionika 9, 26–32 (translated from Russian).

Pershin, S.V., 1983. Hydrobionic features for optimizing the exterior forms of the propeller system in cetaceans. Bionika 17, 13–24 (translated from Russian).

Pershin, S.V., 1988. Fundamentals of Hydrobionics. Sudostroyeniye Publication, Leningrad (translated from Russian).

Petersen, C.G.J., 1925. The motion of whales during swimming. Nature 116, 327–329.

Pettigrew, J.B., 1893. Animal Locomotion. D. Appleton, New York.

Pilleri, G., 1976. Comparative study of the skin and general myology of *Platanista indi* and *Delphinus delphis* in relation to hydrodynamics and behaviour. Invest. Cetacea 6, 89–127.

Pilleri, G., Gihr, M., Purves, P.E., Zbinden, K., Kraus, C., 1976. On the behaviour, bioacoustics and functional morphology of the Indus River dolphin (*Platanista indi* Blyth, 1859). Invest. Cetacea 6, 11–141.

Platzer, M.F., Lai, J.C.S., Dohring, C.M., 1998. Flow separation control by means of flapping foils. In: Meng, J.C.S. (Ed.), Proceedings of the International Symposium on Seawater Drag Reduction. Newport, Rhode Island, pp. 471–478.

Ponganis, P.J., Pierce, R.W., 1978. Muscle metabolic profiles and fiber-type composition in some marine mammals. Comp. Biochem. Physiol. B: Biochem. 59, 99–102.

Purves, P.E., 1963. Locomotion in whales. Nature 197, 334–337.

Purves, P.E., 1969. The structure of the flukes in relation to laminar flow in cetaceans. Z. Saugetierkd 34, 1–8.

Purves, P.E., Dudok van Heel, W.H., Jonk, A., 1975. Locomotion in dolphins part I: hydrodynamic experiments on a model of the bottle-nosed dolphin, *Tursiops truncatus*, (Mont.). Aquat. Mamm. 3, 5–31.

Pyatetskiy, V.Y., Kayan, V.P., 1975. On kinematics of swimming of bottlenose dolphin. Bionika 9, 41–45 (translated from Russian).

Reid, K., Mann, J., Weiner, J.R., Hecker, N., 1995. Infant development in two aquarium bottlenose dolphins. Zoo Biol. 14, 135–147.

Ridgway, S.H., Carder, D.A., 1993. Features of dolphin skin with potential hydrodynamic importance. IEEE Eng. Med. Biol. 12, 83–88.

Ridgway, S.H., Howard, R., 1979. Dolphin lung collapse and intramuscular circulation during free diving: evidence from nitrogen washout. Science 206, 1182–1183.

Ridgway, S.H., Scronce, B.L., Kanwisher, J., 1969. Respiration and deep diving in the bottlenose porpoise. Science 166, 1651–1654.

Rohr, J.J., Fish, F.E., 2004. Strouhal numbers and optimization of swimming by odontocete cetaceans. J. Exp. Biol. 207, 1633–1642.

Rohr, J.J., Fish, F.E., Gilpatrick, J.W., 2002. Maximum swim speeds of captive and free ranging delphinids: critical analysis of extraordinary performance. Mar. Mamm. Sci. 18, 1–19.

Rohr, J., Hendricks, E., Quigley, L., Fish, F., Gilpatrick, J., Scardino-Ludwig, J., 1998. Observations of dolphin swimming speed and Strouhal number. In: SPAWARS System Center Technical Report 1769, San Diego, CA.

Rohr, J., Latz, M.I., Fallon, S., Nauen, J.C., Hendricks, E., 1998. Experimental approaches towards interpreting dolphin-stimulated bioluminescence. J. Exp. Biol. 201, 1447–1460.

Romanenko, E.V., 1976. Acoustics and hydrodynamics of certain marine animals. Sov. Phys. Acoust. 22, 357–358 (translated from Russian).

Romanenko, E.V., 1981. Distribution of dynamic pressure over the body of an actively swimming dolphin. Sov. Phys. Dokl. 26, 1037–1038 (translated from Russian).

Romanenko, E.V., 1995. Swimming of dolphins: experiments and modelling. In: Ellington, C.P., Pedley, T.J. (Eds.), Biological Fluid Dynamics. The Company of Biologists, Cambridge, pp. 21–33.

Romanenko, E.V., Yanov, V.G., 1973. Experimental results of studying the hydrodynamics of dolphins. Bionika 7, 52–56 (translated from Russian).

Rosen, M.W., 1962. Experiments with Swimming Fish and Dolphins. ASME Pap. N. 61-WA-203.

Rosen, M.W., 1963. Flow Visualization Experiments with a Dolphin. NOTS Tech. Publ. 3065.

Rosen, M.W., Cornford, N.E., 1971. Fluid friction of fish slimes. Nature 234, 49–51.

Russell, M., Vailas, A.C., Edgerton, V.R., Sasson, C., Nekooi, A., Durivage, J., 1983. Dolphin tension matrix components. Physiologist 26, A–26.

Saadat, M., Fish, F.E., Domel, A., Di Santo, V., Lauder, G.V., Haj-Hariri, H., 2017. On the rules for aquatic locomotion. Phys. Rev. Fluids 2, 083102.

Scholander, P.F., 1959a. Wave-riding dolphins: how do they do it? Science 129, 1085–1087.

Scholander, P.F., 1959b. Wave-riding dolphins. Science 130, 1658.

Shoele, K., Zhu, Q., 2015. Drafting mechanisms between a dolphin mother and calf. J. Theor. Biol. 382, 363–377.

Shoemaker, P.A., Ridgway, S.H., 1991. Cutaneous ridges in odontocetes. Mar. Mamm. Sci. 7, 66–74.

Similä, T., 1997. Sonar observations of killer whales (*Orcinus orca*) feeding on herring schools. Aquat. Mamm. 23 (2), 119–126.

Similä, T., Ugarte, F., 1993. Surface and underwater observations of cooperatively feeding killer whales in northern Norway. Can. J. Zool. 71, 1494–1499.

Skrovan, R.C., Williams, T.M., Berry, P.S., Moore, P.W., Davis, R.W., 1999. The diving physiology of bottlenose dolphins (*Tursiops truncatus*) II: biomechanics and changes in buoyancy at depth. J. Exp. Biol. 202, 2749–2761.

Slijper, E.J., 1961. Locomotion and locomotory organs in whales and dolphins (Cetacea). Symp. Zool. Soc. Lond. 5, 77–94.

Slijper, E.J., 1979. Whales. Cornell University Press, Ithaca, New York.

Smith, G.J.D., Browne, K.W., Gaskin, D.E., 1976. Functional myology of the harbour porpoise, *Phocoena phocoena* (L.). Can. J. Zool. 54, 716–729.

Smith, T.G., Siniff, D.B., Reichle, R., Stone, S., 1981. Coordinated behavior of killer whales, *Orcinus orca*, hunting a crabeater seal, *Lobodon carcinophagus*. Can. J. Zool. 59, 1185–1189.

Smits, A.J., 2019. Undulatory and oscillatory swimming. J. Fluid Mech. 874, P1.

Sokolov, W., 1960. Some similarities and dissimilarities in the structure of the skin among the members of the suborders Odontoceti and Mystacoceti (Cetacea). Nature 185, 745–747.

Sokolov, V., Bulina, I., Rodionov, V., 1969. Interaction of dolphin epidermis with flow boundary layer. Nature 222, 267–268.

Strickler, T.L., 1980. The axial musculature of *Pontoporia blainvillei*, with comments on the organization of this system and its effect on fluke-stroke dynamics in the cetacea. Am. J. Anat. 157, 49–59.

Sun, Q., Morikawa, H., Kobayashi, S., Ueda, K., Miyahara, H., Nakashima, M., 2010a. Structure and mechanical properties on tail flukes of dolphin. J. Aero Aqua Bio-Mech. 1, 45–50.

Sun, Q., Morikawa, H., Kobayashi, S., Ueda, K., Miyahara, H., Nakashima, M., 2010b. Structure and bending properties of central part of tail fin of dolphin. J. Biomech. Sci. Eng. 5, 388–398.

Suzuki, A., Tsuchiya, T., Takahashi, Y., Tamate, H., 1983. Histochemical properties of myofibers in longissimus muscle of common dolphins (*Delphinus delphis*). Acta Histochem. Cytochem. 16, 223–231.

Tanaka, H., Li, G., Uchida, Y., Nakamura, M., Ikeda, T., Liu, H., 2019. Measurement of time-varying kinematics of a dolphin in burst accelerating swimming. PLoS One 14, e0210860.

Tavolga, M.C., Essapian, F.S., 1957. The behavior of the bottle-nosed dolphin (*Tursiops truncatus*): mating, pregnancy, parturition and mother-infant behavior. Zoologica 42, 11–31.

Toedt, M.E., Reuss, L.E., Dillaman, R.M., Pabst, D.A., 1997. Collagen and elastin arrangement in the blubber of common dolphin (*Delphinus delphis*). Am. Zool. 37, 56A.

Torres, L.G., Read, A.J., 2009. Where to catch a fish? The influence of foraging tactics on the ecology of bottlenose dolphins (*Tursiops truncatus*) in Florida Bay, Florida. Mar. Mamm. Sci. 25, 797–815.

Triantafyllou, M.S., Triantafyllou, G.S., 1995. An efficient swimming machine. Sci. Am. 272, 64–69.

Triantafyllou, M.S., Triantafyllou, G.S., Gopalkrishnan, R., 1991. Wake mechanics for thrust generation in oscillating foils. Phys. Fluids A 3, 2835–2837.

Triantafyllou, G.S., Triantafyllou, M.S., Grosenbaugh, M.A., 1993. Optimal thrust development in oscillating foils with application to fish propulsion. J. Fluids Struct. 7, 205–224.

Triantafyllou, M.S., Winey, N., Trakht, Y., Elhassid, R., Yoerger, D., 2020. Biomimetic design of dorsal fins for AUVs to enhance maneuverability. Bioinspir. Biomim. 15, 035003.

Tsai, W.-L., 1998. Dorsoventral Bending of the Tail and Functional Morphology of the Caudal Vertebrae in the Bottlenose Dolphin, *Tursiops truncatus*. Master's thesis, Oklahoma State University, Stillwater OK.

Uskova, Y.T., Shmyrev, A.N., Rayevskiy, V.S., Bogdanova, L.N., Momot, L.N., Belyayev, V.V., Uskov, I.A., 1983. The nature and hydrodynamic activity of dolphin eye secretions. Bionika 17, 72–75 (translated from Russian).

Vasilevskaya, G.I., 1974. Structural features of the delphinid pectoral flippers. Bionika 8, 127–132 (translated from Russian).

Vaughn, R.L., Muzi, E., Richardson, J.L., Würsig, B., 2011. Dolphin bait-balling behaviors in relation to prey ball escape behaviors. Ethology 117, 859–871.

Videler, J., Kamermans, P., 1985. Differences between upstroke and downstroke in swimming dolphins. J. Exp. Biol. 119, 265–274.

Vogel, S., 1994. Life in Moving Fluids. Princeton University Press, Princeton.

Wainwright, D.K., Fish, F.E., Ingersoll, S., Williams, T.M., St. Leger, J., Smits, A.J., Lauder, G.V., 2019. How smooth is a dolphin? The ridged skin of odontocetes. Biol. Lett. 15, 20190103.

Wainwright, S.A., Pabst, D.A., Brodie, P.F., 1985. Form and possible function of the collagen layer underlying cetacean blubber. Am. Zool. 25, 146A.

Walker, J.A., 2000. Does a rigid body limit maneuverability? J. Exp. Biol. 203, 3391–3396.

Walsh, M.J., 1990. Riblets. Prog. Astronaut. Aeronaut. 123, 203–261.

Watson, A.G., Fordyce, R.E., 1993. Skeleton of two minke whales, *Balaenoptera acutorostrata*, stranded on the south-east coast of New Zealand. N. Z. Nat. Sci. 20, 1–14.

Webb, P.W., 1975. Hydrodynamics and energetics of fish propulsion. Bull. Fish. Res. Board Can. 190, 1–158.

Webb, P.W., 1988. Simple physical principles and vertebrate aquatic locomotion. Am. Zool. 28, 709–725.

Webb, P.W., 2004. Maneuverability-general issues. IEEE J. Ocean. Eng. 29, 547–555.

Webb, P.W., 2006. Stability and maneuverability. In: Shadwick, R.E., Lauder, G.V. (Eds.), Fish Biomechanics. Academic Press, Amsterdam, pp. 281–332.

Weber, P.W., Howle, L.E., Murray, M.M., Fish, F.E., 2009. Lift and drag performance of odontocete cetacean flippers. J. Exp. Biol. 212, 2149–2158.

Weber, P.W., Murray, M.M., Howle, L.E., Fish, F.E., 2009. Comparison of real and idealized cetacean flippers. Bioinspir. Biomim. 4, 046001.

Weihs, D., 1972. Semi-infinite vortex trails, and their relation to oscillating airfoils. J. Fluid Mech. 54, 679–690.

Weihs, D., 1973. Hydromechanics of fish schooling. Nature 241, 290–291.

Weihs, D., 1981. Effect of swimming path curvature on the energetics of fish. Fish. Bull. U.S. 79, 171–176.

Weihs, D., 1989. Design features and mechanics of axial locomotion in fish. Am. Zool. 29, 151–160.

Weihs, D., 1993. Stability of aquatic animal locomotion. Contemp. Math. 141, 443–461.

Weihs, D., 2002. Dynamics of dolphin porpoising revisited. Integr. Comp. Biol. 42, 1071–1078.

Weihs, D., 2004. The hydrodynamics of dolphin drafting. J. Biol. 3, 1–16.

Weihs, D., Webb, P.W., 1983. Optimization of locomotion. In: Webb, P.W., Weihs, D. (Eds.), Fish Biomechanics. Praeger, New York, pp. 339–371.

Williams, T.M. 1987. Approaches for the study of exercise physiology and hydrodynamics in marine mammals. In Approaches to Marine Mammal Energetics, pp. 127-145, A.C. Huntley, D.P. Costa, G.A.J. Worthy, and M.A. Castellini, Eds. Spec. Publ. Soc. Mar. Mamm. No. 1.

Williams, T.M., 2001. Intermittent swimming by mammals: a strategy for increasing energetic efficiency during diving. Am. Zool. 41, 166–176.

Williams, T.M., Davis, R.W., Fuiman, L.A., Francis, J., Le, B.J., Horning, M., Calambokidis, J., Croll, D.A., 2000. Sink or swim: strategies for cost-efficient diving by marine mammals. Science 288, 133–136.

Williams, T.M., Friedl, W.A., Fong, M.L., Yamada, R.M., Sedivy, P., Haun, J.E., 1992. Travel at low energetic cost by swimming and wave-riding bottlenose dolphins. Nature 355, 821–823.

Williams, T.M., Friedl, W.A., Haun, J.E., 1993. The physiology of bottlenose dolphins (*Tursiops truncatus*): heart rate, metabolic rate and plasma lactate concentration during exercise. J. Exp. Biol. 179, 31–46.

Williams, T.M., Shippee, S.F., Rothe, M.J., 1996. Strategies for reducing foraging costs in dolphins. In: Greenstreet, S.P.R., Tasker, M.L. (Eds.), Aquatic Predators and their Prey. Fishing News Books, Oxford, pp. 4–9.

Winn, H.E., Olla, B.L., 1979. Behavior of Marine Animals. vol. 3 Plenum Press, New York.

Woodcock, A.H., 1948. The swimming of dolphins. Nature 161, 602.

Woodcock, A.H., McBride, A.F., 1951. Wave-riding dolphins. J. Exp. Biol. 28, 215–217.

Wu, T.Y., 1971. Hydrodynamics of swimming propulsion. Part 2. Some optimum shape problems. J. Fluid Mech. 46, 521–544.

Würsig, B., Würsig, M., 1979. Behavior and ecology of the bottlenose dolphin, *Tursiops truncatus*, in the South Atlantic. Fish. Bull. 77, 399–412.

Yanov, V.G., 1990. Kinematics of dolphins: new experimental results. Sov. Phys. Dokl. 35, 921–923 (translated from Russian).

Yanov, V.G., 1991. System-functional organization of the swim kinematics of a dolphin. Dokl. Akad. Nauk SSSR 317, 1089–1093 (translated from Russian).

Yasui, W.Y., 1980. Morphometrics, Hydrodynamics and Energetics of Locomotion for a Small Cetacean, *Phocoena phocoena* (L.). M. Sc. thesis, University of Guelph, Ontario.

Yates, G.T., 1983. Hydromechanics of body and caudal fin propulsion. In: Webb, P.W., Weihs, D. (Eds.), Fish Biomechanics. Praeger, New York, pp. 177–213.

Yuen, H.S.H., 1961. Bow wave riding of dolphins. Science 134, 1011–1012.

CHAPTER

5

Cardiovascular physiology in dolphins and other cetaceans

Paul J. Ponganis[a] *and Birgitte I. McDonald*[b]

[a]Scripps Institution of Oceanography, La Jolla, CA, United States [b]Moss Landing Marine Laboratories at San Jose State University, Moss Landing, CA, United States

Abstract

The cardiovascular system in cetaceans follows the general mammalian model but is custom designed for individual species. Notable features include (a) a compliant, elastic ascending aorta and aortic arch, (b) the retia mirabilia, (c) venous plexuses in the head and tracheo-bronchial tree, (d) prominent epidural veins, (e) diaphragmatic caval slings, and (f) thermoregulatory plexuses. During high-speed surface and subsurface swimming of more active porpoises and dolphins, high heart rates, enlarged hearts, and higher affinities of hemoglobin for oxygen (O_2) are postulated to maximize lung-to-muscle O_2 transport just as during exercise of terrestrial mammals. However, during dives, the cardiovascular dive response (bradycardia and vasoconstriction) is variable and serves to regulate the depletion of O_2 stores according to the nature of a given dive. Depth, dive duration, exercise, lung volume reflexes, and cognitive control have all been postulated to influence heart rate during the dive.

This chapter examines the anatomy and physiology of the cardiovascular system of the bottlenose dolphin (*Tursiops truncatus*), with comparison to other cetaceans and with particular emphasis on cardiovascular responses during diving. After review of basic cardiovascular structure and function in these animals, we focus on the nature of the cardiovascular dive response (the decrease in heart rate [bradycardia] and increase in peripheral vasoconstriction associated with the breath hold). Here, we also consider neuroregulation of the dive response during the dive, the effect of the dive response on oxygen (O_2) store management during the dive, and the effects of sustained high heart rates during the postdive surface interval. We do not review the contributions of the cardiovascular system to respiratory gas exchange and thermoregulation as these topics are addressed in other chapters. As one of the primary functions of the cardiovascular system is blood O_2 transport from the lungs to tissues, we conclude with a consideration of hemoglobin (Hb) and the O_2-Hb dissociation curve in cetaceans.

Cardiovascular anatomy and function

Heart—Anatomy

The hearts of bottlenose dolphins and other cetaceans follow the four-chamber mammalian design. Blood returns from the body via the anterior and posterior venae cavae into the right atrium and then enters the right ventricle from where it is pumped via the pulmonary artery to the lungs for gas exchange. Blood returns from the lungs through the pulmonary veins to the left atrium and enters the left ventricle from which it is pumped into the aorta and out to body organs and tissues. The valves of the heart (tricuspid, pulmonary, mitral, and aortic) are also characteristics of mammals. The foramen ovale and ductus arteriosus, essential components of the fetal circulation, both close after birth as in other mammals (Slijper, 1961, 1962).

In contrast to terrestrial mammals, but similar to seals, the cetacean heart is dorso-ventrally flattened and broad (Race et al., 1959; Truex et al., 1961; Slijper, 1962; Rowlatt and Gaskin, 1975; Rowlatt, 1981; Ochrymowych and Lambertsen, 1984; Bisaillon et al., 1987; Tarpley et al., 1993; Drabek and Burns, 2002). Although the heart can be almost

The Physiology of Dolphins
https://doi.org/10.1016/B978-0-323-90516-9.00011-7

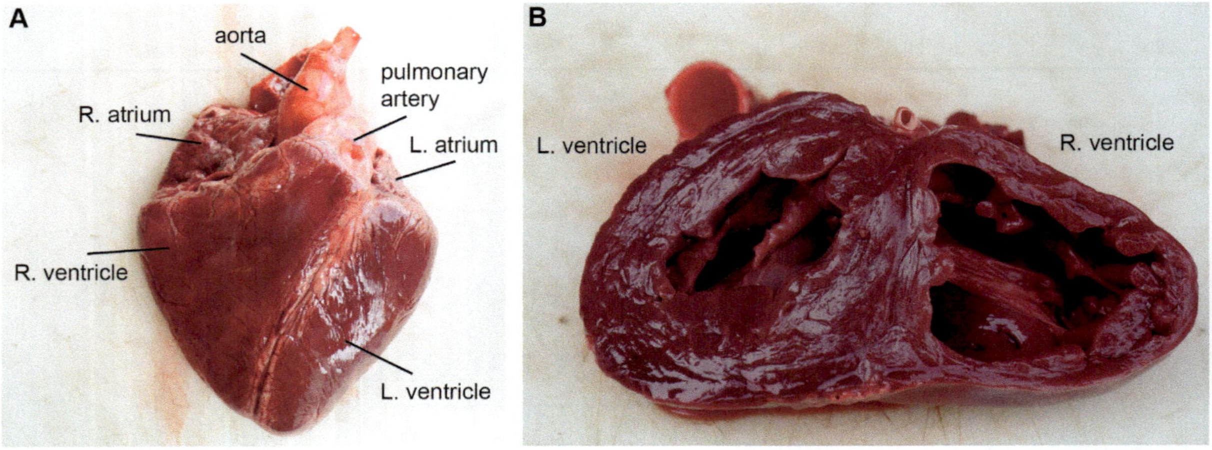

FIG. 1 Heart of the harbor porpoise (*Phocoena phocoena*). (A) Ventral view of the four-chambered heart, pulmonary artery, and aorta. (B) Cross section of the heart at the level of the ventricles. The thicker-walled left ventricle weighs about twice as much as the right ventricle (Rowlatt and Gaskin, 1975). *With permission from Robin Dunkin.*

oval in deeper divers and in larger cetacean species, the hearts of dolphins and porpoises have the more typical triangular shape of many mammals (Fig. 1). The endocardium (inner surface) of the ventricles of the bottlenose dolphin has been reported to be thicker and more elastic than in other mammals (Simpson and Gardner, 1972). The multiple anastomoses between the right and left coronary arterial trees that have been observed in several cetacean species are absent in the bottlenose dolphin (Race et al., 1959; Truex et al., 1961; Sommer et al., 1968; Cave, 1977; Ochrymowych and Lambertsen, 1984; Bisaillon et al., 1988; Picco et al., 2020). Such coronary anastomoses also do not occur in the harbor porpoise (*Phocoena phocoena*) (Rowlatt and Gaskin, 1975). The coronary arterial interconnections in deep divers such as the sperm whale (*Physeter macrocephalus*) are thought to optimize blood flow and O_2 delivery to the heart during long dives.

The Purkinje fibers are specialized, rapidly conducting cells of the cardiac electrical conduction system located beneath the endocardium of the ventricles. These fibers allow for uniform contraction of the ventricular walls, and they have been noted to be large in bottlenose dolphins, as well as other cetaceans (White and Kerr, 1915–1917; Simpson and Gardner, 1972; Pfeiffer, 1990; James et al., 1995; Ono et al., 2009; Aoki et al., 2017). The activation pattern and possibly the distribution pattern of these fibers in the ventricular walls have been postulated to contribute to the shape and duration of the electrocardiogram (ECG) signal in cetaceans (King et al., 1953; Kanwisher and Senft, 1960; Hamlin and Scher, 1961; Hamlin et al., 1970, 1972; Rowlatt and Gaskin, 1975; Meijler et al., 1992; Harms et al., 2013; Storlund et al., 2021). This type B ECG pattern, which is found in ungulates and pinnipeds in addition to cetaceans, contrasts with the type A pattern of primates, cats, and dogs. It has also been reported that Purkinje-like fibers, Todd fibers, occur in the atrial wall of the white-beaked dolphin (*Lagenorhynchus albirostris*) (Van Nie, 1987). It was postulated that such fast-conducting fibers might facilitate transmission of electrical signals from the atrium to the ventricle, and allow for rapid transitions between slow heart rates during dives and fast heart rates at the surface.

The position of the heart in the cetacean thorax is more rostral than in other mammals (Cozzi et al., 2017). The heart is just caudal to the axillary line between the foreflippers, essentially at the cranial end of the thorax. The heart lies ventral to the lungs and along the sternum. The shape and position of the heart have been postulated to accommodate better the compression of the chest by hydrostatic pressure at depth and/or to allow for a very convex, rostral curvature of the diaphragm during rapid exhalations at the surface (Drabek, 1975; Slijper, 1962; Cozzi et al., 2017).

Heart size

The relative size of the heart of the bottlenose dolphin, about 0.5% of body mass (Ridgway and Johnston, 1966; Ridgway and Kohin, 1995; Turner et al., 2006), is characteristic of most mammals and other cetaceans (Tables 1 and 2). A relative heart mass of 0.93% for the bottlenose dolphin (Slijper, 1962) has often been cited in the literature; however, it appears this measurement was from a single juvenile dolphin (Sergeant and Brodie, 1969). Based upon the results and sample sizes from the four studies for bottlenose dolphins in Table 1, a relative heart mass near 0.5% appears to be the most accurate number.

TABLE 1 Relative heart masses (M_H/M_B) of dolphins and porpoises, expressed as percentage (%) of body mass.

Species	M_H/M_B (%) Mean	M_H/M_B (%) Range	n	Reference
Bottlenose dolphin *Tursiops truncatus*	0.54	0.54–0.56	4	Ridgway and Johnston (1966)
	0.45	0.35–0.61	20	Ridgway and Kohin (1995)
	0.55		36	Turner et al. (2006)
	0.56 Female 0.64 Male		8 11	Mallette et al. (2016)
	0.51		50	Sergeant (1969)
	0.5		1	Irving et al. (1941a)
	0.93		1	Slijper (1962)
Indo-Pacific bottlenose dolphin *Tursiops aduncus*	0.6		131	Plon et al. (2012)
Pacific white-sided dolphin *Lagenorhynchus obliquidens*	0.85	0.70–0.92	5	Ridgway and Johnston (1966)
	0.75	0.72–0.78	4	Ridgway and Kohin (1995)
	0.72	0.67–0.84	5	Ponganis et al. (2022)
Peale's dolphin *Lagenorhynchus australis*	1.14		1	Goodall et al. (1997)
Dall's porpoise *Phocoenoides dalli*	1.31	1.25–1.34	4	Ridgway and Johnston (1966)
	0.99	0.57–1.34	35	Ridgway and Kohin (1995)
	0.89	0.68–1.20	36	Ponganis et al. (2022)
Northern right whale dolphin *Lissodelphis borealis*	0.82	0.70–0.93	4	Ponganis et al. (2022)
Harbor porpoise *Phocoena phocoena*	0.50 Female 0.53 Male		20 20	Rowlatt and Gaskin (1975)
	0.69 Female 0.69 Male		44 8	McLellan et al. (2002)
	0.8		97	Lockyer et al. (2003)
	0.84	0.65–1.22	27	Sergeant (1969)
Spotted dolphin *Stenella attenuata*	0.42	0.37–0.51	11	Perrin and Roberts (1972)
Spinner dolphin *Stenella longirostris*	0.46	0.37–0.52	6	Perrin and Roberts (1972)
Striped dolphin *Stenella coeruleoalba*	0.85	0.62–1.15	11	Miyazaki et al. (1981)
Indo-Pacific humpback dolphin *Sousa chinensis*	0.52		350	Plon et al. (2012)
Long-beaked common dolphin *Delphinus capensis*	0.62		308	Plon et al. (2012)
Common dolphin *Delphinus delphis*	0.48		50	Sergeant (1969)
Rough-toothed dolphin *Steno bredanensis*		0.5–0.8	15	Miyazaki and Perrin (1994)

Continued

TABLE 1 Relative heart masses (M_H/M_B) of dolphins and porpoises, expressed as percentage (%) of body mass—cont'd

Species	M_H/M_B (%) Mean	M_H/M_B (%) Range	n	Reference
Francisciana dolphin *Pontoporia blainvillei*	0.61 Female 0.61 Male		16 21	Weber Rosas et al. (2007)
	0.51 Female 0.49 Male	0.49–0.52 0.48–0.49	5 2	Kamiya and Yamasaki (1974)
Guiana dolphin *Sotalia guianensis*	0.56 Female 0.62 Male		12 12	Weber Rosas et al. (2007)
Tucuxi *Sotalia fluviatilis*	0.45 Female 0.43 Male		6 5	Da Silva and Best (1994)
Amazon River dolphin *Inia geoffrensis*	0.35 Female 0.33 Male	0.24–0.47 0.29–0.37	4 5	Best and Da Silva (1989)
Ganges River dolphin *Platanista gangetica*	0.26 Female 0.24 Male	0.24–0.28 0.22–0.25	2 2	Kamiya and Yamasaki (1974)

Most mammals are typically in the 0.6% range with more athletic species such as dogs, horses, and antelopes in the 0.8%–1.2% range (Schneider et al., 1964; Snow, 1985; Woodall, 1992; Williams et al., 2015a). Exceptional greyhounds and racehorses have values as high as 1.7%–2% (Poole and Erickson, 2011). High values of 1.19%–1.55% also occur in smaller, mammals with high resting metabolic rates, such as bats, shrews, and Arctic hares (Bishop, 1997). Sloths have among the lowest (0.24%) relative heart masses reported in mammals (Bishop, 1997). In other marine mammals, typical relative heart masses are 0.5%–0.8%, 0.73%, and 0.3% in pinnipeds (seals, furs seals, and walrus), sea otters, and manatees, respectively (Crile and Quiring, 1940; Bryden, 1972; Morejohn et al., 1975; Bisaillon, 1982; Drabek and Burns, 2002). Abbreviations: M_B, body mass; M_H, heart mass; n, sample size.
Based on supplemental files of Ponganis et al. (2022).

TABLE 2 Relative heart masses (M_H/M_B) of larger odontocetes and baleen whales, expressed as percentage (%) of body mass.

Species	M_H/M_B (%) Mean	M_H/M_B (%) Range	n	Reference
Beluga *Delphinapterus leucas*	0.56	0.56–0.57	2	Crile and Quiring (1940)
	0.57	0.37–0.69	20	Sergeant and Brodie (1969)
Narwhal *Monodon monoceros*	0.59 Female 0.51 Male	0.54–0.65 0.48–0.57	4 3	Hay and Mansfield (1989)
Long-finned pilot whale *Globicephala melas*		0.50–0.59[a] 0.39–0.45[a]		Bloch et al. (1993), Cowan (1966), Bloch et al. (1993), Lockyer (1993)
	0.44		1	Sergeant and Brodie (1969)
Killer whale *Orcinus orca*	0.64	0.50–0.73	3	Sergeant (1969)
Pygmy sperm whale *Kogia breviceps*	0.33		1	Sergeant (1969)
Sperm whale *Physeter macrocephalus*	0.38		25	McAlpine (1985)
	0.34			Slijper (1962)
	0.47		1	Race et al. (1959)
	0.32		1	Quiring (1943)
Baird's beaked whale *Berardius bairdii*	0.63		1	Sergeant (1969)
Sei whale *Balaenoptera borealis*	0.52		27	McAlpine (1985)

TABLE 2 Relative heart masses (M_H/M_B) of larger odontocetes and baleen whales, expressed as percentage (%) of body mass—cont'd

Species	M_H/M_B (%) Mean	M_H/M_B (%) Range	n	Reference
	0.40	0.34–0.49	17	Sergeant (1969)
Fin whale *Balaenoptera physalus*	0.52		33	McAlpine (1985)
	0.55			Slijper (1962)
	0.4			Nishiwaki (1950)
	0.64		1	Quiring (1943)
Blue whale *Balaenoptera musculus*	0.55			Slijper (1962)
Minke whale *Balaenoptera acutorostrata*	0.5			Slijper (1962)
Bryde's whale *Balaenoptera brydei*	0.46	0.23–0.68	20	Sergeant (1969)
Humpback whale *Megaptera novaeangliae*	0.5			Slijper (1962)
	0.50	0.45–0.56	3	Quiring (1943)
Pacific right whale *Eubalaena sieboldii*	0.56	0.43–0.73	6	Omura et al. (1969)
Bowhead whale *Balaena mysticetus*	0.6	0.3–1.2	9	Tarpley et al. (1993)
Gray whale *Eschrichtius robustus*	0.55		1	Sergeant (1969)

Most values for these larger cetaceans are at the lower end of the typical mammalian range (see Table 1 legend). Abbreviations: M_B, body mass; M_H, heart mass; n, sample size.

[a] *Estimates for pilot whales were based on measurements of heart mass and body length (Cowan, 1966) and two different body length-body mass relationships (Bloch et al., 1993; Lockyer, 1993).*

Although heart mass scales allometrically over a wide range of body masses in mammals (Brody, 1945; Stahl, 1967), other factors such as metabolic demand (including aerobic exercise capacity and thermoregulation) can contribute to relative heart masses that are larger than the typical 0.5%–0.6% value of most mammals (Bishop, 1997; Williams et al., 2015a). Common mammalian examples of more athletic species with larger relative heart masses (0.8%–1.2%) include dogs, horses, and antelopes (Schneider et al., 1964; Snow, 1985; Woodall, 1992). Even higher relative heart masses of 1.2%–1.5% occur in smaller mammals, such as bats, shrews, and Arctic hares, which have high metabolic rates due to size, activities, and/or a cold environment (Bishop, 1997). As far as we know, the highest relative heart masses in mammals are 1.7%–2% in the most elite greyhounds and racehorses (Poole and Erickson, 2011).

In regard to the range of relative heart masses in cetaceans, the relative heart masses of bottlenose dolphins, Pacific white-sided dolphins (*Lagenorhynchus obliquidens*), and Dall's porpoises (*Phocoenoides dalli*), three species of similar body mass, were 0.45%–0.55%, 0.75%–0.86%, and 0.99%–1.31% (Ridgway and Johnston, 1966; Ridgway and Kohin, 1995; Turner et al., 2006). In other words, for a Dall's porpoise and bottlenose dolphin of the same body mass, the heart of the Dall's porpoise is about twice as large as that of the dolphin. Based on the higher swim speeds, larger muscle mass, thinner blubber layer, and suspected deeper-diving activity of Dall's porpoises, Ridgway and Johnston (1966) suggested that a relatively larger heart in Dall's porpoises supported a higher metabolic rate and faster/more efficient gas exchange and restoration of larger blood and muscle O_2 stores.

Given the data in Tables 1 and 2 and the positive correlation of aerobic exercise capacity and maximal O_2 consumption with relative heart mass (Bishop, 1997; Williams et al., 2015a), Dall's porpoise is, by far, the most "elite" athlete among the cetaceans. This species would be predicted to be capable of prolonged, elevated aerobic activity, similar to that of many horses and dogs, although probably less than the most elite racehorses and greyhounds. As suggested by Ridgway and Johnston (1966) and Williams et al. (2015a), enlarged hearts appear to occur in those cetacean species with high-speed, intense surface-swimming activities. Certainly, Dall's porpoise fits this picture with its high surface swim speeds and "rooster tailing" through the water, as well as with its stocky build, and increased muscle mass, blood volume, Hb concentration, and Hb-O_2 affinity (Ridgway, 1966; Ridgway and Johnston, 1966; Horvath et al., 1968; Jefferson, 1987, 1988). The larger relative heart masses of the striped dolphin (*Stenella coeruleoalba*) and Pacific white-sided dolphin (Table 1), although lower than that of Dall's porpoise, also stand out among cetaceans, and suggest that they too are capable of sustained, intense aerobic activity.

The harbor porpoise is a cetacean with high resting and field metabolic rates (Kanwisher and Sundnes, 1965; Rojano-Doñate et al., 2018). It is among the smallest of cetaceans, lives in cold waters, and has exceptionally high foraging rates (Wisniewska et al., 2016). A larger relative heart mass (0.7%–0.8%, Table 1) is consistent with the higher metabolic rate and thermoregulatory requirements of this small-bodied species. One study had lower (0.5%) relative heart mass values; it is unclear if the heart preservation technique, body weight estimation formula, or differences in body mass composition (blubber thickness) may have accounted for this difference. Heart size also varies with physiological condition, such as athletic training or pregnancy (Hill and Olson, 2008; Savu et al., 2012). It should not be surprising that the highest relative heart masses (0.86%) in harbor porpoises were reported in pregnant animals (McLellan et al., 2002). Again, these findings support Ridgway and Johnston's original suggestion that relatively larger hearts in some cetaceans support higher metabolic demands, whether those elevations in metabolic rate are due to high swim speeds, deep dives, or thermoregulatory needs. At the other end of the spectrum, the Ganges River dolphin (*Platanista gangetica*) has a relative heart mass of 0.25%, similar to those of manatees and sloths, and about half that of the bottlenose dolphin (Table 1). On the basis of its small relative heart size, this small river dolphin would be expected to have the lowest metabolic demands and smallest aerobic exercise capacity among the cetacean species in Tables 1 and 2.

Exceptionally, large relative heart masses do not occur in many deep-diving cetaceans, such as the beluga (*Delphinapterus leucas*), Baird's beaked whale (*Berardius bairdii*), and sperm whale (Table 2) or even in the long-finned pilot whale (*Globicephala melas*), which has been postulated to have a high-diving metabolic rate due to high transit swim speeds (Aoki et al., 2017). Apparently, an increase in relative heart mass is not essential in all species either to meet metabolic demand during the dive or to allow for rapid, efficient gas exchange at the surface between dives. Indeed, heart rate, pulmonary blood flow, and lung ventilation requirements after the dive are probably most dependent on the magnitude of O_2 depletion and carbon dioxide (CO_2) accumulation during the dive. These requirements, in turn, are dependent on dive duration, dive metabolic rate, and the size of blood and muscle oxygen stores (i.e., blood volume, Hb concentration, muscle mass, and myoglobin concentration).

It is also important to remember that body composition (% blubber, % muscle, and blood volume) varies among these many species and can even vary seasonally within a species (fasting, migration, and location). These differences in body composition and body mass may contribute to differences in overall metabolic needs and relative heart masses. The sperm whale is a good example; its relative heart mass is low, near 0.3% (Table 2). This whale may either be an exceptionally efficient swimmer, and/or its size, low muscle mass (22%–30% of body mass), high blubber content (33%–40% of body mass), and large spermaceti organ (~6%–7% of body mass) may result in a relatively low mass-specific metabolic demand (Omura, 1950; Lockyer, 1976, 1991; Clarke, 1978). The technical difficulty of accurate total body weight determinations in large whales may also contribute to inaccuracy.

Returning to consideration of the bottlenose dolphin's unremarkable relative heart mass of 0.5%, and the relationship of aerobic exercise capacity to relative heart mass, the aerobic scope (the ratio of maximum O_2 consumption to resting O_2 consumption) of the bottlenose dolphin would be expected to be "typical" of most mammals. Namely, as in most "typical" mammals, maximum O_2 consumption would be about 12 × the allometrically predicted resting O_2 consumption, and far less than the 30-fold increase observed in elite mammalian athletes such as horses and dogs (Taylor et al., 1987). Indeed, maximum O_2 consumption rates of two exercising bottlenose dolphins were 7–11 × the allometrically predicted resting O_2 consumption rate (Williams et al., 1993).

Hemodynamics: Cardiac output and stroke volume

The design and size of the heart in combination with heart rate are prime determinants of the ability of the heart to pump blood throughout the body. Cardiac output, the flow rate at which blood is pumped to the body, is determined by heart rate and stroke volume (the volume of blood ejected from the heart during a heartbeat) (Hellsten and Nyberg, 2016). In comparative physiology, these indices of cardiac function are often expressed relative to body mass, i.e., cardiac output in $mL\,kg^{-1}\,s^{-1}$, and stroke volume in $mL\,kg^{-1}$. In the bottlenose dolphin, based on a relative heart mass typical of most mammals, the cardiac output and stroke volume of the heart should be in the range expected for a mammal of its size.

There have been few measurements of cardiac output and stroke volume in dolphins (Sommer et al., 1968; Miedler et al., 2015; Fahlman et al., 2019, 2020b). As outlined in Table 3, stroke volumes of anesthetized bottlenose dolphins and

TABLE 3 Cardiac outputs (CO) and stroke volumes (SV) of bottlenose dolphins.

Animal	Condition	HR (beats min^{-1}) (measured)	CO (mL $kg^{-1}s^{-1}$) (measured)	SV (mL kg^{-1}) (measured)	CO (mL $kg^{-1}s^{-1}$) (allometric)	SV (mL kg^{-1}) (allometric)	Reference
Dolphin 175 kg (mean)	At rest	41	0.5	0.8	1.1	0.9	Miedler et al. (2015)
	1-min postexercise	84	1.7	1.3			
Dolphin 169 kg (mean)	Apnea at rest (spontaneous)	44	0.8	1.1	1.1	0.9	Fahlman et al. (2019)
	Postapneas (stationary)	78	1.8	1.3			
Dolphin 80 kg	First anesthesia	84	1.1	1.1	1.3	0.8	Sommer et al. (1968)
Dolphin 80 kg	Second anesthesia	120	1.1	0.6	1.3	0.8	Sommer et al. (1968)
Dolphin 70 kg	Anesthesia	110	0.8	0.4	1.3	0.8	Sommer et al. (1968)
Dolphin 114 kg	Anesthesia	140	1.8	0.8	1.2	0.8	Sommer et al. (1968)
Beluga 162 kg	Apnea at rest (spontaneous)	38	1.9	3.1	1.2	0.8	Fahlman et al. (2020b)
False killer whale 520 kg	Apnea at rest (spontaneous)	29	0.6	1.2	0.9	0.9	Fahlman et al. (2020b)
Human 70 kg	At rest	80	1.2	0.9	1.3	0.8	Sommer et al. (1968)
Pony 171 kg (mean)	At rest	39	1.5	2.4	1.1	0.9	Taylor et al. (1987)
	At $\dot{V}_{O2max}$	215	8.9	2.5			

Data were collected at rest, after 1.5-min exercise sessions, after stationary apneas at the surface, and during anesthesia. The allometrically predicted values at rest are also provided for each dolphin. Data for a young beluga and an adult false killer whale are also included (Fahlman et al., 2020b). To facilitate comparison to terrestrial mammals, corresponding values are provided for humans and horses. Measurements were made with different techniques: echocardiography (Miedler et al., 2015; Fahlman et al., 2019), indocyanine green dilution (Sommer et al., 1968), and the Fick technique (Taylor et al., 1987). In the awake dolphins, CO and SV reported by the authors in the table were from early apnea when heart rates (HR) were constant because both HR and SV varied with respirations. For example, the highest individual postapneic CO was 2.6 mL kg^{-1} s^{-1} with a stroke volume of 1.8 mL kg^{-1} (Fahlman et al., 2019). In dolphins at rest and under anesthesia, and in humans, the measured SVs were close to allometrically predicted values (Holt et al., 1968) as would be expected since relative heart mass is about 0.5%, the typical value for most mammals (Table 1). In contrast, COs in dolphins during anesthesia and rest periods were less than, although, often close to predicted values (Stahl, 1967) because of the time of measurement during apnea, and especially, during anesthesia, because of potential effects of the anesthetic, body positioning (out of water also), and effects of positive pressure ventilation. COs in dolphins after exercise or after surface breath holds were above the allometrically predicted value. In contrast to the dolphins and humans, CO and SV of ponies were greater than predicted allometrically. This is attributed to the larger relative heart masses (0.8%–1.2%) of most horses (Snow, 1985). In contrast to the variability of SV with breathing in dolphins, SV remains constant in the pony (Taylor et al., 1987) even when exercising at maximum oxygen consumption ($\dot{V}_{O2max}$). CO and SV have not been measured in dolphins during surface swimming or during dives to depth. Additional data are now also available from a beluga and a false killer whale. The large SV of the beluga was attributed to the size of the heart in this young animal. In adult belugas, however, relative heart mass is similar to that of bottlenose dolphins (Table 2).

dolphins at rest are close to those predicted by allometric equations (Holt et al., 1968). This is expected since the allometric equations are based on large numbers of mammals, and, as reviewed earlier, the relative heart mass of the bottlenose dolphin is typical of most mammals (Table 1).

In the awake dolphins (Miedler et al., 2015; Fahlman et al., 2019), both heart rate and stroke volume increased with respirations and decreased during the breath hold. To report data at a constant heart rate, the cardiac outputs and stroke volumes summarized by the authors were measured during early apnea. Consequently, cardiac outputs measured in early apneas of dolphins (Table 3) can be less than the allometrically predicted value since the allometric equations are based on continuously breathing mammals in a steady state with relatively stable heart rates both during and between respirations (Stahl, 1967; Fahlman et al., 2020b). In contrast to the resting state, cardiac outputs and stroke volumes of the dolphins after exercise or long breath holds were elevated above the allometrically predicted resting values (Table 3). For the sake of comparison, it should also be noted that the pony has a resting cardiac output and stroke volume that are greater than calculated by allometric equations (Table 3). This is again consistent with heart size; as reviewed earlier, most horses have a relative heart size about twice the typical mammalian value (Crile and Quiring, 1940; Snow, 1985).

The change in stroke volume with respirations in the dolphin is not unexpected. Apneic-to-eupneic changes in stroke volume occur in seals (Ponganis et al., 1990). Changes in stroke volume can also occur during exercise in some mammals, including humans (Åstrand et al., 1964; Vella and Robergs, 2005). In contrast, stroke volume in the horse remains constant (Table 3), even with exercise at maximum O_2 consumption (Taylor et al., 1987). Stroke volume can vary because of changes in venous return and optimal filling of the heart, and to changes in the contractility of ventricular muscle (sympathetic nervous system activation) (Hellsten and Nyberg, 2016). Overall, although studies are limited, the cardiac output and stroke volume of the heart of the bottlenose dolphin at rest are as predicted for a mammal of its size. From the available data, it is expected that increased heart rates during surface swims or during surface intervals between dives will also be associated with increased stroke volumes.

Hemodynamics: Heart function and blood pressure

Investigation of cardiac physiology and hemodynamics in the dolphin has only been conducted in one cardiac catheterization study in which the bottlenose dolphins were under anesthesia (Sommer et al., 1968). As the authors remarked, positive pressure ventilation, anesthetic drug side effects, body position, out-of-water effects on the body, and even occult disease may have affected results. The so-called filling pressures of the heart (right atrial pressure, left atrial pressure, and pulmonary capillary wedge pressure) were occasionally equivalent to human values but were often elevated, possibly secondary to the mode of ventilation (per the authors). In addition, mean pulmonary artery pressures were high, up to 4× the human reference values, while mean aortic pressures were about 1.5 × the human value (Sommer et al., 1968). With the available data, pulmonary vascular resistance and systemic vascular resistance, indices of the impedance or afterload that the right and left ventricles must work against, can be calculated (Kaplan, 1979; Kwan et al., 2019). In both cases, the values in the anesthetized dolphins were extreme, with pulmonary vascular resistance as high as 679 dynes s cm^{-5} (upper normal limit in humans: 250), and systemic vascular resistance as high as 2570 dynes s cm^{-5} (upper normal limit in humans: 1200) (Kaplan, 1979; Kwan et al., 2019). However, despite the high vascular resistances, most cardiac output measurements were equivalent to or greater than the predicted resting level (Table 3). From the perspective of cardiac function, that is remarkable. Clearly, further hemodynamic investigation of dolphins in the awake state would benefit our understanding of the roles of intracardiac, pulmonary artery, and systemic arterial pressures in cardiac function of these animals.

The vasculature

The arterial and venous systems of the bottlenose dolphin and other cetaceans are remarkable for many features (see Fig. 2). Readers are referred to the literature for detailed descriptions and excellent overviews of vascular anatomy (Slijper, 1962; Rommel et al., 2006; Costidis and Rommel, 2012, 2016b; Cozzi et al., 2017). Most studies have focused

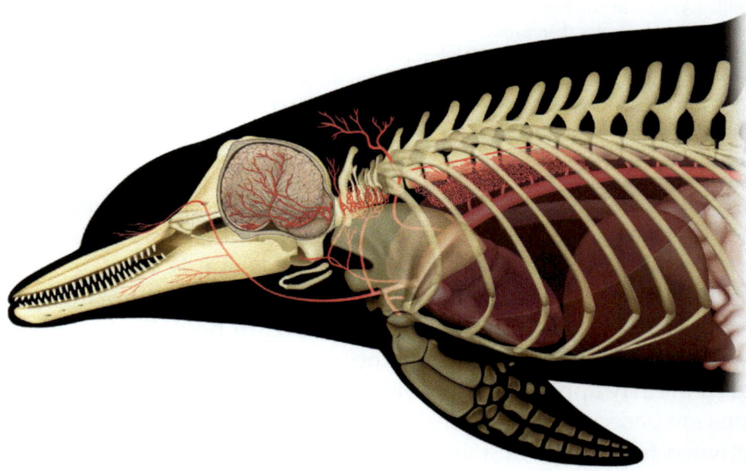

FIG. 2 Cervical and thoracic retia mirabilia of the bottlenose dolphin, *Tursiops truncatus*. Branches of the brachiocephalic trunk, internal thoracic arteries, and intercostal arteries extend from the aorta to form interbranching arterial networks of the retia, which eventually enter the intervertebral canal to form the spinal meningeal arteries, the major blood supply to the brain. The internal carotid artery is vestigial. The external carotid artery extends to tissues associated with the lower jaw and rostrum. For review, see Cozzi et al. (2017). *Illustration by Uko Gorter.*

on the systemic arterial and venous systems. As far as we know, anatomical features of the pulmonary arteries and veins have not been extensively examined.

However, the pulmonary vasculature has some unique features. In contrast to terrestrial mammals, the septae of the lung's alveoli in cetaceans have a double capillary bed, one on each side of the septal wall (for review, see Piscitelli et al., 2013). In addition, there are occasional transeptal anastomoses between the pulmonary artery and vein (Piscitelli et al., 2013). Lastly, smooth muscle sphincters have been observed around pulmonary veins in the lungs of cetaceans, including the bottlenose dolphin (Slijper, 1962; Simpson and Gardner, 1972). Interestingly, sphincters around branches of the portal vein within the liver have been noted in bottlenose dolphins and several other, although not all, cetaceans (Slijper, 1962; Simpson and Gardner, 1972; Hilton and Gaskin, 1978). The function of these sphincters is unknown. Pulmonary vein sphincters have the potential to affect flow in the transeptal pulmonary arterio-venous anastomoses mentioned above and to contribute to postulated regulation of ventilation-perfusion matching in the dolphin lung during dives (Garcia Párraga et al., 2018).

Major structural features of the arterial system in cetaceans include (a) modifications of the compliance and elasticity of the ascending aorta, aortic arch, and distal aorta; (b) the thoracic and vertebral retia mirabilia that form the major blood supply to the brain; (c) the presence of both a countercurrent arrangement of arteries and veins, as well as superficial arterio-venous anastomoses in the flippers, fluke, and dorsal fin for thermoregulation (Scholander and Schevill, 1955; Elsner et al., 1974, 2004; Plön et al., 2019); and (d) intra-abdominal countercurrent vascular networks for thermoregulation of reproductive organs (Rommel et al., 1992, 1993, 1994; Pabst et al., 1995).

The venous system is notable for extensive venous plexuses in the head and in the tracheal/bronchial walls, the presence of large veins (epidural veins) within the posterior vertebral canal, and the presence of a diaphragmatic muscular sling or sphincter around the posterior vena cava (Slijper, 1962; Cozzi et al., 2005; Ninomiya et al., 2005; Costidis and Rommel, 2012; Lillie et al., 2018). The spleen, which can function as a large storage organ for red blood cells in some seals, is small in the bottlenose dolphin and other cetaceans (Slijper, 1958, 1962; Castellini and Castellini, 1993; Haldiman and Tarpley, 1993; E Silva et al., 2014). In this chapter, we review the aorta and the retia, briefly consider the venous plexuses and epidural veins, and conclude with the hemodynamics of blood flow in the posterior vena cava during fluking. Thermoregulation and the associated design/function of more peripheral vascular networks are examined in Chapter 3: Thermoregulation.

Aorta

The aorta and its tributaries, in addition to acting as a vascular conduit, contribute to the regulation of the magnitude and distribution of blood flow to organs and tissues, to management of blood pressure, and to optimization of the function of the heart. As we will review later, during the bradycardia (decreased heart rate) and decreased cardiac output of the dive response, it is the contraction of smooth muscle in arterial walls that constricts arteries, maintains blood pressure, and regulates peripheral blood flow (Irving, 1938; Scholander, 1940; Irving et al., 1941b; Grinnell et al., 1942). Equally important are the structure and biomechanical properties of the aorta that facilitate cardiac function and blood flow during the long pauses between the slow heart beats of intense dive responses (Drabek, 1977; Rhode et al., 1986; Shadwick and Gosline, 1994). Modifications of the aorta occur not only in cetaceans but also in pinnipeds. A highly compliant, elastic expansion of the ascending aorta, the so-called aortic bulb, is especially prominent in deep-diving seals (Drabek, 1975; Drabek and Burns, 2002).

Before consideration of the aorta in bottlenose dolphins and other odontocetes, we first review the anatomy of the aorta in baleen whales. The fin whale (*Balaenoptera physalus*) is a good example because of the anatomical and biomechanical studies of Shadwick and Gosline (Shadwick and Gosline, 1994; Gosline and Shadwick, 1996; Shadwick, 1999), and because of the Krogh Principle that, for a given problem or question, there will be an animal in which it is best studied (Krogh, 1929). The large body size of baleen whales in combination with the dive response should result in exceptionally low heart rates during dives and in optimal design of the aorta to support cardiac function and blood flow under those conditions.

Shadwick and Gosline found that the fin whale had an enlarged, highly compliant, and elastic aortic arch. In comparison with the thoracic aorta, the arch had a higher elastin to collagen ratio, larger diameter, and a different arrangement of lamellar units (layers of elastin, collagen, and smooth muscle). The thoracic and abdominal sections of the aorta were rigid (Shadwick and Gosline, 1994; Lillie et al., 2013). Shadwick and Gosline hypothesized that the arch acted as an elastic reservoir (windkessel) to maintain blood pressure and flow during the pause between heart beats (diastole, the relaxation phase of the heartbeat). During ventricular contraction (systole, the ejection phase of the heartbeat), the compliant arch expands to accommodate the stroke volume of the heart, and then, during diastole, it gradually contracts due to its elastic fibers to maintain blood pressure and peripheral flow. In the presence of widespread vasoconstriction at low heart rates, the compliant aortic arch also serves to decrease the workload and O_2 consumption of the heart. In addition, because coronary blood flow only occurs during diastole, maintenance of diastolic blood pressure is especially important for perfusion of the coronary arteries. An enlarged ascending aorta or aortic arch has also been found in three other baleen whales, the minke whale (*Balaenoptera acutorostrata*, Ochrymowych and Lambertsen, 1984), the bowhead whale (*Balaena mysticetus*, Tarpley et al., 1997; Castellini and Ponganis, 2020), and the gray whale (*Eschrichtius robustus*, Elsner et al., 1966a). Enlargement of the ascending aorta is absent or only slightly increased in Bryde's whale (*Balaenoptera brydei*, Melnikov, 1994, 1997).

Based upon the wavelengths of pressure waves generated by the heart at different heart rates, Shadwick and Gosline have also postulated that the dimensions and biomechanical properties of the arch and of the rigid descending aorta of the fin whale decrease aortic impedance and allow for a more efficient circulation during the higher heart rates of the surface period. The aortic impedance is decreased through the destructive interference of outgoing and reflected pressure waves at the higher heart rates expected in the fin whale at the surface (above 20 beats min^{-1} (bpm)). Recently, documented heart rates in the blue whale (*Balaenoptera musculus*) support these hypotheses on the function of the aortic arch in baleen whales (Goldbogen et al., 2019).

Such quantitative biomechanical studies of the ascending aorta and aortic arch have not been performed in odontocetes. An expansion of the ascending aorta has been mentioned in summary reports of smaller to medium-size odontocetes, including Pacific bottlenose dolphins, pilot whales, and Cuvier's beaked whale (*Ziphius cavirostris*) (Elsner et al., 1966a). Ridgway also reported enlargements of the aorta in every cetacean he dissected or necropsied (Ridgway, 1972).

However, no aortic dilatation was observed in dissections of harbor porpoises or in angiographic studies of Atlantic bottlenose dolphins (Viamonte et al., 1968; Rowlatt and Gaskin, 1975). Viamonte et al. reported that the only aortic dilatation observed was associated with the sinuses of Valsalva in the aortic root. These three sinuses (spaces) bulge outward as they extend from the bases of the aortic valve leaflets to the aortic wall; the ostia (openings) of the right and left coronary artery are located in two of the sinuses. Typically, in humans, the diameter of the aorta at this level is about 1.5 × that of the aortic valve (Campens et al., 2014). In the Atlantic bottlenose dolphin, based on echocardiographic measurements of the diameters of the ascending aorta and aortic valve, the ascending aorta is also about 1.5 × the diameter of the aortic valve (Chetboul et al., 2012; Miedler et al., 2015; Fahlman et al., 2019, 2020b). Melnikov (1997) also reported that the ascending aorta was not enlarged in Dall's porpoise, the Pacific white-sided dolphin, and the spinner dolphin. In view of all these findings, it appears that any dilatation of the ascending aorta of the Atlantic bottlenose dolphin is not especially prominent. Despite the lack of significant enlargement, the ascending aorta of bottlenose dolphins has been described as elastic and thickened (Simpson and Gardner, 1972), which would support some function as an elastic reservoir.

Enlargement of the ascending aorta in odontocetes has also been suggested to be proportional to diving capacity (Galantsev, 1991) (Ridgway, personal communication). Enlargements of the ascending aorta have been reported or mentioned in deeper-diving odontocetes, including sperm whales (Melnikov, 1997), pilot whales (Jackson, 1845; Elsner et al., 1966a), and Cuvier's beaked whale (Elsner et al., 1966a; Díaz-Delgado et al., 2016). Certainly, the plastic cast of the ascending aorta of the beluga appears quite prominent in Vogl and Fisher's monograph on retia in the beluga and narwhal (*Monodon monoceros*) (Vogl and Fisher, 1982). In the sperm whale, Melnikov (1997) described the ascending aorta as a bulbous expansion. He found that the diameter of the ascending aorta was 1.5 × the diameter of the aorta at the level of the coronary ostia (i.e., the sinuses of Valsalva) and that large diameter was actually on the aortic arch after the take-off of the right brachiocephalic trunk. So, there is definite evidence of enlargement of the ascending aorta in some deep-diving odontocetes. There are also reports of high elastin content in the ascending aorta of a beluga (Simpson and Gardner, 1972) and higher elastin/collagen ratios in the proximal versus distal aorta in three delphinid species (Mompeó et al., 2020). In conclusion, although aortic structure and biomechanical properties have not been examined as extensively in odontocetes as in mysticetes, we suspect that a compliant, elastic ascending aorta exists even in the bottlenose dolphin and that the elastic reservoir—windkessel function is more developed in deeper, longer-duration divers, and, especially, in larger animals with baseline slow heart rates due to large body mass. Furthermore, anatomic and biomechanical research is needed to evaluate the hypothesis of an aortic arch windkessel in odontocetes and determine how blood pressure and flow are maintained during the long pauses between slow heartbeats.

Retia mirabilia

A rete mirabile ("wonderful net") has been described as the extension of a single artery into multiple branches that eventually coalesce back into a single artery (Cozzi et al., 2017). It is a complex structure (Figs. 2 and 3) that can be associated with a corresponding network of interbranching veins, and it is not exclusive to cetaceans. In the bottlenose dolphin, the largest rete mirabile is the thoracospinal rete. It is primarily arterial and extends from the aorta via

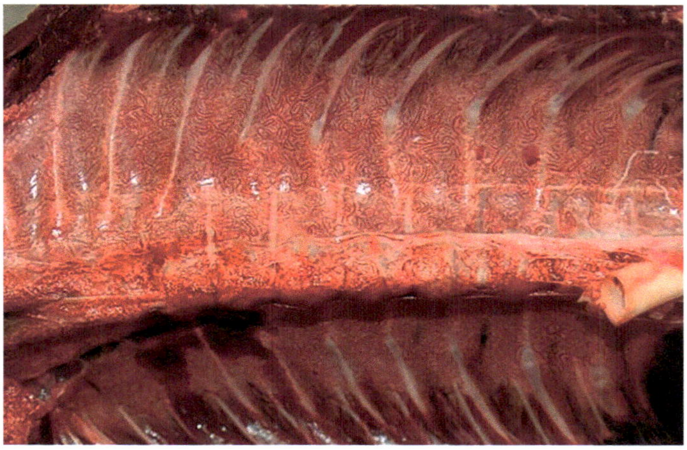

FIG. 3 Thoracic rete mirabile of the common dolphin, *Delphinus delphis*. In this view of the dorsal surface of the thoracic cavity, the rete mirabile is visible between the ribs, extending from the vertebral column toward the periphery. Most of the thoracic aorta had been removed; a transected segment remained attached to the vertebrae near the diaphragmatic end of the thorax. *Credit: P. Ponganis.*

intercostal and dorsal thoracic arteries into the intervertebral canal where it continues cranially to form the spinal meningeal arteries and the primary blood supply to the brain (Galliano et al., 1966; Nagel et al., 1968; Viamonte et al., 1968; McFarland et al., 1979). In the bottlenose dolphin, this rete extends the entire length of the thorax into the skull (Fig. 2). In the literature, the thoracospinal rete complex is also often described as two retia, the thoracic rete and the spinal rete. In addition, there is a contiguous, more cranial segment, the cervical rete (see Fig. 2).

Thoracospinal retia also occur in other cetaceans and are primarily composed of arterial blood vessels. Slijper (1962) described these retia to be most extensive in dolphins and sperm whales. Melnikov (1997) considered the thoracospinal rete of the sperm whale to be the most well-developed among cetaceans; it begins in the midlumbar region. In contrast, the rete was less developed in ziphiids and rorqual whales; in these whales, it only begins in the mid-thorax region (Slijper, 1962). The anatomy of the thoracospinal rete is exceptionally well described in the narwhal and beluga (Vogl and Fisher, 1981a,b, 1982; Vogl et al., 1981). In these two monodontids, the thoracospinal rete forms the blood supply to both the brain and spinal cord, is predominantly arterial and associated with fat, and has muscular arterial walls, but is poorly innervated. The cervical rete of the bowhead whale also appears to be primarily arterial (Pfeiffer and Kinkead, 1990). For detailed descriptions of vascular supply and structure of the thoracospinal retia, readers are referred to McFarland et al. (1979), Vogl and Fisher (1982), Melnikov (1997), and Costidis and Rommel (2016a,b). Excellent overview summaries are provided in Rommel et al. (2006) and Cozzi et al. (2017). In summary, the major anatomical features of the thoracospinal retia mirabilia in cetaceans are that the retia are predominantly arterial, not well-innervated, embedded in fat and are the primary blood supply to the brain.

The thoracospinal rete has been postulated to have many different functions, including windkessel functions, engorgement to prevent "lung squeeze" at depth, thermoregulation, and modification of the composition of blood, including the trapping of air bubbles (Slijper, 1962; Hui, 1975; Vogl and Fisher, 1982). One effect of the rete is a dampening of the pulse pressure in the spinal meningeal artery; mean blood pressure was the same as in the aorta, but there was no systolic peak or diastolic trough in the pressure profile (Nagel et al., 1968). Although blood flow in the spinal meningeal artery was nonpulsatile in that study, blood does, nonetheless, flow to the brain, and, in fact, blood flow can even selectively decrease in one cerebral hemisphere versus the other during unihemispheric sleep in dolphins (Ridgway et al., 2006).

Recent hypotheses on the function of the thoracospinal rete have included nitrogen absorption, a supplemental blood storage function, and dampening of venous pressure waves in the large veins (epidural veins) within the vertebral canals of cetaceans. In an investigation of the harbor porpoise thorocospinal rete (Blix et al., 2013), it was found that arteries in the central portion of the rete became quite thin-walled and were, as in other species, embedded in fat. Based on the fivefold greater solubility of nitrogen in fat than in water and the thin walls of the retinal vessels, the authors postulated that the fat acted as a localized nitrogen sink to lower elevations of blood nitrogen during dives and minimize the risk of decompression sickness. We also note that lowering of high nitrogen levels in the blood supply to the brain could also minimize the risk of nitrogen narcosis during dives.

In another recent study, a numerical model was constructed for heart rate, stroke volume, and cardiac output during dives in bottlenose dolphins (Bonato et al., 2019). For heart rates less than 10 bpm, the model resulted in a cardiac output that was too low to support adequate blood flow to the brain and heart, the two most O_2-dependent organs in the body. On this basis, it was suggested that transfer of blood from the spinal rete could supplement blood flow to adequate levels during the severe bradycardia. The model, of course, is subject to assumptions about stroke volumes and heart rates during the dives and is dependent on the postulated transfer of intra-abdominal pressures to the thorax to compress thoracic vessels and propel additional blood from the rete to the brain. In addition, as far as we know, a heart rate of less than 10 bpm has not been reported in a bottlenose dolphin, either during trained dives in a tank as long as 4.5 min, or during dives at sea to as deep as 210 m and almost 6-min duration (Elsner et al., 1966b; Ridgway, 1986; Williams et al., 1999; Houser et al., 2010; Noren et al., 2012).

A third paper that will be reviewed later in the chapter modeled changes in intra-abdominal pressure and venous pressure profiles due to fluking (Lillie et al., 2018). In that study, it was postulated that the spinal rete may protect the brain against pressure oscillations in the epidural veins within the vertebral canal. These hypotheses have been further developed in a more recent paper (Lillie et al., 2022).

Other retia of variable size and structure have been described throughout the cetacean body, including an ophthalmic rete, and retia in the pterygoid sinus, in mandibular fat pads, in the nasal plug/passages of the head, in the neck, along the esophagus, within the thorax, in the lumbar region (between the lumbar vertebrae and the hypaxial muscles), and in the pelvis (Slijper, 1962; McFarland et al., 1979; Vogl and Fisher, 1982; Melnikov, 1997; Piscitelli et al., 2010; Costidis and Rommel, 2012; Ninomiya et al., 2014; Costidis and Rommel, 2016b). Many of these retia, such as the ophthalmic rete, are also associated with complex venous networks (plexuses) and are probably involved in thermoregulation. In addition, in the head, it has been suggested that these vascular networks may (a) allow for gas exchange and nitrogen absorption from blood into associated fat in pterygoid sinus and mandibular fat, (b) affect acoustic properties of the intermandibular fat body, (c) engorge to prevent barotrauma in the pterygoid air sinus and nasal passages, and (d) engorge in the nasal plug to promote sealing of the plug in the blowhole during a dive (Costidis and Rommel, 2012, 2016a,b). In the abdominal cavity, extensive arterial and venous networks are associated with the uterus/ovaries and the testes in odontocetes ranging from bottlenose dolphins to sperm whales (Rommel et al., 1992, 1993; Melnikov, 1997). These vascular networks associated with the reproductive organs often consist of parallel arteries associated with veins or venous plexuses in a countercurrent flow arrangement that can contribute to thermoregulation. See Section "Thermal effects of exercise and diving," and Fig. 10 in Chapter 3 for further review of these structures.

Venous plexuses and epidural veins

Extensive venous plexuses throughout the body, including those described in the prior paragraph, have long been noted in many cetaceans, including the bottlenose dolphin (Slijper, 1962). As already mentioned, many of these venous networks, dependent on their anatomy and location, may contribute to thermoregulation with influx of cooler blood from the skin and appendages, and efflux of heat away from central organs (see Chapter 3 for further review). Engorgement of large venous plexuses at depth is also an important mechanism for baroprotection in the cranial air sinuses (see prior paragraph) and in the middle ear (Sassu and Cozzi, 2007). The increase in venous volume decreases the fixed volume within a rigid bony cavity and allows for compression of air to smaller volumes without mechanical disruption of tissues.

Venous plexuses are also considered to play a role in baroprotection in the respiratory system with transfer of peripheral blood into the thorax and allowance of greater compression of air volume without associated tissue damage. It has been proposed that engorgement of the large submucosal venous plexuses found in the tracheae and bronchi of cetaceans contribute to such baroprotection (Slijper, 1962; Leith, 1989; Cozzi et al., 2005; Ninomiya et al., 2005; Davenport et al., 2009, 2013). The extensive development of the thoracic rete within the thorax in deep-diving cetaceans also raises the possibility that these structures, although arterial, may contribute to baroprotection (Slijper, 1962; Melnikov, 1997). Volume of the intrathoracic retial tissue of deeper-diving kogiid whales is twice that in the bottlenose dolphin (Piscitelli et al., 2010). In addition, in a modeling study of the fin whale, the estimated volume of blood in the rete was adequate for pressure equilibration (Lillie et al., 2013). Another structure that may contribute to the equilibration of pressure by engorgement within the thorax is the pericardial venous plexus (Harrison and Tomlinson, 1956; Piscitelli et al., 2013).

In seals, there are large hepatic sinuses that can store blood and that empty into the vena cava (Harrison and Tomlinson, 1956; Elsner et al., 1964). From available anatomical studies, cetaceans do not appear to have these large hepatic sinuses (Harrison and Tomlinson, 1956; Barnett et al., 1958; Slijper, 1962). In the northern elephant seal, for example, the posterior vena cava and hepatic sinuses can store up to 20% of the blood volume (Elsner et al., 1964). Slijper (1962) did report that the hepatic veins were enlarged in odontocetes, but not in mysticetes. The relative size of these hepatic veins and their role in blood storage in cetaceans is unclear. Recently, large central veins and vascular sinuses have been observed in the liver tissue of several odontocete species, including the bottlenose dolphin

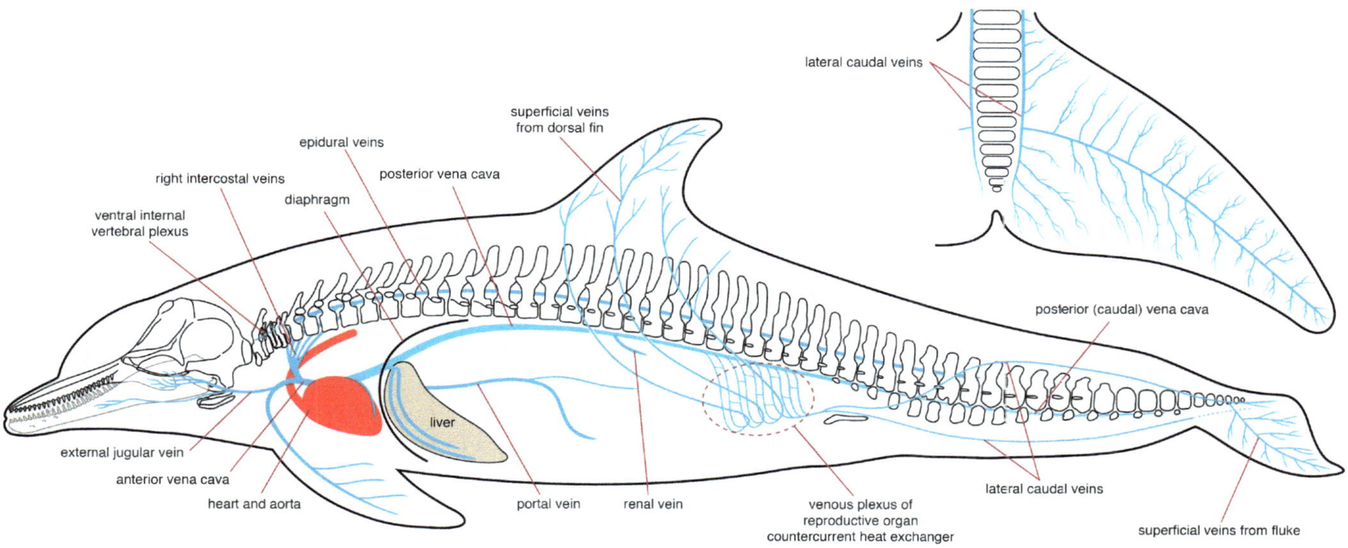

FIG. 4 Venous anatomy in the bottlenose dolphin. The major components of the venous system include the anterior and posterior venae cavae, the paired epidural veins inside the vertebral canal, countercurrent heat exchangers (see Chapter 3: Thermoregulation) in the dorsal fin, flippers, and fluke and around the reproductive organs, and venous plexuses of the head and abdomen (partially illustrated). The anterior vena cava receives venous inflow from the epidural veins via the large intercostal veins, and from the cranial plexi, and central veins from heat exchangers in the flippers and dorsal fin. The epidural veins receive inflow from the brain and dorsal musculature. There are small communicating veins (not illustrated) in the lumbar region between the epidural veins and posterior vena cava. The posterior vena cava receives input from the kidneys, muscle, multiple abdominal structures, including the hepatic veins, and various venous plexi associated with countercurrent heat exchangers in the fluke and in the abdomen. Posterior to the renal vein, the vena cava is paired and has two channels. The posterior vena cava travels inside a diaphragmatic sling as it enters the thorax through the caval hiatus. The portal vein conducts blood from the gastrointestinal tract, spleen, pancreas, and gall bladder to the liver. The upper right insert is a schematic diagram of the superficial venous network of the fluke. Flow through superficial veins in the fluke and dorsal fin returns cooled blood to the abdomen via the lumbo-caudal venous plexus and then into the abdominal heat exchangers around the reproductive organs. Central veins from the fin and fluke heat exchangers (not illustrated) return warm blood more directly to the vena cava. See Harrison and Tomlinson (1956), Rommel et al. (1992, 1993, 1994, 2006), and Lillie et al. (2018) for detailed reviews. *Modified from Harrison and Tomlinson (1956), Rommel et al. (1992, 1993, 1994, 2006), and Lillie et al. (2018) and suggestions from S.A. Rommel, A.M. Costidis, and D.A. Pabst. Illustration by Uko Gorter.*

(Cozzi et al., 2017). The portal vein sphincters described earlier in this section were also seen. These authors have suggested that there may be complex regulation of liver blood flow in these animals during diving. Furthermore, investigation would be beneficial.

Another striking feature of the cetacean venous system is the presence of large epidural veins located ventral to the spinal cord inside the vertebral canal (Harrison and Tomlinson, 1956; Slijper, 1962). These veins are valveless and communicate with the anterior and posterior venae cavae via intercostal and lumbar veins (Fig. 4). In the anterior portion of the vertebral canal, a venous rete drains blood from the brain and eventually links with the epidural veins in the mid—thorax region to empty into the anterior vena cava (Harrison and Tomlinson, 1956). Posteriorly, the epidural veins receive blood from the dorsal musculature through small veins that enter the vertebral canal through the ligamentum flavum (the dorsal ligament between adjacent vertebrae). Blood from the tail fluke does not drain into the epidural veins; rather, blood drains from the tail either via lateral veins beneath the skin or the chevron veins at the bases of the vertebrae into the vena cava. Readers are again referred to Chapter 3 for further review.

The volume of the vasculature within the vertebral canal is markedly increased in deeper-diving cetaceans (Rowlands et al., 2021). In beaked whales, extreme deep divers, vascular volume was 22–35 times greater than spinal cord volume in contrast to the 6–10-fold value in shallower diving dolphins. Remarkably, vascular volume in Cuvier's beaked whale occupied 96% of the volume of the vertebral canal.

Posterior vena caval hemodynamics during fluking

In seals, a striated muscle sphincter around the posterior vena cava at the level of the diaphragm is considered to regulate venous return to the heart (Elsner et al., 1971; Ronald et al., 1977). Such a distinct sphincter around the vena cava does not occur in cetaceans, but there have been observations of diaphragmatic "slings" at the caval hiatus (the opening in the diaphragm through which the vena cava travels, Fig. 4), reports of sphincter-like arrangements of muscle in the diaphragm around the caval hiatus, as well as mention of the absence of a sphincter (Harrison and Tomlinson, 1956; Slijper, 1962; Rowlatt and Gaskin, 1975; Hilton and Gaskin, 1978). This question of a diaphragmatic sphincter in cetaceans led to a recent examination of sphincters in cetaceans and to investigation of the role of such sphincters in the hemodynamics of the posterior vena cava (Lillie et al., 2018).

Lillie et al. (2018) found that diaphragmatic sphincters (annular or V-shaped arrangements of muscle fibers around the caval hiatus) existed in all seven cetacean species examined (harbor porpoise, Dall's porpoise, Pacific white-sided dolphin, beluga, killer whale (*Orcinus orca*), fin whale, and minke whale). In addition, there did not appear to be any structural differences in the sphincters among the odontocetes. This was not expected as the authors had hypothesized that higher intra-abdominal pressures during fluking in faster-swimming cetaceans would require more well-developed sphincters to prevent surges or oscillations in venous return to the heart.

Previously, the authors had found that faster swimmers such as Dall's porpoises, Pacific white-sided dolphins, and harbor porpoises had higher diaphragmatic subserosal collagen contents than the slower cetacean species, and that pinnipeds lacked such collagen in their diaphragms (Lillie et al., 2017). From the correlation of diaphragm collagen content with swim speed, they hypothesized that this increase in collagen content reinforced the diaphragm against deformation due to high intra-abdominal pressures generated by fluking. Pinnipeds, on the other hand, lacked collagen because they did not experience elevated intra-abdominal pressures since they did not fluke. Even in phocid seals, the lateral movements of hindflippers would not generate the intra-abdominal pressures developed in cetaceans by compression of the abdominal cavity during fluking.

These findings led to new hypotheses as to (a) posterior vena caval hemodynamics during fluking, (b) the function of the diaphragmatic sphincter in cetaceans, (c) another role for the intravertebral rete, and (d) differences in the function of the diaphragmatic sphincter in cetaceans versus phocid seals. Basically, the authors examined and modeled how elevations in intra-abdominal pressure during fluking might influence posterior vena caval inflow into the thorax, as well as retrograde flow from the posterior vena cava through the lumbar communicating veins into the epidural veins.

In a hemodynamic model of vena caval blood flow into the thorax, blood flow into the thorax oscillated with each stroke of the tail if intra-abdominal pressure was low (Fig. 5). The return of blood to the heart was irregular with a surge

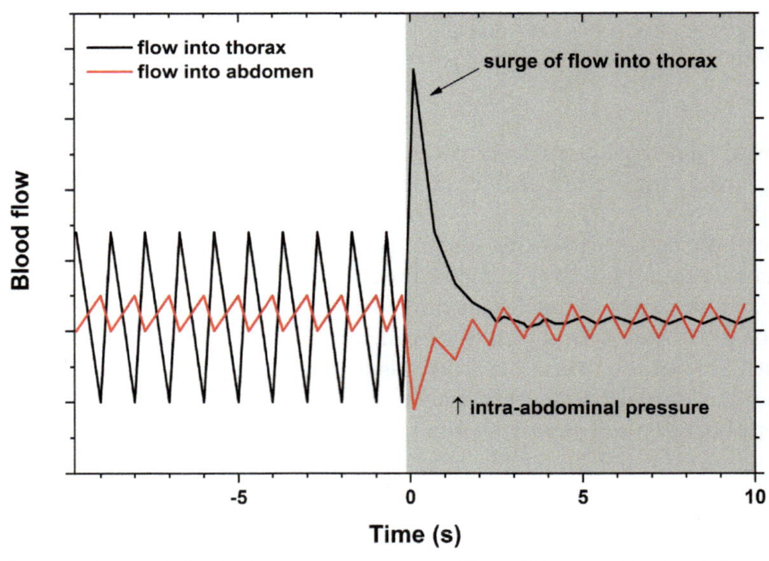

FIG. 5 Hypothesized effect of increased intra-abdominal pressure on vena caval blood flow entering the abdomen from the periphery and exiting the abdomen into the thorax. In the model, fluking is continuous, and intra-abdominal pressure and vena caval blood flow into the thorax oscillate due to fluking if intra-abdominal pressures are low. A caval sphincter could minimize oscillations in venous return to the heart under these conditions. If greater intra-abdominal pressures occur during fluking (time 0), the intra-abdominal segment of the vena cava will compress and dampen the flow oscillations into the thorax. This process has been described as a vascular waterfall or Starling resistor effect. Blood flow into the thorax and heart would be more constant at the higher intra-abdominal pressure even without a sphincter. However, at the onset of an increase in intra-abdominal pressure, there is a transient forward surge of flow into the thorax, and also a backward surge out of the abdomen, most probably into the epidural veins. Here again, a diaphragmatic caval sphincter could minimize this transient forward surge, preventing overdistention of the heart. In regard to retrograde flow into epidural veins, the communicating veins between the vena cava and epidural veins are small, thus offering resistance to such flow. In addition, the spinal/cranial retia mirabilia may protect against any pulsatility or transient retrograde surge of blood via the epidural veins toward the brain. *Modified from Lillie et al. (2018).*

of blood with each stroke. Under these conditions, transient contractions of a caval sphincter could minimize surges into the thorax, providing a more constant return of blood to the heart. However, if increased intra-abdominal pressure (time 0, Fig. 5) occurred with fluking, compression of the intra-abdominal segment of the vena cava dampened the flow oscillations into the thorax. At first glance, this might seem to obviate the need for a diaphragmatic sphincter to regulate venous return to the heart under these conditions. However, there was a large surge of blood flow into the thorax at the initial increase in intra-abdominal pressure (time 0, Fig. 5), and it is this surge that the authors postulate the sphincter contraction can regulate to control blood flow into the thorax. Hence, a diaphragmatic sphincter is present in all cetaceans regardless of swim speed and diaphragm collagen contents. However, the sphincter need not be more developed in faster-swimming species because high intra-abdominal pressure minimizes any surge of flow into the thorax. Equally significant at the onset of the elevated intra-abdominal pressure (time 0, Fig. 5) was the retrograde surge of blood out of the vena cava into the epidural veins and toward the brain. The authors postulate that any retrograde blood flow or pressure oscillations experienced by the brain are minimized by the small size of the lumbar communicating veins and, importantly, by the retia.

In contrast to cetaceans, pinnipeds do not fluke, yet the phocid seals, especially, have well-developed vena caval sphincters. In addition, as reviewed earlier, the phocids can store blood, which sometimes can be completely oxygenated (Meir et al., 2009), in their large hepatic sinuses and posterior venae cavae. So, in phocids in contrast to cetaceans, Lillie et al. (2018) postulated that intra-abdominal pressures may fluctuate with lateral undulations of the hindflippers, but that such intra-abdominal pressures were low, and not elevated. Under those conditions, there would be a continual oscillation in flow into the thorax that could be minimized by transient contractions of a sphincter. In phocids, regulation of venous return to the heart with a vena caval sphincter is postulated not only to optimize cardiac filling from the large venous reservoir behind the diaphragm but also to meter out the large blood O_2 store contained in the hepatic sinuses and posterior venae cavae.

Cardiovascular function during diving

In this section, we first review the cardiovascular dive response and examine the heart rate profiles of some cetaceans during dives. We conclude with consideration of the implications of these heart rate profiles for peripheral blood flow, O_2 store management, and respiratory gas uptake, distribution, and removal during dives.

The dive response

The cardiovascular dive response, which consists of the bradycardia (decreased heart rate) and peripheral vasoconstriction associated with the breath hold, underlies diving physiology and the management of O_2 stores during dives (Ponganis et al., 2011). This response results in the redistribution of blood flow to organs and tissues and the conservation of the blood O_2 store (Scholander, 1940; Irving et al., 1941b; Scholander et al., 1942; Elsner et al., 1966a). In forcibly submerged seals, the dive response was extreme with blood flow essentially restricted to the brain and heart (Scholander, 1940; Scholander et al., 1942; Zapol et al., 1979; Blix et al., 1983). However, in free-diving seals, the bradycardia was often less intense and variable, with the preservation of digestive, renal, and hepatic function during aerobic dives (Jones et al., 1973; Davis et al., 1983; Thompson and Fedak, 1993; Andrews et al., 1997; Davis, 2014).

The intensity and pattern of heart rate reduction may be influenced by many potential factors, including exercise, depth, lung volume changes, and volitional control. Readers are referred to several recent papers and reviews for consideration of the neuroregulation and neuropathways involved in the dive response (Davis and Williams, 2012; Panneton, 2013; Williams et al., 2015b; Elmegaard et al., 2016, 2019; Ponganis et al., 2017; Kaczmarek et al., 2018; Panneton and Gan, 2020; Fahlman et al., 2020a). Because of the complexity of peripheral blood flow measurements, heart rate has often been used as an index of the intensity of the dive response in free-diving and free-swimming animals.

Early studies of heart rates in small- and medium-sized cetaceans

Some of the earliest heart rate studies in cetaceans were conducted with long break-away ECG leads and a benchtop recorder in free-swimming dolphins and porpoises (Scholander, 1940; Irving et al., 1941a). In fragmentary heart rate records, Irving and Scholander observed that both the intensity and pattern of bradycardia were variable during short dives of bottlenose dolphins, but that the reduction in heart rate was minimal in rapidly breathing, surface-swimming harbor porpoises.

In the early 1960s, Elsner and coworkers applied the long ECG lead technique to a Pacific bottlenose dolphin and found that apneic heart rates during spontaneous swimming ranged between 40 and 60 bpm, about a 50% reduction from heart rates associated with breathing (eupnea) (Elsner et al., 1966b). During a trained, stationary, 4.7-min dive, heart rate declined to about 10 bpm, gradually rose to about 30 bpm by the end of the dive, and reached 120 bpm during rapid postdive respirations with minimal decreases in heart rate during the short apneic intervals between breaths.

During the 1960s–70s, Ridgway and Kanwisher utilized both the long ECG lead approach and underwater telemetry to examine heart rate in a variety of species during rest (with slow respiratory rates) and during shallow swimming/dives (Kanwisher and Sundnes, 1965; Ridgway, 1972, 1986; Kanwisher and Ridgway, 1983). They documented that, under these conditions, the oscillation in heart rate occurred with every breath (respiratory arrhythmia) and that the change in heart rate was similar to that reported by Elsner et al. (1966b). Regardless of breath-hold duration, heart rate decreased during apnea by about 50% from the faster heart rate (tachycardia) during eupnea. It was also noted again that after longer dives, the postdive tachycardia was higher and remained high throughout the rapid respiratory cycles until exhaled CO_2 levels returned to baseline. Species included killer whales, pilot whales, belugas, bottlenose dolphins, common dolphins, Pacific white-sided dolphins, Dall's porpoises, harbor porpoises, and Amazon River dolphins (Spencer et al., 1967; Ridgway, 1972). Among mysticetes, a respiratory arrhythmia heart rate pattern was documented in a young gray whale much later with use of a Holter monitor (Ponganis and Kooyman, 1999).

Characteristics of the respiratory arrhythmia heart rate profile in dolphins have been examined in recent post–breath hold investigations of trained animals. These studies suggested that the peak heart rate and magnitude of the respiratory arrhythmia after the breath hold may be related to respiratory rate and other respiratory variables, including tidal volume (Cauture et al., 2019; Blawas et al., 2021a). Similar to heart rate at rest, the peak and minimum heart rates of the respiratory arrhythmia were inversely related to body mass (Blawas et al., 2021b).

Heart rate profiles during diving

Advances in biologging technology have now allowed the documentation of heart profiles during trained and natural dives in several cetacean species. In more recent studies of cetaceans under managed care, the general heart rate profile described above has been confirmed in bottlenose dolphins, Risso's dolphins (*Grampus griseus*), harbor porpoises, killer whales, belugas, and pilot whales during trained stationary breath holds and spontaneous dives of three to 4-min duration, as well as in bottlenose dolphins during trained dives at sea (Williams et al., 1999, 2015b; Noren et al., 2004, 2012; Houser et al., 2010; Davis and Williams, 2012; Elmegaard et al., 2016; Suzuki et al., 2017; McDonald et al., 2018; Bickett et al., 2019; Aoki et al., 2021; Blawas et al., 2021a). Application of the equipment and techniques developed with animals under managed care has now allowed initial investigations of heart rates of wild cetaceans at sea in three species: the narwhal, harbor porpoise, and blue whale (Williams et al., 2017; Goldbogen et al., 2019; McDonald et al., 2021).

During a dive, heart rate is, of course, dependent on body mass, as well as on the nature of a given dive. Heart rate at rest declines as body mass increases (Stahl, 1967). This relationship was reflected in the range of dive heart rate profiles for cetaceans ranging from a 54 kg harbor porpoise to a 70,000 kg blue whale in Fig. 6. In addition to body mass, dive heart rates vary among and within dives because of a variety of factors, including dive duration and depth, dive function (foraging, travel or sleep), locomotory activity, changes in dive expectations, and stressors (Davis and Williams, 2012; Noren et al., 2012; Williams et al., 2015b, 2017; Elmegaard et al., 2016, 2019; Goldbogen et al., 2019; Fahlman et al., 2020a).

Most of what we know about the factors that regulate of heart rate during dives of cetaceans is from trained animals, primarily bottlenose dolphins. Exercise is one such factor. In deeper dives, trained bottlenose dolphins exhibited heart rate profiles that were similar to profiles observed in diving seals and sea lions: rapidly decreasing from predive values of 100–120 beats min^{-1} to lows of 20–30 beats min^{-1}, before increasing again as the dolphin started ascent (Williams et al., 1999; Houser et al., 2010) (Fig. 7). Based on the similarity to heart rate profiles previously documented in shallow, stationary dives (Elsner et al., 1966b), Williams et al. (1999) concluded that the dive response during deep dives dominated over the typical mammalian exercise response of increased heart rate and muscle blood flow.

However, more recently, a correlation between stroke rate and heart rate in short shallow dives led to the hypothesis that bottlenose dolphins exhibit an exercise modulated dive heart rate response (Davis and Williams, 2012; Noren et al., 2012; Williams et al., 2015b). Trained harbor porpoises also exhibited an exercise modulated heart rate response

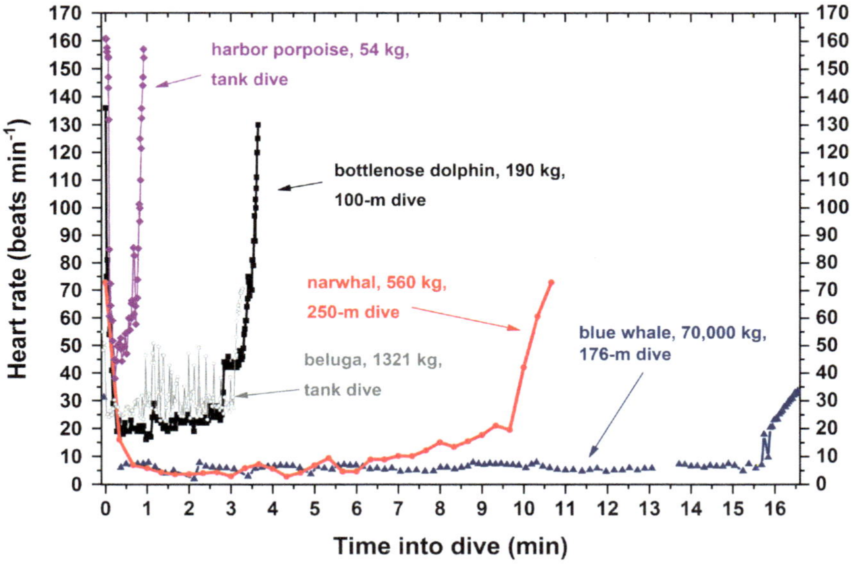

FIG. 6 Heart rate profiles of five cetaceans during spontaneous dives. The common features of the dive heart rate profile include the overall pattern, the variability in beat-to-beat heart rates, and the general effects of body mass on heart rate. Heart rate typically decreases during descent to a level determined by the nature of a dive and then increases during ascent. Pre- and postdive heart rates are elevated and are higher in smaller animals due to the allometric relationship of body mass and heart rate. The intensity and pattern of the bradycardia during the bottom phase of the dive are a function of both the nature of the dive and body mass. For instance, it is expected that heart rate of a beluga during a deep, long dive would be much slower and probably less variable than a shallow, leisurely dive in a tank. *Modified from McDonald et al. (2018), Houser et al. (2010), Bickett et al. (2019), Williams et al. (2017), and Goldbogen et al. (2019).*

during dives that varied in duration and activity (McDonald et al., 2018). The exercise modulated dive response has been proposed to be key to maximizing aerobic dive duration through the efficient balancing of blood and muscle O_2 store depletion (Davis and Kanatous, 1999; Davis and Williams, 2012; Davis, 2014; Williams et al., 2015a,b). In this model, increased heart rates associated with increased stroke rates have been postulated to increase muscle blood flow and O_2 delivery, thus supplementing and prolonging aerobic metabolism in exercising muscle and potentially resulting in simultaneous depletion of blood and muscle O_2 stores.

Measurements of heart rates in wild cetaceans have demonstrated that exercise modulation of heart rate is probably variable and, again, dependent on the nature of a dive (Figs. 6 and 8). In narwhals, heart rate was extremely low and did not correlate with stroke rate during dives immediately postrelease (Williams et al., 2017). In contrast, during dives several hours after release, heart rate was higher (though still low) and it correlated with stroke rate. In the shallow-diving harbor porpoise (Fig. 8), the bradycardia was moderate, relatively stable, and appeared independent of dive type, dive duration, and activity (McDonald et al., 2021). In these spontaneously diving, foraging, and highly active animals, heart rate was not linked with activity. In a blue whale (Fig. 8), heart rates were consistently below allometrically predicted resting heart rates in both foraging and nonforaging dives (Goldbogen et al., 2019). However, heart rate did transiently increase during feeding lunges, consistent with an increase in heart rate because of exercise. These three investigations are a beginning. Furthermore, evaluation of the regulation of heart rate in cetaceans at sea and better understanding of the role of the cardiovascular response in foraging ecology and stress responses require more research and further advances in technology.

Heart rate profiles during postdive intervals and during high-speed surface travel

We have already mentioned that the typical apneic decline in heart rate during slow breathing at rest did not occur during the rapid breathing of the immediate postdive period of the bottlenose dolphin or during the rapid breathing of a fast surface-swimming harbor porpoise. Sustained high heart rates presumably optimize gas exchange during these conditions of high respiratory activity (Fahlman et al., 2018).

As illustrated in Fig. 7A and B, high heart rates were maintained during surface intervals between serial 100-m dives of a trained bottlenose dolphin while diving heart rates declined to about 20 bpm (Houser et al., 2010). After the dive

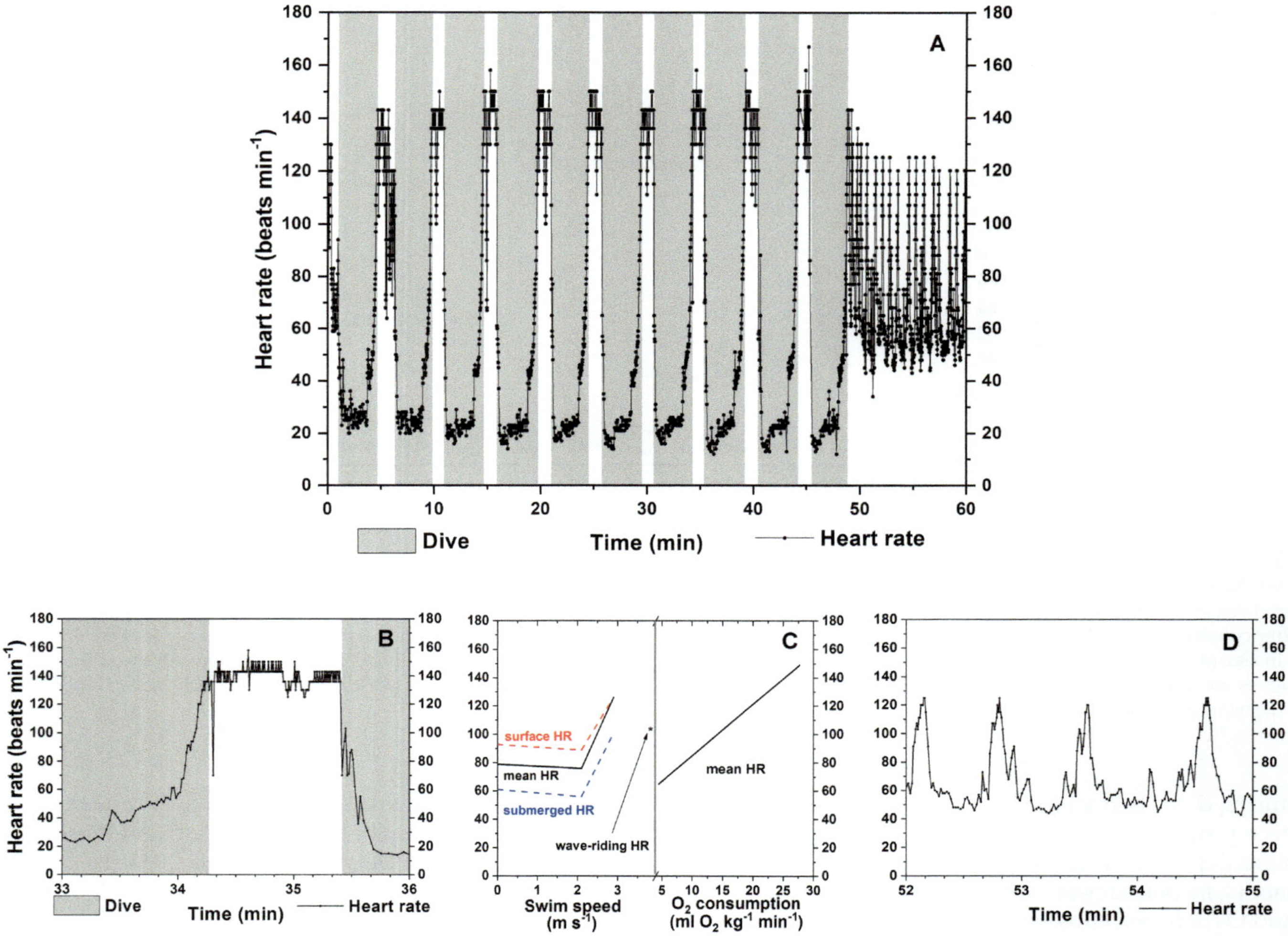

FIG. 7 Heart rates in the bottlenose dolphin. (A) Heart rate profiles of a bottlenose dolphin during serial trained dives to 100-m depth and during the post-dive-bout period when the dolphin underwent ultrasound exams while beached. Heart rate was 140–150 beats min^{-1} (bpm) during surface intervals between dives, reached 30–40 bpm during the bottom phases of the dives, and demonstrated a consistent increase during ascent. (B) 3-min heart rate profile from A illustrates the variability in heart rate both during the dives and during the surface interval. However, despite respirations during the surface interval, the typical respiratory sinus arrhythmia (elevated heart rate with inspiration, decreased heart rate during the breath hold) was not evident. Heart rate remained high throughout the surface interval. (C) During surface swimming of a bottlenose dolphin, the difference in heart rate between inspiration and the breath hold (sinus arrhythmia) gradually disappeared as swim speed increased. Both surface and submerged heart rates increased with swim speed and merged at the highest swim speeds. Heart rate increased with exertion in exercise studies. At maximum oxygen consumption, heart rate was 140 bpm, similar to that during surface intervals between dives (A, B). During wave-riding, heart rate was lower at higher speeds, consistent with a lower metabolic rate during wave-riding. (D) 3-min heart rate profile during post-dive-bout period in A demonstrated a typical sinus arrhythmia pattern associated with breathing in contrast to that during the surface intervals between dives. *Modified from Houser et al. (2010) and Williams et al. (1992, 1993).*

session, similar to the earlier observations of Elsner and Ridgway, eupneic and apneic heart rates gradually declined, and the typical 50% difference between apneic and eupneic heart rate returned (Fig. 7A and D). Sustained high heart rates during surface intervals between dives have also been documented in the harbor porpoise, one of the smallest cetaceans, and in the blue whale, the largest (McDonald et al., 2018, 2021; Goldbogen et al., 2019).

In addition to Scholander's early observations in the harbor porpoise, current evidence also indicates that high heart rates are maintained during high-speed surface travel of dolphins. Elevated eupneic and apneic heart rates merged in trained bottlenose dolphins both during high-speed surface swimming at sea and during maximal exercise studies where dolphins swam against a load cell (Fig. 7C) (Williams et al., 1992, 1993). In contrast to diving, the physiology of exercise during high-speed surface travel of dolphins and porpoises appears more similar to that of terrestrial

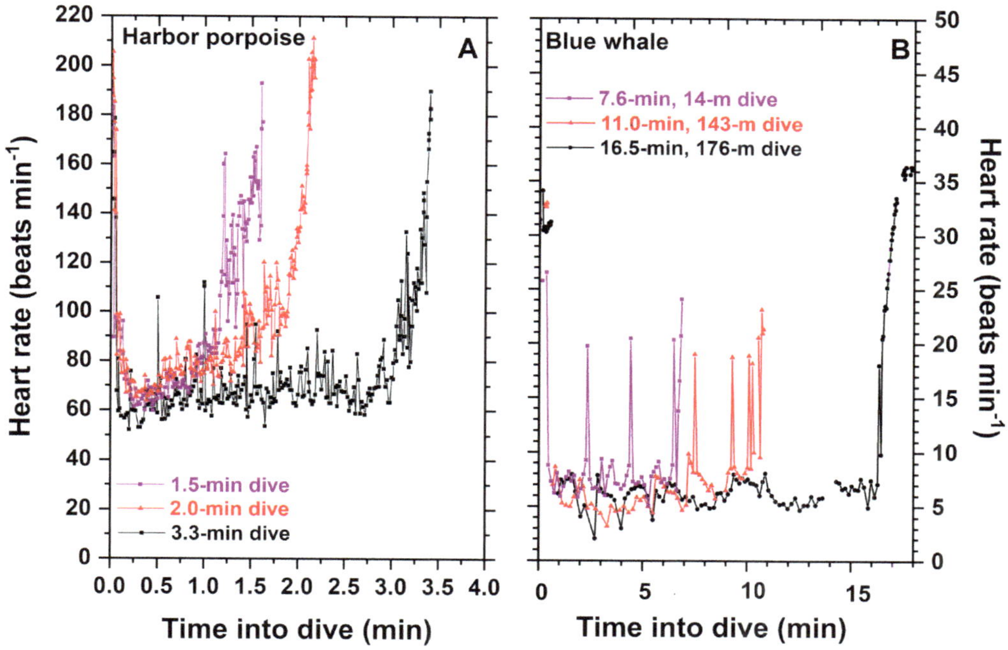

FIG. 8 Harbor porpoise and blue whale heart rates during dives at sea. Heart rate in diving cetaceans is affected by both body mass and the nature of a dive as evidenced by these heart rate profiles of one of the smallest porpoises and the largest whale. (A) Heart rate in the harbor porpoise ranges from 50 to 80 beats min^{-1} (bpm) during the bottom phase of the dive and is near 190–220 bpm while at the surface. (B) In contrast, in the blue whale, heart rate is typically 2–10 bpm during the dive and reaches 25–35 bpm on the surface. There is considerable variation in heart rate in both species during the dive, often including abrupt one to three-beat oscillations in heart rate. Feeding lunges in blue whales were typically associated with transient increases in heart rate of two to five bpm (most evident in the longest dive). All dives were associated with an increase in heart rate during ascent. Gaps in the blue whale heart rate profiles were secondary to artifact in the ECG record. *Modified from McDonald et al. (2021) and Goldbogen et al. (2019).*

mammals (Taylor et al., 1987) with maximization of O_2 transport from lungs to muscle supported by high heart rates and respiratory rates. The increased heart masses of faster-swimming species, and even the increased affinities of Hb for O_2 in the faster species are consistent with this hypothesis (Ridgway and Johnston, 1966; Horvath et al., 1968).

Heart rate and oxygen store management

Review of the role of heart rate and the dive response in O_2 store management and diving physiology must take into consideration the species and type of dive performed. The bottlenose dolphin and many shallow-diving odontocetes have larger lung masses (possibly a greater respiratory O_2 store) than deeper-diving species (Piscitelli et al., 2010). From available data, the coastal form of the bottlenose dolphin also has a lower Hb content, smaller blood volume, lower myoglobin concentration, and smaller relative muscle mass (Ridgway and Johnston, 1966; Mallette et al., 2016; Pabst et al., 2016). Based on (a) the coastal bottlenose dolphin's lower mass-specific O_2 store with a greater percentage of O_2 in the lungs, and (b) its predominant shallow (<20 m) diving (above the expected depth of alveolar collapse and lack of pulmonary gas exchange (Ridgway and Howard, 1979; Mate et al., 1995)), a less intense bradycardia and greater transport of O_2 from the lung to muscle would be expected during shallow dives and subsurface swims. Indeed, the very low exhaled lung O_2 fractions of a bottlenose dolphin trained to actively swim between divers at 20-m depth, as well as the elevated heart rates during trained underwater swims at shallow depths, suggest that this is the case (Ridgway et al., 1969; Noren et al., 2012). On the other hand, during deep dives near or below the depth of alveolar collapse, one would expect a more intense bradycardia, decreased muscle blood flow, and greater reliance on myoglobin-bound O_2 stores in muscle. This is again supported by the lower heart rates during deeper dives of bottlenose dolphins (Williams et al., 1999; Houser et al., 2010).

Different heart rate profiles will affect cardiac output and degree of peripheral vasoconstriction during different types of dives. These changes in the magnitude and distribution of blood flow will not only affect O_2 store management but will also affect the uptake and distribution of other respiratory gases (CO_2, nitrogen). Readers are referred to Chapter 6: Respiratory physiology in the dolphin and other whales and Chapter 7: Diving physiology in dolphins and human for further review of respiratory and diving physiology.

Blood O_2 transport

One of the primary functions of the cardiovascular system is the transport of O_2 to tissues. The higher the blood O_2 content is, the greater the delivery of O_2 to the periphery with every heartbeat is. Here, we review the primary determinants of blood O_2 content, hemoglobin (Hb) concentration, and the oxygen-hemoglobin (O_2-Hb) dissociation curve. Compared with the Hb-bound O_2, the amount of dissolved O_2 is quite low. Readers are referred to Chapter 6 for the details of the uptake and distribution of respiratory gases.

Hemoglobin concentration

Hemoglobin concentrations in bottlenose dolphins ranged from about 14–15 g dL^{-1} in coastal forms to 18.5 g dL^{-1} in deeper diving, offshore forms (Ridgway and Johnston, 1966; Duffield et al., 1983). For comparison, human standard values are 12–16 g dL^{-1} (Dominelli et al., 2021). In general, Hb concentrations were higher in cetaceans that dive deeper and/or that have higher metabolic rates whether because of higher routine swim speeds, foraging activity, or thermoregulatory demand (Ridgway and Johnston, 1966). Although available data are limited and sometimes conflicting (Table 4), the highest values (20–23 g dL^{-1}) among odontocetes occurred in documented or suspected deep-diving

TABLE 4 Hemoglobin concentration ([Hb]) and P_{50} (partial pressure of O_2 at which Hb is 50% saturated) in dolphins and other cetaceans.

Species	[Hb] ($g\,dL^{-1}$)	P_{50} (mmHg)	Reference
Bottlenose dolphin (coastal) *Tursiops truncatus*	14.4	24.6, 26.8	Horvath et al. (1968), Ridgway and Johnston (1966), Lenfant (1969)
Bottlenose dolphin (offshore) *Tursiops truncatus gilli*	18.5	26.0	Duffield et al. (1983), Lenfant (1969)
Common dolphin *Delphinus delphis*	18	21.6	Horvath et al. (1968), Sharp et al. (2014)
Pacific white-sided dolphin *Lagenorhynchus obliquidens*	17.0	20.2, 24.8	Horvath et al. (1968), Lenfant (1969), Ridgway and Johnston (1966)
Dall's porpoise *Phocoenoides dalli*	20.3	19.1	Horvath et al. (1968), Ridgway and Johnston (1966)
Beluga *Delphinapterus leucas*	20.9–23.0	24.2	Choy et al. (2019), Dhindsa et al. (1974), Ridgway et al. (1984)
Harbor porpoise *Phocoena phocoena*	19, 20	24.5	Green and Redfield (1933), Koopman et al. (1999), Reed et al. (2000)
Amazon dolphin *Inia geoffrensis*	14.3	25	Lenfant (1969)
Sperm whale *Physeter macrocephalus*	15.1–15.8, 22	26.5	Lenfant (1969), Green and Redfield (1933), Tawara (1950)
Gray whale *Eschrichtius robustus*	14.0	30.0	Lenfant (1969)
Risso's dolphin *Grampus griseus*	21.4,	30.5	Lenfant (1969)
Killer whale *Orcinus orca*	16.0–18.0	30.7, 25.2	Dhindsa et al. (1974), Lenfant et al. (1968), Ridgway (1972)
Pilot whale *Globicephala macrorhynchus*, *G. melas*	15.1–16.5, 19.6	31	Lenfant (1969), Medway and Moldovan (1966), Ridgway et al. (1970), Velten et al. (2013)
Rough-toothed dolphin *Steno bredanensis*	13.7–15.5		Manire et al. (2018)
Yangtze finless porpoise *Neophocaena asiaeorientalis*	14.3–17.1		Nabi et al. (2017)

TABLE 4 Hemoglobin concentration ([Hb]) and P_{50} (partial pressure of O_2 at which Hb is 50% saturated) in dolphins and other cetaceans—cont'd

Species	[Hb] ($g\,dL^{-1}$)	P_{50} (mmHg)	Reference
Spotted dolphin *Stenella attenuata*	17		St. Aubin et al. (2013)
Spinner dolphin *Stenella longirostris*	18		Suarez et al. (2019)
Commerson's dolphin *Cephalorhynchus commersoni*	18		Hedrick and Duffield (1991)
Fraser's dolphin *Lagenodelphis hosei*	20.2–22.1		Clark et al. (2002), Rhinehart et al. (1995)
Striped dolphin *Stenella coeruleoalba*	20.9		Gales (1992)
Narwhal *Monodon monoceros*	22.5		Vogl and Fisher (1982)
Beaked whale *Mesoplodon* sp.	22.9		Velten et al. (2013)
Blue whale *Balaenoptera musculus*	9.6		Laurie (1933)
Fin whale *Balaenoptera physalus*	9.6		Laurie (1933)
Sei whale *Balaenoptera borealis*	15.6		Tawara (1950)
Bowhead whale *Balaena mysticetus*	20.5		Castellini et al. (2006)

Note: 1 mm Hg = 0.133 kPa.

species such as beaked whales, sperm whales, narwhals, belugas, Dall's porpoises, Risso's dolphins, Fraser's dolphins (*Lagenodelphis hosei*), and striped dolphins. The high Hb content of the harbor porpoise has been attributed to its small size and thermoregulatory metabolic demands. Hemoglobin concentration in pilot whales, relatively active deep divers (Aguilar De Soto et al., 2008; Aoki et al., 2017), was high in one study but was similar to that of the coastal bottlenose dolphin in three other studies (Table 4). In comparison, much higher Hb concentrations (25–27 $g\,dL^{-1}$) have been reported in some phocid seals (Ponganis, 2011).

Relatively, low Hb concentrations of 10–15 $g\,dL^{-1}$ have been reported in several species of rorqual whales, while bowhead whale Hb content was near 20 $g\,dL^{-1}$ (see Table 4). Very few data were available for mysticetes. However, these differences in Hb concentrations between rorquals and the bowhead paralleled similar differences in the concentrations of muscle myoglobin, the other major O_2-binding protein in the body (Tawara, 1950; Lawrie, 1953; Hochachka and Foreman, 1993; Noren and Williams, 2000; Cartwright et al., 2016).

O_2-Hb dissociation curves

A single Hb molecule can bind up to four O_2 molecules; at full saturation of Hb, there are 1.34 mL $O_2\,g^{-1}$ Hb in blood (Dominelli et al., 2021). The sigmoidally shaped O_2-Hb dissociation curve reveals the degree of Hb saturation as a function of the partial pressure of O_2 (P_{O_2}) in blood (Fig. 8). The P_{50} (P_{O_2} at which Hb is 50% saturated) is an index of the affinity of Hb for O_2; as affinity increases, the P_{50} decreases (Dominelli et al., 2021). The O_2 affinity of Hb can reversibly change as a function of blood temperature, CO_2, and pH, as well as by changes in intracellular organophosphates (Storz, 2016). Hemoglobin's affinity for O_2 decreases with increased temperature or acidity, and increases with lowered temperature and alkalosis. Thus, the O_2-Hb dissociation curve can shift to the left (increased affinity, lower P_{50}) or to the right (decreased affinity, higher P_{50}). Decreased Hb-O_2 affinity promotes the release of O_2 from Hb in the periphery, while increased affinity enhances uptake of O_2 by Hb in the lung. In many mammals and birds exposed to chronic

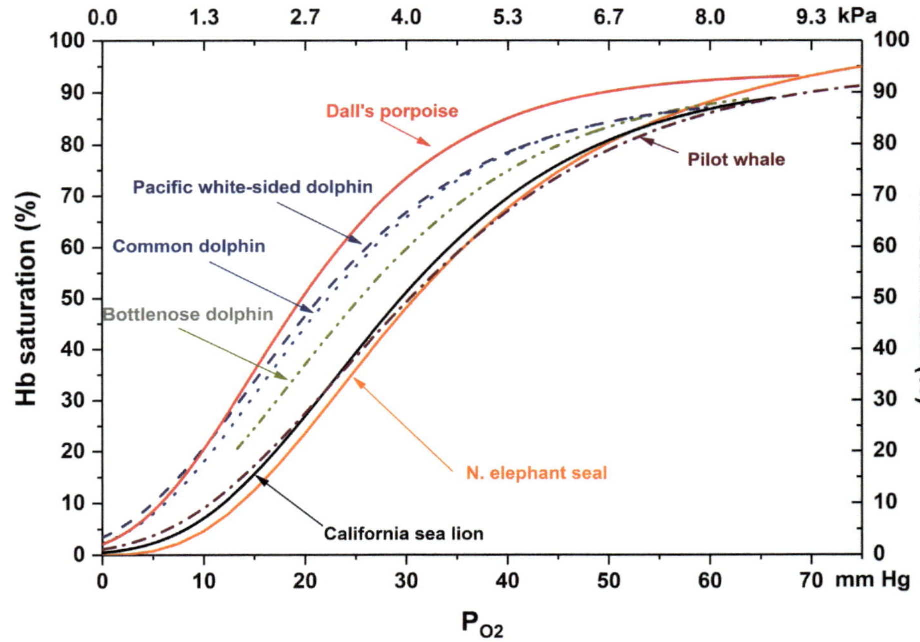

FIG. 9 Increased hemoglobin-oxygen (Hb-O_2) affinities in four small pelagic cetaceans. Increased affinities of Hb for O_2 in the dolphins and porpoises are indicated by the left shift of their oxygen-hemoglobin (O_2-Hb) dissociation curves relative to those of the pilot whale, the sea lion, and the premier pinniped diver, the elephant seal (Horvath et al., 1968; Lenfant, 1969; Meir et al., 2009; McDonald and Ponganis, 2013). See Table 4 for a comparison of oxygen affinities of Hb of other cetaceans. Abbreviations: *Hb*, hemoglobin; P_{O_2}, partial pressure of O_2. *Modified from Horvath et al. (1968), Lenfant (1969), McDonald and Ponganis (2013), and Meir et al. (2009).*

hypoxia (i.e., high altitude or burrows), Hb-O_2 affinity is elevated secondary to genetic mutations and changes in the primary structure of Hb (Storz et al., 2010; Storz, 2016; Storz and Scott, 2019).

Relative to other cetaceans, humans, and several pinnipeds, Hb-O_2 affinity is markedly increased (P_{50} decreased) in several small, fast-swimming pelagic cetaceans (see Table 4, Fig. 9). The P_{50} was 19–22 mmHg (2.53–2.93 kPa) in Dall's porpoise, the Pacific white-sided dolphin, and the common dolphin (Horvath et al., 1968). The P_{50} of the bottlenose dolphin (24 mmHg, 3.20 kPa) was intermediate, but still less than the standard value of 27 mmHg (3.60 kPa) in humans (Dominelli et al., 2021) and less than the P_{50}s of sea lions and elephant seals (see Fig. 9). In larger dolphins and whales, the P_{50} was equivalent or greater than that in the bottlenose dolphin and sometimes even greater than in humans (see Table 4, Fig. 8). In other words, in contrast to the relationship of Hb affinity and body mass in terrestrial mammals (Schmidt-Neilsen and Larimer, 1958; Lahiri, 1975; Snyder, 1983), the affinity of Hb was less in the larger whales than in the smaller dolphin species. In the only study of Hb-O_2 affinity in a beaked whale, the O_2 affinities of isolated Hb fractions were considered slightly higher than in human Hb; P_{50} values were not specifically reported (Vedvick and Itano, 1976).

Increased Hb-O_2 affinity is considered advantageous in hypoxic situations in that it enhances blood O_2 uptake in the lung and increases Hb saturation at low P_{O_2} values (Dominelli et al., 2021). Shifts in Hb-O_2 affinity due to changes in pH, CO_2, and temperature in the periphery contribute to adequate O_2 delivery (Schumacker et al., 1985, 1987; Dominelli et al., 2021). The range of Hb-O_2 affinities in marine mammals, including deep divers, suggests that an elevation in Hb-O_2 affinity as high as in Dall's porpoise is not essential to diving capacity.

It has been suggested that elevated Hb-O_2 affinity facilitates O_2 uptake from the relatively large lung O_2 stores of shallow-diving dolphins and porpoises (Snyder, 1983). In addition to a role during shallow dives, the increased Hb-O_2 affinity may also be important in facilitating O_2 uptake during the rapid subsurface swimming activities of Dall's porpoise and other small pelagic dolphins.

Summary

The design of the cardiovascular system in cetaceans follows the general mammalian model but is modified according to the needs of a given species. Remarkable structures, developed to varying degrees in different species, include

(a) a compliant, elastic ascending aorta and aortic arch; (b) the retia mirabilia; (c) venous plexuses in the head and tracheo-bronchial tree; (d) prominent epidural veins; (e) diaphragmatic caval slings; and (f) thermoregulatory plexuses. During high-speed surface and subsurface swimming of more active porpoises and dolphins, high heart rates, enlarged hearts, and higher Hb-O_2 affinities are postulated to maximize lung-to-muscle O_2 transport just as during exercise of terrestrial mammals. However, during dives, the cardiovascular dive response (bradycardia and vasoconstriction) is variable and serves to regulate the depletion of O_2 stores according to the nature of a given dive. Depth, dive duration, exercise, lung volume reflexes, and cognitive control have all been postulated to influence heart rate during the dive.

References

Aguilar De Soto, N., Johnson, M.P., Madsen, P.T., Diaz, F., Dominguez, I., Brito, A., Tyack, P., 2008. Cheetahs of the deep sea: deep foraging sprints in short-finned pilot whales off Tenerife (Canary Islands). J. Anim. Ecol. 77, 936–947.

Andrews, R.D., Jones, D.R., Williams, J.D., Thorson, P.H., Oliver, G.W., Costa, D.P., Le Boeuf, B.J., 1997. Heart rates of northern elephant seals diving at sea and resting on the beach. J. Exp. Biol. 200, 2083–2095.

Aoki, K., Sato, K., Isojunno, S., Narazaki, T., Miller, P.J.O., 2017. High diving metabolic rate indicated by high-speed transit to depth in negatively buoyant long-finned pilot whales. J. Exp. Biol. 220, 3802–3811.

Aoki, K., Watanabe, Y., Inamori, D., Funasaka, N., Sakamoto, K.Q., 2021. Towards non-invasive heart rate monitoring in free-ranging cetaceans: a unipolar suction cup tag measured the heart rate of trained Risso's dolphins. Proc. R. Soc. Lond. B 376, 20200225.

Åstrand, P.-O., Cuddy, T.E., Saltin, B., Stenberg, J., 1964. Cardiac output during submaximal and maximal work. J. Appl. Physiol. 19, 268–274.

Barnett, C.H., Harrison, R.J., Tomlinson, J.D.W., 1958. Variations in the venous systems of mammals. Biol. Rev. 33, 442–487.

Best, R.C., Da Silva, V.M.F., 1989. Amazon river dolphin, boto *Inia geoffrensis* (de Blainville, 1817). In: Ridgway, S.H., Harrison, R.J. (Eds.), Handbook of Marine Mammals. River Dolphins and the Larger Toothed Whales, vol. 4. Academic Press, San Diego, CA, pp. 1–24.

Bickett, N.J., Tift, M.S., St. Leger, J., Ponganis, P.J., 2019. Heart rates, heart rate profiles, and electrocardiograms in three killer whales, a beluga, and a pilot whale: an exploratory investigation. Mar. Mamm. Sci. 35, 1112–1132.

Bisaillon, A., 1982. Notes on the anatomy of the heart in the harp seal (*Pagophilus groenlandicus* Erxleben, 1777). Cells Tissues Organs 114, 177–184.

Bisaillon, A., Martineau, D., St-Pierre, M.A., 1987. Anatomy of the heart of the beluga whale (*Delphinapterus leucas*). J. Morphol. 191, 89–100.

Bisaillon, A., Martineau, D., St-Pierre, M.A., Beland, P., 1988. Arterial and venous vasculature of the heart of the beluga whale (*Delphinapterus leucas*). J. Morphol. 195, 305–312.

Bishop, C.M., 1997. Heart mass and the maximum cardiac output of birds and mammals: implications for estimating the maximum aerobic power input of flying animals. Philos. Trans. R. Soc. Lond. B. 352, 447–456.

Blawas, A.M., Nowacek, D.P., Allen, A.S., Rocho-Levine, J., Fahlman, A., 2021a. Respiratory sinus arrhythmia and submersion bradycardia in bottlenose dolphins (*Tursiops truncatus*). J. Exp. Biol. 224, jeb234096.

Blawas, A.M., Nowacek, D.P., Rocho-Levine, J., Robeck, T.R., Fahlman, A., 2021b. Scaling of heart rate with breathing frequency and body mass in cetaceans. Philos. Trans. R. Soc. Lond. B. 376, 20200223.

Blix, A.S., Elsner, R., Kjekshus, J.K., 1983. Cardiac output and its distribution through capillaries and A-V shunts in diving seals. Acta Physiol. Scand. 118, 109–116.

Blix, A.S., Walloe, L., Messelt, E., 2013. On how whales avoid decompression sickness and why they sometimes strand. J. Exp. Biol. 216, 3385–3387.

Bloch, D., Lockyer, C., Zachariassen, M., 1993. Age and Growth Parameters of the Long-Finned Pilot Whale off the Faroe Islands. International Whaling Commission Report, Special Issue, 14, pp. 163–208.

Bonato, M., Bagnoli, P., Centelleghe, C., Maric, M., Brocca, G., Mazzariol, S., Cozzi, B., 2019. Dynamics of blood circulation during diving in the bottlenose dolphin (*Tursiops truncatus*): the role of the retia mirabilia. J. Exp. Biol. 222, jeb198457.

Brody, S., 1945. Bioenergetics and Growth: With Special Reference to the Efficiency Complex in Domestic Animals. Reinhold Publishing Corporation, Oxford.

Bryden, M.M., 1972. Growth and development of marine mammals. In: Harrison, R.J. (Ed.), Functional Anatomy of Marine Mammals, I. Academic Press, London, pp. 1–79.

Campens, L., Demulier, L., De Groote, K., Vandekerckhove, K., De Wolf, D., Roman, M.J., Devereux, R.B., De Paepe, A., De Backer, J., 2014. Reference values for echocardiographic assessment of the diameter of the aortic root and ascending aorta spanning all age categories. Am. J. Cardiol. 114, 914–920.

Cartwright, R., Newton, C., West, K.M., Rice, J., Niemeyer, M., Burek, K., Wilson, A., Wall, A.N., Remoinda-Bennett, J., Tejeda, A., Messi, S., Marcial-Hernandez, L., 2016. Tracking the development of muscular myoglobin stores in mysticete calves. PLoS One 11 (1), e0145893.

Castellini, J.M., Castellini, M.A., 1993. Estimation of splenic volume and its relationship to long-duration apnea in seals. Physiol. Zool. 66, 619–627.

Castellini, M.A., Ponganis, P.J., 2020. Cardiovascular and pulmonary systems. In: George, J.C., Thewissen, J.G.M. (Eds.), The Bowhead Whale *Balaena mysticetus*: Biology and Human Interactions. Academic Press, London, pp. 225–236.

Castellini, M.A., Elsner, R., Baskurt, O.K., Wenby, R.A., Meiselman, H.J., 2006. Blood rheology of Weddell seals and bowhead whales. Biorheology 43, 57–69.

Cauture, F., Sterba-Boatwright, B., Rocho-Levine, J., Harms, C., Miedler, S., Fahlman, A., 2019. Using respiratory sinus arrhythmia to estimate inspired tidal volume in the bottlenose dolphin (*Tursiops truncatus*). Front. Physiol. 10.

Cave, A.J.E., 1977. The coronary vasculature of the bottlenose dolphins (*Tursiops truncatus*). In: Harrison, R.J. (Ed.), Functional Anatomy of Marine Mammals. Academic Press, London, pp. 199–215.

Chetboul, V., Lichtenberger, J., Mellin, M., Mercera, B., Hoffmann, A.-C., Chaix, G., Trehiou-Sechi, E., Misbach, C., Petit, A., Lefebvre, H.P., Gaide, N., Tissier, R., Delfour, F., 2012. Within-day and between-day variability of transthoracic anatomic M-mode echocardiography in the awake bottlenose dolphin (*Tursiops truncatus*). J. Vet. Cardiol. 14, 511–518.

Choy, E.S., Campbell, K.L., Berenbrink, M., Roth, J.D., Loseto, L.L., 2019. Body condition impacts blood and muscle oxygen storage capacity of free-living beluga whales (*Delphinapterus leucas*). J. Exp. Biol. 222, jeb191916.

Clark, L.S., Cowan, D.F., Worthy, G.J., Haubold, E.M., 2002. An anatomical and pathological examination of the first recorded stranding of a Fraser's dolphin (*Lagenodelphis hosei*) in the Northwestern Gulf of Mexico. Gulf Mexico Sci. 20, 38.

Clarke, M.R., 1978. Structure and proportions of the spermaceti organ in the sperm whale. J. Mar. Biol. Assoc. 58, 1–17.

Costidis, A., Rommel, S.A., 2012. Vascularization of air sinuses and fat bodies in the head of the Bottlenose dolphin (*Tursiops truncatus*): morphological implications on physiology. Front. Physiol. 3.

Costidis, A.M., Rommel, S.A., 2016a. The extracranial arterial system in the heads of beaked whales, with implications on diving physiology and pathogenesis. J. Morphol. 277, 5–33.

Costidis, A.M., Rommel, S.A., 2016b. The extracranial venous system in the heads of beaked whales, with implications on diving physiology and pathogenesis. J. Morphol. 277, 34–64.

Cowan, D.F., 1966. Observations on the pilot whale *Globicephala melaena*: organ weight and growth. Anat. Rec. 155, 623–628.

Cozzi, B., Bagnoli, P., Acocella, F., Costantino, M.L., 2005. Structure and biomechanical properties of the trachea of the striped dolphin *Stenella coeruleoalba*: evidence for evolutionary adaptations to diving. Anat. Rec. 284, 500–510.

Cozzi, B., Huggenberger, S., Oelschläger, H., 2017. Dolphin Anatomy. Academic Press, London.

Crile, G., Quiring, D.P., 1940. A record of the body weight and certain organ and gland weights of 3690 animals. Ohio J. Sci. 40, 219–259.

Da Silva, V.M.F., Best, R.C., 1994. *Sotalia fluviatilis* (Gervais, 1853). In: Ridgway, S.H., Harrsion, R. (Eds.), Handbook of Marine Mammals. The First Book of Dolphins, vol. 5. Academic Press, San Diego, pp. 43–69.

Davenport, J., Fraher, J., Fitzgerald, E., McLaughlin, P., Doyle, T., Harman, L., Cuffe, T., Dockery, P., 2009. Ontogenetic changes in tracheal structure facilitate deep dives and cold water foraging in adult leatherback sea turtles. J. Exp. Biol. 212, 3440–3447.

Davenport, J., Cotter, L., Rogan, E., Kelliher, D., Murphy, C., 2013. Structure, material characteristics and function of the upper respiratory tract of the pygmy sperm whale. J. Exp. Biol. 216, 4639–4646.

Davis, R.W., 2014. A review of the multi-level adaptations for maximizing aerobic dive duration in marine mammals: from biochemistry to behavior. J. Comp. Physiol. B. 184, 23–53.

Davis, R.W., Kanatous, S.B., 1999. Convective oxygen transport and tissue oxygen consumption in Weddell seals during aerobic dives. J. Exp. Biol. 202, 1091–1113.

Davis, R.W., Williams, T.M., 2012. The marine mammal dive response is exercise modulated to maximize aerobic dive duration. J. Comp. Physiol. A. 198, 583–591.

Davis, R.W., Castellini, M.A., Kooyman, G.L., Maue, R., 1983. Renal GFR and hepatic blood flow during voluntary diving in Weddell seals. Am. J. Physiol. 245, R743–R748.

Dhindsa, D.S., Metcalfe, J., Hoversland, A.S., Hartman, R.A., 1974. Comparative studies of the respiratory functions of mammalian blood X. Killer whale (*Orcinus orca* linnaeus) and beluga whale (*Delphinapterus leucas*). Respir. Physiol. 20, 93–103.

Díaz-Delgado, J., Fernández, A., Xuriach, A., Sierra, E., Bernaldo De Quirós, Y., Mompeo, B., Pérez, L., Andrada, M., Marigo, J., Catão-Dias, J.L., Groch, K.R., Edwards, J.F., Arbelo, M., 2016. Verminous arteritis due to *Crassicauda sp.* in Cuvier's beaked whales (*Ziphius cavirostris*). Vet. Pathol. 53, 1233–1240.

Dominelli, P.B., Wiggins, C.C., Roy, T.K., Secomb, T.W., Curry, T.B., Joyner, M.J., 2021. The oxygen cascade during exercise in health and disease. Mayo Clin. Proc. 96, 1017–1032.

Drabek, C.M., 1975. Some anatomical aspects of the cardiovascular system of Antarctic seals and their possible functional significance in diving. J. Morphol. 145, 85–105.

Drabek, C.M., 1977. Some anatomical and functional aspects of seal hearts and aortae. In: Harrison, J.R. (Ed.), Functional Anatomy of Marine Mammals. Academic Press, San Francisco, CA, pp. 217–234.

Drabek, C.M., Burns, J.M., 2002. Heart and aorta morphology of the deep-diving hooded seal (*Cystophora cristata*). Can. J. Zool. 80, 2030–2036.

Duffield, D.A., Ridgway, S.H., Cornell, L., 1983. Hematology distinguishes coastal and offshore forms of dolphins (*Tursiops*). Can. J. Zool. 61, 930–933.

E Silva, F.M.D.O., Carvalho, V.L., Guimarães, J.P., Vergara-Parente, J.E., De Meirelles, A.C.O., Marmontel, M., Miglino, M.A., 2014. Accessory spleen in cetaceans and its relevance as a secondary lymphoid organ. Zoomorphology 133, 343–350.

Elmegaard, S.E., Johnson, M., Madsen, P.T., McDonald, B.I., 2016. Cognitive control of heart rate in diving harbor porpoises. Curr. Biol. 26, R1167–R1176.

Elmegaard, S.L., McDonald, B.I., Madsen, P.T., 2019. Drivers of the dive response in trained harbour porpoises (*Phocoena phocoena*). J. Exp. Biol. 222, jeb208637.

Elsner, R.W., Scholander, P.F., Craig, A.B., Dimond, E.G., Irving, L., Pilson, M., Johansen, K., Bradstreet, E., 1964. A venous blood oxygen reservoir in the diving elephant seal. Physiologist 7, 124.

Elsner, R., Franklin, D.L., Van Citters, R.L., Kenney, D.W., 1966a. Cardiovascular defense against asphyxia. Science 153, 941–949.

Elsner, R., Kenney, D.W., Burgess, K., 1966b. Bradycardia in the trained dolphin. Nature 212, 407.

Elsner, R., Hanafee, W.N., Hammond, D.D., 1971. Angiography of the inferior vena cava of the harbor seal during simulated diving. Am. J. Physiol. 220, 1155–1157.

Elsner, R., Pirie, J., Kenney, D.D., Schemmer, S., 1974. Functional circulatory anatomy of the cetacean appendages. In: Harrison, R.J. (Ed.), Functional Anatomy of Marine Mammals. vol. 2. Academic Press, New York, pp. 143–159.

Elsner, R., Geoerge, J.C., O'Hara, T., 2004. Vasomotor responses of isolated peripheral blood vessels from bowhead whales: thermoregulatory implications. Mar. Mamm. Sci. 20, 546–553.

Fahlman, A., Jensen, F.H., Tyack, P.L., Wells, R.S., 2018. Modeling tissue and blood gas kinetics in coastal and offshore common bottlenose dolphins, *Tursiops truncatus*. Front. Physiol. 9.

Fahlman, A., Miedler, S., Rocho-Levine, J., Jabois, A., Arenarez, J., Marti-Bonmati, L., García-Párraga, D., Cauture, F., 2019. Re-evaluating the significance of the dive response during voluntary surface apneas in the bottlenose dolphin, *Tursiops truncatus*. Sci. Rep. 9, 8613.

Fahlman, A., Cozzi, B., Manley, M., Jabas, S., Malik, M., Blawas, A., Janik, V.M., 2020a. Conditioned variation in heart rate during static breath-holds in the bottlenose dolphin (*Tursiops truncatus*). Front. Physiol. 11.

Fahlman, A., Miedler, S., Marti-Bonmati, L., Ferrero Fernandez, D., Muñoz Caballero, P., Arenarez, J., Rocho-Levine, J., Robeck, T., Blawas, A., 2020b. Cardiorespiratory coupling in cetaceans; a physiological strategy to improve gas exchange? J. Exp. Biol. 223, jeb226365.

Galantsev, V.P., 1991. Adaptational changes in the venous system of diving mammals. Can. J. Zool. 69, 414–419.

Gales, N.J., 1992. Mass stranding of striped dolphin, *Stenella coeruleoalba*, at Augusta, Western Australia: notes on clinical pathology and general observations. J. Wildl. Dis. 28, 651–655.

Galliano, R.E., Morgane, P.J., McFarland, W.L., Nagel, E.L., Catherman, R.L., 1966. The anatomy of the cervicothoracic arterial system in the bottlenose dolphin (*Tursiops truncatus*) with a surgical approach suitable for guided angiography. Anat. Rec. 155, 325–337.

Garcia Párraga, D., Moore, M., Fahlman, A., 2018. Pulmonary ventilation–perfusion mismatch: a novel hypothesis for how diving vertebrates may avoid the bends. Proc. R. Soc. Lond. B 285, 20180482.

Goldbogen, J.A., Cade, D.E., Calambokidis, J., Czapanskiy, M.F., Fahlbusch, J., Friedlaender, A.S., Gough, W.T., Kahane-Rapport, S.R., Savoca, M.S., Ponganis, K.V., Ponganis, P.J., 2019. Extreme bradycardia and tachycardia in the world's largest animal. Proc. Natl. Acad. Sci. U. S. A. 116, 25329–25332.

Goodall, R.N.P., Norris, K.S., Schevill, W.E., Fraga, F., Praderi, R., Iniguez Jr., M.A., De Haro, J.C., 1997. Review and Update on the Biology of Peale's Dolphin, *Lagenorhynchus australis*. Reports International Whaling Commission, 47, pp. 777–796.

Gosline, J.M., Shadwick, R.E., 1996. The mechanical properties of fin whale arteries are explained by novel connective tissue designs. J. Exp. Biol. 199, 985–997.

Green, A.A., Redfield, A.C., 1933. On the respiratory function of the blood of the porpoise. Biol. Bull. 64, 44–52.

Grinnell, S.W., Irving, L., Scholander, P.F., 1942. Experiments on the relation between blood flow and heart rate in the living seal. J. Cell. Comp. Physiol. 19, 341–350.

Haldiman, J.T., Tarpley, R.J., 1993. Anatomy and physiology. In: Burns, J.J., Montague, J.J., Cowles, C.J. (Eds.), The Bowhead Whale. Allen Press, Lawrence, KS, pp. 71–150.

Hamlin, R.L., Scher, A.M., 1961. Ventricular activation process and genesis of QRS complex in the goat. Am. J. Phys. 200, 223–228.

Hamlin, R.L., Jackson, R.F., Himes, J.A., Pipers, F.S., Townsend, A.C., 1970. Electrocardiogram of the bottle-nosed dolphin (*Tursiops truncatus*). Am. J. Vet. Res. 31, 501–505.

Hamlin, R.L., Ridgway, S.H., Gilmartin, W.G., 1972. Electrocardiograms of pinnipeds. Am. J. Vet. Res. 33, 867–875.

Harms, C.A., Jensen, E.D., Townsend, F.I., Hansen, L.J., Schwacke, L.H., Rowles, T.K., 2013. Electrocardiograms of bottlenose dolphins (*Tursiops truncatus*) out of water: habituated collection versus wild postcapture animals. J. Zoo Wildl. Med. 44, 972–981.

Harrison, R.J., Tomlinson, J.D.W., 1956. Observations on the venous system in certain Pinnipedia and Cetacea. Proc. Zool. Soc. London 126, 205–233.

Hay, K.A., Mansfield, A.W., 1989. Narwhal *Monodon monoceros* Linnaeus, 1758. In: Ridgway, S.H., Harrison, R.J. (Eds.), Handbook of Marine Mammals 4 River Dolphins and Larger Toothed Whales. Academic Press, San Diego, CA, pp. 145–176.

Hedrick, M.S., Duffield, D.A., 1991. Haematological and rheological characteristics of blood in seven marine mammal species: physiological implications for diving behaviour. J. Zool. 225, 273–283.

Hellsten, Y., Nyberg, M., 2016. Cardiovascular adaptations to exercise training. Compr. Physiol. 6, 1–32.

Hill, J.A., Olson, E.N., 2008. Cardiac plasticity. N. Engl. J. Med. 358, 1370–1380.

Hilton, J.W., Gaskin, D.E., 1978. Comparative volumes and vascular microanatomy of the intrahepatic venous system of the harbour porpoise, *Phocoena phocoena* (L.). Can. J. Zool. 56, 2292–2298.

Hochachka, P.W., Foreman, R.A., 1993. Phocid and cetacean blueprints of muscle metabolism. Can. J. Zool. 71, 2089–2098.

Holt, J.P., Rhode, E.A., Kines, H., 1968. Ventricular volumes and body weight in mammals. Am. J. Physiol. 215, 704–715.

Horvath, S.M., Chiodi, H., Ridgway, S.H., Azar Jr., S., 1968. Respiratory and electrophoretic characteristics of hemoglobin of porpoises and sea lion. Comp. Biochem. Physiol. 24, 1027–1033.

Houser, D.S., Dankiewicz-Talmadge, L.A., Stockard, T.K., Ponganis, P.J., 2010. Investigation of the potential for vascular bubble formation in a repetitively diving dolphin. J. Exp. Biol. 213, 52–62.

Hui, C., 1975. Thoracic collapse as affected by the retia thoracica in the dolphin. Respiration 25, 63–70.

Irving, L., 1938. Changes in the blood flow through the brain and muscles during the arrest of breathing. Am. J. Physiol. 122, 207–214.

Irving, L., Scholander, P.F., Grinnell, S.W., 1941a. The respiration of the porpoise, *Tursiops truncatus*. J. Cell. Comp. Physiol. 17, 145–168.

Irving, L., Scholander, P.F., Grinnell, S.W., 1941b. Significance of the heart rate to the diving ability of seals. J. Cell. Comp. Physiol. 18, 283–297.

Jackson, J.B.S., 1845. Dissections of a spermaceti whale and three other cetaceans. Boston J. Nat. Hist. 5, 10–171.

James, T.N., Kawamura, K., Meijler, F.L., Yamamoto, S., Terasaki, F., Hayashi, T., 1995. Anatomy of the sinus node, av node, and his bundle of the heart of the sperm whale (*Physeter macrocephalus*), with a note on the absence of an os cordis. Anat. Rec. 242, 355–373.

Jefferson, T.A., 1987. A study of the behavior of Dall's porpoise (*Phocoenoides dalli*) in the Johnstone Strait, British Columbia. Can. J. Zool. 65, 736–744.

Jefferson, T.A., 1988. Phocoenoides dalli. Mamm. Species 319, 1–7.

Jones, D.R., Fisher, H.D., McTaggart, S., West, N.H., 1973. Heart rate during breath-holding and diving in the unrestrained harbor seal (*Phoca vitulina richardi*). Can. J. Zool. 51, 671–680.

Kaczmarek, J., Reichmuth, C., McDonald, B.I., Kristensen, J.H., Larson, J., Johansson, F., Sullivan, J.L., Madsen, P.T., 2018. Drivers of the dive response in pinnipeds; apnea, submergence or temperature? J. Exp. Biol. 221.

Kamiya, T., Yamasaki, F., 1974. Organ Weights of *Pontoporia blainvillei* and *Platanista gangetica* (Platanistidae). Scientific Reports of the Whales Research Institute, 25, pp. 265–270.

Kanwisher, J.W., Ridgway, S.H., 1983. The physiological ecology of whales and porpoises. Sci. Am. 248, 110–120.

Kanwisher, J.W., Senft, A.W., 1960. Physiological measurements on a live whale. Science 131, 1379–1380.

Kanwisher, J.W., Sundnes, G., 1965. Physiology of a small cetacean. Hvalrådets skr. 48, 45–53.

Kaplan, J.A., 1979. Cardiac Anesthesia. Grune and Stratton, New York.

King, R.L., Jenks, J.L., White, P.D., 1953. The electrocardiogram of a beluga whale. Circulation 8, 387–393.

Koopman, H.N., Westgate, A.J., Read, A.J., 1999. Hematology values of wild harbor porpoises (*Phocoena phocoena*) from the Bay of Fundy, Canada. Mar. Mamm. Sci. 15, 52–64.

Krogh, A., 1929. Progress of physiology. Am. J. Physiol. 90, 243–251.

Kwan, W.C., Shavelle, D.M., Laughrun, D.R., 2019. Pulmonary vascular resistance index: getting the units right and why it matters. Clin. Cardiol. 42, 334–338.
Lahiri, S., 1975. Blood oxygen affinity and alveolar ventilation in relation in body weight in mammals. Am. J. Phys. 229, 529–536.
Laurie, A.H., 1933. Some aspects of respiration in blue and fin whales. Discovery Rep. 7, 363–406.
Lawrie, R., 1953. The activity of the cytochrome system in muscle and its relation to myoglobin. Biochem. J. 55, 298–305.
Leith, D.E., 1989. Adaptations to deep breath-hold diving: respiratory and circulatory mechanics. Undersea Biomed. Res. 16, 345–353.
Lenfant, C., 1969. Physiological properties of blood of marine mammals. In: Anderson, H.T. (Ed.), The Biology of Marine Mammals. Academic Press, New York, pp. 95–116.
Lenfant, C., Kenney, D.W., Aucutt, C., 1968. Respiratory function in the killer whale (*Orcinus orca* Linnaeus). Am. J. Physiol. 215.
Lillie, M.A., Piscitelli, M.A., Vogl, A.W., Gosline, J.M., Shadwick, R.E., 2013. Cardiovascular design in fin whales: high-stiffness arteries protect against adverse pressure gradients at depth. J. Exp. Biol. 216, 2548–2563.
Lillie, M.A., Vogl, A.W., Gerard, S.G., Raverty, S., Shadwick, R.E., 2022. Retia mirabilia: Protecting the cetacean brain from locomotion-generated blood pressure pulses. Science 377, 1452–1456.
Lillie, M.A., Vogl, A.W., Raverty, S., Haulena, M., McLellan, W.A., Stenson, G.B., Shadwick, R.E., 2017. Controlling thoracic pressures in cetaceans during a breath-hold dive: importance of the diaphragm. J. Exp. Biol. 220, 3464–3477.
Lillie, M.A., Vogl, A.W., Raverty, S., Haulena, M., McLellan, W.A., Stenson, G.B., Shadwick, R.E., 2018. The caval sphincter in cetaceans and its predicted role in controlling venous flow during a dive. J. Exp. Biol. 221.
Lockyer, C., 1976. Body weights of some species of large whales. J. Cons. Int. Explor. Mer 36, 259–273.
Lockyer, C., 1991. Body composition of the sperm whale, *Physeter catodon*, with special reference to the possible function of fat depots. Rit Fiskideildar 12, 1–24.
Lockyer, C., 1993. Seasonal Changes in Body Fat Condition of Northeast Atlantic Pilot Whales, and Their Biological Significance. International Whaling Commission Report, Special Issue, 14, pp. 325–350.
Lockyer, C., Heide-Jørgensen, M.P., Jensen, J., Walton, M.J., 2003. Life history and ecology of harbour porpoises (*Phocoena phocoena*) from West Greenland. NAMMCO Sci. Publ. 5, 177–194.
Mallette, S.D., McLellan, W.A., Scharf, F.S., Koopman, H.N., Barco, S.G., Wells, R.S., Ann Pabst, D., 2016. Ontogenetic allometry and body composition of the common bottlenose dolphin (*Tursiops truncatus*) from the U.S. mid-Atlantic. Mar. Mamm. Sci. 32, 86–121.
Manire, C.A., Reiber, C.M., Gaspar, C., Rhinehart, H.L., Byrd, L., Sweeney, J., West, K.L., 2018. Blood chemistry and hematology values in healthy and rehabilitated rough-toothed dolphins (*Steno bredanensis*). J. Wildl. Dis. 54, 1–13.
Mate, B.R., Rossbach, K.A., Nieukirk, S.L., Wells, R.S., Irvine, A.B., Scott, M.D., Read, A.J., 1995. Satellite-monitored movements and dive behavior of a bottlenose dolphin (*Tursiops truncatus*) in Tampa Bay, Florida. Mar. Mamm. Sci. 11, 452–463.
McAlpine, D.F., 1985. Size and growth of heart, liver, and kidneys in North Atlantic fin (*Balaenoptera physalus*), sei (*B. borealis*), and sperm (*Physeter macrocephalus*) whales. Can. J. Zool. 63, 1402–1409.
McDonald, B.I., Ponganis, P.J., 2013. Insights from venous oxygen profiles: oxygen utilization and management in diving California sea lions. J. Exp. Biol. 216, 3332–3341.
McDonald, B.I., Johnson, M., Madsen, P.T., 2018. Dive heart rate in harbour porpoises is influenced by exercise and expectations. J. Exp. Biol. 221.
McDonald, B.I., Elmegaard, S., Johnson, M., Wisniewska, D., Rojano-Doñate, L., Siebert, U., Teilmann, J., Madsen, P., 2021. High heart rates in hunting harbour porpoises. Proc. R. Soc. Lond. B 288, 20211596.
McFarland, W.L., Jacobs, M.S., Morgane, P.J., 1979. Blood supply to the brain of the dolphin, *Tursiops truncatus*, with comparative observations on special aspects of the cerebrovascular supply of other vertebrates. Neurosci. Biobehav. Res. 3, 193.
McLellan, W.A., Koopman, H.N., Rommel, S.A., Read, A.J., Potter, C.W., Nicolas, J.R., Westgate, A.J., Pabst, D.A., 2002. Ontogenetic allometry and body composition of harbour porpoises (*Phocoena phocoena*, L.) from the western North Atlantic. J. Zool. 257, 457–471.
Medway, W., Moldovan, F., 1966. Blood studies on the North Atlantic pilot (pothead) whale, *Globicephala melaena* (Traill, 1809). Physiol. Zool. 39, 110–116.
Meijler, F.L., Wittkampf, F.H., Brennen, M.K.R., Baker, V., Wassenar, C., Baken, E.E., 1992. Electrocardiogram of the humpback whale (*Megaptera novaeangliae*), with specific reference to atrioventricular transmission and ventricular excitation. J. Am. Coll. Cardiol. 20, 475–479.
Meir, J.U., Champagne, C.D., Costa, D.P., Williams, C.L., Ponganis, P.J., 2009. Extreme hypoxemic tolerance and blood oxygen depletion in diving elephant seals. Am. J. Physiol. Regul. Integr. Comp. Physiol. 297, R927–R939.
Melnikov, V.V., 1994. The arterial systems of the head and the anterior truncal part of *Balaenoptera edeni*. In: Investigations on Cetacea. vol. 25, pp. 309–318.
Melnikov, V.V., 1997. The arterial system of the sperm whale (*Physeter macrocephalus*). J. Morphol. 234, 37–50.
Miedler, S., Fahlman, A., Valls Torres, M., Álvaro Álvarez, T., Garcia-Parraga, D., 2015. Evaluating cardiac physiology through echocardiography in bottlenose dolphins: using stroke volume and cardiac output to estimate systolic left ventricular function during rest and following exercise. J. Exp. Biol. 218, 3604–3610.
Miyazaki, N., Perrin, W.F., 1994. Rough-toothed dolphin *Steno bredanensis* (Lesson, 1828). In: Ridgway, S.H., Harrison, R.J. (Eds.), Handbook of Marine Mammals. vol. 5. Academic Press, San Diego, CA, pp. 1–20.
Miyazaki, N., Fujise, Y., Fujiyama, T., 1981. Body and Organ Weight of Striped and Spotted Dolphins off the Pacific Coast of Japan. Scientific Reports Whales Research Institute, 33, pp. 27–67.
Mompeó, B., Pérez, L., Fernández, A., Saavedra, P., Rivero, M., Arbelo, M., Arregui, M., Suárez-Santana, C., Bernaldo-De-Quiros, Y., 2020. Morphological structure of the aortic wall in three Delphinid species with shallow or intermediate diving habits: evidence for diving adaptation. J. Morphol. 281, 377–387.
Morejohn, G.V., Ames, J.A., Lewis, D.B., 1975. Post Mortem Studies of Sea Otters, *Enhydra lutris* L., in California. Marine Resources Technical Report 30, California Department of Fish and Game.
Nabi, G., Hao, Y., Zeng, X., Jinsong, Z., McLaughlin, R.W., Wang, D., 2017. Hematologic and biochemical differences between two free ranging Yangtze finless porpoise populations: the implications of habitat. PLoS One 12, e0188570.
Nagel, E.L., Morgane, P.J., McFarland, W.L., Galliano, R.E., 1968. Rete mirabile of dolphin: its pressure-damping effect on cerebral circulation. Science 161, 898–900.

Ninomiya, H., Inomata, T., Shirouzu, H., Katsumata, E., 2005. Microanatomy of the terminal air spaces of Baird's beaked whale (*Berardius bairdii*) lungs. J. Vet. Med. Sci. 67, 473–479.

Ninomiya, H., Imamura, E., Inomata, T., 2014. Comparative anatomy of the ophthalmic rete and its relationship to ocular blood flow in three species of marine mammal. Vet. Ophthalmol. 17, 100–105.

Nishiwaki, M., 1950. On the Body Weight of Whales. Scientific Reports Whales Research Institute, 4, pp. 184–209.

Noren, S.R., Williams, T.M., 2000. Body size and skeletal muscle myoglobin of cetaceans: adaptations for maximizing dive duration. Comp. Biochem. Physiol. A: Physiol. 126, 181–191.

Noren, S.R., Cuccurullo, V., Williams, T.M., 2004. The development of diving bradycardia in bottlenose dolphins (*Tursiops truncatus*). J. Comp. Physiol. B Biochem. Syst. Environ. Physiol. 174, 139–147.

Noren, S.R., Kendall, T., Cuccurullo, V., Williams, T.M., 2012. The dive response redefined: underwater behavior influences cardiac variability in freely diving dolphins. J. Exp. Biol. 215, 2735–2741.

Ochrymowych, C., Lambertsen, R.H., 1984. Anatomy and vasculature of a minke whale heart. Am. J. Anat. 169, 165–175.

Omura, H., 1950. On the Body Weight of Sperm and Sei Whales Located in the Adjacent Waters of Japan. Scientific Reports Whales Research Institute, 4, pp. 1–13.

Omura, H., Ohsumi, S., Nemoto, T., Nasu, K., Kasuya, T., 1969. Black Right Whales in the North Pacific. Scientific Reports Whales Research Institute, 21, pp. 1–78.

Ono, N., Yamaguchi, T., Ishikawa, H., Arakawa, M., Takahashi, N., Saikawa, T., Shimada, T., 2009. Morphological varieties of the Purkinje fiber network in mammalian hearts, as revealed by light and electron microscopy. Arch. Histol. Cytol. 72, 139–149.

Pabst, D.A., Rommel, S.A., McLellan, W.A., Williams, T.M., Rowles, T.K., 1995. Thermoregulation of the intra-abdominal testes of the bottlenose dolphin (*Tursiops truncatus*) during exercise. J. Exp. Biol. 198, 221–226.

Pabst, D.A., McLellan, W.A., Rommel, S.A., 2016. How to build a deep diver: the extreme morphology of mesoplodonts. Integr. Comp. Biol. 56, 1337–1348.

Panneton, W.M., 2013. The mammallian diving response: an enigmatic reflex to preserve life? Physiology 28, 284–297.

Panneton, W.M., Gan, Q., 2020. The mammalian diving response: inroads to its neural control. Front. Neurosci. 14.

Perrin, W.F., Roberts, E.L., 1972. Organ weights of non-captive porpoise (*Stenella* spp.). Bull. South. Calif. Acad. Sci. 71, 19–32.

Pfeiffer, C.J., 1990. Observations on the ultrastructural morphology of the bowhead whale (*Balaena mysticetus*) heart. J. Zoo Wildl. Med. 21, 48–55.

Pfeiffer, C.J., Kinkead, T.P., 1990. Microanatomy of retia mirabilia of bowhead whale foramen magnum and mandibular foramen. Cells Tissues Organs 139, 141–150.

Picco, N., Coton, F.E., Fosser, G., Guerrero-Rojas, E.A., Belerenian, G., Abuin, G., 2020. Coronary anatomy of Commerson' s dolphin (*Cephalorhynchus commersonii*). Int. J. Morphol. 38, 7–12.

Piscitelli, M., McLellan, W., Rommel, S., Blum, J., Barco, S., Pabst, D.A., 2010. Lung size and thoracic morphology in shallow (*Tursiops truncatus*) and deep (*Kogia* spp.) diving cetaceans. J. Morphol. 271, 654–673.

Piscitelli, M.A., Raverty, S.A., Lillie, M.A., Shadwick, R.E., 2013. A review of cetacean lung morphology and mechanics. J. Morphol. 274, 1425–1440.

Plon, S., Albrecht, K.H., Cliff, G., Froneman, P.W., 2012. Organ weights of three dolphin species (*Sousa chinensis*, *Tursiops aduncus* and *Delphinus capensis*) from South Africa: implications for ecological adaptation? J. Cetacean Res. Manag. 12, 265–276.

Plön, S., Frainer, G., Wedderburn-Maxwell, A., Cliff, G., Huggenberger, S., 2019. Dorsal fin and hump vascular anatomy in the Indo-Pacific humpback dolphin (*Sousa plumbea*) and the Indo-Pacific bottlenose dolphin (*Tursiops aduncus*). Mar. Mamm. Sci. 35, 684–695.

Ponganis, P.J., 2011. Diving mammals. In: Terjung, R. (Ed.), Comprehensive Physiology. John Wiley & Sons, Inc, pp. 517–535.

Ponganis, P.J., Fujise, Y., Miyazaki, N., 2022. Morphology and physiology in some small pelagic cetaceans: Is Dall's porpoise a deep diver and a thoroughbred of the sea? Mar. Mamm. Sci. 38, 1442–1469.

Ponganis, P.J., Kooyman, G.L., 1999. Heart rate and electrocardiogram characteristics of a young California gray whale (*Eschrichtius robustus*). Mar. Mamm. Sci. 15, 1198–1207.

Ponganis, P.J., Kooyman, G.L., Zornow, M.H., Castellini, M.A., Croll, D.A., 1990. Cardiac output and stroke volume in swimming harbor seals. J. Comp. Physiol. 160B, 473–482.

Ponganis, P.J., Meir, J.U., Williams, C.L., 2011. In pursuit of Irving and Scholander: a review of oxygen store management in seals and penguins. J. Exp. Biol. 214, 3325–3339.

Ponganis, P.J., McDonald, B.I., Tift, M.S., Williams, C.L., 2017. Heart rate regulation in diving sea lions: the vagus nerve rules. J. Exp. Biol. 220, 1372–1381.

Poole, D.C., Erickson, H.H., 2011. Highly athletic terrestrial mammals: horses and dogs. Compr. Physiol. 1, 1–37.

Quiring, D.P., 1943. Weight data on five whales. J. Mammal. 24, 39–45.

Race, G.J.M.D., Edwards, W.L.J.M.D., Halden, E.R.M.D., Wilson, H.E.M.D., Luibel, F.J.M.D., 1959. A large whale heart. Circulation 19, 928–932.

Reed, J.Z., Chambers, C., Hunter, C.J., Lockyer, C., Kastelein, R., Fedak, M.A., Boutilier, R.G., 2000. Gas exchange and heart rate in the harbour porpoise, *Phocoena phocoena*. J. Comp. Physiol. B 170, 1–10.

Rhinehart, H., Gorzelany, J.F., Wells, R.S., 1995. Blood Values from 23 Live-Stranded Fraser's Dolphins (*Lagenodelphis hosei*). International Association of Aquatic Animal Medicine—Archives.

Rhode, E.A., Elsner, R., Peterson, T.M., Campbell, K.B., Spangler, W., 1986. Pressure-volume characteristics of aortas of harbor and Weddell seals. Am. J. Physiol. 251, R174–R180.

Ridgway, S.H., 1966. Dall porpoise *Phocoenoides dalli* (True): observations in captivity and at sea. In: Norsk Hvalfangst-Tidende. vol. 55, pp. 97–110.

Ridgway, S.H., 1972. Mammals of the Sea Biology and Medicine. Charles C Thomas, Springfield, IL.

Ridgway, S.H., 1986. Diving by cetaceans. In: Brubakk, A.O., Kanwisher, J.W., Sundnes, G. (Eds.), Diving in Animals and Man. Royal Norwegian Society of Science and Letters, Trondheim, pp. 33–62.

Ridgway, S.H., Howard, R., 1979. Dolphin lung collapse and intramuscular circulation during free diving: evidence from nitrogen washout. Science 206, 1182–1183.

Ridgway, S.H., Johnston, D.G., 1966. Blood oxygen and ecology of porpoises of three genera. Science 151, 456–458.

Ridgway, S.H., Kohin, S., 1995. The relationship between heart mass and body mass for three cetacean genera: narrow allometry demonstrates interspecific differences. Mar. Mamm. Sci. 11, 72–80.

Ridgway, S.H., Scronce, B.L., Kanwisher, J., 1969. Respiration and deep diving in the bottlenose porpoise. Science 166, 1651–1654.

Ridgway, S.H., Simpson, J.G., Patton, G.S., Gilmartin, W.G., 1970. Hematological findings in some small cetaceans. J. Am. Vet. Assoc. 157, 566–575.

Ridgway, S.H., Bowers, C.A., Miller, D., Schultz, M.L., Jacobs, C.A., Dooley, C.A., 1984. Diving and blood oxygen in the white whale. Can. J. Zool. 62, 2349–2351.

Ridgway, S., Houser, D., Finneran, J., Carder, D., Keogh, M., Van Bonn, W., Smith, C., Scadeng, M., Dubowitz, D., Mattrey, R., Hoh, C., 2006. Functional imaging of dolphin brain metabolism and blood flow. J. Exp. Biol. 209, 2902–2910.

Rojano-Doñate, L., McDonald, B.I., Wisniewska, D.M., Johnson, M., Teilmann, J., Wahlberg, M., Højer-Kristensen, J., Madsen, P.T., 2018. High field metabolic rates of wild harbour porpoises. J. Exp. Biol. 221, jeb185827.

Rommel, S.A., Pabst, D.A., McLellan, W.A., Mead, J.A., Potter, C.W., 1992. Anatomical evidence for a countercurrent heat exchanger associated with dolphin testes. Anat. Rec. 232, 150–156.

Rommel, S.A., Ann Pabst, D., McLellan, W.A., 1993. Functional morphology of the vascular plexuses associated with the cetacean uterus. Anat. Rec. 237, 538–546.

Rommel, S.A., Pabst, D.A., McLellan, W.A., Williams, T.M., Friedl, W.A., 1994. Temperature regulation of the testes of the bottlenose dolphin (*Tursiops truncatus*): evidence from colonic temperatures. J. Comp. Physiol. B. 164, 130–134.

Rommel, S.A., Costidis, A.M., Fernandez, A., Jepson, P.D., Pabst, A.D., McLellan, W.A., Houser, D.S., Cranford, T.W., Van Helden, A.L., Allen, D.M., Barros, N.B., 2006. Elements of beaked whale anatomy and diving physiology and some hypothetical causes of sonar-related stranding. J. Cetacean Res. Manag. 7, 189–209.

Ronald, K., McCarter, R., Selley, L.J., 1977. Venous circulation of the Harp seal (*Pagophilus groenlandicus*). In: Harrison, R.J. (Ed.), Functional Anatomy of Marine Mammals. Academic Press, New York, pp. 235–270.

Rowlands, C.E., McLellan, W.A., Rommel, S.A., Costidis, A.M., Yopak, K.E., Koopman, H.N., Glandon, H.L., Ann Pabst, D., 2021. Comparative morphology of the spinal cord and associated vasculature in shallow versus deep diving cetaceans. J. Morphol. 282, 1415–1431.

Rowlatt, U., 1981. The cardiac ventricles of a baleen whale (*Balaenoptera acutorostrata*: Minke whale) and a toothed whale (*Hyperoodon ampullatus*: Bottlenose whale). J. Morphol. 168, 85–96.

Rowlatt, U., Gaskin, D.E., 1975. Functional anatomy of the heart of the harbor porpoise, *Phocaena phocaena*. J. Morphol. 146, 479–493.

Sassu, R., Cozzi, B., 2007. The external and middle ear of the striped dolphin *Stenella coeruleoalba* (Meyen 1833). Anat. Histol. Embryol. 36, 197–201.

Savu, O., Jurcuţ, R., Giuşcă, S., Mieghem, T.V., Gussi, I., Popescu, B.A., Ginghină, C., Rademakers, F., Deprest, J., Voigt, J.-U., 2012. Morphological and functional adaptation of the maternal heart during pregnancy. Circ. Cardiovasc. Imaging 5, 289–297.

Schmidt-Neilsen, K., Larimer, J.L., 1958. Oxygen dissociation curves of mammalian blood in relation to body size. Am. J. Phys. 195, 424–428.

Schneider, H.P., Truex, R.C., Knowles, J.O., 1964. Comparative observations of the hearts of mongrel and greyhound dogs. Anat. Rec. 149, 173–180.

Scholander, P.F., 1940. Experimental investigations on the respiratory function in diving mammals and birds. Hvalrådets skr. 22, 1–131.

Scholander, P.F., Schevill, W.E., 1955. Counter-current vascular heat exchange in the fins of whales. J. Appl. Physiol. 8, 279–282.

Scholander, P.F., Irving, L., Grinnell, S.W., 1942. Aerobic and anaerobic changes in seal muscle during diving. J. Biol. Chem. 142, 431–440.

Schumacker, P.T., Suggett, A.J., Wagner, P.D., West, J.B., 1985. Role of hemoglobin P_{50} in O_2 transport during normoxic and hypoxic exercise in the dog. J. Appl. Physiol. 59, 749–757.

Schumacker, P.T., Rowland, J., Saltz, S., Nelson, D.P., Wood, L.D., 1987. Effects of hyperthermia and hypothermia on oxygen extraction by tissues during hypovolemia. J. Appl. Physiol. 63, 1246–1252.

Sergeant, D.E., 1969. Feeding Rates of Cetacea. Report on Norwegian fishery and marine investigations, 15, pp. 246–258.

Sergeant, D.E., Brodie, P.F., 1969. Body size in white whales, *Delphinapterus leucas*. J. Fish. Res. Board Can. 26, 2561–2580.

Shadwick, R.E., 1999. Mechanical design in arteries. J. Exp. Biol. 202, 3305–3313.

Shadwick, R.E., Gosline, J.M., 1994. Arterial mechanics in the fin whale suggest a unique hemodynamic design. Am. J. Physiol. 267, R805–R818.

Sharp, S.M., Knoll, J.S., Moore, M.J., Moore, K.M., Harry, C.T., Hoppe, J.M., Niemeyer, M.E., Robinson, I., Rose, K.S., Brian Sharp, W., Rotstein, D., 2014. Hematological, biochemical, and morphological parameters as prognostic indicators for stranded common dolphins (*Delphinus delphis*) from Cape Cod, Massachusetts, U.S.A. Mar. Mamm. Sci. 30, 864–887.

Simpson, J.G., Gardner, M.B., 1972. Comparative microscopic anatomy of selected marine mammals. In: Ridgway, S.H. (Ed.), Mammals of the Sea. Biology and Medicine. C.C. Thomas, Springfield, IL, pp. 229–301.

Slijper, E.J., 1958. Organ weights and symmetry problems in porpoises and seals. Arch. Neerland Zool. 13, 97–113.

Slijper, E.J., 1961. Foramen ovale and ductus arteriosus Botalli in aquatic mammals. Mammalia 25, 528–570.

Slijper, E.J., 1962. Whales. Hutchinson and Co, London.

Snow, D.H., 1985. The horse and dog, elite athletes—why and how? Proc. Nutr. Soc. 44, 267–272.

Snyder, G.K., 1983. Respiratory adaptations in diving mammals. Respir. Physiol. 54, 269–294.

Sommer, L.S., McFarland, W.L., Galliano, R.E., Nagel, E.L., Morgane, P.J., 1968. Hemodynamic and coronary angiographic studies in the bottlenose dolphin (*Tursiops truncatus*). Am. J. Physiol. 215, 1498–1505.

Spencer, M.P., Gornall, T.A., Poulter, T.C., 1967. Respiratory and cardiac activity of killer whales. J. Appl. Physiol. 22, 974–981.

St. Aubin, D.J., Forney, K.A., Chivers, S.J., Scott, M.D., Danil, K., Romano, T.A., Wells, R.S., Gulland, F.M.D., 2013. Hematological, serum, and plasma chemical constituents in pantropical spotted dolphins (*Stenella attenuata*) following chase, encirclement, and tagging. Mar. Mamm. Sci. 29, 14–35.

Stahl, W.R., 1967. Scaling of respiratory variables in mammals. J. Appl. Physiol. 22, 453–460.

Storlund, R.L., Rosen, D.A.S., Trites, A.W., 2021. Electrocardiographic scaling reveals differences in electrocardiogram interval durations between marine and terrestrial mammals. Front. Physiol. 12.

Storz, J.F., 2016. Hemoglobin–oxygen affinity in high-altitude vertebrates: is there evidence for an adaptive trend? J. Exp. Biol. 219, 3190–3203.

Storz, J.F., Scott, G.R., 2019. Life ascending: mechanism and process in physiological adaptation to high-altitude hypoxia. Annu. Rev. Ecol. Evol. Syst. 50, 503–526.

Storz, J.F., Runck, A.M., Moriyama, H., Weber, R.E., Fago, A., 2010. Genetic differences in hemoglobin function between highland and lowland deer mice. J. Exp. Biol. 213, 2565–2574.

Suarez, L.J.A., Paz, J.I., Balquin, J.T., Aragones, L.V., 2019. Haematological and serum biochemical reference values extrapolated from rehabilitated stranded spinner dolphins (*Stenella longirostris*). Philipp. J. Vet. Med. 56, 114–121.

Suzuki, M., Tomoshige, M., Ito, M., Koga, S., Yanagisawa, M., Bungo, T., Makiguchi, Y., 2017. Increase in serum noradrenaline concentration by short dives with bradycardia in Indo-Pacific bottlenose dolphin *Tursiops aduncus*. Gen. Comp. Endocrinol. 248, 1–4.

Tarpley, R.J., Hillmann, D.J., Henk, W.G., 1993. Observations of the External Morphology and Coronary Vasculature of the Heart of the Bowhead Whale *Balaena mysticetus*. Report to the Department of Wildlife Management Northslope Borough. Barrow AK.

Tarpley, R.J., Hillmann, D.J., Henk, W.G., George, J.C., 1997. Observations on the external morphology and coronary vasculature of a fetal heart of the bowhead whale, *Balaena mysticetus*. Anat. Rec. 247, 556–581.

Tawara, T., 1950. On the Respiratory Pigments of Whale (Studies on Whale Blood II). Scientific Reports Whales Research Institute, 3, pp. 95–101.

Taylor, C.R., Karas, R.H., Weibel, E.R., Hoppeler, H., 1987. Adaptive variation in the mammalian respiratory system in relation to energetic demand. II. Reaching the limits to oxygen flow. Respir. Physiol. 69, 7–26.

Thompson, D., Fedak, M.A., 1993. Cardiac responses of grey seals during diving at sea. J. Exp. Biol. 174, 139–164.

Truex, R.C., Nolan, F.G., Schneider, H.P., Perlmutter, H.I., 1961. Anatomy and pathology of the whale heart with special reference to the coronary circulation. Anat. Rec. 141, 325–353.

Turner, J.P., Clark, L.S., Haubold, E.M., Worthy, G.A.J., Cowan, D.F., 2006. Organ weights and growth profiles in bottlenose dolphins (*Tursiops truncatus*) from the northwestern Gulf of Mexico. Aquat. Mamm. 32, 46–57.

Van Nie, C.J., 1987. Todd fibres—Purkinje fibres—in the wall of the right atrium in dolphins (their structure and hypothetical function). Aquat. Mamm. 13, 103–104.

Vedvick, T.S., Itano, H.A., 1976. Partial characterization of the hemoglobin from Hubbs' beaked whale (*Mesoplodon carlhubbsi*). Comp. Biochem. Physiol. B: Biochem. 55, 65–68.

Vella, C.A., Robergs, R.A., 2005. A review of the stroke volume response to upright exercise in healthy subjects. Br. J. Sports Med. 39, 190–195.

Velten, B.P., Dillaman, R.M., Kinsey, S.T., McLellan, W.A., Pabst, D.A., 2013. Novel locomotor muscle design in extreme deep-diving whales. J. Exp. Biol. 216, 1862–1871.

Viamonte, M., Morgane, P., Galliano, R., Nagel, E., McFarland, W., 1968. Angiography in the living dolphin and observations on blood supply to the brain. Am. J. Physiol. 214, 1225–1249.

Vogl, A.W., Fisher, H.D., 1981a. Arterial circulation of the spinal cord and brain in the Monodontidae (order Cetacea). J. Morphol. 170, 171–180.

Vogl, A.W., Fisher, H.D., 1981b. The internal carotid artery does not directly supply the brain in the Monodontidae (order Cetacea). J. Morphol. 170, 207–214.

Vogl, A.W., Fisher, H.D., 1982. Arterial retia related to supply of the central nervous system in two small toothed whales—Narwhal (*Monodon monoceros*) and Beluga (*Delphinapterus leucas*). J. Morphol. 174, 41–56.

Vogl, A.W., Todd, M.E., Fisher, H.D., 1981. An ultrastructural and fluorescence histochemical investigation of the innervation of retial arteries in *Monodon monoceros*. J. Morphol. 168, 109–119.

Weber Rosas, F.C., De Oliveira, M.R., De Mattos, G.E., 2007. Organ weights of *Pontoporia blainvillei* and *Sotalia guianensis* (Cetacea: Pontoporiidae and Delphinidae). Marine Biological Association of the United Kingdom, J. Mar. Biol. Assoc. U. K. 87, 825–828.

White, P.D., Kerr, W.J., 1915–1917. The heart of the sperm whale with special reference to the A-V conduction system. Heart 6, 207–216.

Williams, T.M., Friedl, W.A., Fong, M.L., Yamada, R.M., Sedivy, P., Haun, J.E., 1992. Travel at low energetic cost by swimming and wave-riding bottlenose dolphins. Nature 355, 821–823.

Williams, T.M., Friedl, W.A., Haun, J.E., 1993. The physiology of bottlenose dolphins (*Tursiops truncatus*): heart rate, metabolic rate and plasma lactate concentration during exercise. J. Exp. Biol. 179, 31–46.

Williams, T.M., Haun, J.E., Friedl, W.A., 1999. The diving physiology of bottlenose dolphins (*Tursiops truncatus*). I. Balancing the demands of exercise for energy conservation at depth. J. Exp. Biol. 202, 2739–2748.

Williams, T.M., Bengston, P., Steller, D.L., Croll, D.A., Davis, R.W., 2015a. The healthy heart: lessons from nature's elite athletes. Physiology 30, 349–357.

Williams, T.M., Fuiman, L.A., Kendall, T., Berry, P., Richter, B., Noren, S.R., Thometz, N., Shattock, M.J., Farrell, E., Stamper, A.M., Davis, R.W., 2015b. Exercise at depth alters bradycardia and incidence of cardiac anomalies in deep-diving marine mammals. Nat. Commun. 6, 6055.

Williams, T.M., Blackwell, S.B., Richter, B., Sinding, M.-H.S., Heide-Jørgensen, M.P., 2017. Paradoxical escape responses by narwhals (*Monodon monoceros*). Science 358, 1328–1331.

Wisniewska, D.M., Johnson, M., Teilmann, J., Rojano-Doñate, L., Shearer, J., Sveegard, S., Miller, L.A., Siebert, U., Madsen, P.T., 2016. Ultra-high foraging rates of harbor porpoises make them vulnerable to anthropogenic disturbance. Curr. Biol. 26, 1–6.

Woodall, P.F., 1992. Relative heart weights of some African antelope. J. Zool. 228, 666–669.

Zapol, W.M., Liggins, G.C., Schneider, R.C., Qvist, J., Snider, M.T., Creasy, R.K., Hochachka, P.W., 1979. Regional blood flow during simulated diving in the conscious Weddell seal. J. Appl. Physiol. 47, 968–973.

CHAPTER

6

Respiratory physiology in the dolphin and other whales[*]

Marina Piscitelli-Doshkov[a,b], *Gerald L. Kooyman*[c], *and Andreas Fahlman*[d,e,f]

[a]Department of Zoology, The University of British Columbia, Vancouver, BC, Canada [b]Department of Husbandry, North Carolina Aquarium on Roanoke Island, Manteo, NC, United States [c]Scripps Institution of Oceanography, La Jolla, CA, United States [d]Fundación Oceanogràfic de la Comunitat Valenciana, Valencia, Spain [e]Global Diving Research SL., Valencia, Spain [f]Kolmården Wildlife Park, Kolmården, Sweden

Abstract

Bottlenose dolphins, like all mammals, depend on the respiratory system to bring oxygen into the body and remove the carbon dioxide produced by aerobic metabolism. In marine mammals, such as dolphins, that divide their time at the surface exchanging gases through the lungs and spending considerable time underwater holding their breath, the lungs have specialized adaptations. For example, the hydrostatic pressure of water increases with depth, causing compression of air-filled spaces resulting in lung compression and eventually alveolar collapse. This process affects the function of the respiratory system and gas exchange during diving. In addition, dolphins need to complete a respiratory cycle during the short period in the air when porpoising, which has resulted in fast and high respiratory flow rates. Thus, life in the ocean has resulted in several adaptations to obtain sufficient oxygen to perform dives.

Introduction

Function of the respiratory system

The function of the cardiorespiratory system is to bring air from the atmosphere to the cells, where oxygen is used and converted to water, energy, and carbon dioxide. The mammalian respiratory system acts like a bellow that brings in fresh air and removes air exchanged with the blood in the alveoli. Thus, the inhalation rate of air into the alveoli determines the rate of oxygen consumption. Therefore, anything limiting the respiratory capacity (e.g., tidal volume, breathing frequency, and the ability to exchange gas with the blood) can limit the overall metabolic rate and exercise capacity (Table 1).

The respiratory system of marine mammals, especially bottlenose dolphins (*Tursiops truncatus*) and other cetaceans, has evolved for the specialized function of life at the surface and underwater. For example, in addition to functioning like land mammals, the respiratory system of dolphins is capable of high respiratory flow during exhalation at the surface (Kooyman and Cornell, 1981; Fahlman et al., 2015) and lung collapse under high ambient pressure as the animal descends in the water column (Ridgway et al., 1969; Moore et al., 2011). These respiratory adaptations are the result of anatomical differences (see Section "Anatomy and structure"). The mammalian respiratory system is divided into the upper conducting airways and the lower respiratory airways (terminal bronchi through alveoli). In this chapter, we will explain the anatomy and function of humans to highlight the differences observed in marine mammals, especially dolphins. For example, in marine mammals, differences in alveolar structure are apparent between groups (Piscitelli et al., 2013; Otero-Sabio et al., 2021) but are particularly striking compared with humans (Fig. 1). In addition, the respiratory mechanics and physiology of dolphins and other marine mammals have been investigated in excised tissues, carcasses, and live animals (see Sections "Functional studies in dead animals"

[*]This book has a companion website hosting complementary materials. Visit this URL to access it: https://www.elsevier.com/books-and-journals/book-companion/9780323905169.

The Physiology of Dolphins
https://doi.org/10.1016/B978-0-323-90516-9.00005-1

TABLE 1 Tidal volume (V_T), breathing frequency (f_R), minute ventilation ($\dot{V}_E$), and maximum expiratory and inspiratory flow during rest in a number of marine mammal species.

Species	Weight (kg)	V_T (L)	f_R (breaths min^{-1})	$\dot{V}_E$ (L min^{-1})	Respiratory durations (ms)		Maximum respiratory flow (L s^{-1})		Reference
		Rest	Rest	Rest	Expiration	Inspiration	Expiration	Inspiration	
Grey seal (*Halichoerus grypus*)[a]	160–250	6.3	19.4	123			9.7	8.4	Reed et al. (1994)
Weddell seal (*Leptonychotes weddellii*)	280–430	4.0–10.0	4.0–10.0	32.4			5.8–7.0	4.6–6.3	Kooyman et al. (1971), Falke et al. (2008)
Harbor porpoise (*Phocoena phocoena*)[a]	28	1.0–1.1	4.7–5.3	4.6–5.8			4.9–6.0	4.0–5.3	Reed et al. (2000)
Pacific bottlenose dolphin (*Tursiops truncatus gilli*)	285	26.4					115–162	45–56	Kooyman and Cornell (1981)
Atlantic bottlenose dolphin (*Tursiops truncatus*)	140–250	5–10	3.4		0.26–0.31	0.43–0.66	20–140	15–33	Irving et al. (1941), Ridgway et al. (1969), and Fahlman et al. (2015)
California sea lion (*Zalophus californianus*)	32–46	1.2–2	16.4–18.1	21.3-22.9	0.8	1.0	8.6–14	5	Kerem et al. (1975), Matthews (1977)
Gray whale (*Eschrichtius robustus*)	1500–6000[b]	38–212	1–2	23–303	0.3–1.2		101–202	116–176	Wahrenbrock et al. (1974), Kooyman et al. (1975)
Killer whale (*Orcinus orca*)	1090[c]–3700	46–218	1				180	61	Spencer et al. (1967), Kasting et al. (1989)
Beluga whale (*Delphinapterus leucas*)	324–640	14.0–12.6			0.52	0.89–0.98	58.5 ± 23.6	28.2 ± 10.0	Kasting et al. (1989), Fahlman et al. (2019)
Pilot whale (*Globicephala scammoni*)	450	9–39.5	1–2						Olsen et al. (1969)
Fin whale (*Balaenoptera physalus*)	≈40,000		0.5–1		0.73 ± 0.15	0.91 ± 0.23			Lafortuna et al. (2003)
Minke whale (*Balaenoptera acutorostrata*)	1840–5740		0.48–1.35						Blix and Folkow (1995)
Pacific walrus (*Odobenus rosmarus divergens*)	640–841	8–15			1.30 ± 0.33	1.96 ± 0.56	>45	11	Borque-Espinosa et al. (2021)
South American sea lion (*Otaria flavescens*)	94–286	2.4–7.6			2.43 ± 0.62	1.89 ± 0.23	3.6–9.1	3.1–7.0	Fahlman and Madigan (2016)
Human (*Homo sapiens*)	70	0.6	18	11.3			7–10	5	Stahl (1967), Jordanoglou and Pride (1968), Knudson et al. (1976)

The range of maximal flows between different individuals is given when available.
[a] *Data collected during surface periods between breath-holds/apneas.*
[b] *Body mass of wild animals was estimated from length (4.8–5.8 m estimated to be approximately 1500–2300 kg (Krogh, 1929).*
[c] *Beached female.*

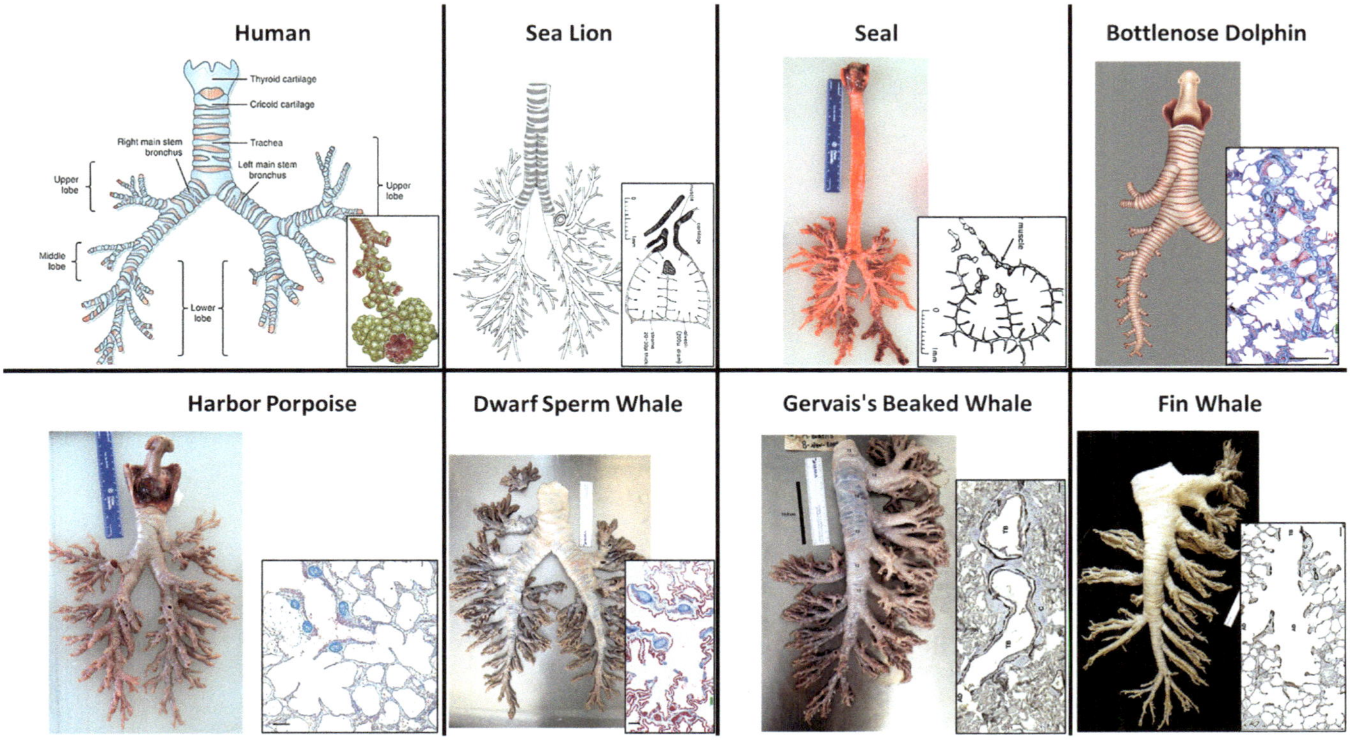

FIG. 1 Comparative gross anatomy of conducting airways in the respiratory system of a human compared to pinnipeds and cetaceans. Insets are close-up illustrations of the terminal (gas-exchanging) airways. Note the long trachea of the pinnipeds compared to the cetacean species. Only the left main bronchus of the beaked and fin whale conducting airways is shown. Inset scale bars = 1 mL for harbor seal (*Phoca vitulina*) and California sea lion (*Zalophus californianus*), 700 μm for dolphin, 100 μm for all others. The dolphin, porpoise, and dwarf sperm whale stained with Masson's trichrome. Mayer hematoxylin counterstained was used for the beaked whale and fin whale. (References used to illustrate the respiratory trees: Denison and Kooyman, 1973; Nakakuki, 1975; Nakakuki, 1993; Netter, 2006; Otero-Sabio et al., 2021). Stranded animal specimens: SWT009 *Kogia breviceps*, VAQS20091107 *Mesoplodon bidens*, KLC355 *Phoca vitulina*, WAM654 *Balenoptera physalus*, DFO9373 *Phocoena phocoena*, CLP001 *Kogia breviceps*, WAM647 *Tursiops truncatus*, and CAHA 536 *Phocoena phocoena*.

and "Functional studies in live animals"). Throughout this chapter, we will describe these differences and how they support diving adaptations. First, however, to better understand these adaptations, the following sections will present a basic introduction to the respiratory system's different volumes (see Section "Volumes of the respiratory system") and mechanics (see Section "Respiratory mechanics").

Volumes of the respiratory system

The anatomy of the mammalian lung recruits different alveolar regions that are functionally important for gas exchange. These volumes include the inspiratory and expiratory reserve volumes, the tidal volume, and the residual volume (Table 1). The sum of the tidal volume and inspiratory reserve volume is the inspiratory capacity, whereas the functional residual capacity is the sum of the expiratory reserve volume and residual volume (Fig. 2). The tidal volume is the air inhaled and exhaled during a normal breath. The tidal volume is only a fraction of the total lung capacity. Both humans and animals can increase the tidal volume during exercise up to the maximal volume that can be inhaled and exhaled, which is the vital capacity. Thus, the vital capacity is the sum of the expiratory and inspiratory reserve volumes and the tidal volume and is determined by the residual volume. The residual volume is the amount of air that remains in the airways and alveoli after maximal exhalation. Thus, the total lung capacity is the total volume of the lung and is the sum of the vital capacity and the residual volume (Fig. 2). In humans, the functional residual capacity is approximately 45%–50% of the total lung capacity, whereas the residual volume is between 18% and 25% and is determined by the flexibility of the chest (Stocks and Quanjer, 1995).

Respiratory mechanics

Two critical definitions in respiratory physiology are minute and alveolar ventilation. The former is a measure of the total volume of air brought into the airways over a period of time, whereas the latter is an index of the effective

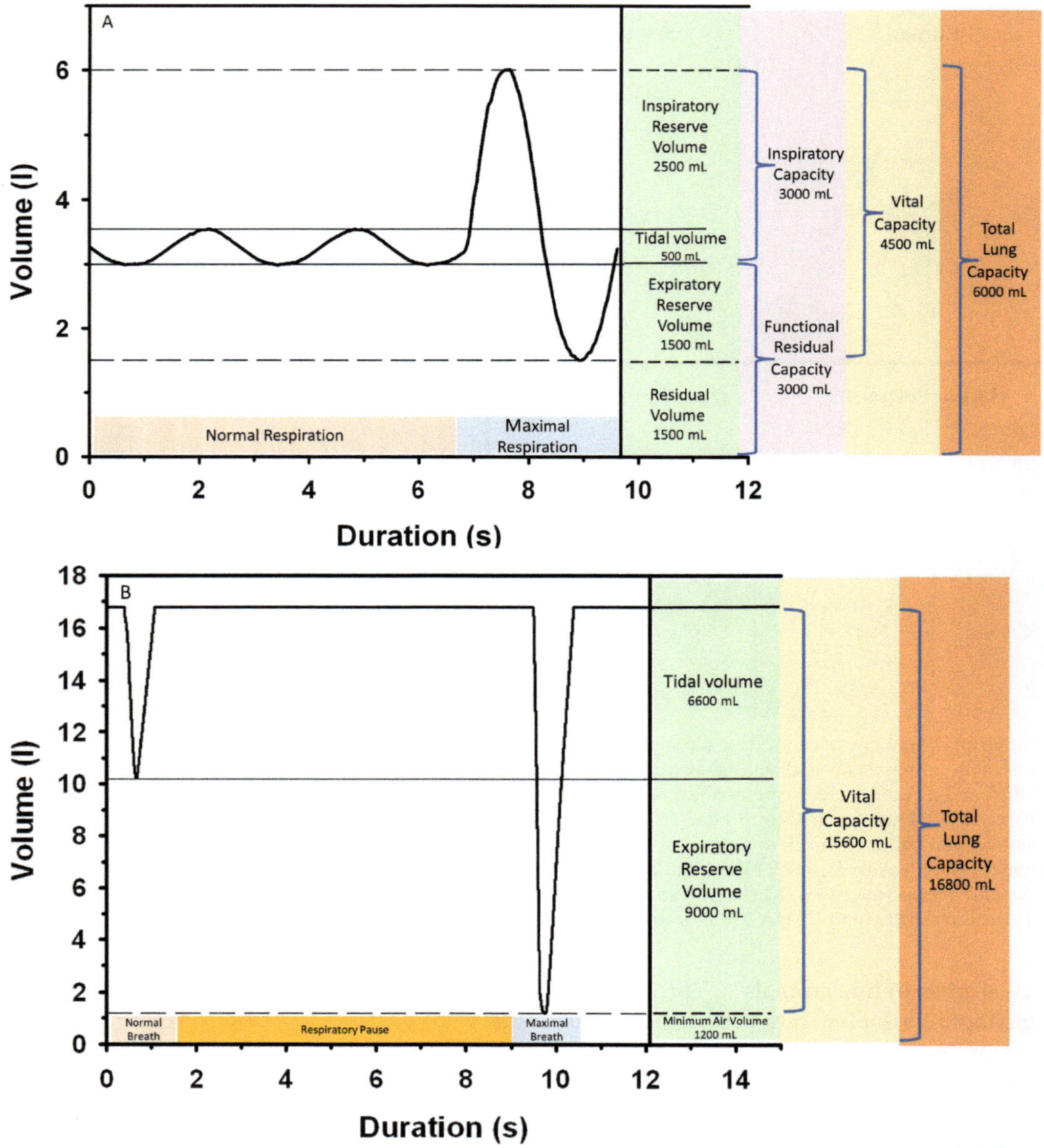

FIG. 2 Representative spirometry data illustrating lung volumes in a healthy adult. (A) human (body mass 70 kg) and (B) bottlenose dolphin (*Tursiops truncatus*, body mass 190 kg). Volumes within the green box are measured during spirometry, whereas capacities in the yellow and orange boxes are calculated. The figure highlights the terrestrial versus aquatic breathing strategy. The latter begins with exhalation from a filled lung, followed by an immediate inhalation and a respiratory pause. The pause length depends on the species, but in dolphins, the pause ranges from 10 to 30 s—consequently, dolphins, and most other marine mammals, "breath-hold" between breaths.

ventilation, or the volume available for gas exchange. Humans have voluntary control over breathing and can change the volume of air brought into the alveoli by increasing the tidal volume up to the vital capacity (Fig. 2) (West, 2012). The minute ventilation is the product of the tidal volume and breathing frequency (Table 1). Changing minute ventilation can occur by changing the breathing frequency, the number of breaths per minute, or the tidal volume. Consequently, a human or any other mammal can increase the amount of air or oxygen brought into the respiratory system by increasing the breathing frequency or tidal volume. When it comes to gas exchange, it is the alveolar ventilation that is important and it is this quantity that provides changes in available oxygen for metabolism. However, the relative importance of increasing breathing frequency or tidal volume is functionally very different, as they affect the alveolar ventilation differently. As the conducting airways, i.e., the pharynx, larynx, trachea, and bronchi, do not participate in gas exchange, the fresh air brought into this area is not available for exchange with the pulmonary blood. Located at the end of the conducting airways is the gas exchange surface of a mammalian lung. As the conducting airways do not participate in gas exchange, they are called "dead space" and "dead space volume." Consequently, the volume of new, fresh air brought into the alveoli during each breath is the tidal volume minus the dead space volume. Thus, the volume of fresh air brought into the alveoli is the tidal volume minus the dead space volume times the breathing frequency. The effect is that a change in breathing frequency or tidal volume does not have the same impact on alveolar ventilation.

This more significant effect on alveolar ventilation based on tidal volume results from the reduced dead space ventilation when changing the volume rather than the frequency. Dead space volume and dead space ventilation affect alveolar ventilation. In other words, an increase in breathing frequency or tidal volume, which result in the same minute ventilation, will result in different alveolar ventilation. For example, humans' average tidal volume and breathing frequencies are approximately 0.5 L and 12 breaths min^{-1}, respectively, resulting in minute ventilation of 6 L min^{-1} (West, 2012). In humans, the dead space is approximately 0.15 L; therefore, the alveolar ventilation is around 4.2 L. Let us assume that exercise results in a doubling of the minute ventilation, either by doubling in tidal volume or breathing frequency. In both instances, the minute ventilation doubles to 12 L min^{-1}. Breathing frequency doubling from 4.2 to 8.4 L min^{-1} results in a 100% increase in minute ventilation. However, the doubling in tidal volume from 0.5–1 L results in a 143% increase in alveolar ventilation, i.e., 10.2 L, while the doubling in breathing frequency results in a doubling of the alveolar ventilation (8.4 L). Similarly, a reduction in minute ventilation to 3 L min^{-1}, either by a decrease in breathing frequency or tidal volume, results in a 50% (via breathing frequency) and 71% (via tidal volume) reduction in alveolar ventilation, respectively. Consequently, the breathing strategy significantly alters the effective ventilation.

With this basic pattern of respiratory volumes and mechanics explained, we will now describe how the dolphin's system is modified both anatomically and functionally to meet the needs of an animal that is so briefly at the air/water interface.

Anatomy and structure

Dolphins, like all mammals, depend on the respiratory system to bring oxygen into the body for gas exchange and to produce vocalizations. However, unlike typical terrestrial mammals, dolphins have adapted their respiratory system to control buoyancy, accomplish fast ventilatory events at the water surface, and exchange gas while diving below the water surface (Fahlman et al., 2017; Piscitelli et al., 2010). Both the upper and lower airways have specific adaptations and functions in the dolphin compared with other marine mammals (Figs. 1 and 3). The respiratory system is divided into the upper conducting airways and the lower respiratory airways (terminal bronchi through alveoli) in all mammals.

Air flow pathway and branching pattern

The pathway of airflow through the respiratory system is modified in cetaceans. In dolphins, the air pathway starts with a single blowhole ("nose") on top of the head that facilitates rapid breathing at the surface with a set of four paired, nasal air sacs that function in producing sound. The pathway then follows with a short, paired nasal passage, where a bony septum separates the nasal cavity into right and left bony internal nares. Mead (1975) comprehensively reviewed the differences among delphinid nasal passages. In odontocetes (toothed whales, including the dolphin), bony internal nares end with the nasopharynx, a very short, hollow, tube-shaped airspace. In dolphins, the nasopharynx is approximately 5 cm long. The larynx, or "goose beak," is inserted into the distal nasopharynx cavity and is held in place by the palatopharyngeal sphincter muscle. The larynx connects to a short trachea, which divides early into a right accessory bronchus that enters the right lung lobe and ends with the bifurcation at the carina of each primary bronchus to each uni-lobed lung (Fig. 1). This uni-lobed structure to the lung with no apparent tissue fissures is very different from the typical mammalian pattern (including humans, pinnipeds [except for elephant seals, *Mirounga* spp.], and sea otters, *Enhydra lutris*, where lungs are multi-lobed with apparent lobar fissures in the tissue. The major and minor bronchi divide throughout the lung parenchyma in an asymmetric monopodial branching pattern, typical of non-human terrestrial mammals (Nakakuki, 1993; Phalen and Oldham, 1983; Patra, 1986; Wang and Kraman, 2004), to the level of the terminal bronchi. The terminal respiratory airways participate in gas exchange. Unlike the monopodial conducting airways, the branching pattern of these terminal airways symmetrically (dichotomously) divides terminating at the alveolus. The symmetric versus asymmetric dividing airways have implications for airflow mechanics and airway resistance that have not been investigated in marine mammals (Kabilan et al., 1985; Cohen and Briant, 1989; Asgharian et al., 2016). Although the terminal airways' basic structure and branching pattern are like terrestrial mammals, there are many microstructural adaptations in this region, in both cetaceans and pinnipeds, discussed below.

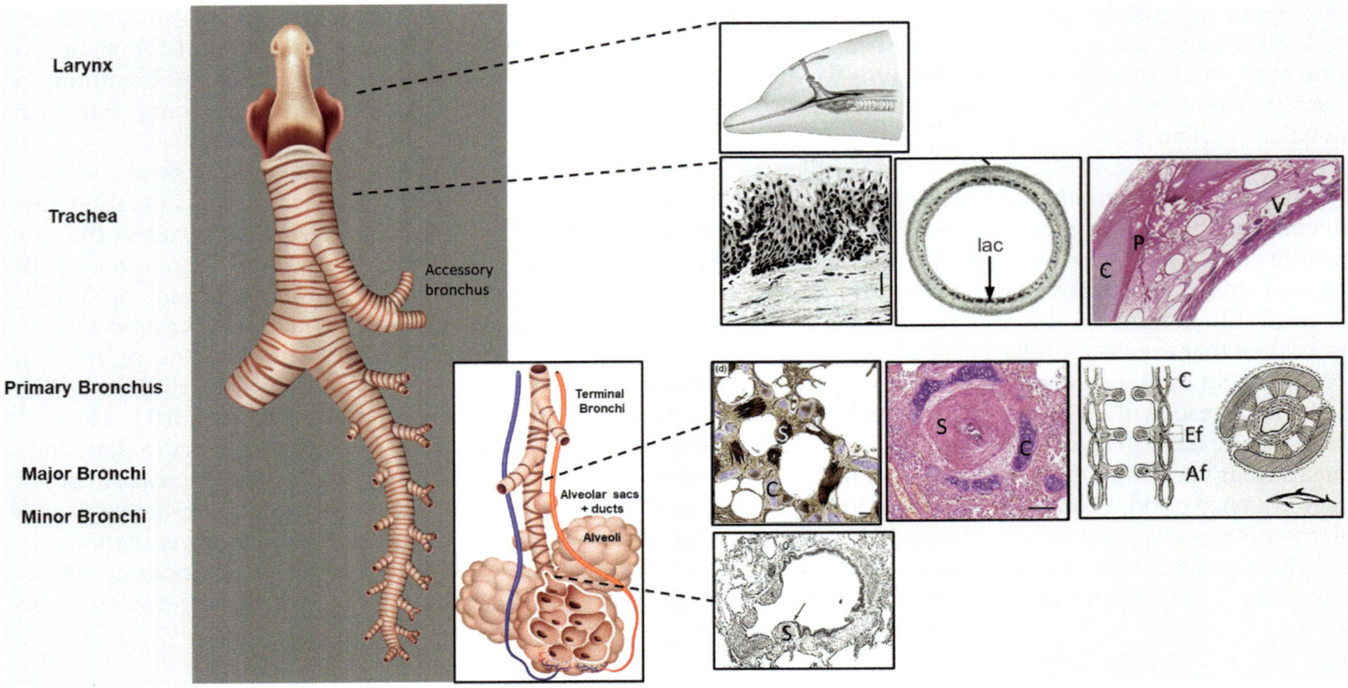

FIG. 3 Respiratory system adaptations in the dolphin at the level of the larynx, trachea, terminal bronchi, and alveolar structures. The larynx is adapted to fit tightly into the nasopharynx, completely separating the oral cavity from the nasal cavity. Tracheal adaptations include regions of transitional epithelium with a higher density of underlying submucosal blood vessels. Terminal airway adaptations include reinforcing the airways with cartilage and a series of robust myoelastic sphincters (Insets are from the following references: Ridgway, 1972; Fanning and Harrison, 1974; Slijper, 1979; Ballarin et al., 2018; Denk et al., 2020; Otero-Sabio et al., 2021). Af, annular fibers; C, cartilage; Ef, elastic fibers; lac, lacunae; P, perichondrium; S, sphincter; V, venule.

The lung

The overall elasticity appear different between species on gross and histological examination of cetacean lungs. In the shallow-diving delphinids and phocoenids, the lungs are resilient and spongy with an enhanced smooth muscle-elastin network throughout the terminal airways and a relatively thick pleura that is well vascularized with a robust elastic layer (Wislocki, 1929; Bélanger, 1940). These adaptations presumably aid in increasing elastic recoil of the lung, leading to faster respiratory flow and thus an overall rapid ventilatory event. In contrast, slow-ventilating mysticetes (fin (*Balaenoptera physalus*) and minke whales (*Balaenoptera acutorostrata*)) and deep-diving cetaceans, such as beaked whales (Ziphiidae) and sperm whales (*Physeter macrocephalus*), lack the terminal airway enhancement in smooth muscle–elastin sphincters (Otero-Sabio et al., 2021). Visually, the lungs in these species appear to lack the spongy resilience and appear more collapsed (Fig. 4), presumably resulting in more compliant lungs adapted for compression at depth with a lower minimum air volume. However, no published mechanical studies have tested this hypothesis.

Furthermore, the lungs of shallow-diving dolphins are relatively larger ($100\,mL\,kg^{-1}$) compared with humans ($70–80\,mL\,kg^{-1}$) and deeper-diving cetaceans ($30\,mL\,kg^{-1}$). Because the lungs are larger, they are better adapted for routine positive buoyancy during shallow diving, similar to that of sea otters, and fast ventilation under constant exercise load (Fig. 5). In contrast, as air-filled spaces compress with increasing pressure at deeper depths, the lungs of deeper-diving species have a reduced role in oxygen storage and buoyancy control compared with shallower-diving species. As a result, the lung size of deeper divers is overall smaller, in both mass and lung volume, than shallow-diving species. Because the lungs of deeper divers are smaller and these species "live" at deeper depths, the alveoli spend a significant amount of time collapsed. Deep-diving cetaceans do not rely on their lungs at extreme depths for oxygen storage nor buoyancy control the way shallow-diving species do; instead, they increase oxygen storage in muscle and blood compartments (reviewed in Ponganis, 2015).

Larynx and trachea

The larynx and trachea are adapted for ventilation in the aquatic environment and act as an air storage site during diving. The stiff, elbowed-shaped larynx and associated set of complex muscles maintain the larynx in a tightly shut

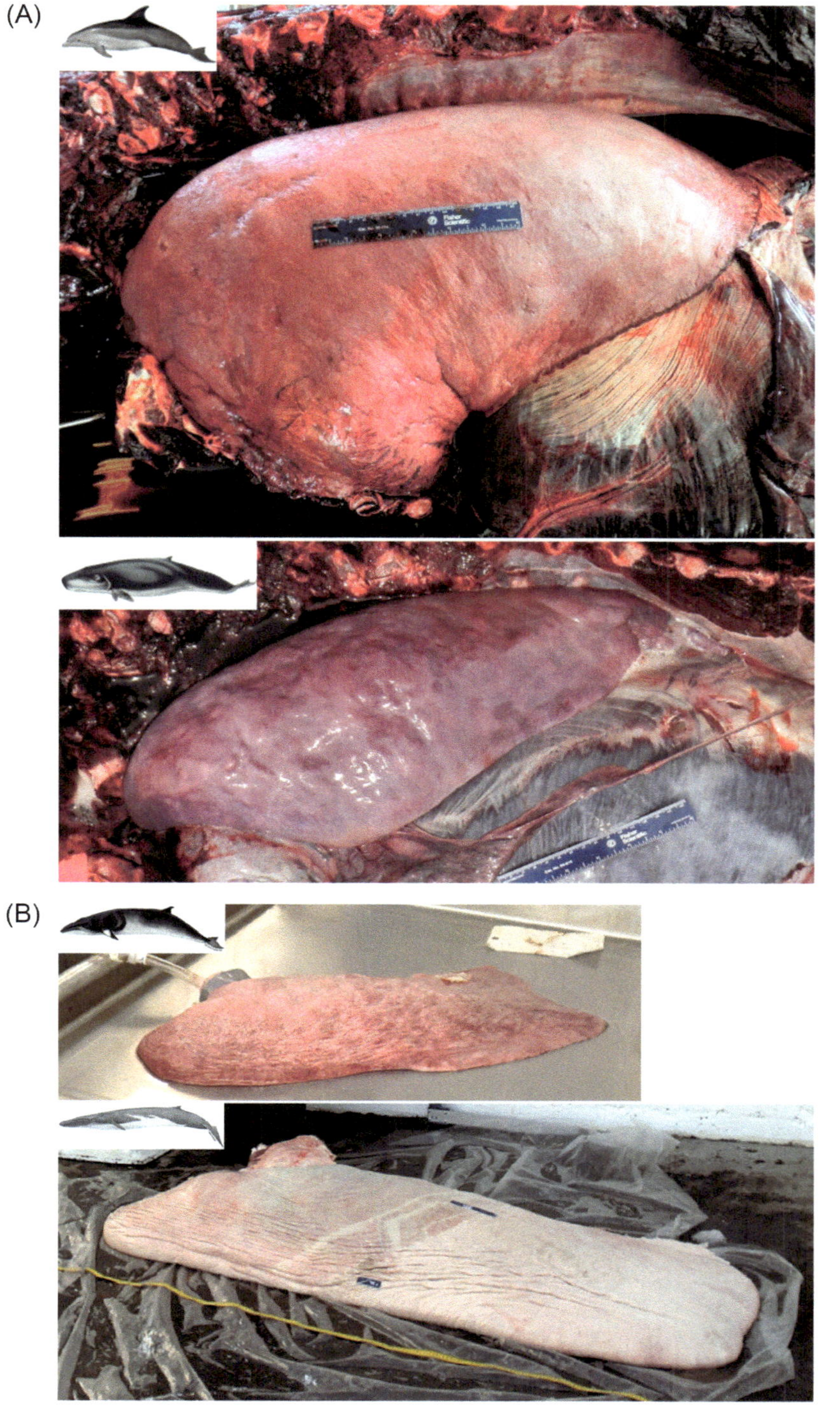

FIG. 4 Gross external photographs of two distinct types of cetacean lungs: (A) spongy bottlenose dolphin lung (*Tursiops truncatus*, top) compared with (B) collapsed lung of a dwarf sperm whale lung (*Kogia sima*), minke whale (*Balenoptera acutorostrata*), and fin whale (*Balenoptera physalus*). Note the crenated (wrinkly) pleura of the collapsed lungs, while the bottlenose dolphin is more spongy or resilient with higher elastic recoil or outward force. As a result, more air is trapped in the turgid lung and will float if placed in water, whereas the collapsed lung contains less air and is less buoyant in water. The terminal airways of bottlenose dolphin lungs also possess a robust, smooth muscle-elastin network (robust myoelastic sphincters). Specimens: CAHA067Tt, CAHA065 Ks, VGT235Ba, and W66Bp.

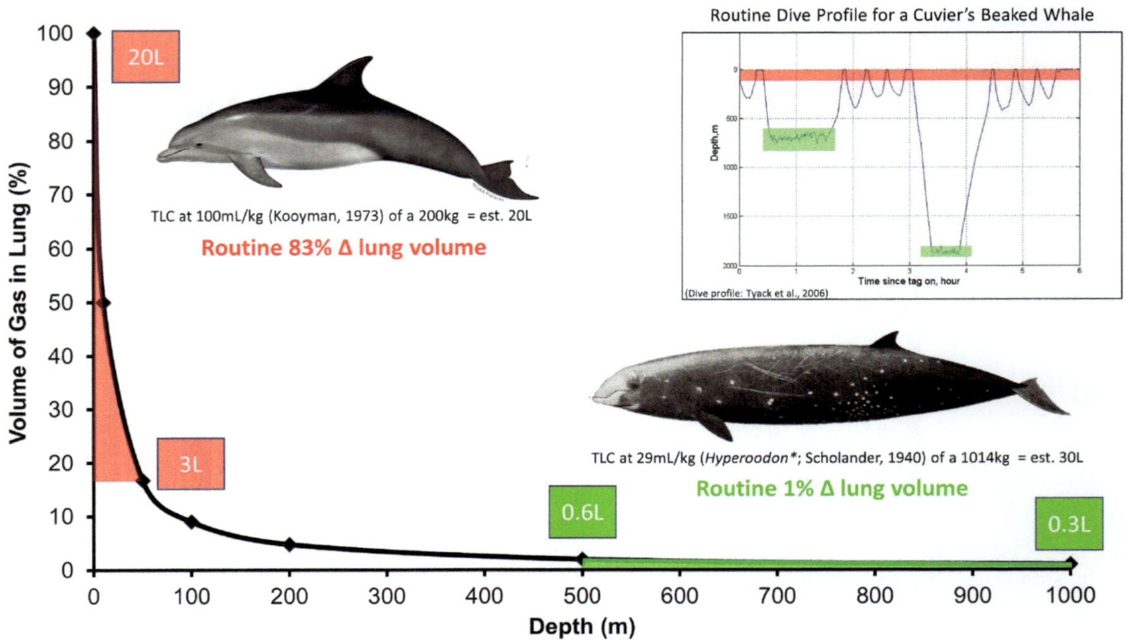

FIG. 5 The routine diving depth and associated changes in lung volume compared between a bottlenose dolphin (*Tursiops truncatus*, in red), and a Cuvier's beaked whale (*Ziphius cavirostris*, in green). Near-shore bottlenose dolphins are shallow, dynamic divers that routinely experience positive buoyancy and a routine 83% change in lung volume while submerged. In contrast, *Ziphius cavirostris* currently holds the mammalian record for the absolute deepest and most prolonged dives, at 2992 m and 222 min (Schorr et al., 2014; Quick et al., 2020). On average, deep-diving beaked whales quickly obtain deep depths (>1000 m) and "live" at these depths up to an hour (Tyack et al., 2006; Quick et al., 2020), where they will routinely experience negative buoyancy and essentially a completely collapsed respiratory system. References used to compile figure: (Scholander, 1940; Kooyman, 1973; Tyack et al., 2006) (* denotes that we know no comparable data yet available for *Z. cavirostris*).

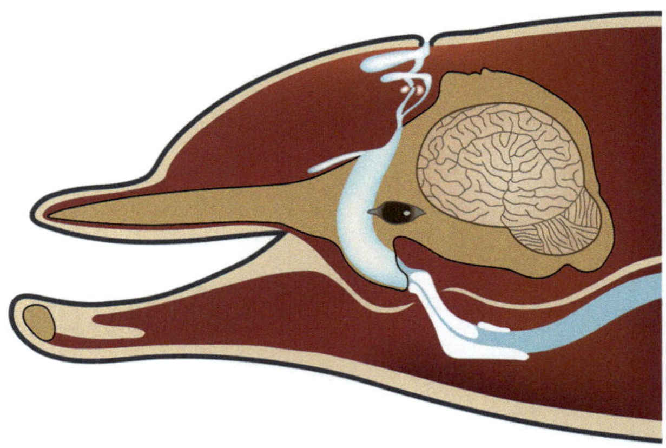

FIG. 6 Upper conducting respiratory system (larynx in white) in relation to the paranasal air sacs near the blowhole. The upper conducting respiratory system is the site of air storage at depth for recirculation in the cranial air spaces to perform sound production. *Illustration by Uko Gorter.*

position at a 45–80 degrees angle into the nasopharynx of the skull. The muscles act on the larynx during ventilation to actively open it to allow airflow. This design assures that water cannot enter the respiratory tract through the blowhole during diving. This design also allows the esophagus to bifurcate around the goose beak so that simultaneous swallowing and breathing can occur while also assuring that any seawater swallowed has no anatomical means of entering the respiratory tract through the esophagus. The larynx is a stiff structure and has been hypothesized to be the site of air storage at depth, well away from the gas-exchanging structures, while remaining near cranial air sacs (most cetaceans) (Fig. 6), laryngeal air sac (mysticetes) (Gandilhon et al., 2015), and tracheal air sac (ribbon seal, *Histriophoca fasciata*) (Burns, 1981) for sound production. It is worth noting that these air sacs, used

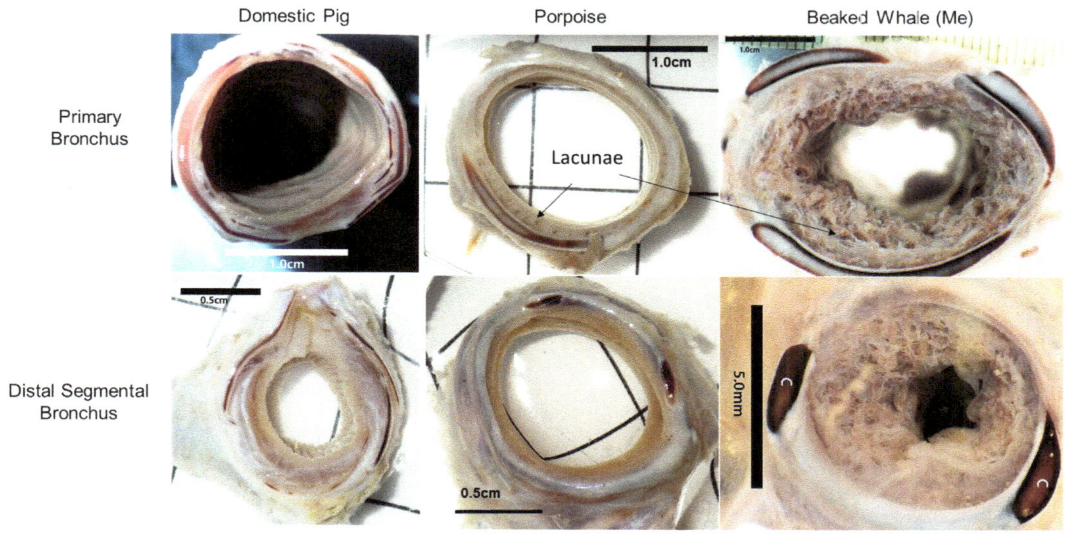

FIG. 7 Peribronchial vascular plexus in domestic pig (*Sus scrofa*), shallow-diving harbor porpoise (*Phocoena phocoena*), and deeper-diving Gervais's beaked whale (*Mesoplodon europaeus*). C, cartilage. Specimens: Wilbur001 Pig, DFO9373 Pp, KLC038 Me.

for sound production, are not encased completely by bone so that they can accommodate the air volume changes caused by pressure changes during diving.

In many cetaceans, the trachea's submucosal lumen lining consists of a plexus of arteriole–venous lacunae and transitional epithelium (Wislocki, 1929; Fanning and Whitting, 1969; Fanning and Harrison, 1974; Fanning, 1977; Cozzi et al., 2005; Smodlaka et al., 2006; Davenport et al., 2013), similar in structure to the corpus cavernosum (Bagnoli et al., 2011). This structure may become blood-engorged at deeper depths to prevent transluminal pressure differences, which may be critically important to avoid tracheobronchial barotrauma. In deeper-diving beaked whales, similar erectile vascular structures, corpus cavernosum nasalis, are also found lining the nasal passage (Costidis and Rommel, 2016). This extensive arterial plexus can fill the soft tissue in the intranasal and palatopharyngeal regions. In bottlenose dolphins and Baird's beaked whale (*Berardius bairdii*), the lacunae of blood vessels in the submucosa extend from the extrapulmonary bronchi to the terminal bronchi and vary significantly in thickness (Wislocki, 1929; Fanning and Harrison, 1974; Ninomiya et al., 2005). In general, the lacunae are situated between the elastic membranes, which enclose the hyaline tracheal cartilage and the smooth muscle layer (Fanning, 1977). These lacunae are supplied by pulmonary veins and surround the pulmonary arteries up to the terminal airways. In theory, these structures can become blood-engorged to reduce the gas volume of the airway lumen. This blood engorgement and reduction in gas volume would decrease the airway's functional compliance (e.g., in the trachea or bronchi), reduce the dead space (e.g., in the nasal cavity), and protect the airway against barotrauma during diving. Comparison of this adaptation (Fig. 7) shows that the dolphin version is unimpressive compared with a deep-diving beaked whale or *Kogia* spp., which lends support to this being a diving adaptation to manage dead air volumes and airway compliance throughout all the conducting airways.

Bronchial airways

In cetaceans, the short trachea and major and minor bronchi are reinforced with complete anastomosing cartilage rings that extend through the terminal airways to the mouth of alveolar structures. Similarly, in sea lion lungs (a typical otariid seal that is a fast swimmer and breather but not as extreme as the dolphin), there are thick and almost contiguous rings of cartilage that reinforce the walls of the last two or three generations of terminal bronchi (no respiratory bronchioles) extending to the mouth of the alveolar sacs (Denison et al., 1971; Denison and Kooyman, 1973). Fanning proposed that the terminal bronchi in dolphins are respiratory in function (Fanning, 1977) based on a very thin epithelium and close association with capillaries, which would enhance gas-exchanging capabilities during a breath-hold. The final ring of cartilage on each route forms the entrance to alveolar sacs consisting of a cluster of approximately 50 alveoli (Fig. 1, California sea lion, *Zalophus californianus* inset). The mechanics of the cartilage appears to vary in flexibility or compliance between cetacean and pinniped species and across age classes, with the deep-diving beaked whale possessing one of the stiffest tracheas among marine mammals (Moore et al., 2014).

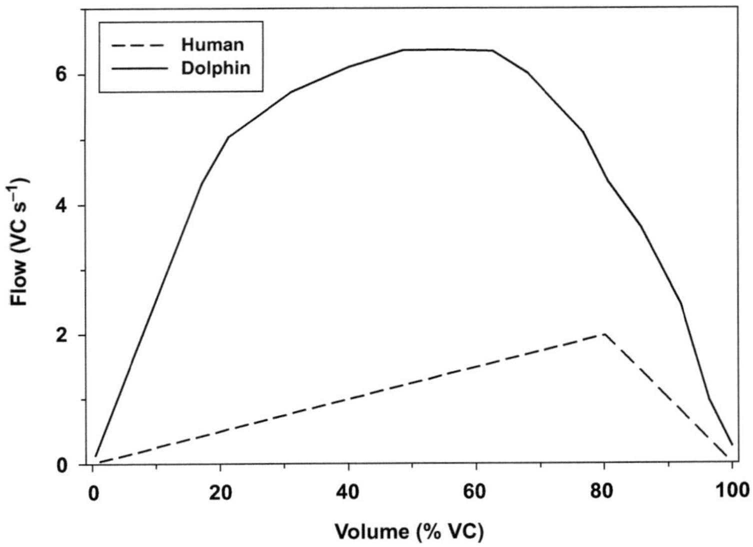

FIG. 8 Maximum expiratory flow rate comparing human (*Homo sapiens*, maximal respiratory maneuver) and a trained Pacific bottlenose dolphin (*Tursiops truncatus gilli*). *Data reproduced from Fahlman et al. (2015), Kooyman and Cornell (1981), and Kooyman and Sinnett (1979).*

During the brief time at the water's surface, high respiratory flow rates are achieved due to the robustness of the cartilage in the small terminal airways, which allows dolphins to exchange a large vital capacity (with a low residual volume). In addition, the cartilage reinforces the distal airways, decreasing compliance and thus preventing premature collapse and maintaining patency of the terminal airspaces, as observed by the plateau on the flow-volume curve (Fig. 8). However, during a dive, these same distal airways undergo a graded collapse in such a way that pushes the air away from gas-exchanging structures and into stiffer conducting airways of the upper respiratory tree (Scholander, 1940). In sea lion (Otariidae) lungs, Denison and Kooyman (1973) confirmed a hierarchy of compliance of the alveolar and bronchial airways (where proximal airways were more rigid and more distal gas-exchanging airways were more compliant). Furthermore, they concluded that the reduced compliance did not have to be very significant for graded compression collapse to occur, i.e., collapse begins with the alveoli. In contrast, in humans, the equivalent of terminal bronchi is respiratory bronchioles that lack cartilage altogether, are very compliant, and undergo collapse at high effort (observed as the steep decline on the flow-volume curve, Fig. 8). In humans, maximal expiratory flow occurs at a specific lung volume, and it decreases quickly, causing the flow-volume curve for exhalation to resemble a triangle. The relative compliance of these terminal bronchi and the graded collapse of gas-exchanging airways is essential in preventing diving-related diseases such as decompression sickness.

In the terminal bronchi of delphinids (dolphins) and beluga whales (*Delphinapterus leucas*), very strong, robust myoelastic (smooth muscle-elastin) sphincters (MES) are prominent. The mechanics of the myoelastic sphincters have not been studied, and the function of the sphincters remains unknown. However, in mammals, smooth muscle is known to regulate airway diameter. It may also control the regional ventilation distribution under different physiological conditions (e.g., rest versus exercise) (Cieri, 2019). Piscitelli et al. (2013) reviewed the multiple hypothesized functions of MES from published literature. A noteworthy observation during excised lung inflation experiments was that fresh lungs, collected within 1–2h of death, completely resisted any insertion of air via pump because the sphincters were so tightly constricted (*Stenella* spp. taken in fisheries, Kooyman, *personal observation*; and beluga whales taken during Inuit harvest, Piscitelli, *personal observation*). However, once a few hours passed (probably allowing adenosine triphosphate (ATP) depletion of the sphincters), the lungs could be inflated for mechanical experiments. The terminal bronchi in the shallow-diving harbor porpoise (*Phocoena phocoena*) also possess myoelastic sphincters, but they are not as prominent or robust as the delphinids (Kooyman and Sinnett, 1979). In shallow-diving mysticetes and deeper-diving beaked (Ziphiidae) and sperm whales (Physeteroidea), the terminal bronchi possess longitudinal smooth muscle; however, in the deep divers, the terminal bronchi are enhanced with vasculature in the submucosa (Figs. 1 and 7). Myoelastic sphincters have also been observed in the shallow-diving Amazon river dolphin (*Inia geoffrensis*), tucuxi (*Sotalia fluviatilis*), and La Plata dolphin (*Pontoporia blainvillei*) [robust MES in boto and tucuxi (Santos Júnior et al., 2017); less thick and less robust MES in La Plata dolphin (*Pontoporia blainvillei*) (Crespo and De Cidre, 2005)]. Smooth muscle is present in the ringed seal (*Pusa hispida*), California sea lion, and Weddell seal

(*Leptonychotes weddellii*) in the terminal airways, but this differs from the myoelastic sphincters seen in odontocetes (Denison and Kooyman, 1973; Boyd, 1975; Smodlaka et al., 2006). In these pinniped species, there is a smooth muscle knob at the terminal bronchi opening into the alveolar ducts. In contrast, the presence of a series of MES in the terminal airways appears to be consistently found in more recently evolved odontocetes with enhanced ventilatory flow rates and relatively shallow divers where buoyancy control is critical (Fig. 5) (delphinids, phocoenids, river dolphins, and beluga) (reviewed in Piscitelli et al., 2013).

Alveolar structures

In dolphins, the alveolar sacs are large and polygonal in shape. All alveolar structures (ducts, sacs, and alveolus) contain sparse smooth muscle fibers, enhanced with elastin fibers in the alveolar wall with thicker alveolar septum than terrestrial mammals and humans. Mysticetes especially have enlarged alveolar sacs and ducts with large elastin bundles throughout all the alveolar structures (Murata, 1951). Sea lions (*Zalophus*), also fast swimmers and ventilators, but not as extreme as dolphins, also have cartilage-reinforced terminal airways. However, their alveolar sacs do not have alveolar ducts. Instead, they are arranged in pairs invested by a thick stroma, which leads to a cluster of about 50 alveoli that are approximately 200–250 μm in diameter (Denison et al., 1971; Denison and Kooyman, 1973).

The enhancement in elastin fibers and bundles observed in cetacean and pinniped alveolar structures gives this gas exchange region of the lung both the ability to collapse to a low lung volume while diving and enhances the evacuation of air at high flow rates by collapsing to low residual volume during ventilation (Denison et al., 1971; Denison and Kooyman, 1973). This enhancement in elastin fibers leads to highly compliant lungs (Fig. 9) (more so than in terrestrial

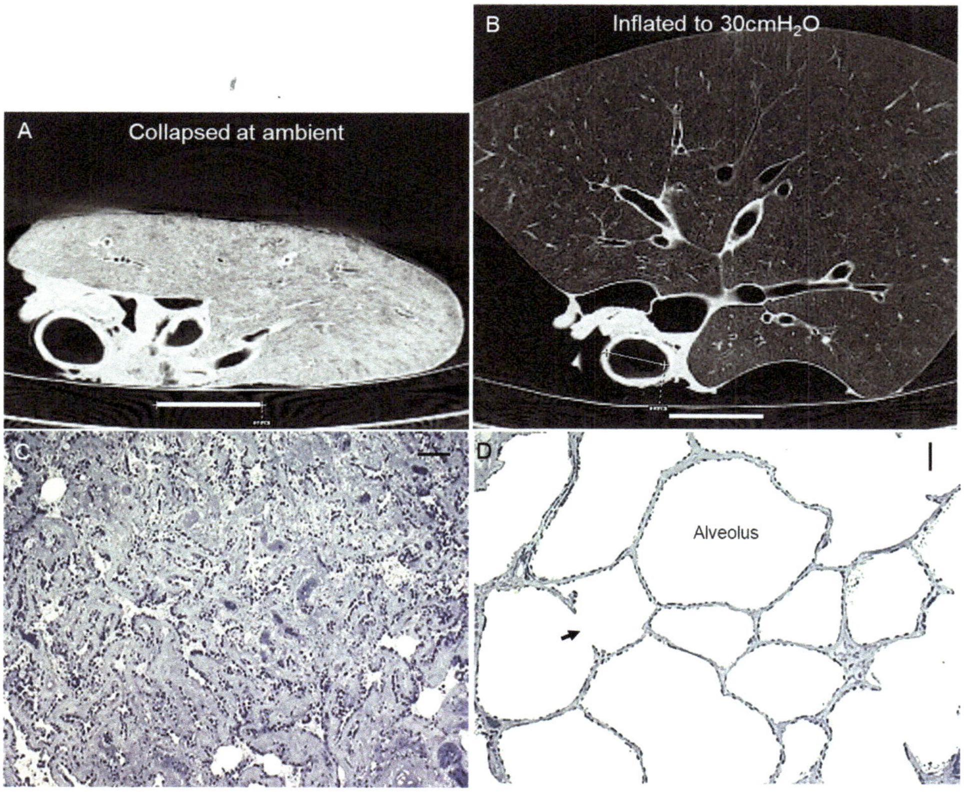

FIG. 9 Images taken from a computer tomography (CT) scan of an excised lung from an adult beluga whale (*Delphinapterus leucas*) (A) collapsed at ambient pressure and (B) inflated to total lung capacity (airway pressure of 30 cmH_2O). Beluga lungs can collapse or deflate down to low residual volume. Black represents air in these CT images. The pictures scaled to each other. Light micrographs (at 10× magnification) of lung tissue from an adult beluga whale in two physiological states: (C) alveoli collapsed (folded upon each other) at ambient pressure; and (D) alveoli inflated to full capacity (30 cmH_2O airway pressure). Note in (D) that the arrow denotes Kohn's pore connecting two alveoli. When beluga lungs collapse, such as during diving, little air volume is left inside the respiratory (gas-exchanging) airways and lower conducting bronchi. Note in (A) some of the larger conducting airways remain patent while the rest of the lung collapses. Similarly, lungs can deflate to low residual volumes during expiration and thus maximize air exchange with each breath. Performed CT scans with the help of Dean Malpas, Canada Diagnostics, and assistance from R. Shadwick. Specimens: A, B: HI-2012-01; C, D: HI-2013-03.

mammals or humans) and an enhancement in the amount of air that can be exchanged. While at the same time allowing the lungs of dolphins and sea lions to empty more readily so little air remains in the lung. The minimal air volume, which is the air left in the lung when relaxed, is only about 7% of a dolphin's estimated total lung capacity compared with a human, which is 25% of total lung capacity. Thus, in the dolphin, the tidal volume is similar to the vital capacity (90% of total lung capacity). Thus, the dolphin can exchange more air from its lungs than terrestrial mammals in any given ventilatory event.

A recent study using lungs of an unknown inflation state found that the alveolar lengths and diameters measured in bottlenose dolphins and striped dolphins were significantly smaller than fin and blue whales (*Balaenoptera musculus*). This observation supports the hypothesis that smaller-bodied cetaceans possess a higher oxygen consumption and higher metabolic rates (Otero-Sabio et al., 2021). However, because a standard inflation state was not used, results are still questionable, and this topic warrants further investigation in marine mammals using a standard pressure. Few studies have taken on this challenge to preserve cetacean lungs in a standard (at total capacity, 30 cm H_2O) inflated state for microscopy (Fanning, 1977) bottlenose dolphin; (Kooyman and Sinnett, 1979) harbor porpoise and spinner dolphin (*Stenella longirostris*); (Piscitelli, unpublished observation Figs. 9 and 10) beluga and fin whale (*Balaenoptera physalus*); (Nicholson, 1984) Australian sea lions; (Denison and Kooyman, 1973) harbor seals (*Phoca vitulina*), elephant seals, and Weddell seals). This is an area of active research, and previous publications lack an explanation of the functional consequences. This technique is valuable for evaluating the size of the terminal airways across species using a standard stereological method to observe how alveolar size varies with body size, which is essential for evaluating enhancements in gas exchange and differences in oxygen consumption and metabolic rates (Tenney and Remmers, 1963).

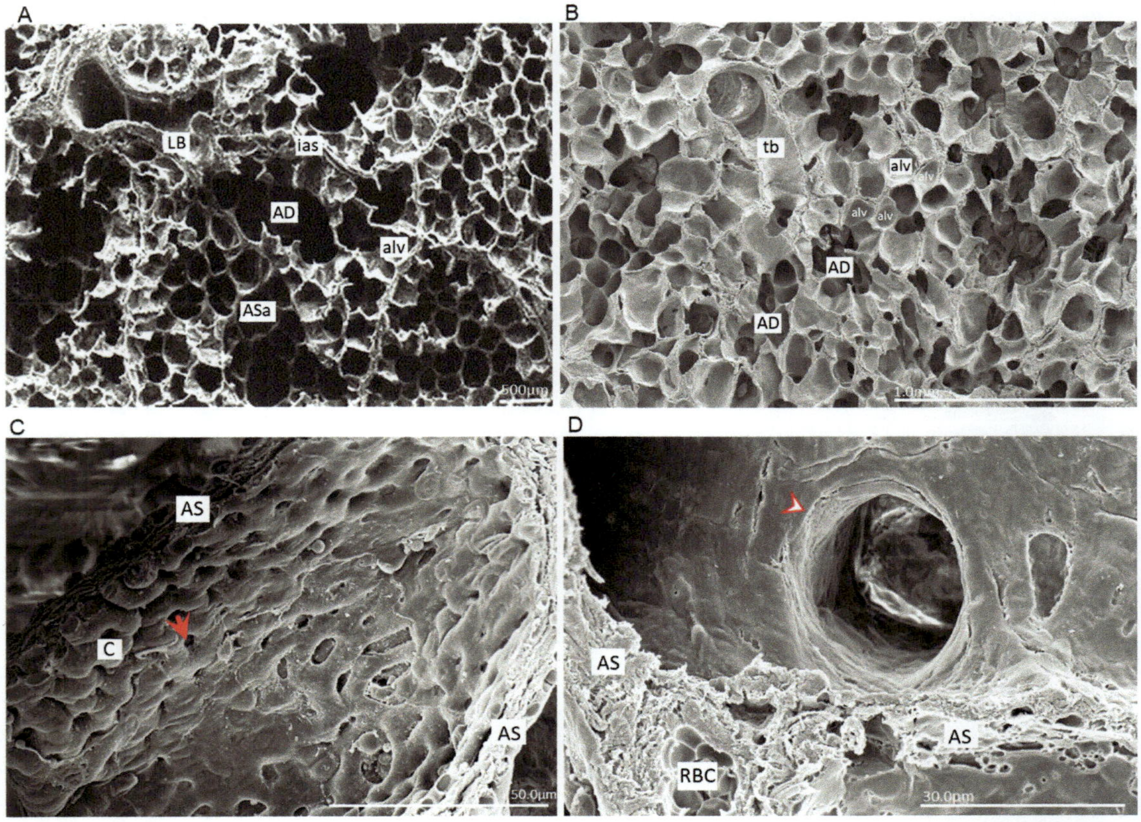

FIG. 10 Scanning electron microscopy (SEM) comparing alveolar morphology in (A) an Australian sea lion (*Neophoca cinerea*) compared to (B, C, D) an adult beluga whale (*Delphinapterus leucas*). (A) Sea lion lung parenchyma shows the division of lobular bronchus to supply individual acini. The thick stroma dividing the terminal airways and alveoli is apparent in the sea lion lung and not in the beluga. (B) Inflated terminal airways (0.5 mm in diameter), including terminal bronchi, alveolar ducts, and alveoli. (C) A bulging capillary bed on the surface of an alveolus. (C) and (D) frequent pores of Kohn visible (red arrowheads) perforating the thin alveolar septum of the alveolar wall. (D) The pore is approximately 24 μ in diameter and lined with type I alveolar epithelial cells with raised borders between type I cells extending down into the pore. Preserved lung tissue was at full capacity (air pressure of 30 cmH_2O). alv, alveolus; AD, alveolar duct; AS, thick and thin alveolar septa; ASa, alveolar sacs; ias, interacinar septa; LB, lobular bronchus; tb, terminal bronchi; C, capillary bed; RBC, red blood cell (Supplementary Video 1 in the online version at https://doi.org/10.1016/B978-0-323-90516-9.00005-1). (A) is from Nicholson, 1984. Beluga lung specimens: HI-2014-03 and HI-2014-08.

Pores of Kohn allow collateral ventilation between two groups of alveoli fed by different terminal bronchi and may also be a pathway for bacterial infections (such as pneumonia) to spread (Desplechain et al., 1983). They give the lung lobes an open structure or a collateral pathway that air can move through the lung without passing through the conducting airways. Until recently, no study has observed pores of Kohn in the alveolar wall of marine mammals [bottlenose dolphin (Fanning, 1977), Australian sea lion (*Neophoca cinerea*) (Nicholson, 1984), tucuxi and boto river dolphins (Santos Júnior et al., 2017)]. However, the presence of pores may depend on the animal's age or location of sampling; at least this variation has been observed in humans (Terry and Traystman, 2016). In addition, there has been little research examining marine mammal lungs using scanning electron microscopy (SEM) to observe alveolar spaces. We present the first evidence of pores of Kohn in beluga lungs that was 24 μm in diameter (Figs. 9 and 10). Collateral ventilation tends to occur in less lobulated lungs, and the evidence for collateral ventilation has been shown in a bottlenose dolphin (García-Párraga et al., 2018). With only a right and a left lung lobe in cetaceans, such a system would benefit a diver to prevent barotrauma. For example, if a region was obstructed, expanding air during ascent could escape via another route and not result in overexpansion and alveolar rupture. Another benefit has recently been proposed in the *selective gas exchange* hypothesis (Fahlman et al., 2021), where the alveolar collapse would be heterogeneous (García-Párraga et al., 2018). Thus, each lobe would have a collapsed or an open region. By directing blood flow to either area, the increased/decreased solubility (at varying pressures) would allow selective gas exchange (García-Párraga et al., 2018; Fahlman et al., 2021). Furthermore, support for the selective gas exchange hypothesis is that the diving-related heart rate response is conditioned in dolphins (Elsner et al., 1966; Fahlman et al., 2020), suggesting that there is an ability to actively alter gas exchange while diving. Such a hypothesis would help explain how alveolar tissue stress associated with repeated compression/reinflation during diving could result in failure of this mechanism and increase uptake of nitrogen, which may result in the formation of gas emboli and gas embolic pathology.

Functional studies in dead animals

Excised tissue

Conversely, physics determines much of respiratory function and capacity. For example, driving gas exchange is partial pressures and not the number of moles of gas, e.g., content. The partial pressure of a gas depends on the content and gas solubility. The "solubility" is the same for all gases in the gas phase and follows the ideal gas law. Thus, in a volume of gas, the number of moles and partial pressure are directly related and determined by Dalton's law. However, in a liquid or tissue, the solubility varies greatly between gases and the type of liquid or tissue. Thus, in a fluid or tissue, the content of oxygen or carbon dioxide is highly different, which results in different partial pressures for the same number of molecules of different gases. For example, carbon dioxide is approximately 35 times more soluble than oxygen in water. For this reason, a similar change in partial pressure requires 35 times more molecules of carbon dioxide than oxygen.

Similarly, limiting lung capacity is the size of the lungs and their "structural properties." Think of the structural properties as describing how flexible or compliant the lungs and the chest are. We predict that the structural properties of shallow-diving dolphins are different from those that are diving deep, as pressure will have different effects on the respiratory system. For example, a smaller lung and a stiffer upper airway, and a more compliant alveolar space will result in faster compression and alveolar collapse, which ends gas exchange. On the contrary, a larger lung with more compliant airways and stiffer alveolar space will collapse at a deeper depth.

Consequently, these properties are vital to understanding how gas exchange varies during diving. Unfortunately, measuring the structural properties of live animals has been challenging, and few studies have done so. An alternative approach has been to assess the compliance of excised tissues, which has helped define, or at least estimate, the total lung capacity, the minimum air volume, and the compliance of the lungs and conducting airways (Goldbogen et al., 2015; Fahlman et al., 2011, 2014; Denison and Kooyman, 1973; Kooyman and Sinnett, 1979).

In excised tissues, pressure-volume experiments can estimate lung compliance. Compliance is measured as the volume expansion per unit of pressure change and visualized as the steepness on pressure-volume curves. These pressure-volume experiments, known as static experiments (see Section "Lung function testing and respiratory health"), are performed by stepwise injections of a small volume of air into the lung until reaching a predetermined transpulmonary pressure. The steepness of these curves provides an estimate of the flexibility or compliance, where a steeper curve represents increased compliance. Differences in compliance between species provide scientists with a better understanding of how the lung will compress during diving, and more importantly, understand how the gas exchange will be affected. For example, changes in gas exchange with depth will affect the amount of nitrogen taken up from the lungs during a dive, which will affect the risk of gas bubble formation during ascent.

The pressure-volume experiments also estimate the total lung capacity, defined as the volume of air that results in a change in a predetermined transpulmonary pressure. Many studies use the volume reached at a transpulmonary pressure of 30 cmH_2O as the estimated total lung capacity. These lung inflation experiments have determined the estimated total lung capacity for several marine mammals as $TLC_{est} = 0.135 \times M_B^{0.92}$, where TLC_{est} is the estimated total lung capacity, M_b is the body mass in kg, and estimated total lung capacity is in liters (Kooyman, 1973; Piscitelli et al., 2010; Fahlman et al., 2011). Although this equation reliably predicts total lung capacity for a range of shallow and deep-diving marine mammals, there are some notable exceptions. Specifically, the sperm whale (*Physeter macrocephalus*), northern bottlenose whale (*Hyperoodon ampullatus*), pygmy sperm whale (*Kogia breviceps*), and dwarf sperm whale (*Kogia sima*) all have a considerably smaller mass-specific total lung capacity than estimated by the equation (Piscitelli et al., 2010). However, the volume of the conducting airways (dead space air volume) complicates this estimate. CT may provide new capabilities to estimate the relative volume of the conducting airways and the alveolar space more accurately in future studies (Fig. 11).

As previously mentioned, the size of the lung also affects at what depth the alveoli collapse, and a smaller lung volume would result in alveolar collapse and cessation of gas exchange at a shallower depth. This size difference may explain why the lungs of deep-diving cetaceans have been smaller compared with land mammals (Piscitelli et al., 2010). However, alveolar collapse would also prevent oxygen and carbon dioxide exchange and minimize the lung's use as an oxygen store. The reduced lung oxygen store appears compensated for in deep-diving cetaceans. They tend to have greater blood volume with a high hematocrit and hemoglobin and a larger than average relative muscle mass and muscle myoglobin concentration (Pabst et al., 2016). Near-shore bottlenose dolphins are usually shallow divers and thus appear to have a lung capacity that the estimated total lung capacity equation can calculate. However, there are off-shore populations of bottlenose dolphins that dive deep (>1000 m), and their tidal volume does not appear to be different from near-shore, shallow-diving bottlenose dolphins (Klatsky et al., 2007; Fahlman et al., 2018, 2023). There are no differences in lung mass between near-shore and off-shore bottlenose dolphins (Piscitelli et al., 2010), suggesting that small lung size is not a general adaptation to prevent nitrogen uptake.

Cadavers

Although studies of excised tissues provide valuable data on the structural properties, these studies do not provide insight into respiratory system function while breathing normally or while exposed to pressure. For example, excised tissues cannot answer how much the alveoli and airways compress when inside a compliant or stiff chest during a dive. Experiments on seal cadavers showed that the chest wall was very flexible and did not provide much resistance to compress. In the dolphin, on the other hand, the force to inflate the chest and lung was considerable. On the contrary, the dolphin chest had considerable elastic recoil and emptied to low volumes when relaxed. The dolphin chest wall is similar to a rubber band stretching, which requires work, returning to its original size when released. Thus, the lungs empty to low volumes in both the seals and dolphins, so they are essentially gas-free when relaxed (Leith, 1976; Kooyman and Sinnett, 1979; Fahlman et al., 2014, 2018). See Section "Functional studies in live animals": the lungs and pressure for further discussion on thoracic collapse at depth.

CT imaging of whole cadavers placed in a fiberglass hyperbaric chamber examined the effects of pressure on lung function. Common dolphin (*Delphinus delphis*) cadavers obtained from fisheries bycatch had their airways intubated, allowing the inflation of the lungs with a known volume of air. Both the conducting airways and alveoli compressed with increasing pressure, and alveolar collapse and termination of gas exchange likely did not happen until a depth of at least 100 m (Fig. 12) (Moore et al., 2011; Denk et al., 2020). These results agreed with chamber studies on live seals and sea lions where venous and arterial gas content were measured, and the pulmonary shunt was calculated at different pressures (Kooyman and Sinnett, 1982). In the common dolphin, the alveolar space and conducting airways compress under pressure from the alveoli to the trachea (caudally to cranial) (Denk et al., 2020) (Fig. 12). Thus, the chamber experiments on cadavers provided additional information on how the respiratory system responds under pressure. However, as pointed out by the authors, postmortem animals do not have cardiovascular tone and function, affecting lung function and mechanics. In live human breath-hold divers, the trachea undergoes compression of the posterior wall when under pressure with a decreased tracheal diameter and average 50% reduction in air volume (Lindholm and Nyrén, 2005). Thus, tracheal compression occurs in both dolphins and humans.

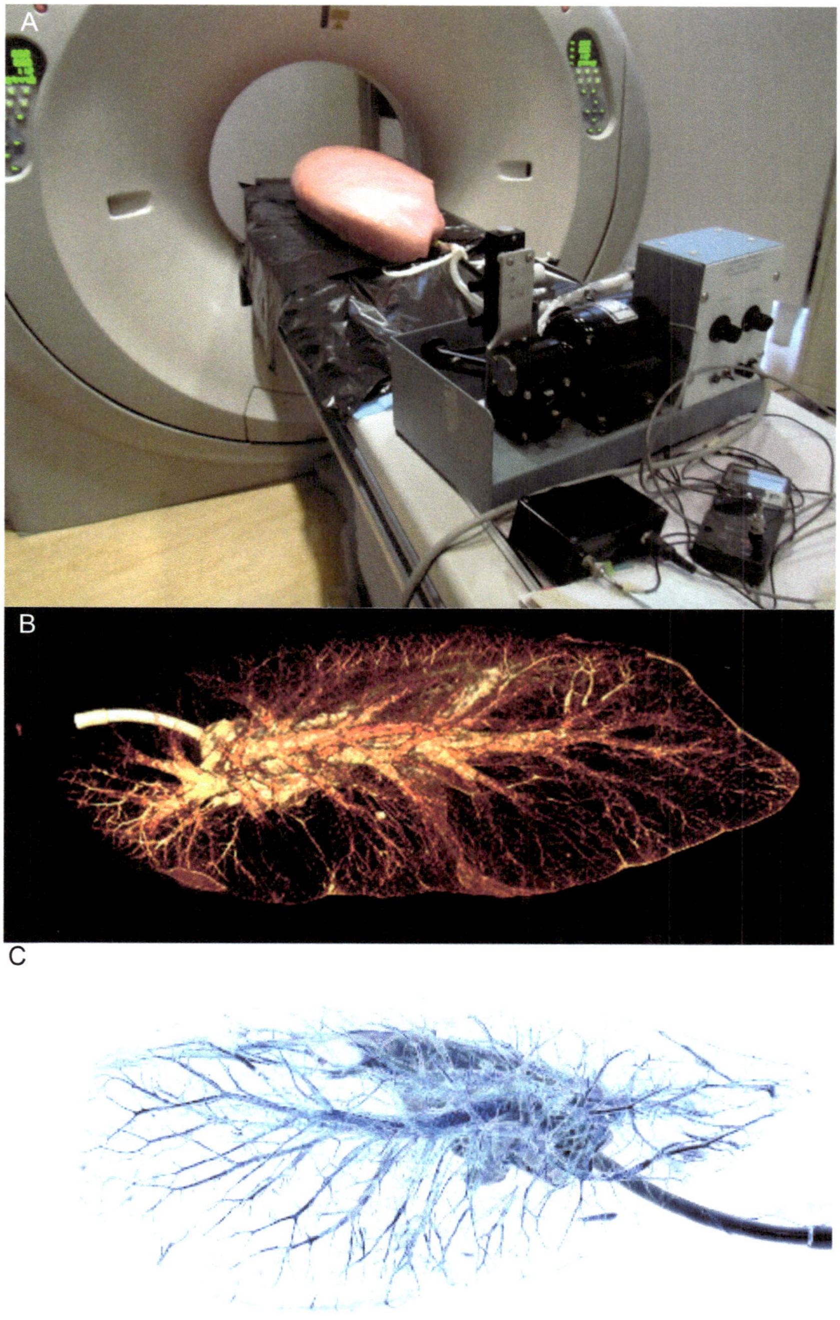

FIG. 11 A computed tomography (CT) scan of lungs at full inflation (30 cmH_20). (A) Image of the inflated beluga (*Delphinapterus leucas*) lung lying on the bed of a CT scanner (Supplemental Video 2 in the online version at https://doi.org/10.1016/B978-0-323-90516-9.00005-1). (B) 3D reconstruction of the airways in a minke whale (*Balaenoptera acutorostrata*) lung. (C) 3D reconstruction of a killer whale (*Orcinus orca*) lung (Supplementary Video 3 in the online version at https://doi.org/10.1016/B978-0-323-90516-9.00005-1). The gold-white features on the external lung are lymphatic tissues (such as lymph nodes and vessels). The internal gold and blue highlighting in the 3D reconstructions show the monopodial branching pattern and robust nature of the cartilage throughout all the airways. CT scans and reconstructions were performed with the help of Dean Malpas, Canada Diagnostics, and with assistance from R. Shadwick. Specimens: HI-2012-01 Dl, VGT235 Ba, MBH010713 Oo. *Reproduced from Goldbogen, J.A., Shadwick, R.E., Lillie, M.A., Piscitelli, M.A., Potvin, J., Pyenson, N.D. et al., 2015. Using morphology to infer physiology: case studies on rorqual whales (Balaenopteridae). Can. J. Zool. 93(9), 687–700.*

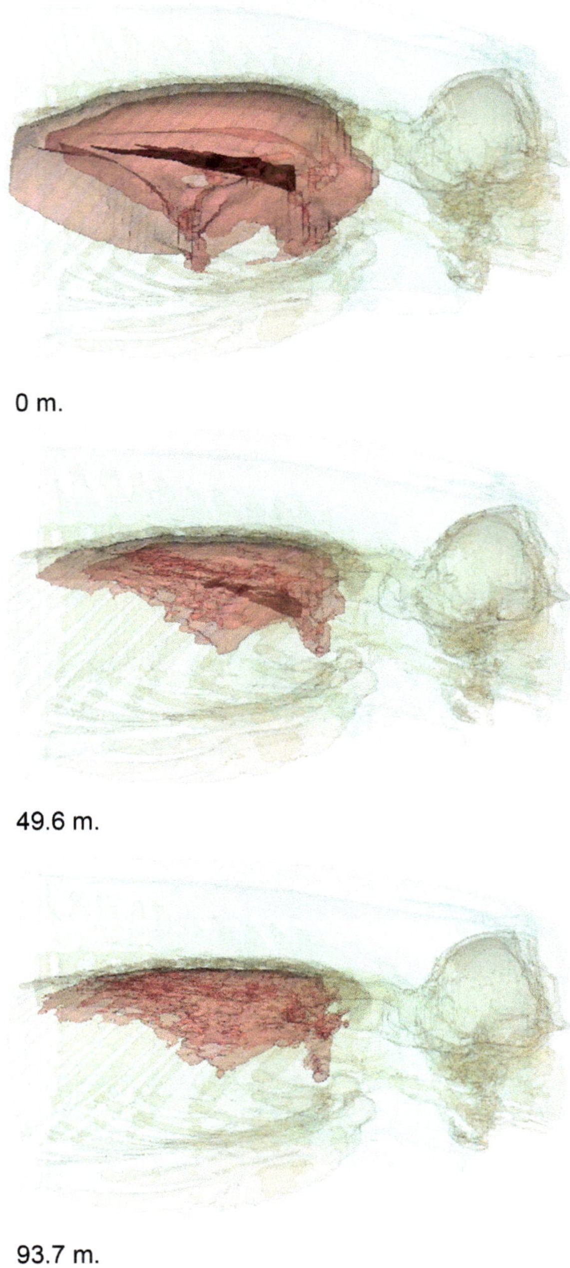

FIG. 12 Images from computed tomography of a deceased common dolphin (*Delphinus delphis*) pressurized to 49.6 and 93.7 m. The black linear shape in the lung represents the air within the mainstem bronchi. The lung and mainstem bronchi compress caudally and ultimately disappear with increasing pressure. The lung volume changed the most between the surface (0 m, upper panel) and 49.6 m (middle panel) treatments. *Reproduced from Denk et al. (2020).*

Functional studies in live animals

The breathing strategy of cetaceans is very different from that of land mammals. Short, forceful breaths and a low breathing frequency suggest that cetaceans have larger tidal volumes than land mammals. Measurements of tidal volume have confirmed this. As early as the 1940s, smaller cetaceans, such as the harbor porpoise and bottlenose dolphin, were restrained and a flow meter was placed above the blowhole (Scholander, 1940; Irving et al., 1941). Stress may have an impact on respiration in restrained animals. One way to minimize stress during respiration studies is to work with trained animals. Trained animals are willing to participate in research activities. Although this controlled environment may introduce different biases, it allows measurements to be collected that would be difficult to obtain from wild animals (Pryor, 2019; Ramirez, 2019). Thus, studying trained animals provides an opportunity to get measures that may be difficult or ethically challenging to collect from wild animals (see Section "Future possibilities").

Lung function testing in live cetaceans

In the late 1960s, the first lung function measurements in live cetaceans were performed in a female killer whale (*Orcinus orca*) and a trained pilot whale (*Globicephala macrorhynchus* [then called *scammoni*]) (Spencer et al., 1967; Olsen et al., 1969). In both the pilot whale and killer whale, the average tidal volume as a percentage of the total lung capacity (killer whale: 55%, range: 36%–55%; pilot whale: range: 20%–88%) was much larger as compared with humans (about 20%–25% of total lung capacity). The largest breath measured in the pilot whale was 88% of the total lung capacity, suggesting that the residual volume was as low as 12% of the total lung capacity. The compliance of the respiratory system, which is a measure of how flexible the lung is, was measured in a number of pinnipeds and the bottlenose dolphins and pilot whale and was at least twice as compliant as a human lung (Olsen et al., 1969; Leith, 1976; Fahlman et al., 2011, 2014, 2015) (Table 2). In both the killer whale and pilot whale, the breath duration was about 1 s, and the maximal expiratory flow in the killer whale reached up to 180 L s^{-1} or about 12 times higher than humans (Spencer et al., 1967; Olsen et al., 1969; West, 2012). Lung function measurements performed in a trained bottlenose dolphin confirmed an exceptional respiratory capacity that allows the exchange of almost the entire lung volume within a second (Kooyman and Cornell, 1981). In addition, the study on the bottlenose dolphin also showed that the expiratory flow, unlike in humans, appears to be effort-dependent (Kooyman and Cornell, 1981). This observation means cetaceans can maintain a high respiratory flow over all lung volumes. In contrast, the highest respiratory flow occurs at a specific lung volume in a human and then decreases either at higher or lower volumes (Fig. 8). This maintenance of a high flow rate allows the dolphins to rapidly complete a breath, even when the tidal volume is close to total lung capacity.

Minimum air volume, residual volume, and total lung capacity

As explained, lung function and mechanics measurements are vital to understand how gas exchange changes during diving. In addition, predictions of human disturbances, such as climate change, may impact diving and foraging capacity. However, such measurements have been challenging as equipment for these measurements has not been readily available. Custom-made spirometers for dolphins, introduced by Kooyman and Cornell (1981), allow

TABLE 2 Values of specific lung compliance (sC_L) for a range of marine mammal species.

Species	sCL (cmH_2O^{-1})	References
Harbor seal (*Phoca vitulina*)	1.48 ± 0.03	Fahlman et al. (2011, 2014)
Northern elephant seal (*Mirounga angustirostris*)	0.68 ± 0.20	Fahlman et al. (2014)
Steller sea lion (*Eumetopias jubatus*)	0.25 ± 0.05	Fahlman et al. (2014)
California sea lion (*Zalophus californianus*)	0.92 ± 0.16	Fahlman et al. (2014)
Northern fur seal (*Callorhinus ursinus*)	1.58	Fahlman et al. (2014)
South American sea lion (*Otaria flavescens*)	0.41 ± 0.10	Fahlman and Madigan (2016)
Walrus (*Odobenus rosmarus divergens*)	0.61	Fahlman et al. (2015)
Bottlenose dolphin (*Tursiops truncatus*)	0.31 ± 0.04	Fahlman et al. (2015)
Pilot whale (*Globicephala scammoni*)	0.13	Olsen et al. (1969)
Human (*Homo sapiens*)	0.08–0.12	Stahl (1967), Stocks and Quanjer (1995), Galetke et al. (2007), West (2012)

sC_L is computed as the measured lung compliance divided by the minimum air volume.

measurement of the tidal volume, functional residual capacity, and inspiratory reserve capacity, but not the residual volume. In humans, body plethysmography or a gas dilution test measures residual volume. Body plethysmography is not straightforward for dolphins, but a helium gas dilution determined residual volume in a pilot whale (Olsen et al., 1969). This study confirmed that the functional total lung capacity is similar to that measured in excised lungs and that the residual volume is smaller compared with land mammals.

Although the lung capacity of dolphins has been estimated using excised tissue, this has not been measured in live dolphins. Spirometry measurements suggest that the vital capacity is close to the estimated total lung capacity from excised tissues, which agrees with the results in the pilot whale. These results also agree with data from excised lungs that suggest that the minimum air volume, as determined by water displacement, is about 7% (Kooyman and Sinnett, 1979; Piscitelli et al., 2010; Fahlman et al., 2011). Consequently, dolphins, and likely also other cetaceans, can exchange almost all the gas volume inside the lung during a single breath. This concept has important implications for gas exchange in a species that spend considerable time underwater. Imagine that you have been diving for a few minutes to catch your food. Now you are back at the surface to replenish the oxygen consumed and remove the carbon dioxide produced from aerobic metabolism during the dive. If you have ever held your breath, you know how the urge to breathe makes you breathe faster and deeper when you return to the surface. You do this to increase the minute and alveolar ventilation, which helps bring more oxygen into the body and also helps remove the carbon dioxide. As summarized earlier, the critical measure for gas exchange is the alveolar ventilation, which is the minute ventilation minus the dead space ventilation (see Section "Introduction"). The alveolar ventilation increases more with a change in tidal volume than breathing frequency. Consequently, the low minimum air volume/residual volume in dolphins allows them to take deep breaths and exchange almost the entire lung full, which then helps improve the alveolar ventilation. However, dolphins can exchange nearly the whole lung air volume during a single breath, but they also do this with extremely high respiratory flows.

Respiratory flow

Dolphins swimming or porpoising must expire and inspire during the very short duration their blowhole is in the air. Like most marine mammals, dolphins breathe by first exhaling and then inhaling (the aquatic breathing strategy). The exhalation is audible, and during deep breaths, it sounds like a small explosion or like puncturing a balloon (Supplementary Video 4 in the online version at https://doi.org/10.1016/B978-0-323-90516-9.00005-1). Much of the spout of a whale is because the exhalation begins before the blowhole is above water, leaving more time for the slower inhalation, which must occur before submersion. During voluntary breaths, exhalation is primarily passive, driven by the elastic recoil of the chest. However, during maximal respiratory efforts, the flow increases with the help of the respiratory (intercostal) muscles and diaphragm, enabling extreme respiratory flows (Cotten et al., 2008; Fahlman et al., 2015, 2017). Consequently, active exhalation recruits air from the expiratory reserve volume (Fig. 2), which increases both expiratory and inspiratory volumes for the next breath. The maximal effort represents a flow rate of approximately 10.3 vital capacities per second (Fig. 8 shows a breath for a forced breath to 6.5 vital capacities per second in the harbor porpoise) or between three and five times higher than the respiratory flow capacity of humans (Kooyman and Cornell, 1981). Similar respiratory capacities have also been measured in the pilot whale (Olsen et al., 1969) and the beluga (Fahlman et al., 2019) and make cetaceans the respiratory flow rate champions of all mammals studied.

Inspiratory flows are not as impressive, where in the bottlenose dolphin and beluga, they are generally only 25%–33% of the expiratory flow. Inhalation is an active phase and requires muscle effort to expand the lungs and the chest. Imagine the dolphin's chest as it inhales similar to a rubber band stretching. The energy used to stretch the rubber band (expanding the chest) releases as the rubber band (or chest) recoils inwards. A higher expiratory flow than inspiratory is typical for mammals, with active inhalation driven by muscle work. In contrast, exhalation is driven by passive elastic recoil of the chest and active muscle work. The respiratory flow has only been measured on dolphins at rest. While swimming, there is likely coupling between locomotion and breathing, which potentially will result in even higher respiratory flows.

Measurements of respiratory flow in wild cetaceans are few, and in most instances, rely on respiratory proxies or stranded animals (Spencer et al., 1967; Sumich, 2001). However, direct measurements of respiratory flow have been made for wild, juvenile gray whales (*Eschrichtius robustus*) in a shallow enclosed area in Baja, California. These measurements suggested that while resting at the surface ("logging"), whales seldom breathe maximally and that the range of respiratory flows and tidal volumes is likely much lower (see Section "Future possibilities").

The lungs and pressure

During breath-hold diving, dolphins are exposed to the increasing pressure as they descend deeper into the water column. Boyle's law applies here and states that at a constant temperature, the pressure (P) of a given quantity of gas varies inversely with its volume (V):

$$P \bullet V = k \quad (1)$$

where k is a constant. For the same gas after a change in either the volume or pressure from time 1 to time 2, Boyle's law can be expressed as follows:

$$P_1 \bullet V_1 = P_2 \bullet V_2 \quad (2)$$

Or, in words, the pressure and volume at time 1 ($P_1 \bullet V_1$) are equal to the pressure and volume at time 2 ($P_2 \bullet V_2$). So, what happens in your ears as you dive, or in the airplane as you ascend, is that the gas volume changes, pulling/pushing on the surrounding structures, which causes discomfort until the pressure inside the ears is equalized (opening the Eustachian tube to allow more air into the inner ear).

In breath-hold divers, the respiratory system undergoes similar changes, and with increasing pressure, the respiratory gas volume decreases. At the surface, the ambient pressure is 1 ATA (Table 3), and for every 10 m in depth, the pressure increases by 1 ATA, for a total pressure of 2 ATA at 10 m. Assuming that the volume is 10 L at the surface, we can rearrange the equation for Boyle's law and determine the volume ($V_2 = P_1 \bullet V_1 \bullet P_2^{-1}) = 1 \bullet 10/2 = 5$ L). Thus, a gas volume will halve at 10 m and be a third of surface volume at 20 m (3 ATA).

As the lungs' gas volume changes, so do the pressure inside the lungs so that it is equal to the ambient (outside) pressure. The increased pressure results in increased gas partial pressure, which has several effects. First, it affects gas exchange, or the diffusion rate, according to Fick's law of diffusion:

$$\text{Diffusion rate} = \Delta P \bullet \text{SA} \bullet D \bullet T^{-1} \quad (3)$$

where ΔP is the partial pressure difference inside the alveoli and the pulmonary capillary, SA is the surface area available for diffusion, D is a diffusion rate coefficient specific for each gas and membrane, and T is the thickness of the membrane through which diffusion happens. As the ambient pressure increases during descent, the pressure inside the alveoli increases as the lungs compress (Wilmshurst, 1998). This compression increases the partial pressure of oxygen, nitrogen, and carbon dioxide, resulting in a higher diffusion rate. However, as the lungs compress, the surface area for diffusion decreases. It is also likely that the thickness of the membrane increases, thereby reducing the diffusion rate, resulting in what is commonly called a pulmonary shunt (Bostrom et al., 2008). The pulmonary shunt continues to increase until all the alveoli have collapsed, at which time gas exchange ceases. Alveolar compression and collapse were theoretically described by Scholander as early as 1940, then called lung collapse, and then predicted theoretically (Bostrom et al., 2008; Fahlman et al., 2009), and also shown experimentally in both pinnipeds and dolphins in several seminal studies (Ridgway and Howard, 1979; Kooyman and Sinnett, 1982; Falke et al., 1985; McDonald and Ponganis, 2012).

A combination of behavior and respiratory adaptations affects the respiratory system compressing and collapsing. Imagine the mammalian lungs as a very soft and pliable (compliant) balloon (the alveoli) attached to a stiff (noncompliant) tube (the conducting airways), which we will call the Scholander balloon pipe model of the lung. The balloon pipe is filled with a total of 10 L of air with 9 L (V_B) in the balloon and 1 L in the stiff airways (V_A). First, let us assume that the conducting airways are relatively incompressible, and pressure will mainly cause compression of the balloon. As you lower this balloon pipe into the water, the pressure will compress the volume of the balloon (alveoli), and gas pushes into the upper airways. We can use a slightly different form of Boyle's law to determine the depth (pressure) where the alveoli collapse. In this case, we have the following equation:

$$V_{B2} = (V_{B1} + V_A) \bullet P_1 \bullet P_2^{-1} - V_A \quad (4)$$

TABLE 3 Pressure units as kilopascal (kPa), atmosphere absolute (ATA), bar, millimeter of mercury (mmHg), centimeter of water (cmH_2O), and pounds per square inch (psi).

kPa[a]	ATA	bar	mmHg torr^{-1}	cmH_2O	psi
1	0.00987	0.01	7.5	10.2	0.145
101.3	1	1.013	760	1033	14.7

[a] *SI unit which is the scientific standard.*

which is a special case of Boyle's law earlier described. The balloon (alveoli) collapses when $V_{B2} = 0$, which happens when $P_2 = (V_{B1} + V_A) \bullet P_1 \bullet V_A^{-1}$. In other words, the balloon, or alveoli, collapses at a depth when the air volume of the lungs (airways and alveoli) is equal to the upper airways, or all the gas pushed into the upper airways (Bostrom et al., 2008). However, as the conducting airways are also collapsible, as shown in experiments in live pinnipeds (Kooyman et al., 1970), this relationship is more complicated as now the conducting airways will also compress. Consequently, the way that the conducting airways and alveoli compress during diving depends on the structural properties of the different parts of the respiratory system (Bostrom et al., 2008). Species with very stiff conducting airways will experience alveolar collapse and cessation of gas exchange at deeper depths than those with flexible upper airways, with all other things being equal. Beaked whales, the champion deep-diving cetaceans, appear to have very stiff conducting airways, possibly as an anatomical adaptation to help the alveoli collapse at shallower depths (Moore et al., 2014).

The residual volume is considerable in humans, and the stiff chest resists further volume reduction. The increasing pressure as the human diver descends without a reduction in volume results in negative pressure within the thorax, called lung squeeze (Lundgren and Miller, 1999). If the diver continues to descend and the lung squeeze increases it will eventually result in pulmonary edema and force fluids into the alveoli (Wilmshurst, 1998). One effect of the negative pressure is that blood tends to be drawn into the chest. Intrathoracic blood pooling has been reported in human breath-hold divers (Craig, 1968; Schaefer et al., 1968). Both the dive response and negative intrathoracic pressure may help redistribute blood into the thoracic cavity, which may help compress the alveoli and prevent barotrauma (Craig, 1968; Schaefer et al., 1968). In a human, an estimated 1.4–1.7 L of blood can be redistributed and fill the thorax (Schaefer et al., 1968). This volume represents 25%–30% of an average adult human's total blood volume (approximately 5.5 L) (Saladin, 2007).

On the other hand, marine mammals have several anatomical adaptations to prevent lung squeeze. A chest with considerable elastic recoil in the dolphin leaves little air in the relaxed respiratory system (Leith, 1976; Fahlman et al., 2014, 2018). The chest of the dolphin has double-jointed ribs that allow the chest to compress to very low alveolar volumes (Fig. 12) (Piscitelli et al., 2010). This anatomical adaptation was photographed in a bottlenose dolphin (thought to dive on full inspiration) diving to 300 meters, showing the compression of the chest (Fig. 13) (Ridgway et al., 1969).

FIG. 13 Photograph of "Tuffy," a trained US Navy bottlenose dolphin (*Tursiops truncatus*), at 300 m depth. Compression of the chest wall began around the first 10 m of the dive and continued to 70 m. Divers observing from depth described the compression as just caudal to the pectoral flipper. *Image from Ridgway et al. (1969).*

Data in freely diving Weddell seals suggests chest compression may be relatively limited (Falke et al., 2008). One study demonstrated that the seal's thoracic circumference decreased by less than 2% when measured at 196 m. This study hypothesized that vascular structures within the thorax might become engorged with blood, similar to intrathoracic blood pooling in humans, to help compress the alveolar space and prevent negative intrathoracic pressures (Falke et al., 2008). Phocid seals possess elaborate venous structures within their thoracic cavity that would allow either intrathoracic blood pooling or a thoracic (respiratory) pump that would allow blood engorgement (Harrison and Tomlinson, 1956; Fanning and Harrison, 1974; Ponganis et al., 2006; Falke et al., 2008).

The thoracic (respiratory) pump is another mechanism that may aid the flow of venous blood from the abdominal to the thoracic cavity during ventilation (Lindholm and Nyrén, 2005; Ponganis et al., 2006; Saladin, 2007). During inhalation, the thoracic cavity expands, and the internal pressure drops. Simultaneously, the caudal movement of the diaphragm raises the pressure in the abdominal cavity. This rise in abdominal cavity pressure results in a negative pressure differential pulling the blood cranially, via the caudal vena cava, from the abdominal to the thoracic cavity (Saladin, 2007). For those marine mammals that inhale before diving, like cetaceans, the thoracic pump may bring venous blood into the thorax just before diving, followed by intrathoracic blood pooling as part of lung compression and reduction in peripheral blood flow. Lillie et al. (2018) described the presence of a caval sphincter in shallow-diving dolphins that may regulate blood flow from the abdominal cavity into the thoracic cavity (see also Chapter 5: Cardiovascular physiology in dolphins and other cetaceans). However, no study has mechanically investigated the caval sphincter in other cetacean species.

These two vascular responses, well-documented in humans, suggest that deeper-diving cetaceans, which likely have larger blood volumes than shallow divers (Ridgway and Johnston, 1966; Ridgway and Harrison, 1986), may redirect a large amount of blood into the thoracic cavity during a dive. If this hypothesis is true, then one would expect a deeper-diving cetacean to possess more voluminous vascular structures within its thoracic cavity (Melnikov, 1997), as well as mechanisms of blood movement from the periphery to the thorax. Scholander (1940) referred to vascular structures that could dynamically change shape as "swelling" and "displacement" organs. Several vascular structures within the cetacean thoracic cavity may function in this manner (see Chapter 5 for further discussion). These structures include the thoracic arterial rete, intrathoracic venous appendages, and the pericardial venous plexus (Scholander, 1940; Harrison and Tomlinson, 1956; Ridgway, 1972; Hui, 1975; Vogl and Fisher, 1982; Melnikov, 1997; Ninomiya et al., 2005; Ponganis et al., 2006; Costidis and Rommel, 2012).

The thoracic arterial rete lies dorsally between the ribs. It extends into the neural canal to form the epidural rete (Walmsley, 1938; Ridgway and Johnston, 1966; Ridgway, 1972; Vogl and Fisher, 1982; Costidis et al., 2009; Lillie et al., 2017). This rete might affect the degree and pattern of thoracic collapse in a shallow-diving common dolphin (Hui, 1975). However, Scholander (1940) hypothesized that the volume of this vascular structure was too small in the shallow-diving harbor porpoise to permit it to function as a significant blood reservoir. Instead, the thoracic arterial rete may act as a pressure dampener to the heart and brain during fluking, which acts like a hydraulic hammer driving blood forward and absorbed by the rete (Lillie et al., 2017). However, it also may have another novel function in deeper-diving cetaceans that stroke to depth in helping collapse the lung. In deeper-diving cetaceans, such as *Kogia* spp., beaked whales, and sperm whales, the thoracic arterial rete is even more extensive and voluminous (Walmsley, 1938) compared to what Scholander (1940) and Hui (1975) found in the shallow-diving dolphins and porpoises. Piscitelli et al. (2010) found the intrathoracic arterial rete was double the volume in a pygmy sperm whale compared to a bottlenose dolphin. In these routine deeper-diving cetaceans, the thoracic arterial rete completely covers the cranial thoracic inlet and extends laterally around the lungs and dorsally into the epaxial muscle and cranially up to the back of the skull around the larynx and trachea (Fig. 14).

Additionally, extensive intrathoracic venous appendages surround the lungs of all deep-diving cetaceans (kogiids, physeterids, and ziphiids) (Costidis et al., 2009). These appendages fill the pleural cavity in the deepest-diving cetaceans, ziphiids (Fig. 15). This study hypothesized that these veins might become engorged with large volumes of blood at depth and potentially function in intrathoracic blood pooling to help collapse the lungs while maintaining pressure equilibrium across the body wall. This concept would be considered a similar but more extreme case of what happens in human breath-hold diving.

Finally, a pericardial venous plexus has been described in the harbor porpoise, harbor seal, grey seal (*Halichoerus grypus*), the elephant seal, and leopard seal (*Hydrurga leptonyx*) (Burne, 1910; Harrison and Tomlinson, 1956; Ponganis et al., 2006). This plexus, a continuous venous ring surrounding the caudal base of the pericardium, connects the caudal vena cava dorsally and the phrenic veins ventrally. It can possess leaf-like projections into the pleural cavity formed by thick-walled veins (0.2–0.3 mL) surrounded by lobules of brown fat (Harrison and Tomlinson, 1956). However, Scholander (1940) hypothesized that these venous plexuses' volume was too small to function as a "swelling" organ

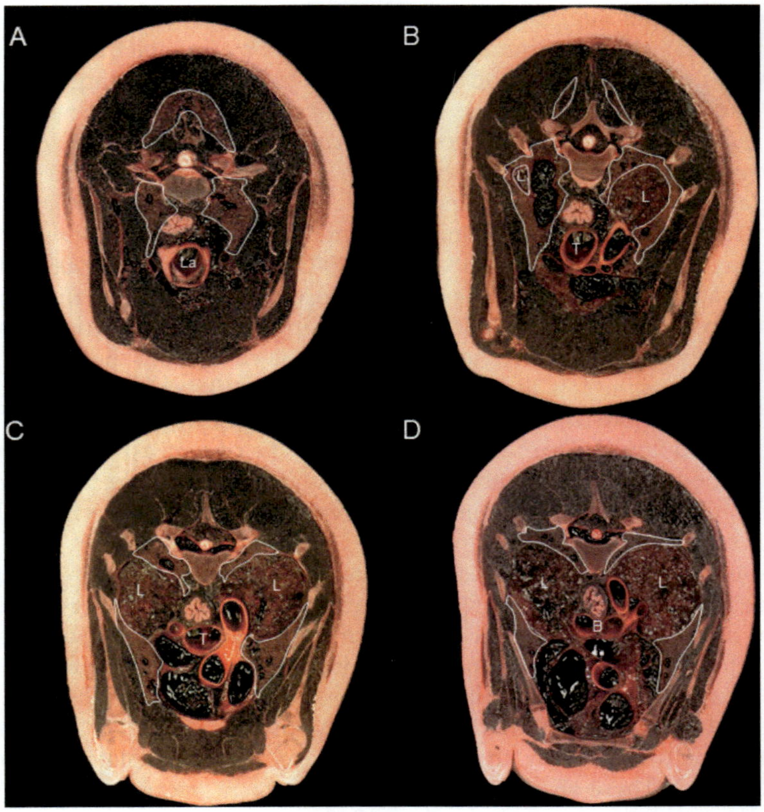

FIG. 14 Cross-sections through a dwarf sperm whale (*Kogia sima*; 160.0 cm in total length) highlight the extent of the thoracic arterial rete (outlined in white). For centimeter comparison to the bottlenose dolphin (*Tursiops truncatus*), see cross-sections of the dolphin in Huggenberger et al. (Huggenberger et al., 2019). L, lung; La, larynx; B, bronchus; T, trachea. Processed cross-sections with assistance from D.A. Pabst, W. McLellan, and S. Rommel. Specimen: VAQS20081002.

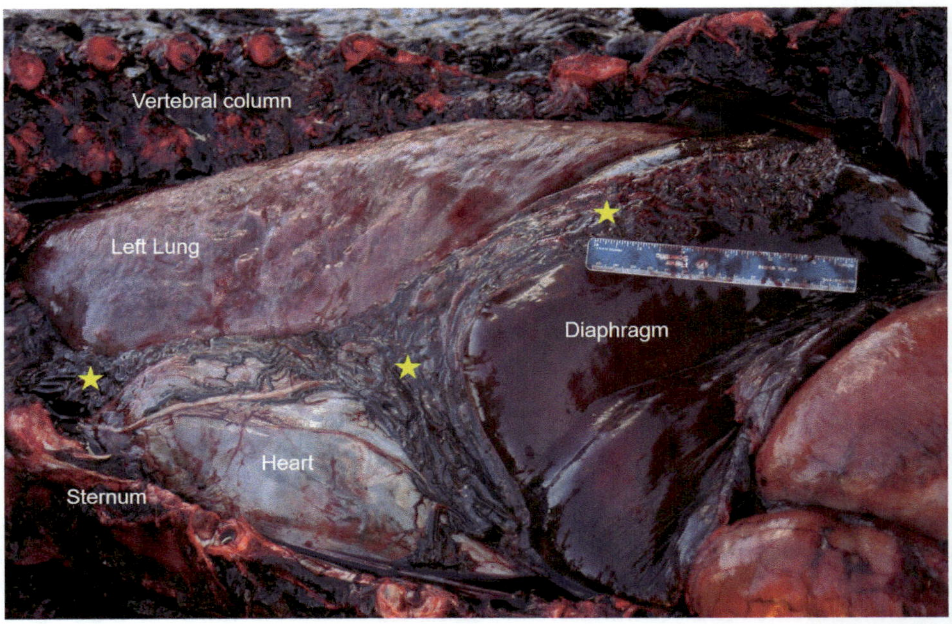

FIG. 15 Engorged intrathoracic venous appendages (denoted by yellow stars) in a Blainville's beaked whale (*Mesoplodon densirostris*). The appendages at their distal cranial and caudal ends are free-floating and unattached. Specimen: GNL053 Md. *Photo courtesy of Gretchen Lovewell.*

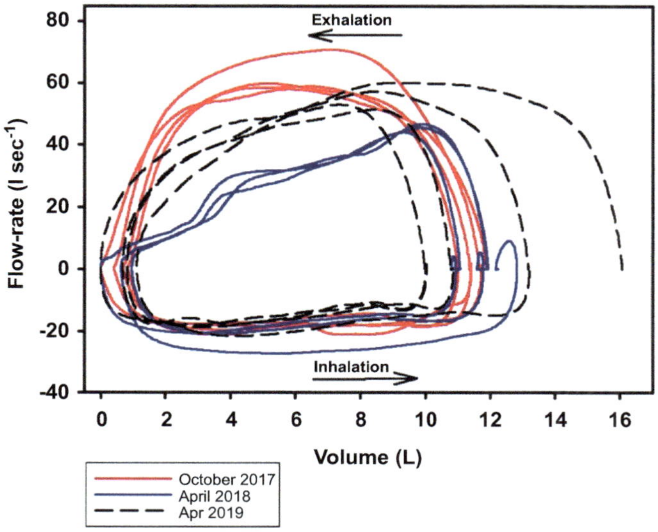

FIG. 16 Flow-volume relationship of a male bottlenose dolphin (*Tursiops truncatus*, age: 6–8 years, body mass: 177 kilograms) when healthy in October 2017, diagnosed with pneumonia in 2018, and after recovery in 2019.

within the thoracic cavity of the harbor porpoise. Thoracic vascular structures may potentially affect thoracic shape change, but the relative size of these thoracic vascular structures in a deep versus shallow-diving cetacean still needs to be investigated.

Lung function testing and respiratory health

Respiratory disease is a significant problem for cetaceans both in the wild and those held under managed care (Sweeney and Ridgway, 1976; Howard et al., 1983; Baker, 1992; Johnson et al., 2009; Bogomolni et al., 2010; Venn-Watson et al., 2012). One problem is that dolphins often do not show symptoms until they are severely affected. Methods to detect lung disease include CT, radiography, ultrasound, cytology, and blood samples (Sweeney and Reddy, 2001; Johnson et al., 2009; Smith et al., 2012; Ivancic et al., 2014; Martony et al., 2017). CT is the preferred diagnostic method, but it is logistically challenging and costly. In addition, there are portable ultrasound and radiography (X-rays) units. Ultrasound can be performed in water but can only image the periphery. Radiography penetrates deeper into the animal but involves beaching it and exposing it to ionizing radiation. Finally, cytology and blood samples provide valuable information about the active disease and the type of infection, e.g., fungal or bacterial. However, none of these modes give information on the lungs' functional aspects of abnormalities or changes. On the other hand, spirometry testing may be a useful diagnostic modality that may help define the functional aspects of respiratory disease.

Spirometry testing is a common diagnostic test used by human doctors, where a patient breathes into a device to measure lung capacity. These tests that assess respiratory capacity and effort allow the physician to evaluate lung health. The patient breathes into a flow meter for the test that measures the respiratory flow during several different maneuvers. For a test of flow restriction, the patient inhales fully to total lung capacity and then exhales forcefully. The results allow the physician to evaluate the lungs for diseases, such as bronchial constriction or emphysema, detected as shape changes in the relationship between respiratory flow and volume (Fig. 16). In dolphins housed in managed care and trained to perform maximal respiratory maneuvers, lung function testing is a useful diagnostic test and evaluates treatment efficacy for lung disease (Borque-Espinosa et al., 2020). It can also provide diagnostic information during voluntary breaths and could be a useful method to help evaluate lung health for wild, stranded animals.

Future possibilities

Direct respiratory capacity measurement in free-swimming dolphins is challenging, but several methods have developed. One such way, named phonospirometry, records the sound of the breath and then estimates the respiratory flow from the sound level (Sumich, 2001). Some limitations with this technique are that it requires validation and calibration for each species. It will also depend on the recording device placement on wild animals as the distance from the blowhole will affect the sound level. In addition, the recording device records the breath's sound in the air; as a result, both the tag and blowhole need to be in the air simultaneously. Finally, as many cetaceans begin to exhale before breaking the water surface, exhalations are often difficult to hear. As a result, most measurements in dolphins have been on inhalation.

Another method proposed is using the heart rate changes associated with breaths, also called respiratory tachycardia. These changes in heart rate associated with respirations are common in all mammals. In humans, the magnitude of the changes is related to the breathing frequency and the tidal volume. Two studies have shown a similar correlation between the magnitude of the heart rate changes and tidal volume in dolphins, which may provide an alternative method to estimate respiratory function (Cauture et al., 2019; Blawas et al., 2021). However, variation in autonomic tone or respiratory disease may alter the respiratory tacgycardia, and this method will require substantial further validation. Advancements in noninvasive physiological sensing technology such as near-infrared spectroscopy may also allow assessing the coupling between heart rate and gas exchange. Pulse oximetry is standard in human medicine to measure the pulse rate and blood oxygen saturation level using near-infrared spectroscopy to shine light through a finger or toe. Recently, this technology has been used to evaluate cerebral blood flow and oxygenation in breath-hold diving humans and seals (McKnight et al., 2019; McKnight et al., 2021). Therefore, near-infrared spectroscopy has the potential to provide a novel view of cardiorespiratory coupling and gas exchange efficiency by measuring heart rate and changes in oxygenation following a breath.

Glossary

Alveolar recruitment The point when collapsed alveoli open up and gas exchange resumes.
Atelectasis Alveolar collapse, resulting in cessation of gas exchange.
Collateral ventilation Ventilatory flow through the lung parenchyma through alternative flow pathways, such as pores of Kohn.
Dead space The volume of air in the respiratory system not participating in gas exchange, e.g., air in the trachea.
Decompression sickness Also called the "bends" or caisson disease; a collection of symptoms observed following a reduction in ambient pressure, which causes bubbles to form in the blood and tissues. In humans, symptoms include dizziness, numbness, fatigue and, in more severe cases, paralysis, problems breathing, and death.
Inert gas narcosis Is caused by the anesthetic effect of lipid-soluble gases at high pressure. In air-breathing divers, the symptoms may ultimately lead to loss of consciousness as pressure increases.
Lung squeeze Pulmonary edema caused by intrathoracic pressures that are lower than environmental pressures during breath-hold diving.
Maximal/forced breath Often used in human lung function testing to assess the maximal capacity of lung function such as VC, PEF, PIF, and airway obstruction. The individual is asked to expire maximally, followed by an inspiration.
Minimum air volume The volume of air left in the relaxed lung.
Respiratory frequency The number of breaths per unit time.
Rete A network of arteries or veins
Tidal volume The volume of air exhaled or inhaled during a normal breath.
Vital capacity The maximal volume of air that can be exchanged in one breath. In marine mammals, VC is close to TLC.

References

Asgharian, B., Miller, F.J., Price, O., Schroeter, J.D., Einstein, D.R., Corley, R.A., et al., 2016. Modeling particle deposition in the pig respiratory tract. J. Aero Sci. 99, 107–124.

Bagnoli, P., Peruffo, A., Costantino, M.L., Cozzi, B., 2011. The upper respiratory tract of dolphins. Ital. J. Anat. Embryol. 116 (3), 159–164.

Baker, J.R., 1992. Causes of mortality and parasites and incidental lesions in dolphins and whales from British waters. Vet. Record. 130, 554–558.

Ballarin, C., Bagnoli, P., Peruffo, A., Cozzi, B., 2018. Vascularization of the trachea in the bottlenose dolphin: comparison with bovine and evidence for evolutionary adaptations to diving. Royal Soc. Open Sci. 5 (4).

Bélanger, L.F., 1940. A study of the histological structure of the respiratory portion of the lungs of aquatic mammals. Am. J. Anat. 67 (3), 437–469.

Blawas, A., Nowacek, D.P., Rocho-Levine, J., Robeck, T., Fahlman, A., 2021. Scaling of heart rate with breathing frequency and body mass in cetaceans. Philos. Trans. R. Soc. Lond. B. 376, 20200223.

Blix, A.S., Folkow, L.P., 1995. Daily energy expenditure in free living minke whales. Acta Physiol. Scand. 153, 61–66.

Bogomolni, A.L., Pugliares, K.R., Sharp, S.M., Patchett, K., Harry, C.T., LaRocque, J.M., et al., 2010. Mortality trends of stranded marine mammals on Cape Cod and southeastern Massachusetts, USA, 2000 to 2006. Dis. Aquat. Org. 88 (2), 143–155.

Borque-Espinosa, A., Burgos, F., Dennison, S., Laughlin, R., Manley, M., Capaccioni, R., Fahlman, A., 2020. Lung function testing as a diagnostic tool to assess respiratory health in bottlenose dolphins, *Tursiops truncatus*. Dis. Aquat. Org. 138, 17–27.

Borque-Espinosa, A, Ferrero-Fernánde, D, Capaccioni Azzati, R, Fahlman, A, 2021. Lung function assessment in the Pacific walrus (*Odobenus rosmarus* divergens) while resting on land and submerged in water. J. Exp. Biol. 224, jeb.227389.

Bostrom, B.L., Fahlman, A., Jones, D.R., 2008. Tracheal compression delays alveolar collapse during deep diving in marine mammals. Res. Physiol. Neurobiol. 161, 298–305.

Boyd, R.B., 1975. A gross and microscopic study of the respiratory anatomy of the Antarctic Weddell seal, *Leptonychotes weddelli*. J. Morphol. 147 (3), 309–336.

Burne, R.H., 1910. Note on the veins of a seal. Proc. Zool. Soc. Lond., 385–387.

Burns J.J. Ribbon Seal, Phoca fasciata. In: Ridgway S.H., Harrison, R.J., editor. Handbook of Marine Mammals. 2. London: Academic Press; 1981. pp. 104-105.

Cauture, F., SterbaBoatwright, B., Miedler, S., Rocho-Levine, J., Harms, C., Fahlman, A., 2019. Using respiratory sinus arrhythmia to estimate inspired tidal volume in the bottlenose dolphin (*Tursiops truncatus*). Front. Physiol. 10, 128.

Cieri, R.L., 2019. Pulmonary smooth muscle in vertebrates: a comparative review of structure and function. Integr. Comp. Biol. 59 (1), 10–28.

Cohen, B.S., Briant, J.K., 1989. Flow distribution in human and canine tracheobronchial airway casts. Health Phys. 57 (Suppl. 1), 21–27.

Costidis, A., Piscitelli, M.A., Rommel, S.A., Harms, C., McLellan, W.A., Rotstein, D., Pabst, D.A. (Eds.), 2009. Elaborate Venous Structures in the Thorax of Physeteriids, Kogiids, and Ziphiids. Abstract, Biennial Meeting of the Society for Marine Mammology. Quebec City.

Costidis, A., Rommel, S.A., 2012. Vascularization of air sinuses and fat bodies in the head of the Bottlenose dolphin (*Tursiops truncatus*): morphological implications on physiology. Front. Physiol. 3 (243), 1–23.

Costidis, A.M., Rommel, S.A., 2016. The extracranial venous system in the heads of beaked whales, with implications on diving physiology and pathogenesis. J. Morphol. 277 (1), 34–64.

Cotten, P.B., Piscitelli, M.A., McLellan, W.A., Rommel, S.A., Dearolf, J.L., Pabst, D.A., 2008. The gross morphology and histochemistry of respiratory muscles in bottlenose dolphins, *Tursiops truncatus*. J. Morphol. 269 (12), 1520–1538.

Cozzi, B., Bagnoli, P., Acocella, F., Constantino, M.L., 2005. Structure and Biomechanical properties of the trachea of the striped dolphin *Stenella coeruleoalba*: evidence for evolutionary adaptations to diving. Anat. Rec. A. 284A, 500–510.

Craig, A.B., 1968. Depth limits of breath hold diving (an example of fennology). Resp. Physiol. 5, 14–22.

Crespo, F.A., De Cidre, L.L., 2005. Functional significance of bronchial sphincters in two Southwestern Atlantic dolphins: *Pontoporia blainvillei* and *Lagenorhynchus obscurus*: a comparative approach. Mammals 69 (2), 233–238.

Davenport, J., Cotter, L., Rogan, E., Kelliher, D., Murphy, C., 2013. Structure, material characteristics and function of the upper respiratory tract of the pygmy sperm whale. J. Exp. Biol. 216, 4639–4646.

Denison, D.M., Kooyman, G.L., 1973. The structure and function of the small airways in pinniped and sea otter lungs. Resp. Physiol. 17, 1–10.

Denison, D.M., Warrell, D.A., West, J.B., 1971. Airway structure and alveolar emptying in the lungs of sea lions and dogs. Resp. Physiol. 13, 253–260.

Denk, M.A., Fahlman, A., Dennison-Gibby, S., Song, Z., Moore, M., 2020. Hyperbaric tracheobronchial compression in cetaceans and pinnipeds. J. Exp. Biol. 223. jeb.217885.

Desplechain, C., Foliguet, B., Barrat, E., Grignon, G., Touati, F., 1983. The pores of Kohn in pulmonary alveoli. Bull. Eur. Phys. Res. 19 (1), 59–68.

Elsner, R., Kenney, D.W., Burgess, K., 1966. Diving bradycardia in the trained dolphin. Nature 212 (5060), 407–408.

Fahlman, A., Cozzi, B., Manley, M., Jabas, S., Malik, M., Blawas, A., Janik, V.M., 2020. Conditioned variation in heart rate during static breath-holds in the bottlenose dolphin (*Tursiops truncatus*). Front. Physiol. 11, 604018.

Fahlman, A., Epple, A., Garcia-Párraga, D.G., Robeck, T., Haulena, M., Piscitelli-Doshkov, M., Brodsky, M., 2019. Characterizing respiratory capacity in belugas (*Delphinapterus leucas*). Resp. Physiol. Neurobiol. 260, 63–69.

Fahlman, A., Hooker, S.K., Olszowka, A., Bostrom, B.L., Jones, D.R., 2009. Estimating the effect of lung collapse and pulmonary shunt on gas exchange during breath-hold diving: the Scholander and Kooyman legacy. Resp. Physiol. Neurobiol. 165, 28–39.

Fahlman, A., Jensen, F., Tyack, P.L., Wells, R., 2018. Modeling tissue and blood gas kinetics in coastal and offshore common Bottlenose dolphins, *Tursiops truncatus*. Front. Physiol. 9 (838), 1–13.

Fahlman, A., Loring, S.H., Ferrigno, M., Moore, C., Early, G., Niemeyer, M., Lentell, B., Wenzel, F., Joy, R., Moore, M.J., 2011. Static inflation and deflation pressure-volume curves from excised lungs of marine mammals. J. Exp. Biol. 214 (22), 3822–3828.

Fahlman, A., Loring, S.H., Johnson, S., Haulena, M., Trites, A.W., Fravel, V.A., van Bonn, W., 2014. Inflation and deflation pressure-volume loops in anesthetized pinnipeds confirms compliant chest and lungs. Front. Physiol. 5 (433), 1–7.

Fahlman, A., Loring, S.H., Levine, G., Rocho-Levine, J., Austin, T., Brodsky, M., 2015. Lung mechanics and pulmonary function testing in cetaceans. J. Exp. Biol. 218, 2030–2038.

Fahlman, A., Madigan, J., 2016. Respiratory function in voluntary participating Patagonia sea lions in sternal recumbency. Front. Physiol. 7, 1–9.

Fahlman, A., McHugh, K., Allen, J., Barleycorn, A., Allen, A., Sweeney, J., Stone, R., Faulkner Trainor, R., Bedford, G., Moore, M.J., Jensen, F.H., Wells, R., 2018. Resting metabolic rate and lung function in wild offshore common bottlenose dolphins, *Tursiops truncatus*, near Bermuda. Front. Physiol. 9 (886).

Fahlman, A., Moore, M.J., Garcia-Parraga, D., 2017. Respiratory function and mechanics in pinnipeds and cetaceans. J. Exp. Biol. 220 (10), 1761–1763.

Fahlman, A., Moore, M.J., Wells, R.S., 2021. How do marine mammals manage and usually avoid gas emboli formation and gas embolic pathology? critical clues from studies of wild dolphins. Front. Mar. Sci. 8. (598633).

Fahlman, A., Tyson Moore, R.B., Stone, R., Sweeney, J., Faulkner Trainor, R., Barleycorn, A., McHugh, K., Allen, J., Wells, R.S., 2023. Deep diving by offshore bottlenose dolphins (*Tursiops spp.*). Mar. Mamm. Sci. 39, 1251–1266.

Falke, K.J., Busch, T., Hoffmann, O., Liggins, G.C.L.J., Mohnhaupt, R., Roberts Jr., J.D., et al., 2008. Breathing pattern, CO2 elimination and the absence of exhaled NO in freely divingWeddell seals. Resp. Physiol. Neurobiol. 162, 85–92.

Falke, K.J., Hill, R.D., Qvist, J., Schneider, R.C., Guppy, M., Liggins, G.C., et al., 1985. Seal lung collapse during free diving: evidence from arterial nitrogen tensions. Science 229, 556–557.

Fanning, J.C., 1977. The Submicroscopic Structure of the Dolphin Lung. PhD thesis, University of Adelaide.

Fanning, J.C., Harrison, R.J., 1974. The structure of the trachea and lungs of the South Australian bottle-nose dolphin. In: Harrison, R.J. (Ed.), Functional Anatomy of Marine Mammals. 2. Academic Press, London, pp. 231–252.

Fanning, J.C., Whitting, H.W., 1969. Observations on the histology of the bronchial tree and lungs of some delphinids. J. Anat. 104, 409.

Galetke, W., Feier, C., Muth, T., Ruehle, K.H., Borsch-Galetke, E., Randerath, W., 2007. Reference values for dynamic and static pulmonary compliance in men. Res. Med. 101, 1783–1789.

Gandilhon, N., Adam, O., Cazau, D., Laitman, J.T., Reidenberg, J.S., 2015. Two new theoretical roles of the laryngeal sac of humpback whales. Mar. Mam. Sci. 31 (2), 774–781.

García-Párraga, D., Moore, M., Fahlman, A., 2018. Pulmonary ventilation–perfusion mismatch: a novel hypothesis for how diving vertebrates may avoid the bends. Proc. R. Soc. Lond. B 285, 20180482.

Goldbogen, J.A., Shadwick, R.E., Lillie, M.A., Piscitelli, M.A., Potvin, J., Pyenson, N.D., et al., 2015. Using morphology to infer physiology: case studies on rorqual whales (Balaenopteridae). Can. J. Zool. 93 (9), 687–700.

Harrison, R.J., Tomlinson, J.D.W., 1956. Observations on the venous system in certain pinnipedia and cetacea. Proc. Zool. Soc. Lond. 126 (2), 205–234.

Howard, E.B., Britt, J.O., Matsumoto, G.K., Itahara, R., Nagano, C.N., 1983. Bacterial diseases. In: Howard, E.B. (Ed.), Pathobiology of Marine Mammal Diseases. CRC Press, BocaRaton, FL, pp. 69–118.

Huggenberger, S., Oelschläger, H., Cozzi, B., 2019. Atlas of the Anatomy of Dolphins and Whales. Academic Press, Amsterdam.

Hui, C.A., 1975. Thoracic collapse as affected by the retia thoracica in the dolphin. Resp. Physiol. 25, 63–70.

Irving, L., Scholander, P.F., Grinnell, S.W., 1941. The respiration of the porpoise, *Tursiops truncatus*. J. Cell Comp. Physiol. 17, 145–168.

Ivancic, M., Solano, M., Smit, C.R., 2014. Computed tomography and cross-sectional anatomy of the thorax of the live bottlenose dolphin (*Tursiops truncatus*). Anat. Rec. 297, 901–915.

Johnson, W.R., Torralba, M., Fair, P.A., Bossart, G.D., Nelson, K.E., Morris, P.J., 2009. Novel diversity of bacterial communities associated with bottlenose dolphin upper respiratory tracts. Environ. Micro Rep. 1 (6), 555–562.

Jordanoglou, J., Pride, N.B., 1968. Factors determining maximum inspiratory flow and maximum expiratory flow of the lung. Thorax 23, 33–37.

Kabilan, S., Lin, C.L., Hoffman, E.A., 1985. Characteristics of airflow in a CT-based ovine lung: a numerical study. J. Appl. Physiol. 102 (4), 1469–1482.

Kasting, N.W., Adderley, S.A.L., Safford, T., Hewlett, K.G., 1989. Thermoregulation in beluga (*Delphinapterus leucas*) and killer (*Orcinus orca*) whales. Physiol. Zool. 62, 687–701.

Kerem, D.H., Kylstra, J.A., Saltzman, H.A., 1975. Respiratory flow rates in the sea lion. Undersea Biomed. Res. 2, 20–27.

Klatsky, L.J., Wells, R.S., Sweeney, J.C., 2007. Offshore bottlenose dolphins (*Tursiops truncatus*): movement and dive behavior near the Bermuda pedestal. J. Mam. 88 (1), 59–66.

Knudson, R.J., Slatin, R.C., Lebowitz, M.D., Burrows, B., 1976. The maximal expiratory flow-volume curve. Am. Rev. Respir. Dis. 113, 587–600.

Kooyman, G.L., 1973. Respiratory adaptations in marine mammals. Am. Zool. 13, 457–468.

Kooyman, G.L., Cornell, L.H., 1981. Flow properties of expiration and inspiration in a trained bottle-nosed porpoise. Physiol. Zool. 54 (1), 55–61.

Kooyman, G.L., Hammond, D.D., Schroede, J.P., 1970. Bronchograms and tracheograms of seals under pressure. Science 169, 82–84.

Kooyman, G.L., Kerem, D.H., Campbell, W.B., Wright, J.J., 1971. Pulmonary function in freely diving Weddell seals, *Leptonychotes weddelli*. Respir. Physiol. 12, 271–282.

Kooyman, G.L., Norris, K.S., Gentry, R.L., 1975. Spout of the gray whale: its physical characteristics. Science 190, 908–910.

Kooyman, G.L., Sinnett, E.E., 1979. Mechanical properties of the harbor porpoise lung, *Phocoena phocoena*. Resp. Physiol. 36 (3), 287–300.

Kooyman, G.L., Sinnett, E.E., 1982. Pulmonary shunts in Harbor seals and sea lions during simulated dives to depth. Physiol. Zool. 55 (1), 105–111.

Lafortuna, C.L., Jahoda, M., Azzellino, A., Saibene, F., Colombini, A., 2003. Locomotor behaviours and respiratory pattern of the Mediterranean fin whale (*Balaenoptera physalus*). Eur. J. Appl. Physiol. 90, 387–395.

Leith, D.E., 1976. Comparative mammalian respiratory mechanics. Physiology 19, 485–510.

Lillie, M.A., Vogl, A.W., Raverty, S., Haulena, M., McLellan, W.A., Stenson, G.B., Shadwick, R.E., 2017. Controlling thoracic pressures in cetaceans during a breath-hold dive: importance of the diaphragm. J. Exp. Biol. 220 (19), 3464–3477.

Lillie, M.A., Vogl, A.W., Raverty, S., Haulena, M., McLellan, W.A., Stenson, G.B., Shadwick, R.E., 2018. The caval sphincter in cetaceans and its predicted role in controlling venous flow during a dive. J. Exp. Biol. 221 (11). jeb177212.

Lindholm, P., Nyrén, S., 2005. Studies on inspiratory and expiratory glossopharyngeal breathing in breath-hold divers employing magnetic resonance imaging and spirometry. Eur. J. Appl. Physiol. 94 (5–6), 646–651.

Lundgren, C.E.G., Miller, J.N., 1999. The Lung at Depth. xvii Marcel Dekker, New York, p. 686.

Martony, M.E., Ivančić, M., Gomez, F.M., Meegan, J.M., Nollens, H.H., Schmitt, T.L., Erlacher-Reid, C.D., Carlin, K.P., Smith, C.R., 2017. Establishing marginal lymph node ultrasonographic characteristics in healthy bottlenose dolphins (*Tursiops truncatus*). J. Zoo Wildlife Med. 48 (4), 961–971.

Matthews, R.C., 1977. Pulmonary Mechanics of California Sea Lions, *Zalophus californianus*. MSc San Diego State University, San Diego.

McDonald, B.I., Ponganis, P.J., 2012. Lung collapse in the diving sea lion: hold the nitrogen and save the oxygen. Biol. Lett. 8 (6), 1047–1049.

McKnight, J.C., Bennett, K.A., Bronkhorst, M., Russell, D.J.F., Balfour, S., Milne, R., Bivins, M., Moss, S.E.W., Colier, W., Hall, A.J., Thompson, D., 2019. Shining new light on mammalian diving physiology using wearable near-infrared spectroscopy. PLoS Biol. 17 (6), e3000306.

McKnight, J.C., Mulder, E., Ruesch, A., Kainerstorfer, J., Wu, J., Hakimi, N., Balfour, S., Bronkhorst, M., Horschig, J.M., Pernett, F., Sato, K., Hastie, G.D., Tyack, P., Schagatay, E., 2021. When the human brain goes diving: using NIRS to measure cerebral and systemic cardiovascular responses to deep, breath-hold diving in elite freedivers. Philos. Trans. R. Soc. Lond. B. 376, 20200349.

Mead, J.G., 1975. Anatomy of the External Nasal Passages and Facial Complex in the Delphinidae (Mammalia: Cetacea). Smithsonian Contributions to Zoology. 1–35.

Melnikov, V.V., 1997. The arterial system of the sperm whale (*Physeter macrocephalus*). J. Morphol. 234 (1), 37–50.

Moore, M.J., Hammar, T., Arruda, J., Cramer, S., Dennison, S., Montie, E., Fahlman, A., 2011. Hyperbaric computed tomographic measurement of lung compression in seals and dolphins. J. Exp. Biol. 214 (Pt 14), 2390–2397.

Moore, C., Moore, M.J., Trumble, S., Niemeyer, M., Lentell, B., McLellan, W., Costidis, A., Fahlman, A., 2014. A comparative analysis of marine mammal tracheas. J. Exp. Biol. 217, 1154–1166.

Murata, T., 1951. Histological studies on the respiratory portions of the lungs of cetacea. Sci. Rep. Whales Res. Inst. 6, 35–47.

Nakakuki, S., 1975. The new interpretation of bronchial tree. Proc. Jpn. Acad. 51 (5), 342–346.

Nakakuki, S., 1993. The bronchial tree and lobular division of the lung of the California sea lion (*Zalophus californianus*). J. Vet. Med. Sci. 55 (4), 669–671.

Netter, F.H., 2006. Atlas of Human Anatomy, fourth ed. Saunders, London.

Nicholson, A., 1984. The Structure-Function Relationship of the Lung of the Australian Sea Lion, *Neophoca cinerea*. PhD thesis, University of Adelaide.

Ninomiya, H., Inomata, T., Shirouzu, H., Katsumata, E., 2005. Microanatomy of the terminal air spaces in Baird's beaked whale (*Berardius bairdii*) lungs. J. Vet. Med. Sci. 67 (5), 473–479.

Olsen, C.R., Hale, F.C., Elsner, R., 1969. Mechanics of ventilation in the pilot whale. Resp. Physiol. 7, 137–149.

Otero-Sabio, C., Centelleghe, C., Corain, L., Graïc, J.-M., Cozzi, B., Rivero, M., Consoli, F., Peruffo, A., 2021. Micro-anatomical, immunohistochemical and morphometric characterization of the terminal portions of the lung in cetaceans. J. Morphol. 282 (2), 291–308.

Pabst, D.A., McLellan, W.A., Rommel, S.A., 2016. How to build a deep diver: the extreme morphology of mesoplodonts. Integr. Comp. Biol. 56 (6), 1337–1348.

Patra, A.L., 1986. Comparative anatomy of mammalian respiratory tracts: the nasopharyngeal region and the tracheobronchial region. J. Toxicol. Environ. Health 17 (2–3), 163–174.
Phalen, R.F., Oldham, M.J., 1983. Tracheobronchial airway structure as revealed by casting techniques. Am. Rev. Respir. Dis. 128 (2 Pt 2), S1–S4.
Piscitelli, M.A., McLellan, W.A., Rommel, S.A., Blum, J.E., Barco, S.G., Pabst, D.A., 2010. Lung size and thoracic morphology in shallow- and deep-diving cetaceans. J. Morphol. 271 (6), 654–673.
Piscitelli, M.A., Raverty, S.A., Lillie, M.A., Shadwick, R.E., 2013. A review of cetacean lung morphology and mechanics. J. Morphol. 274 (12), 1425–1440.
Ponganis, P.J., 2015. Diving Physiology of Marine Mammals and Seabirds. Cambridge University Press, Cornwall, UK.
Ponganis, P.J., Knower Stockard, T., Levenson, D.H., Berg, L., Baranov, E.A., 2006. Cardiac output and muscle blood flow during rest-associated apneas of elephant seals. Comp. Biochem. Physiol. A: Physiol. 144, 105–111.
Ponganis, P.J., Stockard, T.K., Levenson, D.H., Berg, L., Baranov, E.A., 2006. Intravascular pressure profiles in elephant seals: hypotheses on the caval sphincter, extradural vein and venous return to the heart. Comp. Biochem. Physiol. A: Physiol. 145, 123–130.
Pryor, K., 2019. Don't Shoot the Dog!: The Art of Teaching and Training, third ed. Simon & Schuster, Inc., New York, NY.
Quick, N.J., Cioffi, W.R., Shearer, J.M., Fahlman, A., Read, A.J., 2020. Extreme diving in mammals: first estimates of behavioural aerobic dive limits in Cuvier's beaked whales. J. Exp. Biol. 225, jeb222109.
Ramirez, K., 2019. Animal Training: Successful Animal Management Through Positive Reinforcement. First Stone Publishing.
Reed, J.Z., Chambers, C., Fedak, M.A., Butler, P.J., 1994. Gas exchange of captive freely diving grey seals (*Halichoerus grypus*). J. Exp. Biol. 191, 1–18.
Reed, J.Z., Chambers, C., Hunter, C.J., Lockyer, C., Kastelein, R., Fedak, M.A., Boutilier, R.G., 2000. Gas exchange and heart rate in the harbour porpoise, *Phocoena phocoena*. J. Comp. Physiol. B 170, 1–10.
Ridgway, S.H., 1972. Homeostasis in the aquatic environment. In: Ridgway, S.H. (Ed.), Mammals of the Sea: Biology and Medicine. Thomas, Springfield, IL, pp. 590–747.
Ridgway, S.H., Harrison, R.J., 1986. Diving dolphins. In: Bryden, M.M., Harrison, R.H. (Eds.), Research on Dolphins. Oxford Books, Oxford, pp. 33–58.
Ridgway, S.H., Howard, R., 1979. Dolphin lung collapse and intramuscular circulation during free diving: evidence from nitrogen washout. Science 206 (4423), 1182–1183.
Ridgway, S.H., Johnston, D.G., 1966. Blood oxygen and ecology of porpoises of three genera. Science 151 (3709), 456–458.
Ridgway, S.H., Scronce, B.L., Kanwisher, J.W., 1969. Respiration and deep diving in the bottlenose porpoise. Science 166, 1651–1654.
Saladin, K.S., 2007. The circulatory system. In: Anatomy and Physiology: The Unity of Form and Function. McGraw-Hill Companies, Inc, New York, pp. 679–776.
Santos Júnior, L.C., Marmontel, M., Santos, F.G.A., Rici, R.E.G., Miglino, M.A., Carvalho, Y.K., 2017. Posterior respiratory apparatus of Inia geoffrensis and Sotalia fluviatilis : structure and ultrastructure. Int. J. Morph. 35, 1582–1589.
Schaefer, K.E., Allison, R.D., Dougherty Jr., J.H., Carey, C.R., Walker, R., Yost, F., Parker, D., 1968. Pulmonary and circulatory adjustments determining the limits of depths in breathhold diving. Science 162 (3857), 1020–1023.
Scholander, P.F., 1940. Experimental investigations on the respiratory function in diving mammals and birds. Hvalrådets skr. 22, 1–131.
Schorr, G.S., Falcone, E.A., Moretti, D.J., Andrews, R.D., 2014. First long-term behavioral records from Cuvier's beaked whales (*Ziphius cavirostris*) reveal record-breaking dives. PLoS One 9 (3), e92633.
Slijper, E.J., 1979. Whales. Hutchinson and Co, London, p. 511.
Smith, C.R., Solano, M., Lutmerding, B.A., Johnson, S.P., Meegan, J.M., Le-Bert, C.R., Emory-Gomez, F., Cassle, S., Carlin, K., Jensen, E.D., 2012. Pulmonary ultrasound findings in a bottlenose dolphin *Tursiops truncatus* population. Dis. Aquat. Org. 101, 243–255.
Smodlaka, H., Henry, R.W., Reed, R.B., 2006. Microscopic anatomy of the ringed seal (*Phoca hispida*) lower respiratory system. Anat. Histol. Embryol. 35, 35–41.
Spencer, M.P., Thomas, A., Gornall, T.A., Poulter, T.C., 1967. Respiratory and cardiac activity of killer whales. J. Appl. Physiol. 22 (5), 974–981.
Stahl, W.R., 1967. Scaling of respiratory variables in mammals. J. Appl. Physiol. 22, 453–460.
Stocks, J., Quanjer, P.H., 1995. Reference values for residual volume, functional residual capacity and total lung capacity. ATS Workshop on Lung Volume Measurements. Official Statement of The European Respiratory Society. Eur. Res. J. 8, 492–506.
Sumich, J.L., 2001. Direct and indirect measures of oxygen extraction, tidal lung volumes and respiratory rates in a rehabilitating gray whale calf. Aquat. Mamm. 27 (3), 279–283.
Sweeney, J.C., Reddy, M.L., 2001. Cetacean Cytology. In: Dierauf, L.A., FMD, G. (Eds.), CRC Handbook of Marine Mammal Medicine, second ed. CRC Press, pp. 437–446.
Sweeney, J.C., Ridgway, S.H., 1976. Common diseases of small cetaceans. J. Am. Vet. Med. Assoc. 167 (7), 533–540.
Tenney, S.M., Remmers, J.E., 1963. Comparative quantitative morphology of the mammalian lung: diffusing area. Nature 197 (4862), 54–56.
Terry, P.B., Traystman, R.J., 2016. The clinical significance of collateral ventilation. Ann. Am. Thorac. Soc. 13 (12), 2251–2257.
Tyack, P.L., Johnson, M., Aguilar de Soto, N., Sturlese, A., Madsen, P.T., 2006. Extreme diving of beaked whales. J. Exp. Biol. 209, 4238–4253.
Venn-Watson, S., Daniels, R., Smith, C., 2012. Thirty year retrospective evaluation of pneumonia in a bottlenose dolphin *Tursiops truncatus* population. Dis. Aquat. Org. 99, 237–242.
Vogl, A.W., Fisher, H.D., 1982. Arterial retia related to supply of the central nervous system in two small toothed whales—Narwhal (*Monodon monoceros*) and Beluga (*Delphinapterus leucas*). J. Morphol. 174, 41–56.
Wahrenbrock, E.A., Maruscha, G.F., Elsner, R., Kenney, D.W., 1974. Respiration and metabolism in 2 baleen whale calves. Mar. Fish. Rev. 36, 3–9.
Walmsley, R., 1938. Some Observations on the Vascular System of a Female Fetal Finback, p. 70.
West, J.B., 2012. Respiratory Physiology, 9th ed. Wolters Kluwer-Lippincott Williams & Wilkins, Philadelphia.
Wang, P.M., Kraman, S.S., 2004. Fractal branching pattern of the monopodial canine airway. J. Appl. Physiol. 96 (6), 2194–2199.
Wislocki, G.B., 1929. On the structure of the lungs of the porpoise (*Tursiops truncatus*). Am. J. Anat. 44 (1), 47–77.
Wilmshurst, P., 1998. Diving and oxygen. BMJ 317 (7164), 996–999.

CHAPTER

7

Diving physiology in dolphins and human

J. Chris McKnight[a], *Alexander Patrician*[b], *Zeljko Dujic*[c], *Sascha K. Hooker*[a], *and Andreas Fahlman*[d,e,f]

[a]Sea Mammal Research Unit, Scottish Oceans Institute, University of St Andrews, St Andrews, Scotland, United Kingdom [b]Centre for Heart, Lung and Vascular Health, University of British Columbia, Okanagan campus, Kelowna, BC, Canada [c]Department of Integrative Physiology, University of Split School of Medicine, Split, Croatia [d]Fundación Oceanogràfic de la Comunitat Valenciana, Valencia, Spain [e]Kolmården Wildlife Park, Kolmården, Sweden [f]Global Diving Research SL., Valencia, Spain

Abstract

During diving, the body is exposed to a number of environmental stressors that result in physiological responses. Many of these responses are common across both humans and dolphins. One of the best known is the dive response, historically referred to as the "master switch of life," a defense against asphyxia. This is enacted during submersion and consists of a reduction in heart rate and a redistribution of blood to high-priority organs. Increasing hydrostatic pressure during descent compresses all air-filled spaces, resulting in structural changes, and alters blood gas tensions and hemodynamics. In this chapter, we will briefly outline a number of physiological changes associated with diving and compare and contrast these responses between humans and dolphins.

Introduction to diving physiology

A dolphin perspective

Breath-hold diving exposes the mammalian body to multiple extreme environmental stresses. Some of the most challenging are the separation from ambient air and the increasing atmospheric pressure with depth. Submersion in water also affects heat transfer due to the greater thermal conductivity of water (i.e., heat transfer rates increase 24-fold, compared to air temperatures, see Chapter 3: Thermoregulation). For dolphins, diving is a "full-time" activity, and these challenging states constitute "normal" conditions. As such, their physiological adaptations to these challenges have evolved since their ancestral re-entry into the aquatic environment and have been optimized for a lifestyle where almost all of life's activities (feeding, communicating, traveling, breeding, and sleeping) are conducted while submerged in seawater. To paraphrase Fedak and Thompson's (1993) reflection: *dolphins are not surface-dwellers which occasionally leave the surface to forage, but rather are visitors to the surface which live at depth and only periodically return to the surface to breathe.* In this chapter, we hope to explain how this is more than just a semantic distinction.

A human perspective

For humans, breath-hold diving is a "part-time" activity (in other words, humans must breathe and do not live in the water indefinitely) for recreation, competitive sport, military strategy, and even sustenance. It can take a variety of forms, such as underwater rugby/hockey in a pool, spearfishing (performing repetitive shallow dives over many hours), and free-diving. The latter encompasses different disciplines ranging from long static (i.e., stationary) breath-holds, breath-holding while swimming horizontally in a pool (dynamic), and/or diving deep on a static dive line. Since the 1940s, records of performance have shown divers progressively diving deeper (Fitz-Clarke, 2018)—with current depth records using a monofin to achieve 114 and 130 m for female and males, respectively (Fig. 1A). Sustenance diving is typically characteristic of indigenous diving populations, such as the Ama in Japan and Haenyeo in

The Physiology of Dolphins
https://doi.org/10.1016/B978-0-323-90516-9.00003-8

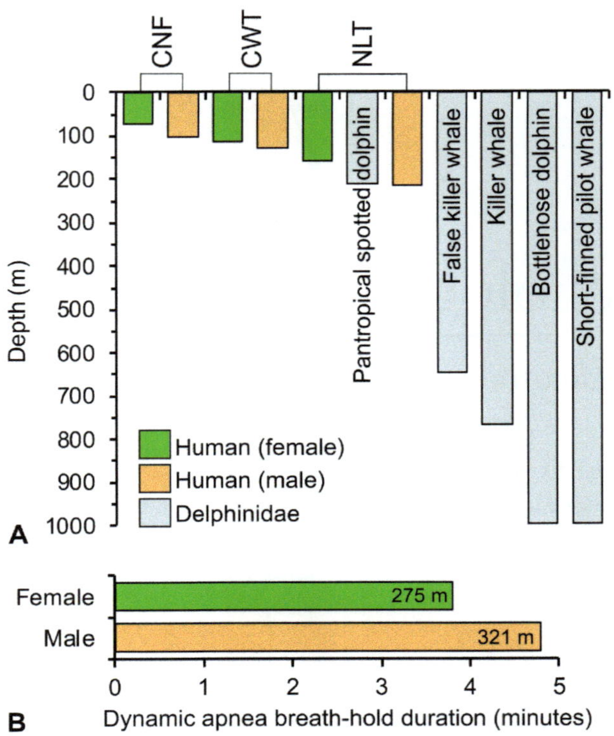

FIG. 1 Comparison of depth records for human and Delphinidae species presented in ascending order of depth (Panel A). Horizontal underwater breath-hold (i.e., dynamic apnea with monofin, DYN) swimming distance and duration records for humans (Panel B). CNF, constant weight without fins; CWT, constant weight with monofin; NLT, no limit discipline. Human depth data from AIDA (2022) and distance and duration data from the top male and female finisher at the 2022 CMAS World Freediving Indoor Championships. Delphinidae maximum dive depths are provided for the pantropical spotted dolphin (*Stenella attenuata*), false killer whale (*Pseudorca crassidens*), bottlenose dolphin (*Tursiops truncatus*) and short-finned pilot whale (*Globicephala macrorhynchus*) based on data from Ropert-Coudert et al. (2018), Fahlman et al. (2023).

Korea (Teruoka, 1932; Rahn and Yokoyama, 1965; Hong and Rahn, 1967; Vaneechoutte et al., 2011), and the Bajau in Indonesia (Abrahamsson and Schagatay, 2014; Ilardo et al., 2018). Despite societal acculturation, these populations still practice breath-hold diving as their primary means of harvesting food. This form of diving has also been performed in coastal countries, such as Croatia, where restrictions on fishing with nets and SCUBA (self-contained underwater breathing apparatus) equipment have established breath-hold diving as a primary fishing modality. Irrespective of the motivations to breath-hold dive, humans, as a species, are less adapted for diving compared to marine species. Indeed, our time in the sea is physiologically challenging, and many of our responses to the challenges of submersion, and our need for equipment, are geared toward simply surviving. Humans are most certainly surface-dwellers, which occasionally leave the surface to forage or to partake in recreational or professional endeavors.

Just as some motivations (i.e., food) for diving in dolphins and humans may overlap, the physical and anaerobic challenges associated with immersion always exist—for example, how is oxygen delivery to vital tissues such as the brain maintained? How do air-filled cavities cope with elevated and rapid hydrostatic pressure changes? How is thermal homeostasis maintained? In this chapter, we will briefly outline a number of physiological changes associated with diving and compare and contrast their responses in humans and dolphins. We will demonstrate how common physiological responses, which are shared among mammals to protect the body, have been accentuated in dolphins to allow them to exploit marine resources without experiencing pathologies often manifested in the human diver.

The dive response

Overview

Immersion and separation from ambient air is perhaps the most immediate limiting factor for a submerged mammal. Without access to air to breathe, aerobic metabolism continuously depletes blood oxygen (O_2) and raises carbon dioxide (CO_2) levels, causing hypoxia (low O_2), hypercapnia (high CO_2), and tissue acidosis (higher acidity). With the cessation of breathing during submersion, all mammals share the same "dive response." This has been referred to as the master switch of life (Scholander, 1963), as it optimizes the use of O_2 and delays the onset of asphyxia. This may even constitute an ancestral response that is "hard-wired" to protect the body from hypoxia and asphyxia rather than an evolutionary adaptation for diving per se (Mottishaw et al., 1999). While this response is autonomic (i.e., involuntary) in humans, marine mammals (e.g., bottlenose dolphin (*Tursiops truncatus*), the harbor porpoise (*Phocoena phocoena*), and the California sea lion (*Zalophus californianus*)) appear to have the ability to regulate heart rate (and blood flow) depending on their planned dive (Elsner et al., 1966; Ridgway et al., 1975; Elmegaard et al., 2016; Kaczmarek et al., 2018; Fahlman et al., 2020b).

The dive response consists of a suite of cardiovascular changes, which allow the diving body to conserve the available O_2 for central organs such as the heart and brain, via a temporary "metabolic retreat." This involves a reduction of the heart rate below resting levels, called bradycardia, and selective distribution of the available O_2 in the blood to tissues that are unable to sustain any prolonged period of hypoxia (i.e., hypoxia-intolerant) or that do not have endogenous O_2 stores. This reduced metabolic state demonstrates some similarities to resting activities such as sleeping and even meditation, although, during diving, the body is not resting. Instead, the reduction in the rate of O_2 consumption, through reduced cardiac work and organ performance, functions to extend dive time and protect the body's tissues from hypoxia, hypercapnia, and blood acidosis. Together with the significant preferential redistribution of blood flow away from tissues which are nonessential to diving (such as the skin and adipose tissue), these responses maintain blood flow to more vital organs (such as the brain) and help reduce the utilization of available O_2, thus extending the dive duration. Redistribution of blood flow, through peripheral vasoconstriction, is a phenomenon demonstrated in both seals (Zapol et al., 1979) and humans (Brick, 1966). This redistribution of blood flow results in hypoxia in perfusion-poor tissues, which in turn result in a decrease in the respiratory rate of the cells and a reduction in the local rate of O_2 consumption (and CO_2 production) (Scholander, 1940; Wheaton and Chandel, 2011). This "master switch of life" is vital in protecting the body's most hypoxia-intolerant tissues and organs, sustaining consciousness, and conserving O_2 required to conduct longer breath-hold durations (see Chapter 5: Cardiovascular physiology in dolphins and other cetaceans).

Facial receptors and cardiovascular responses

In humans, submersion of the face (together with the inhibition of breathing) stimulates receptors around the nose and in the skin of the face (Schuitema and Holm, 1988), which sends afferent signals to the brain. These afferents initiate (1) a vagally mediated increase in parasympathetic nerve activity that reduces the heart rate (Olsen et al., 1962; Craig, 1963) and (2) sympathetically mediated vasoconstriction of peripheral vascular beds, which reduces blood flow to organs that can endure reduced O_2 availability, such as the kidney and skeletal muscles (Fagius and Sundlöf, 1986; Heusser et al., 2009; Kyhl et al., 2016). These combined effects provide the reduction in O_2 consumption during diving, prolonging breath-hold duration and aim to preserve consciousness.

The magnitude of cardiovascular depression in response to diving is variable, augmented by various extrinsic and intrinsic factors. For instance, in humans, facial immersion in cooler water temperatures accentuates the bradycardia (Hurwitz and Furedy, 1986; Schuitema and Holm, 1988; Schagatay and Holm, 1996). Heart rates of 20–30 beats per minute are recorded during dives while immersed in a hyperbaric chamber or during dry breath-holds in the laboratory (Ferrigno et al., 1997; Ferretti, 2001; Bain et al., 2018b), but heart rates as low as 10 beats per minute have been recorded in facial immersion experiments in water at 1°C (Arnold, 1985). Additionally, air temperature appears to interact with the effect of water temperature to influence the magnitude of the human dive response. Immersion experiments showed greater bradycardia with a greater difference between water and air temperature (Schagatay and Holm, 1996).

Depth (i.e., vertical) diving in humans has an additive effect on the magnitude of the dive response. Submersion causes an initial bradycardia, but continued descent further decreases heart rate (especially when descent becomes passive due to negative buoyancy), until minimum heart rate is reached at, or close to, the bottom of the dive (Lemaître et al., 2013). After reaching the deepest part of the dive, the diver must actively swim to ascend. While heart rate remains bradycardic compared to predive values, the increase in metabolic activity and exertion can bring about small increases in heart rate (Lemaître et al., 2013; Mulder et al., 2021). This change in heart rate of human divers is similar to the cardiac variability during ascent in dolphins (Williams et al., 1999, 2015).

Blood pressure

In humans performing dry or facial immersion static apnea, the combined influence of (1) sympathetic excitation and chemoreflex engagement from combined inputs of hypoxia, hypercapnia, and lack of ventilatory rhythm (Heusser et al., 2009; Steinback et al., 2010) and (2) peripheral vasoconstriction likely drive the increases in mean arterial pressure (Fig. 2). Normal mean arterial blood pressure at rest is around 90 mmHg in a healthy young adult, but has been seen to increase up to ~150 mmHg during dry static apneas (Fig. 2, Bain et al., 2016; Bain et al., 2017), during apnea and exercise with face immersion (Bjertnaes et al., 1984), and fully immersed apneas (Ferrigno et al., 1997; Perini et al., 2010). However, blood pressure measurements during immersion are logistically difficult and data acquired during underwater diving are scarce. In an innovative study by Ferrigno et al. (1997), two participants with radial artery catheterization performed breath-holds in a hyperbaric chamber to simulate the hydrostatic pressure of a dive to 50 m. While there were substantial elevations in mean arterial pressure during descent (with occasional systolic peaks in one diver reaching 345 mmHg), mean arterial pressure subsided (remaining higher and lower than baseline in divers one and two, respectively) during the ascent phase of the dive (Ferrigno et al., 1997). While there does appear to be different blood pressure profile trajectories between static/surface apneas (for which blood pressure progressively increases) and deep diving (for which blood pressure inversely mirrors dive profile), differences in subject apnea experience, methodological designs, and acquisition tools make comparisons difficult. Unfortunately, with respect to dolphins, available blood pressure data are lacking to explore their adaptational advantages.

Splenic contraction

An additional component of the dive response seen in both humans and seals, and presumed (although not yet confirmed) to occur in dolphins, is the contraction of the spleen and ejection of stored red blood cells (Schagatay et al., 2001; Thornton et al., 2001; Stewart and McKenzie, 2002). For almost a century, the spleen has been considered to act as a dynamic reservoir, capable of expelling blood into the circulation, resulting in an elevated hematocrit concentration (Barcroft and Poole, 1927). Spleen contraction has been reported in human divers, who, while breath-

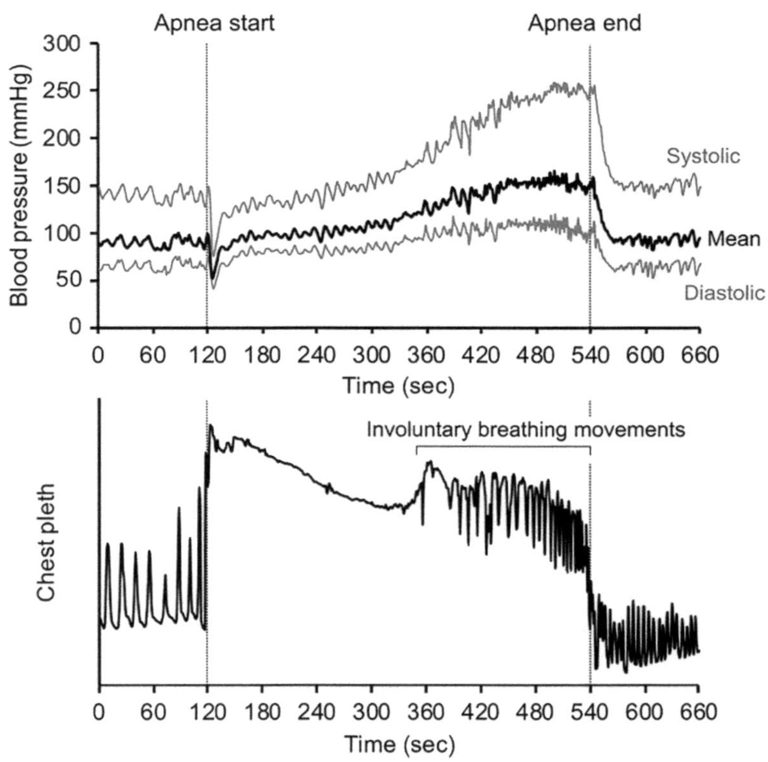

FIG. 2 A 7-min static apnea in a human diver. Continuous blood pressure (via finometer; top panel) and chest plethysmograph (to identify involuntary breathing movements; IBMs; bottom panel). An apnea is divided into an easy-going phase (no IBMs) and struggle phases (with IBMs). Note IBMs progressively intensify with breath-hold duration.

holding, demonstrate a reduction in spleen volume by 18%–35% and an increase in hematocrit of 2%–6% (Schagatay et al., 2001, 2012; Stewart and McKenzie, 2002; Baković et al., 2003). Hypothetically, if there was a 5% increase in hemoglobin concentration from the spleen, this would increase arterial O_2 content by ~5% (assuming all else remains equal; Fig. 4). Larger spleen volumes have been associated with improved apneic performance in competitive divers (Schagatay et al., 2012), and evidence of anatomical and genetic changes in indigenous diving populations (Ama in Japan/Korea and Bajau in Indonesia) also supports the importance of the spleen for apneic diving (Hurford et al., 1990; Ilardo et al., 2018). For example, the Ama have demonstrated a 10.5% increase in hematocrit, following 1 h of repetitive diving to 5–7 m (Hurford et al., 1990). Likewise, the Bajau exhibit larger spleen volumes than demographically matched controls, which have been linked to single nucleoid polymorphisms of the PDE10A gene (Ilardo et al., 2018). Bajau also exhibited single nucleoid polymorphisms to the BDKRB2 gene, which, interestingly, is one of the genes that have been linked to stronger peripheral vasoconstriction during breath-holding (Baranova et al., 2017). With respect to cetaceans, while they generally do not have spleen volumes that are large relative to body mass, cetaceans do have a high prevalence (60%) of one (or more) accessory spleens with similar characteristics to the primary spleen (Tanaka, 1994; Carvalho et al., 2014; Rommel et al., 2018).

Comparative perspectives of the diving response

Just as in the human, the dive response is without a doubt vital to the dolphin's survival while diving. However, for the dolphin spending a considerable time underwater and only intermittently surfacing to breathe, gas management is not just about ensuring there is enough O_2 available to maintain consciousness. Instead, dolphins must use the dive response to both manage the depletion of O_2 and accumulation of metabolic by-products such as CO_2, while also limiting the buildup of nitrogen (N_2). Specifically, dolphins must ensure that there is sufficient available O_2 in the body to support aerobic metabolism where possible. Maintaining aerobic metabolism ensures that dolphins can continue to dive repetitively. Reducing the time spent at the surface is important to maximize the time spent underwater, as required to travel, search for, and catch prey items. Diving can certainly be performed while utilizing anaerobic metabolism in some tissues, but prolonged reliance on anaerobic metabolism eventually results in cellular dysfunction as metabolic end-products accumulate and the supporting glycogen resources deplete. This metabolic *option* is inefficient, as anaerobic metabolism only results in three molecules of adenosine triphosphate (ATP) produced per molecule of sugar, while aerobic metabolism produces 39 ATP. In addition, prolonged anaerobic metabolism eventually forces the animal to stop diving to remove the metabolic by-products (i.e., oxidize lactate) produced. For a wild animal that must breath-hold dive to carry out most of life's activities, this option is not evolutionarily beneficial. Therefore, most diving animals preferentially rely on aerobic metabolism while submerged and stay underwater for durations less than their "aerobic dive limit" (ADL) (Kooyman et al., 1980). ADL is the dive duration associated with the onset of postdive blood lactate accumulation, caused by the necessity for anaerobic metabolism in some body tissues. When diving within the ADL, dolphins can perform repeated dives with short surface intervals. It is likely they only occasionally exceed their aerobic capacity and use anaerobic metabolism sparingly, for instance, to prolong a dive after encountering a high-density aggregation of prey. On these longer dives, anaerobic metabolism should occur in many nonessential tissues, allowing O_2 to be saved for vital tissues that are unable to survive on anaerobic metabolism, such as the brain.

To optimize reliance on aerobic metabolism, despite living mostly without access to ambient air, dolphins are required to balance their short- to medium-term needs efficiently. Maintaining aerobic metabolism can be achieved by limiting dive durations to short dives. During such dives, the dive response is modulated so that circulation to most organs is maintained and O_2 supply is sustained. Therefore, both behavior and physiology, i.e., the dive response, are important to the dolphin to match cardiovascular adjustments to the anticipated behavioral challenge of the next dive (or perhaps sequence of dives)—i.e., "how long will the next dive be and how large a dive response should I have?"

Cognitive control over bradycardia

Dolphins, and other marine mammals, appear to have a conditioned ability to influence the magnitude and rate of change of the dive response—which is not possible for humans. The magnitude of the dive response in dolphins is proportional to the expected length of the dive (Elsner et al., 1966; Fahlman et al., 2019b, 2020b). Similar results have also been reported in other cetaceans and pinnipeds (Jones et al., 1973; Ridgway et al., 1975; Thompson and Fedak, 1993; Jobsis et al., 2001; Elmegaard et al., 2016). Interestingly, when dolphins are able to determine the duration of the apnea, it appears that the reduction in heart rate is minimal (Fahlman et al., 2020b). This suggests that cognition plays a role in the control of the dive response—an exceptional difference between humans and dolphins. Where the dive response in humans is a reflex response, in dolphins, it is, at least partly, cognitively controlled. This ability to vary the dive response is vital to optimize aerobic metabolism as discussed above, ensuring efficient and effective diving patterns, minimizing the time spent at the surface, and maximizing time spent underwater.

Hypoxia and acid-base disturbance

Blood gases in humans

The separation from ambient air is perhaps the most immediate limiting factor for a submerged mammal, and despite the effect of the dive response, hypoxia and tissue acid-base disturbances are an inevitable part of diving. Due to the influence of hydrostatic pressure, the partial pressure of O_2 and CO_2 in the blood (i.e., PO_2 and PCO_2, respectively) is expected to increase in accordance with Boyle's and Henry's law—where the gas is compressed resulting in higher pressure and higher gas solubility. The increasing alveolar PO_2 and PCO_2 result in an elevated diffusion gradient from the lung into the blood, a phenomenon expected to affect the human and the dolphin equally. In two field studies on human divers, the arterial blood gases were measured at 40 m [$n=6$, deep pool (Bosco et al., 2018)] and 60 m [$n=1$, open sea (Scott et al., 2021)]. Both studies showed that arterial blood became hyperoxic (i.e., increased PO_2), during descent. In the pool, arterial PO_2 in 4 (out of 6) divers increased from 94 ± 6 mmHg at the surface to 263 ± 32 mmHg at 40 m, and in open ocean, arterial PO_2 increased from 89 mmHg at the surface to 321 mmHg at 60 m. Unexpectedly, two of the pool divers did not experience hyperoxia at depth (PaO_2 at 40 m was 68 ± 10 mmHg), and the authors postulated that atelectasis (i.e., alveolar collapse) resulted in a right-left intrapulmonary shunt (i.e., blood by-passing the lung), which resulted in their low arterial blood O_2 tensions (Bosco et al., 2018). Interestingly, despite the high solubility of CO_2 in blood and predicted exponential increase in arterial PCO_2 due to hydrostatic loading, arterial PCO_2 across both studies did not exceed 50 mmHg. Even after maximal effort 7.5-min dry static apneas, arterial PCO_2 only increased to 55 mmHg (Fig. 3, Bain et al., 2018b). Whether the solubility of CO_2 buffers the alveolar to arterial PO_2 gradient (Fahlman et al., 2020c), alveolar compression augments pulmonary shunt that reduces gas exchange (Kooyman and Sinnett, 1982; Bostrom et al., 2008; Fahlman et al., 2009), or there exists a buffering capacity to restrain acidosis, these data suggest that the body can mediate the rising arterial CO_2 content.

Blood gases in dolphins

Acquisition of blood samples at depth, from which blood gases can be assessed, has technological and logistical challenges even in controlled human studies. Therefore, blood gas tension and chemistry data for the dolphin have not been regularly measured. Even in seals, which have more accessible blood vessels and have been a more common animal model in physio-logging research, there is limited blood gas data (beyond arterial and venous O_2 tension in the northern elephant seal (*Mirounga angustirostris*) and California sea lion). As such, a comparison of blood gas dynamics in the human and the dolphin is challenging. Nevertheless, a small number of pre- and postdiving measures of blood gas in vertical diving, and cross-sectional measures in static dives do exist for the dolphin (Ridgway et al., 1969; Noren

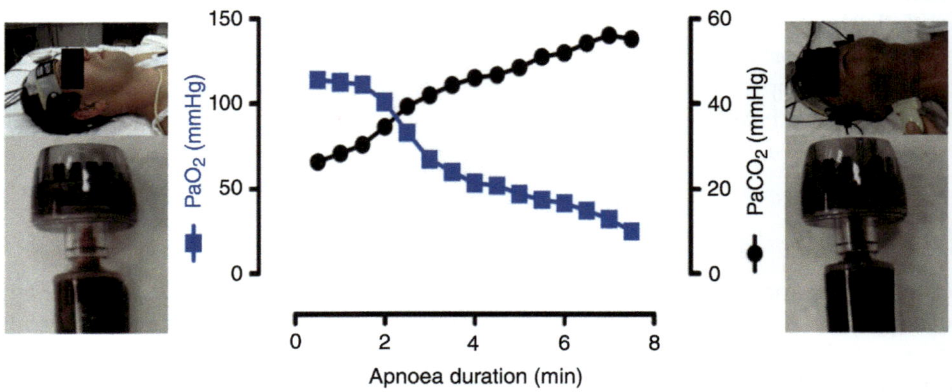

FIG. 3 Arterial blood gases during ~8 min breath-hold in a human. PaO2, arterial partial pressure of oxygen; PaCO2, arterial partial pressure of carbon dioxide. *Reproduced with permission from Bain et al. (2018b).*

et al., 2012; Fahlman et al., 2019a). In the recent study by Fahlman and colleagues, end-expiratory PO_2 following a 4.75-min (284 s) static apnea in a bottlenose dolphin reached 41 mmHg (Fahlman et al., 2019a). In contrast, humans performing static apneas of comparable duration (282 ± 19 s) reported end-expiratory PO_2 of 23.0 ± 0.5 mmHg (Lindholm and Lundgren, 2006).

While there is a paucity of empirical data on blood gas changes during diving in dolphins, it is possible to *hypothesize* that these changes might mirror patterns observed in other marine mammals. Dolphins, like humans and sea lions, inhale before diving. In both the human, as mentioned above, and the sea lion (McDonald and Ponganis, 2013), oxygenation remains high or elevated at depth and mostly shows decline during ascent from depth. In contrast, phocid seals exhale prior to diving (Kooyman and Sinnett, 1982) and their arterial blood O_2 tension declines during the entire duration of the dive. Therefore, as inhalation divers, one could assume that at depth, dolphins are both hyperoxic and hypercapnic, like humans and sea lions.

Involuntary breathing movements

The inevitable changes in blood gas tensions during breath-hold diving are followed by physiological phenomena aimed at removing the body from an asphyxiated state. One such phenomenon in humans is known as involuntary breathing movement (IBMs; Fig. 2). Indeed, if you were to hold your breath for as long as you can endure, you may experience this phenomenon. These IBMs are diaphragmatic contractions, which signal the "physiologic breaking-point," and comprise the struggle phase of an apnea (Schagatay, 2009; Bain et al., 2018b). The intensity of the IBMs increase as the drive to breathe increases, hypoxemia and hypercapnia become more severe and a lack of respiratory rhythm persists. Interestingly, IBMs have been shown to improve cerebral oxygenation (Dujic et al., 2009), especially at higher lung volumes (Palada et al., 2008), which may be critical to slow cerebral hypoxemia.

It is perhaps not surprising that there are no reports of IBMs in dolphins, as apnea is the "normal" respiratory state for dolphins, and this species is not thought to suffer from a physiological break-point and ensuing IBMs. Dolphins have been suggested to have voluntary respiration (Lilly, 1966), but later work proposed that their respiration is both automatic and can be brought under voluntary control just as in other mammals (McCormick, 1969). Nonetheless, the apneustic respiratory mechanism of the dolphin appears to respond differently from the apneustic center in the pons in humans, which may act to terminate IBMs if there is a physiological break-point (Comroe, 1966). Additionally, based on end-expired PO_2 and PCO_2, dolphins live with lower arterial PO_2 and higher arterial PCO_2 and a greater tolerance to elevated CO_2, which in humans triggers the desire to breathe (Ridgway et al., 1969; Fahlman et al., 2015, 2020a; West and Luks, 2016). This blunted hypercapnic ventilatory response may be evidence of a higher chemoreceptor threshold, which results in a higher CO_2 tolerance in the dolphin. In addition, the greater blood CO_2 buffering capacity, resulting from a higher concentration of bicarbonate as compared with terrestrial mammals (Lenfant et al., 1970), means that respiratory acidosis and increasing blood CO_2 tension will be slower to develop.

Loss of consciousness: Hypoxic syncope or shallow-water blackout

If the inevitable changes in blood gases associated with diving exceed extreme thresholds, the arterial O_2 tension may become so low that there can be a loss of consciousness, termed hypoxic syncope. The theoretical lower arterial blood O_2 tensions for consciousness in humans is around 20 mmHg (Ernsting, 1963), which has been supported by data from human breath-hold divers. Lindholm and Lundgren (2006) showed that postapnea end-tidal O_2 partial pressure of 20.3 mmHg (range: 19.6–21.0 mmHg) coincided with a loss of motor control, whereas end-tidal O_2 partial pressure values of only ~3 mmHg higher (mean 23.0 mmHg; range 22.4–23.6 mmHg) did not. In elite apneists under "dry" laboratory conditions, radial artery (Willie et al., 2015) and jugular venous catheterizations (Bain et al., 2016, 2017, 2018a) have demonstrated arterial and central venous PO_2 of 29.6 ± 6.6 mmHg and 25 ± 6 mmHg, respectively, at the end of a breath-hold without syncope (Fig. 3). In an extreme example, and still without syncope, an end-apnea PO_2 tension of 23 mmHg was attained in a diver performing a 435 s breath-hold (Bailey et al., 2017). Fig. 4, which illustrates a hypothetical dive to 100 m, highlights the impact of hydrostatic pressure (particularly on the last 10 m of a dive) on arterial hypoxia and the closeness of such deep diving to the theoretic limit of consciousness.

Dolphins are not reported to suffer mortalities as a result of hypoxic syncope. As "full-time" divers, with a long evolutionary provenance in exploiting an apneic lifestyle, it is perhaps not surprising. For dolphins, a blackout would most likely result in death—and, evolutionarily speaking, this is a definite selection pressure and its avoidance a prerequisite for successful adaptation to an entirely aquatic existence. Nevertheless, how dolphins might avoid hypoxic syncope is interesting. It could be that dolphins avoid exposure to such potentially deleteriously low levels of arterial

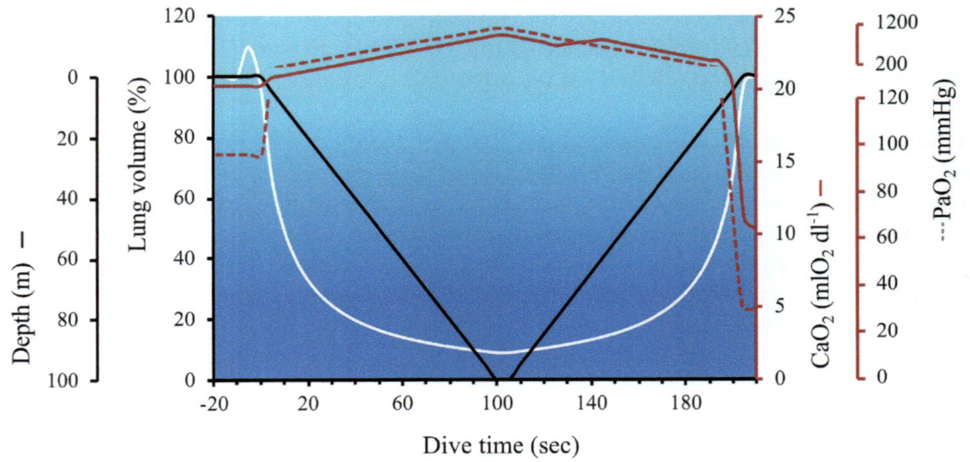

FIG. 4 The impact of hydrostatic pressure (*black*; i.e., depth) on lung volume (*white*) and arterial hypoxia [arterial oxygen content (CaO_2; *dashed red*) and partial pressure of arterial oxygen (PaO_2; *solid red*)] in a human during a simulated dive to 100 m. The increase in lung volume immediately before the start of the dive coincides with lung packing—a maneuver commonly employed by human breath-hold divers to increase the volume of oxygen in the lungs. *Reproduced with permission from Patrician et al. (2021b).*

blood O_2 tensions via sensors, which detect arterial blood O_2 tensions, combined with strict behavioral limits on dive depths and durations. However, considering that some dolphins, such as those in Bermuda, have been recorded to dive to 1000 m (with a total dive time of 13 min, Fahlman et al., 2023), it is likely some animals experience low arterial blood O_2 levels. Cetacean brains have been shown to have higher levels of neuroglobin (Williams et al., 2008; Schneuer et al., 2012), which would help to scavenge reactive O_2 species formed during repeated bouts of ischemia-reperfusion (i.e., an anatomical adaptation to hypoxic exposure). Elevated cerebral glycogen levels along with the astrocyte-neuron lactate shuttle hypothesis (which proposes that the astrocytes increase their rate of glucose uptake, glycolysis, and release of lactate into the extracellular space in response to increased neuron activity), flexible switching between aerobic and anaerobic energy production, and greater tolerance to high lactate concentrations are potential contributing factors (Mason, 2017). Still, the limited data available do not fully explain the dolphin's capacity to maintain neurological performance at low PaO_2.

Pressure effects

Descent into the marine environment comes with the gradual increase in ambient (i.e., hydrostatic) pressure. At the sea surface, all life is exposed to 1 ATA (atmospheres absolute). With increasing depth, the hydrostatic pressure increases by 1 ATA every 10 m (see Chapter 6: Respiratory physiology in the dolphin and other whales) such that at 10 m and 20 m, the hydrostatic pressure is 2 ATA and 3 ATA, respectively. Therefore, descending into the ocean brings with it physiological challenges. However, because humans and dolphins, like all mammals, need to return to the surface to breathe, they are exposed to repeated bouts of elevated pressure, and susceptible to some of the same diving pathologies.

Scholander (1940) proposed that the unusual anatomy of the respiratory system in marine mammals, with a compliant alveolar space and thorax, and stiff conducting airways, might function to reduce pressure-related disease (Scholander, 1940). This balloon-pipe model of the respiratory system has been the main hypothesis as to how dolphins and marine mammals reduce uptake of N_2 and avoid the risks of barotrauma (lung squeeze), N_2 narcosis, and decompression sickness.

Risk of pulmonary barotrauma

Humans have a delicate pulmonary capillary interface (West et al., 1991), and therefore, it is perhaps not surprising that lung injury can occur, due to the cumulative forces of hydrostatic-induced compression and decompression of both the lungs and thoracic cage, centralization of blood volume, hypertension, exertion, and hypercapnic hypoxia. In humans, one such injury, commonly referred to as "lung squeeze," is a form of pulmonary barotrauma that has been extensively reviewed (Ferrigno and Lundgren, 2003; Lindholm and Lundgren, 2009; Dujic and Breskovic, 2012;

Mijacika and Dujic, 2016; Moon et al., 2016; Kumar and Thompson, 2019; Schipke et al., 2019). Lung squeeze manifests shortly after surfacing, is characterized by pulmonary edema and hemoptysis (blood originating from the bronchi or lungs) (Boussuges et al., 1999; Lindholm et al., 2008; Patrician et al., 2021a), and is often associated with (1) productive cough, dyspnea, and chest tightness (Cialoni et al., 2012), (2) decrements in lung function and reduced O_2 saturation (Linér and Johan, 2008), and (3) an impairment in pulmonary gas exchange efficiency (Patrician et al., 2021a). Modeling work suggests that human lungs will begin to compress during a dive to 18m, and the alveoli will be completely collapsed at 235m (Fitz-Clarke, 2007). During ascent, the reduction in pressure results in lung air expansion and alveolar recruitment/opening (see Chapter 6). As the alveoli are opening, there exists potential for the disproportionate coordination between transpulmonary pressure and regional airway surface tensions that can lead to abrupt compensatory airway equalization (i.e., airway/alveolar "popping"). Beyond pulmonary trauma, but still engendered by hydrostatic compression and decompression, is barotrauma of the paranasal sinuses. As depth and therefore hydrostatic pressure increases, sinuses must be equalized or a relative vacuum in the sinus cavity forms. Failure to equalize pressure during descent elicits mucosal congestion, edema, hemorrhagic bullas, and free blood—manifesting in sharp, intense pain of the maxillary, ethmoid, and/or frontal sinuses. This risk increases under various conditions interfering with equilibration of gas pressure in the body's air-containing cavities (e.g., sinus congestion, Eustachian tube blockage, structural anomaly, inflammation, infection).

The anatomical observations made by Scholander (1940) have helped explain how marine mammals avoid pulmonary barotrauma, with either a highly compliant thorax (as in seals), or with a chest with extensive elastic recoil inwards such that a relaxed state results in almost complete emptying of the lung (see Chapter 6). Consequently, without a stiff chest or residual volume, pressure can push the chest inwards until complete atelectasis (alveolar collapse), which helps prevent transthoracic pressure differences and lung squeeze. Initial evidence for chest collapse was presented by Ridgway et al. (1969) in a seminal study where photos was taken of the chest of Tuffy, a bottlenose dolphin, during a dive to 300m, with visible compression of the thoracic cavity (see Fig. 14 in Chapter 6).

Measurements of the thorax in dolphin cadavers (Fahlman et al., 2018) and in live pinnipeds (Leith, 1976; Fahlman et al., 2014) have confirmed that chest compression leaves the alveolar space almost empty of air. The highly compliant alveolar space in the dolphin and other marine mammals (Fahlman et al., 2017b) allows alveolar atelectasis (i.e., airway collapse). It has also been reported that some species of dolphins have submucosal venous lacunae in the airways (Cozzi et al., 2005), a tissue that could fill with blood in the tracheal space, which both reduces the tracheal volume and helps stiffen the wall. This further helps to prevent negative pressure and tissue trauma of the upper airways as the pressure increases during a dive.

Cetaceans have lost the air-filled, bony-walled, noncompliant paranasal sinus system found in humans and other terrestrial mammals. However, cetaceans do have gas-filled sacs akin to sinuses (in which the gas is derived from respiratory air but may vary in composition) that are similar to those of the paranasal system. The cetacean accessory sinus system is unique (Fraser and Purvevs, 1960) in that the unpigmented mucosa-lined structures, which are located on the ventral aspect of the skull, are typically associated with hearing and acoustic isolation of the ears (Houser et al., 2004). The accessory sinus system, which is not completely encased in robust, rigid bony compartments, is presumed to be exposed to the effects of changing barometric pressures encountered during diving. In order to avoid sinus barotrauma, it is assumed that there exists a significant amount of structural flexibility within those structures to compensate for the compression and decompression of the air within these spaces—unlike the much less compliant sinuses of the human. In addition, Fraser and Purves (1960) hypothesized that the elaborate venous plexus in the accessory sinus system could help compensate for lost volume of air during compression, by replacing it with blood to avoid dysbaric trauma to the tissues, as well as facilitating hydrostatic equilibrium of the ears, which are surrounded by the accessory sinus system (Costidis and Rommel, 2012).

Risk of nitrogen narcosis

In his book "Silent World," Jacque Yves Costeau coined the synonym for nitrogen narcosis, "L'ivresse des grandes profondeurs," or "rapture of depths" (Cousteau and Frédéric Dumas, 1953). The narcotic effects of high gas tensions are pronounced in the human diver, and the anesthetic effect of nitrogen greatly affects brain function. Depending on depth, time at depth, and inspired lung volume, the alveolar N_2 partial pressure will be elevated and will facilitate diffusion of N_2 into the blood. Nitrogen narcosis may be relatively common in human during breath-hold dives exceeding ~70–90m, but there are only a few published examples in the literature (Lindholm and Lundgren, 2009; Fitz-Clarke, 2018; Patrician et al., 2021a). Prevalence of reporting is likely impacted due to amnesia (Ferrigno and Lundgren, 2003)—for example, in a recent report by Patrician et al. (2021a) one diver spent 42s beyond 90m (with maximum depth of 117 m). Upon surfacing, the diver did not recall the entire dive, and only slowly began remembering the details of the dive over the subsequent minutes.

Dolphins, and other marine mammals, are not thought to suffer from N_2 narcosis, and Kooyman (1989) suggested that it is unlikely that N_2 levels reach narcotic levels. Still, arterial blood N_2 tensions in a harbor seal (*Phoca vitulina*) exceeded 5 ATA (Kooyman et al., 1972)—levels that would cause severe narcosis in humans. Interestingly, it has been shown that marine mammals appear to have different NMDA (N-methyl-D-aspartate) receptors, which have been suggested to provide resistance toward high-pressure nervous syndrome and possibly also N_2 narcosis (Bliznyuk et al., 2018).

Risk of decompression sickness

In 1908, Haldane first published his seminal work on diving physiology during pressure exposures in a dry pressure chamber, and the first tables to avoid decompression sickness (DCS) were developed (Boycott et al., 1908). Today, it is commonly accepted that DCS, also called "bends," or "caisson disease" is caused by the formation of vascular gas bubbles due to supersaturation in body fluids and tissues when a diver is returning to the surface after a dive (Vann et al., 2011; Pollock and Buteau, 2017). The vascular bubbles do not always result in DCS symptoms, but the probability of symptoms increases with increasing amount of bubbles detected (Sawatzky and Nishi, 1991). Thus, alternative theories were also proposed due to the fact that most SCUBA divers show asymptomatic vascular gas bubbles postdive, whereas DCS is a very rare clinical condition (Ljubkovic et al., 2010, 2011). DCS can also occur in breath-hold divers (commonly referred to as "Taravana," which stems from DCS reports in the indigenous diving populations in Polynesia)—especially those diving repetitively (e.g., spearfishing, safety divers, and/or using underwater scooters) and those performing extreme depths (Cross, 1965; Paulev, 1965; Rahn and Yokoyama, 1965; Schipke et al., 2006; Fitz-Clarke, 2009; Lemaitre et al., 2009; Moon and Gray, 2010; Dujic and Breskovic, 2012). The pathology of DCS, including its manifestation and risk factors, has been extensively reviewed elsewhere (Brubakk and Neuman, 2003); however, in breath-hold divers, symptoms can range from dizziness, nausea, thoracic/skin/joint pain, hemiplegia, paresis, dysarthria, vertigo, and unconsciousness, with short- to long-term prognoses (Cross, 1965; Schipke et al., 2006; Cortegiani et al., 2013; Tetzlaff et al., 2017; Blogg et al., 2023).

Until recently, it was thought that the anatomy of the respiratory system in marine mammals, with stiff airways and compliant alveoli proposed by Scholander (1940), would be sufficient to explain how all marine mammals avoid gas emboli and DCS. However, following naval associated mass-strandings of deep-diving cetaceans (Jepson et al., 2003; Fernández et al., 2005), with symptoms resembling DCS in human divers, a number of studies have shown that marine mammals experience elevated blood and tissue gas tensions that can result in gas emboli, and at times, trauma (Fig. 5, Moore et al., 2009; Van Bonn et al., 2013; Fernández et al., 2017). Theoretical modeling suggested that despite the unusual respiratory anatomy, the repeated dive behavior seen in dolphins would result in blood and tissue N_2 levels that would likely result in gas emboli and a high probability of severe symptoms of DCS (see Figure 3 in Fahlman et al., 2021). Furthermore, work on the cardiorespiratory physiology of cetaceans has attempted to elucidate the mechanism by which cetaceans avoid gas emboli, resulting in the *selective gas exchange* hypothesis (García-Párraga et al., 2018; Fahlman et al., 2021). In short, the lung architecture in the dolphin results in two distinctly different pulmonary regions: one that is open for gas exchange and the other which is collapsed. The conditioned ability to vary heart rate (Fahlman et al., 2020b), and likely blood flow, may enable marine mammals to direct blood either to the region that is open or collapsed and allows selective exchange of gases with different gas solubilities (West, 1962; Farhi and Yokoyama, 1967). This would provide the ability to exchange O_2 and CO_2 while minimizing N_2 exchange (García-Párraga et al., 2018; Fahlman et al., 2021). This hypothesis also explains how stress, such as exposure to sonar, might result in failure of this mechanism causing uptake of N_2 and increasing the risk of gas emboli formation. This provides avenues for new areas of research, offering an alternative explanation for how marine vertebrates can avoid diving-related problems observed in human divers, and how failure of this adaptation may result in diving-related trauma (Frantzis, 1998; Fernández et al., 2005, 2017; García-Párraga et al., 2014; Fahlman et al., 2017a).

Other considerations to minimize metabolic rate

Other factors that affect metabolic rate, and therefore impact on diving, include mechanisms to retain heat balance (i.e., thermoregulation, see Chapter 3) and mechanisms to improve the efficiency of swimming (see Chapter 2: Energetic costs of rest and locomotion in dolphins). For humans, prolonged immersion anywhere other than in the tropics requires a wet suit to protect against the fundamentally different thermal properties of water than air. This challenge is

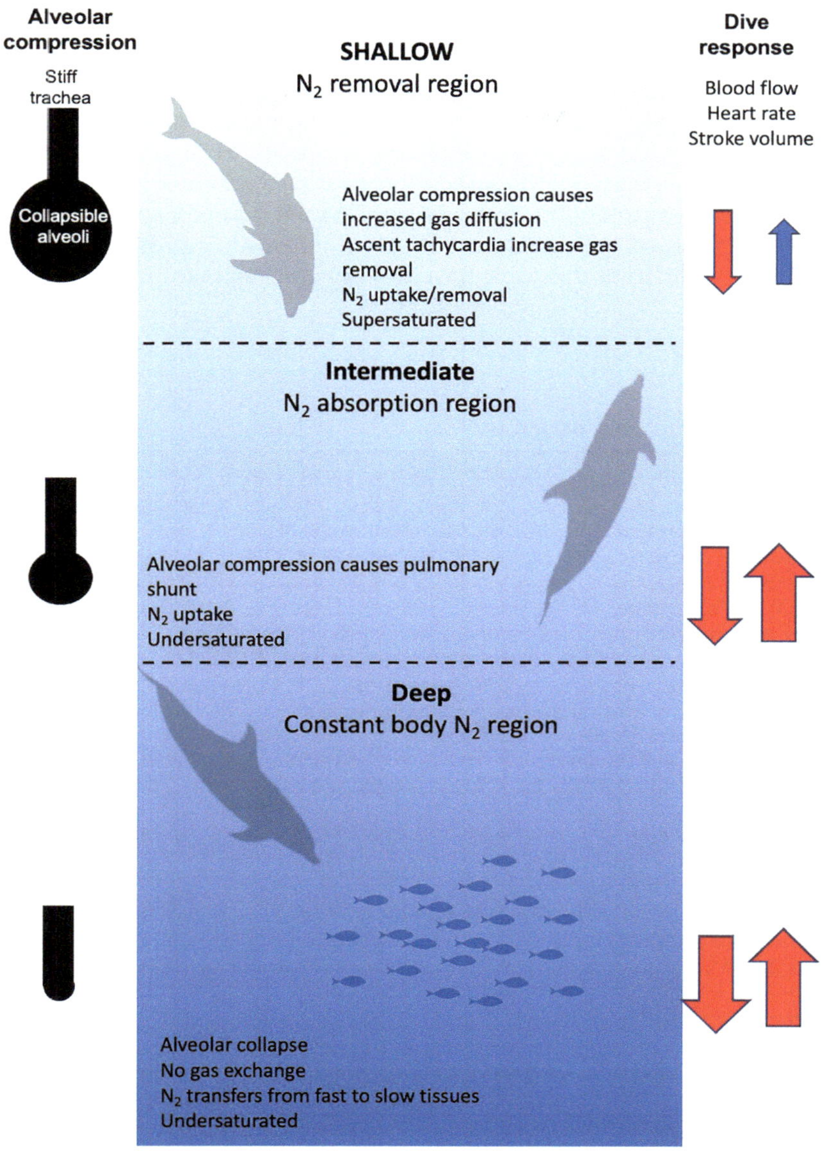

FIG. 5 Risk factors for gas bubble formation, across three generalized regions: shallow (0 to ~30 m), intermediate (~30 to ~200 m), and deep (beyond ~200 m). In the shallow region, tissue and blood PN_2 begins to exceed ambient PN_2, and N_2 is removed. In the intermediate region, ambient PN_2 exceeds tissue and blood PN_2 and N_2 is taken up until the alveoli collapse in the deep region. Following alveolar collapse, N_2 is redistributed from fast tissues to slow tissues. On the left, the balloon/pipe model proposed by Scholander (1940) shows how increasing dive depth causes alveolar compression and collapse. Increasing lung partial pressure initially causes an increase in the gas diffusion rate in the shallow region. In the intermediate region, the compression of the alveoli results in reduction and then cessation of diffusion as the alveoli collapse. On the right, arrows illustrate how diving alters heart rate, stroke volume, and cardiac output (the dive response). During descent (*arrow down*), the heart rate, stroke volume, and cardiac output decreases (*red color*), with a greater reduction with increasing dive duration and depth (*thickness of arrow*). During ascent (*arrow up*), the dive response is maintained until the animal approaches the surface, where the heart rate, stroke volume, and cardiac output increase (*blue colored arrow*) to prediving levels to prepare the animal to rapidly restore the O_2 stores, and remove CO_2 and N_2. *Modified from Fahlman et al. (2021).*

perhaps one of the most difficult faced by mammals making the transition from land to the sea. In retaining the same high body temperatures of terrestrial mammals (including humans), these animals face large thermal gradients and potential heat loss, without adaptations that help minimize this thermal loss.

One of the most notable adaptations recorded in the Ama freedivers was their thermoregulatory adaptation to cold exposure. Historically, the Ama dove year-round in only cotton linens, despite sea temperatures ranging from 27°C in the summer to 10°C in the winter. Early studies concluded that the Ama may have an improved (i.e., reduced) thermal conductance based on findings of improved peripheral vasoconstriction, increased basal metabolic rate in the winter (one of the few clearly recognized and described instances of human metabolic adaptation to regular cold water exposure), and higher shivering threshold (i.e., takes longer to shiver), compared to controls with comparable subcutaneous fat thickness (Rennie et al., 1962; Hong et al., 1986). However, with the introduction of wetsuits (as early as 1977, and more common by 1980), many of these thermoregulatory adaptations against cold in the Ama disappeared and these were no longer evident in the 1980s (Park et al., 1983a,b; Ferretti and Costa, 2003). Unlike the human, where endogenous heat conservation mechanisms are not adequate to resist heat loss in water, dolphins have increased insulation provided by a thick subcutaneous blubber layer (see Chapter 3).

Conclusion

Breath-hold diving exposes all mammals to a number of extreme environmental stresses. Both humans and dolphins show basic mammalian features in their response to diving, but there are stark contrasts between dolphins living permanently at sea, and humans occasionally making forays from the surface into the water. Our understanding of the diving physiology of dolphins nevertheless benefits from studies of diving in more "accessible" human subjects to compare and contrast with the more limited studies available from dolphins and other marine mammals. In this chapter, we have explored similarities and differences between features of the dive response, hemodynamics, and acid-base changes during diving and we have compared strategies used to mediate the effects of hydrostatic pressure on the body. While many differences exist, a comparative exploration of this topic provides novel insights for future work.

References

Abrahamsson, E., Schagatay, E., 2014. A living based on breath-hold diving in the Bajau Laut. Hum. Evol. 29, 171–183.

AIDA, 2022. Association Internationale pour le Développement de l'Apnée. Available: www.aidainternational.org.

Arnold, R.W., 1985. Extremes in human breath hold, facial immersion bradycardia. Undersea Biomed. Res. 12, 183–190.

Bailey, D.M., Willie, C.K., Hoiland, R.L., Bain, A.R., Macleod, D.B., Santoro, M.A., Demasi, D.K., Andrijanic, A., Mijacika, T., Barak, O.F., Dujic, Z., Ainslie, P.N., 2017. Surviving without oxygen: how low can the human brain go? High Alt. Med. Biol. 18, 73–79.

Bain, A.R., Ainslie, P.N., Hoiland, R.L., Barak, O.F., Cavar, M., Drvis, I., Stembridge, M., Macleod, D.M., Bailey, D.M., Dujic, Z., Macleod, D.B., 2016. Cerebral oxidative metabolism is decreased with extreme apnoea in humans; impact of hypercapnia. J. Physiol. 594, 5317–5328.

Bain, A.R., Ainslie, P.N., Barak, O.F., Hoiland, R.L., Drvis, I., Mijacika, T., Bailey, D.M., Santoro, A., Demasi, D.K., Dujic, Z., Macleod, D.B., 2017. Hypercapnia is essential to reduce the cerebral oxidative metabolism during extreme apnea in humans. J. Cereb. Blood Flow Metab. 37, 3231–3242.

Bain, A.R., Ainslie, P.N., Hoiland, R.L., Barak, O.F., Drvis, I., Stembridge, M., Macleod, D.M., Mceneny, J., Stacey, B.S., Tuaillon, E., Marchi, N., De Maudave, A.F., Dujic, Z., Macleod, D.B., Bailey, D.M., 2018a. Competitive apnea and its effect on the human brain: focus on the redox regulation of blood-brain barrier permeability and neuronal-parenchymal integrity. FASEB J. 32, 2305–2314.

Bain, A.R., Drvis, I., Dujic, Z., Macleod, D.B., Ainslie, P.N., 2018b. Physiology of static breath holding in elite apneists. Exp. Physiol. 103, 635–651.

Baković, D., Valic, Z., Eterović, D., Vukovic, I., Obad, A., Marinović-Terzić, I., Dujić, Z., 2003. Spleen volume and blood flow response to repeated breath-hold apneas. J. Appl. Physiol. 95, 1460–1466.

Baranova, T.I., Berlov, D.N., Glotov, O.S., Korf, E.A., Minigalin, A.D., Mitrofanova, A.V., Ahmetov, I.I., Glotov, A.S., 2017. Genetic determination of the vascular reactions in humans in response to the diving reflex. Am. J. Physiol. Heart Circ. Physiol. 312, H622–H631.

Barcroft, J., Poole, L.T., 1927. The blood in the spleen pulp. J. Physiol. 64, 23–29.

Bjertnaes, L., Hauge, A., Kjekshus, J., Søyland, E., 1984. Cardiovascular responses to face immersion and apnea during steady state muscle exercise. A heart catheterization study on humans. Acta Physiol. Scand. 120, 605–612.

Bliznyuk, A., Golan, H., Grossman, Y., 2018. Marine mammals' NMDA receptor structure: possible adaptation to high pressure environment. Front. Physiol. 9, 1633.

Blogg, S.L., Tillmans, F., Lindholm, P., 2023. The risk of decompression illness in breath-hold divers: a systematic review. Diving Hyperb. Med. 53, 31–41.

Bosco, G., Rizzato, A., Martani, L., Schiavo, S., Talamonti, E., Garetto, G., Paganini, M., Camporesi, E.M., Moon, R.E., 2018. Arterial blood gas analysis in breath-hold divers at depth. Front. Physiol. 9, 1558.

Bostrom, B.L., Fahlman, A., Jones, D.R., 2008. Tracheal compression delays alveolar collapse during deep diving in marine mammals. Respir. Physiol. Neurobiol. 161, 298–305.

Boussuges, A., Pinet, C., Thomas, P., Bergmann, E., Sainty, J., Vervloet, D., 1999. Haemoptysis after breath-hold diving. Eur. Respir. J. 13, 697–699.

Boycott, A.E., Damant, G.C.C., Haldane, J.S., 1908. The prevention of decompression-air illness. J. Hyg. 8, 342–443.

Brick, I., 1966. Circulatory responses to immersing the face in water. J. Appl. Physiol. 21, 33–36.

Brubakk, A.O., Neuman, T.S., 2003. Bennett and Elliott's Physiology and Medicine of Diving. Saunders, New York; Edinburgh.

Carvalho, V.L., Guimarães, J.P., Vergara-Parente, J.E., De Meirelles, A.C.O., Marmontel, M., Miglino, M.A., 2014. Accessory spleen in cetaceans and its relevance as a secondary lymphoid organ. Zoomorphology 133, 343–350.

Cialoni, D., Sponsiello, N., Marabotti, C., Marroni, A., Pieri, M., Maggiorelli, F., Tonerini, M., Frammartino, B., 2012. Prevalence of acute respiratory symptoms in breath-hold divers. Undersea Hyperb. Med. 39, 837–844.

Comroe, J.H.J., 1966. Physiology of Respiration. Year Book Medical Publishers Inc, Chicago.

Cortegiani, A., Foresta, G., Strano, G., Strano, M.T., Montalto, F., Garbo, D., Raineri, S.M., 2013. An atypical case of taravana syndrome in a breath-hold underwater fishing champion: a case report. Case Rep. Med. 2013, 939704.

Costidis, A., Rommel, S.A., 2012. Vascularization of air sinuses and fat bodies in the head of the bottlenose dolphin (*Tursiops truncatus*): morphological implications on physiology. Front. Physiol. 3, 243.

Cousteau, J., Frédéric Dumas, F., 1953. The Silent World. Harper & Brothers Publishers, New York.

Cozzi, B., Bagnoli, P., Acocella, F., Costantino, M.L., 2005. Structure and biomechanical properties of the trachea of the striped dolphin *Stenella coeruleoalba*: evidence for evolutionary adaptations to diving. Anat Rec A Discov Mol Cell Evol Biol 284, 500–510.

Craig Jr., A.B., 1963. Heart rate responses to apneic underwater diving and to breath holding in man. J. Appl. Physiol. 18, 854–862.

Cross, E., 1965. Taravana diving syndrome in the tuamotu diver. In: Rahn, H., Yokoyama, T. (Eds.), Physiology of Breath-hold Diving and the Ama of Japan. The National Academies Press, USA, pp. 207–221.

Dujic, Z., Breskovic, T., 2012. Impact of breath holding on cardiovascular respiratory and cerebrovascular health. Sports Med. 42, 459–472.

Dujic, Z., Uglesic, L., Breskovic, T., Valic, Z., Heusser, K., Marinovic, J., Ljubkovic, M., Palada, I., 2009. Involuntary breathing movements improve cerebral oxygenation during apnea struggle phase in elite divers. J. Appl. Physiol. 107, 1840–1846.

Elmegaard, S.L., Johnson, M., Madsen, P.T., McDonald, B.I., 2016. Cognitive control of heart rate in diving harbor porpoises. Curr. Biol. 26, R1167–R1176.

Elsner, R., Kenney, D.W., Burgess, K., 1966. Diving bradycardia in the trained dolphin. Nature 212, 407–408.

Ernsting, J., 1963. The effect of brief profound hypoxia upon the arterial and venous oxygen tensions in man. J. Physiol. 169, 292–311.

Fagius, J., Sundlöf, G., 1986. The diving response in man: effects on sympathetic activity in muscle and skin nerve fascicles. J. Physiol. 377, 429–443.

Fahlman, A., Hooker, S.K., Olszowka, A., Bostrom, B.L., Jones, D.R., 2009. Estimating the effect of lung collapse and pulmonary shunt on gas exchange during breath-hold diving: the Scholander and Kooyman legacy. Respir. Physiol. Neurobiol. 165, 28–39.

Fahlman, A., Loring, S.H., Johnson, S.P., Haulena, M., Trites, A.W., Fravel, V.A., Van Bonn, W.G., 2014. Inflation and deflation pressure-volume loops in anesthetized pinnipeds confirms compliant chest and lungs. Front. Physiol. 5, 433.

Fahlman, A., Loring, S.H., Levine, G., Rocho-Levine, J., Austin, T., Brodsky, M., 2015. Lung mechanics and pulmonary function testing in cetaceans. J. Exp. Biol. 218, 2030–2038.

Fahlman, A., Crespo Picazo, J.-L., Sterba-Boatwright, B., Stacy, B.A., Garcia-Parraga, D., 2017a. Defining risk variables causing gas embolism in loggerhead sea turtles (*Caretta caretta*) caught in trawls and gillnets. Sci. Rep. 7, 2739.

Fahlman, A., Moore, M.J., Garcia-Parraga, D., 2017b. Respiratory function and mechanics in pinnipeds and cetaceans. J. Exp. Biol. 220, 1761–1773.

Fahlman, A., McHugh, K., Allen, J., Barleycorn, A., Allen, A., Sweeney, J., Stone, R., Faulkner Trainor, R., Bedford, G., Moore, M.J., Jensen, F.H., Wells, R., 2018. Resting metabolic rate and lung function in wild offshore common bottlenose dolphins, *Tursiops truncatus*, near Bermuda. Front. Physiol. 9, 886.

Fahlman, A., Brodsky, M., Miedler, S., Dennison, S., Ivančić, M., Levine, G., Rocho-Levine, J., Manley, M., Rocabert, J., Borque Espinosa, A., 2019a. Ventilation and gas exchange before and after voluntary static surface breath-holds in clinically healthy bottlenose dolphins, *Tursiops truncatus*. J. Exp. Biol. 222, 1–9.

Fahlman, A., Miedler, S., Rocho-Levine, J., Jabois, A., Arenarez, J., Marti-Bonmati, L., García-Párraga, D., Cauture, F., 2019b. Re-evaluating the significance of the dive response during voluntary surface apneas in the bottlenose dolphin, *Tursiops truncatus*. Sci. Rep. 9, 8613.

Fahlman, A., Borque-Espinosa, A., Facchin, F., Fernandez, D.F., Caballero, P.M., Haulena, M., Rocho-Levine, J., 2020a. Comparative respiratory physiology in cetaceans. Front. Physiol. 11, 142.

Fahlman, A., Cozzi, B., Manley, M., Jabas, S., Malik, M., Blawas, A., Janik, V.M., 2020b. Conditioned variation in heart rate during static breath-holds in the bottlenose dolphin (*Tursiops truncatus*). Front. Physiol. 11, 604018.

Fahlman, A., Sato, K., Miller, P.J., 2020c. Improving estimates of diving lung volume in air-breathing marine vertebrates. J. Exp. Biol. 223, jeb216846.

Fahlman, A., Moore, M.J., Wells, R.S., 2021. How do Marine mammals manage and usually avoid gas emboli formation and gas embolic pathology? Critical clues from studies of wild dolphins. Front. Mar. Sci. 8. fmars.2021.598633.

Fahlman, A., Tyson Moore, R.B., Stone, R., Sweeney, J., Faulkner Trainor, R., Barleycorn, A., McHugh, K., Allen, J., Wells, R.S., 2023. Deep diving by offshore bottlenose dolphins (*Tursiops spp.*). Mar. Mamm. Sci. 39, 1251–1266.

Farhi, L.E., Yokoyama, T., 1967. Effects of ventilation-perfusion inequality on elimination of inert gases. Respir. Physiol. 3, 12–20.

Fedak, M.A., Thompson, D., 1993. Behavioural and physiological options in diving seals. In: Boyd, I.L. (Ed.), Marine mammals: advances in behavioural and population biology. Symposia of the Zoological Society of London, No. 66, Clarendon Press, pp. 333–348.

Fernández, A., Edwards, J.F., Rodríguez, F., Espinosa De Los Monteros, A., Herráez, P., Castro, P., Jaber, J.R., Martín, V., Arbelo, M., 2005. "Gas and fat embolic syndrome" involving a mass stranding of beaked whales (family Ziphiidae) exposed to anthropogenic sonar signals. Vet. Pathol. 42, 446–457.

Fernández, A., Sierra, E., Díaz-Delgado, J., Sacchini, S., Sánchez-Paz, Y., Suárez-Santana, C., Arregui, M., Arbelo, M., Bernaldo De Quirós, Y., 2017. Deadly acute decompression sickness in Risso's dolphins. Sci. Rep. 7, 13621.

Ferretti, G., 2001. Extreme human breath-hold diving. Eur. J. Appl. Physiol. 84, 254–271.

Ferretti, G., Costa, M., 2003. Diversity in and adaptation to breath-hold diving in humans. Comp. Biochem. Physiol. A: Physiol. 136, 205–213.

Ferrigno, M., Lundgren, C., 2003. Breath-hold diving. In: Brubakk, A.O., Neuman, T.S. (Eds.), Bennett and Elliott's Physiology and Medicine of Diving. Saunders, New York, pp. 153–180.

Ferrigno, M., Ferretti, G., Ellis, A., Warkander, D., Costa, M., Cerretelli, P., Lundgren, C.E., 1997. Cardiovascular changes during deep breath-hold dives in a pressure chamber. J. Appl. Physiol. 83, 1282–1290.

Fitz-Clarke, J.R., 2007. Mechanics of airway and alveolar collapse in human breath-hold diving. Respir. Physiol. Neurobiol. 159, 202–210.

Fitz-Clarke, J.R., 2009. Risk of decompression sickness in extreme human breath-hold diving. Undersea Hyperb. Med. 36, 83–91.

Fitz-Clarke, J.R., 2018. Breath-hold diving. Compr. Physiol. 8, 585.

Frantzis, A., 1998. Does acoustic testing strand whales? Nature 392, 29.

Fraser, F.C., Purves, P.E., 1960. Hearing in cetaceans. Evolution of the accessory air sacs and the structure and function of the outer and middle ear in recent cetaceans. Bull. Brit. Mus. (Nat. Hist.) 7, 1–140.

García-Párraga, D., Crespo-Picazo, J.L., Bernaldo De Quirós, Y., Cervera, V., Martí-Bonmati, L., Díaz-Delgado, J., Arbelo, M., Moore, M.J., Jepson, P.D., Fernández, A., 2014. Decompression sickness ("the bends") in sea turtles. Dis. Aquat. Org. 111, 191–205.

García-Párraga, D., Moore, M., Fahlman, A., 2018. Pulmonary ventilation– perfusion mismatch: a novel hypothesis for how diving vertebrates may avoid the bends. Proc. R. Soc. Lond. B 285, 20180482.

Heusser, K., Dzamonja, G., Tank, J., Palada, I., Valic, Z., Bakovic, D., Obad, A., Ivancev, V., Breskovic, T., Diedrich, A., Joyner, M.J., Luft, F.C., Jordan, J., Dujic, Z., 2009. Cardiovascular regulation during apnea in elite divers. Hypertension 53, 719–724.

Hong, S.K., Rahn, H., 1967. The diving women of Korea and Japan. Sci. Am. 216, 34–43.

Hong, S.K., Rennie, D.W., Park, Y.S., 1986. Cold acclimatization and deacclimatization of Korean women divers. Exerc. Sport Sci. Rev. 14, 231–268.

Houser, D.S., Finneran, J., Carder, D., Van Bonn, W., Smith, C., Hoh, C., Mattrey, R., Ridgway, S., 2004. Structural and functional imaging of bottlenose dolphin (*Tursiops truncatus*) cranial anatomy. J. Exp. Biol. 207, 3657–3665.

Hurford, W.E., Hong, S.K., Park, Y.S., Ahn, D.W., Shiraki, K., Mohri, M., Zapol, W.M., 1990. Splenic contraction during breath-hold diving in the Korean ama. J. Appl. Physiol. 69, 932–936.

Hurwitz, B.E., Furedy, J.J., 1986. The human dive reflex: an experimental, topographical and physiological analysis. Physiol. Behav. 36, 287–294.

Ilardo, M.A., Moltke, I., Korneliussen, T.S., Cheng, J., Stern, A.J., Racimo, F., De Barros Damgaard, P., Sikora, M., Seguin-Orlando, A., Rasmussen, S., Van Den Munckhof, I.C.L., Ter Horst, R., Joosten, L.A.B., Netea, M.G., Salingkat, S., Nielsen, R., Willerslev, E., 2018. Physiological and genetic adaptations to diving in sea nomads. Cell 173, 569–580.e15.

Jepson, P.D., Arbelo, M., Deaville, R., Patterson, I.A., Castro, P., Baker, J.R., Degollada, E., Ross, H.M., Herráez, P., Pocknell, A.M., Rodríguez, F., Howie, F.E., Espinosa, A., Reid, R.J., Jaber, J.R., Martin, V., Cunningham, A.A., Fernández, A., 2003. Gas-bubble lesions in stranded cetaceans. Nature 425, 575–576.

Jobsis, P.D., Ponganis, P.J., Kooyman, G.L., 2001. Effects of training on forced submersion responses in harbor seals. J. Exp. Biol. 204, 3877–3885.

Jones, D.R., Fisher, H.D., Mctaggart, S., West, N.H., 1973. Heart rate during breath-holding and diving in the unrestrained harbor seal. Can. J. Zool. 51, 671–680.

Kaczmarek, J., Reichmuth, C., McDonald, B.I., Kristensen, J.H., Larson, J., Johansson, F., Sullivan, J.L., Madsen, P.T., 2018. Drivers of the dive response in pinnipeds; apnea, submergence or temperature? J. Exp. Biol., jeb.176545.

Kooyman, G.L., 1989. Diverse Divers: Physiology and Behavior. Springer-Verlag, Berlin.

Kooyman, G., Sinnett, E., 1982. Pulmonary shunts in harbor seals and sea lions during simulated dives to depth. Physiol. Zool. 55, 105–111.

Kooyman, G.L., Schroeder, J.P., Denison, D.M., Hammond, D.D., Wright, J.J., Bergman, W.P., 1972. Blood nitrogen tensions of seals during simulated deep dives. Am. J. Phys. 223, 1016–1020.

Kooyman, G., Wahrenbrock, E., Castellini, M., Davis, R., Sinnett, E., 1980. Aerobic and anaerobic metabolism during voluntary diving in Weddell seals: evidence of preferred pathways from blood chemsitry and behavior. J. Comp. Physiol. 138, 335–346.

Kumar, M., Thompson, P.D., 2019. A literature review of immersion pulmonary edema. Phys. Sportsmed. 47, 148–151.

Kyhl, K., Drvis, I., Barak, O., Mijacika, T., Engstrøm, T., Secher, N.H., Dujic, Z., Buca, A., Madsen, P.L., 2016. Organ perfusion during voluntary pulmonary hyperinflation; a magnetic resonance imaging study. Am. J. Physiol. Heart Circ. Physiol. 310, H444–H451.

Leith, D.E., 1976. Comparative mammalian respiratory mechanics. Physiologist 19, 485–510.

Lemaitre, F., Fahlman, A., Gardette, B., Kohshi, K., 2009. Decompression sickness in breath-hold divers: a review. J. Sports Sci. 27, 1519–1534.

Lemaître, F., Lafay, V., Taylor, M., Costalat, G., Gardette, B., 2013. Electrocardiographic aspects of deep dives in elite breath-hold divers. Undersea Hyperb. Med. 40, 145–154.

Lenfant, C., Johansen, K., Torrance, J.D., 1970. Gas transport and oxygen storage capacity in some pinnipeds and the sea otter. Respir. Physiol. 9, 277–286.

Lilly, J.C., 1966. Sonic-ultrasonic emissions of the bottlenose dolphin. In: Norris, K.S. (Ed.), Whales, Dolphins and Porpoises. University of California Press, Berkley, USA, pp. 503–509.

Lindholm, P., Lundgren, C.E., 2006. Alveolar gas composition before and after maximal breath-holds in competitive divers. Undersea Hyperb. Med. 33, 463–467.

Lindholm, P., Lundgren, C.E., 2009. The physiology and pathophysiology of human breath-hold diving. J. Appl. Physiol. 106, 284–292.

Lindholm, P., Ekborn, A., Öberg, D., Gennser, M., 2008. Pulmonary edema and hemoptysis after breath-hold diving at residual volume. J. Appl. Physiol. 104, 912–917.

Linér, M.H., Johan, P.A.A., 2008. Pulmonary edema after competitive breath-hold diving. J. Appl. Physiol. 104, 986–990.

Ljubkovic, M., Marinovic, J., Obad, A., Breskovic, T., Gaustad, S.E., Dujic, Z., 2010. High incidence of venous and arterial gas emboli at rest after trimix diving without protocol violations. J. Appl. Physiol. 109, 1670–1674.

Ljubkovic, M., Dujic, Z., Møllerløkken, A., Bakovic, D., Obad, A., Breskovic, T., Brubakk, A.O., 2011. Venous and arterial bubbles at rest after no-decompression air dives. Med. Sci. Sports Exerc. 43, 990–995.

Mason, S., 2017. Lactate shuttles in neuroenergetics—homeostasis, allostasis and beyond. Front. Neurosci. 11, 43.

McCormick, J.G., 1969. Relationship of sleep, respiration, and anesthesia in the porpoise: a preliminary report. Proc. Natl. Acad. Sci. U. S. A. 62, 697–703.

McDonald, B.I., Ponganis, P.J., 2013. Insights from venous oxygen profiles: oxygen utilization and management in diving California sea lions. J. Exp. Biol. 216, 3332–3341.

Mijacika, T., Dujic, Z., 2016. Sports-related lung injury during breath-hold diving. Eur. Respir. Rev. 25, 506–512.

Moon, R.E., Gray, L.L., 2010. Breath-hold diving and cerebral decompression illness. Undersea Hyperb. Med. 37, 1–5.

Moon, R.E., Martina, S.D., Peacher, D.F., Potter, J.F., Wester, T.E., Cherry, A.D., Natoli, M.J., Otteni, C.E., Kernagis, D.N., White, W.D., Freiberger, J.J., 2016. Swimming-induced pulmonary edema: pathophysiology and risk reduction with sildenafil. Circulation 133, 988–996.

Moore, M.J., Bogomolni, A.L., Dennison, S.E., Early, G., Garner, M.M., Hayward, B.A., Lentell, B.J., Rotstein, D.S., 2009. Gas bubbles in seals, dolphins, and porpoises entangled and drowned at depth in gillnets. Vet. Pathol. 46, 536–547.

Mottishaw, P., Thornton, S., Hochachka, P., 1999. The diving response mechanism and its surprising evolutionary path in seals and sea lions. Am. Zool. 39, 434–450.

Mulder, E., Schagatay, E., Sieber, A., 2021. Using underwater pulse oximetry in freediving to extreme depths to study risk of hypoxic blackout and diving response phases. Front. Physiol. 12, 651128.

Noren, S.R., Kendall, T., Cuccurullo, V., Williams, T.M., 2012. The dive response redefined: underwater behavior influences cardiac variability in freely diving dolphins. J. Exp. Biol. 215, 2735–2741.

Olsen, C.R., Fanestil, D.D., Scholander, P.F., 1962. Some effects of breath holding and apneic underwater diving on cardiac rhythm in man. J. Appl. Physiol. 17, 461–466.

Palada, I., Bakovic, D., Valic, Z., Obad, A., Ivancev, V., Eterovic, D., Shoemaker, J.K., Dujic, Z., 2008. Restoration of hemodynamics in apnea struggle phase in association with involuntary breathing movements. Respir. Physiol. Neurobiol. 161, 174–181.

Park, Y.S., Rahn, H., Lee, I.S., Lee, S.I., Kang, D.H., Hong, S.Y., Hong, S.K., 1983a. Patterns of wet suit diving in Korean women breath-hold divers. Undersea Biomed. Res. 10, 203–215.

Park, Y.S., Rennie, D.W., Lee, I.S., Park, Y.D., Paik, K.S., Kang, D.H., Suh, D.J., Lee, S.H., Hong, S.Y., Hong, S.K., 1983b. Time course of deacclimatization to cold water immersion in Korean women divers. J. Appl. Physiol. 54, 1708–1716.

Patrician, A., Gasho, C., Sprajić, B., Caldwell, H.G., Barak, O., Drvis, I., Dujic, Z., Ainslie, P.N., 2021a. Case studies in physiology: breath-hold diving beyond 100 meters—cardiopulmonary responses in world champion divers. J. Appl. Physiol. 130, 1345–1350.

Patrician, A., Dujić, Ž., Spajić, B., Drviš, I., Ainslie, P.N., 2021b. Breath-hold diving—the physiology of diving deep and returning. Front. Physiol. 12, 639377.

Paulev, P., 1965. Decompression sickness following repeated breath-hold dives. J. Appl. Physiol. 20, 1028–1031.

Perini, R., Gheza, A., Moia, C., Sponsiello, N., Ferretti, G., 2010. Cardiovascular time courses during prolonged immersed static apnoea. Eur. J. Appl. Physiol. 110, 277–283.

Pollock, N.W., Buteau, D., 2017. Updates in decompression illness. Emerg. Med. Clin. North Am. 35, 301–319.

Rahn, H., Yokoyama, T., 1965. Physiology of Breath-Hold Diving and the Ama of Japan. NAS-NRC, Washington, DC, USA.

Rennie, D.W., Covino, B.G., Howell, B.J., Song, S.H., Kang, B.S., Hong, S.K., 1962. Physical insulation of Korean diving women. J. Appl. Physiol. 17, 961–966.

Ridgway, S.H., Scronce, B.L., Kanwisher, J.W., 1969. Respiration and deep diving in the bottlenose porpoise. Science 166, 1651–1654.

Ridgway, S.H., Carder, D.A., Clark, W., 1975. Conditioned bardycardia in the sea lion *Zalophus californianus*. Nature 256, 37–38.

Rommel, S., Costidis, A.M., Lowenstine, L.J., 2018. Anatomy and physiology. In: Gulland, F., Dierauf, L.A., Whitman, K.L. (Eds.), CRC Handbook of Marine Mammal Medicine, third ed. CRC Press, Boca Raton, FL.

Ropert-Coudert, Y., Kato, A., Robbins, A., Grw, H., 2018. The Penguiness Book (Online). World Wide Web electronic publication. Available http://www.penguiness.net.

Sawatzky, K.D., Nishi, R.Y., 1991. Assessment of inter-rater agreement on the grading of intravascular bubble signals. Undersea Biomed. Res. 18, 373–396.

Schagatay, E., 2009. Predicting performance in competitive apnoea diving. Part I: static apnoea. Diving Hyperb. Med. 39, 88–99.

Schagatay, E., Holm, B., 1996. Effects of water and ambient air temperatures on human diving bradycardia. Eur. J. Appl. Physiol. 73, 1–6.

Schagatay, E., Andersson, J.P., Hallén, M., Pålsson, B., 2001. Selected contribution: role of spleen emptying in prolonging apneas in humans. J. Appl. Physiol. 90, 1623–1629.

Schagatay, E., Richardson, M.X., Lodin-Sundström, A., 2012. Size matters: spleen and lung volumes predict performance in human apneic divers. Front. Physiol. 3, 173.

Schipke, J.D., Gams, E., Kallweit, O., 2006. Decompression sickness following breath-hold diving. Res. Sports Med. 14, 163–178.

Schipke, J.D., Lemaitre, F., Cleveland, S., Tetzlaff, K., 2019. Effects of breath-hold deep diving on the pulmonary system. Respiration 97, 476–483.

Schneuer, M., Flachsbarth, S., Czech-Damal, N., Folkow, L., Siebert, U., Burmester, T., 2012. Neuroglobin of seals and whales: evidence for a divergent role in the diving brain. Neuroscience 223, 35–44.

Scholander, P.F., 1940. Experimental investigations on the respiratory function in diving mammals and birds. Hvalrådets skr. 22, 1–131.

Scholander, P.F., 1963. The master switch of life. Sci. Am. 209, 92–106.

Schuitema, K., Holm, B., 1988. The role of different facial areas in eliciting human diving bradycardia. Acta Physiol. Scand. 132, 119–120.

Scott, T., Van Waart, H., Vrijdag, X.C.E., Mullins, D., Mesley, P., Mitchell, S.J., 2021. Arterial blood gas measurements during deep open-water breath-hold dives. J. Appl. Physiol. 130, 1490–1495.

Steinback, C.D., Salmanpour, A., Breskovic, T., Dujic, Z., Shoemaker, J.K., 2010. Sympathetic neural activation: an ordered affair. J. Physiol. 588, 4825–4836.

Stewart, I.B., McKenzie, D.C., 2002. The human spleen during physiological stress. Sports Med. 32, 361–369.

Tanaka, Y., 1994. Microscopy of vascular architecture and arteriovenous communications in the spleen of two odontocetes. J. Morphol. 221, 211–233.

Teruoka, G., 1932. Die Ama und ihre Arbeit. Arbeitsphysiologie 5, 239–251.

Tetzlaff, K., Schöppenthau, H., Schipke, J.D., 2017. Risk of neurological insult in competitive deep breath-hold diving. Int. J. Sports Physiol. Perform. 12, 268–271.

Thompson, D., Fedak, M.A., 1993. Cardiac responses of grey seals during diving at sea. J. Exp. Biol. 174, 139–154.

Thornton, S.J., Spielman, D.M., Pelc, N.J., Block, W.F., Crocker, D.E., Costa, D.P., Leboeuf, B.J., Hochachka, P.W., 2001. Effects of forced diving on the spleen and hepatic sinus in northern elephant seal pups. Proc. Natl. Acad. Sci. U. S. A. 98, 9413–9418.

Van Bonn, W., Dennison, S., Cook, P., Fahlman, A., 2013. Gas bubble disease in the brain of a living California Sea lion (*Zalophus californianus*). Front. Physiol. 4, 5.

Vaneechoutte, M., Kuliukas, A., Verhaegen, M., 2011. Was Man More Aquatic in the Past? Fifty Years after Alister Hardy-Waterside Hypotheses of Human Evolution. U.A.E, Bentham Science Publishers, Sharjah.

Vann, R.D., Butler, F.K., Mitchell, S.J., Moon, R.E., 2011. Decompression illness. Lancet 377, 153–164.

West, J.B., 1962. Regional differences in gas exchange in the lung of erect man. J. Appl. Physiol. 17, 893–898.

West, J.B., Luks, A., 2016. West's Respiratory Physiology: The Essentials. Wolters Kluwer.

West, J.B., Tsukimoto, K., Mathieu-Costello, O., Prediletto, R., 1991. Stress failure in pulmonary capillaries. J. Appl. Physiol. 70, 1731–1742.

Wheaton, W.W., Chandel, N.S., 2011. Hypoxia. 2. Hypoxia regulates cellular metabolism. Am. J. Physiol. 300, C385–C393.

Williams, T.M., Haun, J.E., Friedl, W.A., 1999. The diving physiology of bottlenose dolphins (*Tursiops truncatus*). I. Balancing the demands of exercise for energy conservation at depth. J. Exp. Biol. 202, 2739–2748.

Williams, T.M., Zavanelli, M., Miller, M.A., Goldbeck, R.A., Morledge, M., Casper, D., Pabst, D.A., McLellan, W., Cantin, L.P., Kliger, D.S., 2008. Running, swimming and diving modifies neuroprotecting globins in the mammalian brain. Proc. R. Soc. Lond. B 275, 751–758.

Williams, T.M., Fuiman, L.A., Kendall, T., Berry, P., Richter, B., Noren, S.R., Thometz, N., Shattock, M.J., Farrell, E., Stamper, A.M., Davis, R.W., 2015. Exercise at depth alters bradycardia and incidence of cardiac anomalies in deep-diving marine mammals. Nat. Commun. 6, 6055.

Willie, C.K., Ainslie, P.N., Drvis, I., Macleod, D.B., Bain, A.R., Madden, D., Maslov, P.Z., Dujic, Z., 2015. Regulation of brain blood flow and oxygen delivery in elite breath-hold divers. J. Cereb. Blood Flow Metab. 35, 66–73.

Zapol, W.M., Liggins, G.C., Schneider, R.C., Qvist, J., Snider, M.T., Creasy, R.K., Hochachka, P.W., 1979. Regional blood flow during simulated diving in the conscious Weddell seal. J. Appl. Physiol. 47, 968–973.

C H A P T E R

8

Genetic and molecular adaptations

Ashley M. Blawas[a], *Andreas Fahlman*[b,c,d], *and Jason A. Somarelli*[e]

[a]Division of Marine Science and Conservation, Nicholas School of the Environment, Duke University Marine Laboratory, Beaufort, NC, United States [b]Global Diving Research SL., Valencia, Spain [c]Fundación Oceanogràfic de la Comunitat Valenciana, Valencia, Spain [d]Kolmården Wildlife Park, Kolmården, Sweden [e]Department of Medicine, Duke University Medical Center, Durham, NC, United States

Abstract

Dolphins and other marine mammals possess unique adaptations that allow them to thrive in aquatic environments. While many physiological and anatomical adaptations play a clear role in marine mammal phenotypes, less is known about how genetic and molecular adaptations contribute to these phenotypes. In this chapter, we will explore the current understanding of genetic and molecular adaptations of marine mammals, focusing on adaptations that afford dolphins and other marine mammals an elevated tolerance for hypoxia and ischemia during diving. We will summarize the current knowledge of gene gain/loss events, mutations, and alterations in gene regulatory networks that have occurred in marine mammals, as well as unique patterns of gene expression in this group. These changes will be connected to potential resulting physiological adaptations and will also be compared with other species of physiological relevance, such as those living in chronic hypoxic environments. Based on a review of the available literature, we will speculate how molecular adaptations may have shaped physiological structures and systems to allow marine mammals to thrive in aquatic environments.

Introduction

Dolphins and other marine mammals have undergone unique physiological adaptations to accommodate the challenges of mammalian life in the ocean. These physiological adaptations are fundamentally connected to genetic alterations within their genomes. Indeed, genetic and molecular alterations underpin—at some level—nearly all of the phenotypic adaptations observed in nature; yet, despite the inexorable link between genetics and physiology, it is often difficult to understand clearly how the genetic and molecular composition of a given organism imbues specific physiological traits.

Dolphin phenotypes have evolved to support a fully aquatic lifestyle in marine mammals' return to the sea. Examples of these evolutionary adaptations include a smooth integument that reduces drag during swimming, the ability to echolocate, which permits foraging in water with low visibility, and a dive response that supports oxygen management during diving. Over evolutionary time, the genotypes we observe today have been selected based on the physiological and anatomical phenotypes that enhance the ecological fitness of dolphins in the aquatic environment (Fig. 1). This chapter will combine concepts from ecophysiology, comparative physiology and genomics, and physiological genomics to explore the genetic and molecular adaptations of marine mammals to life in the ocean, with particular attention paid to the bottlenose dolphin (*Tursiops truncatus*).

From genotype to phenotype

What makes a dolphin *a dolphin* is ultimately encoded within its DNA. However, simply connecting dolphin phenotypes to a specific genetic alteration is likely to be highly complex. Even for model organisms classically used in genetics, for which ample genetic and molecular tools exist, the development of genotype-phenotype maps to connect gene-level changes to phenotypic traits requires the combined use of extensive forward genetics, sequencing, and statistical approaches (Bult et al., 2013; Ho and Zhang, 2014; Han et al., 2015; Mackay and Huang, 2018). While efforts in

The Physiology of Dolphins
https://doi.org/10.1016/B978-0-323-90516-9.00013-0

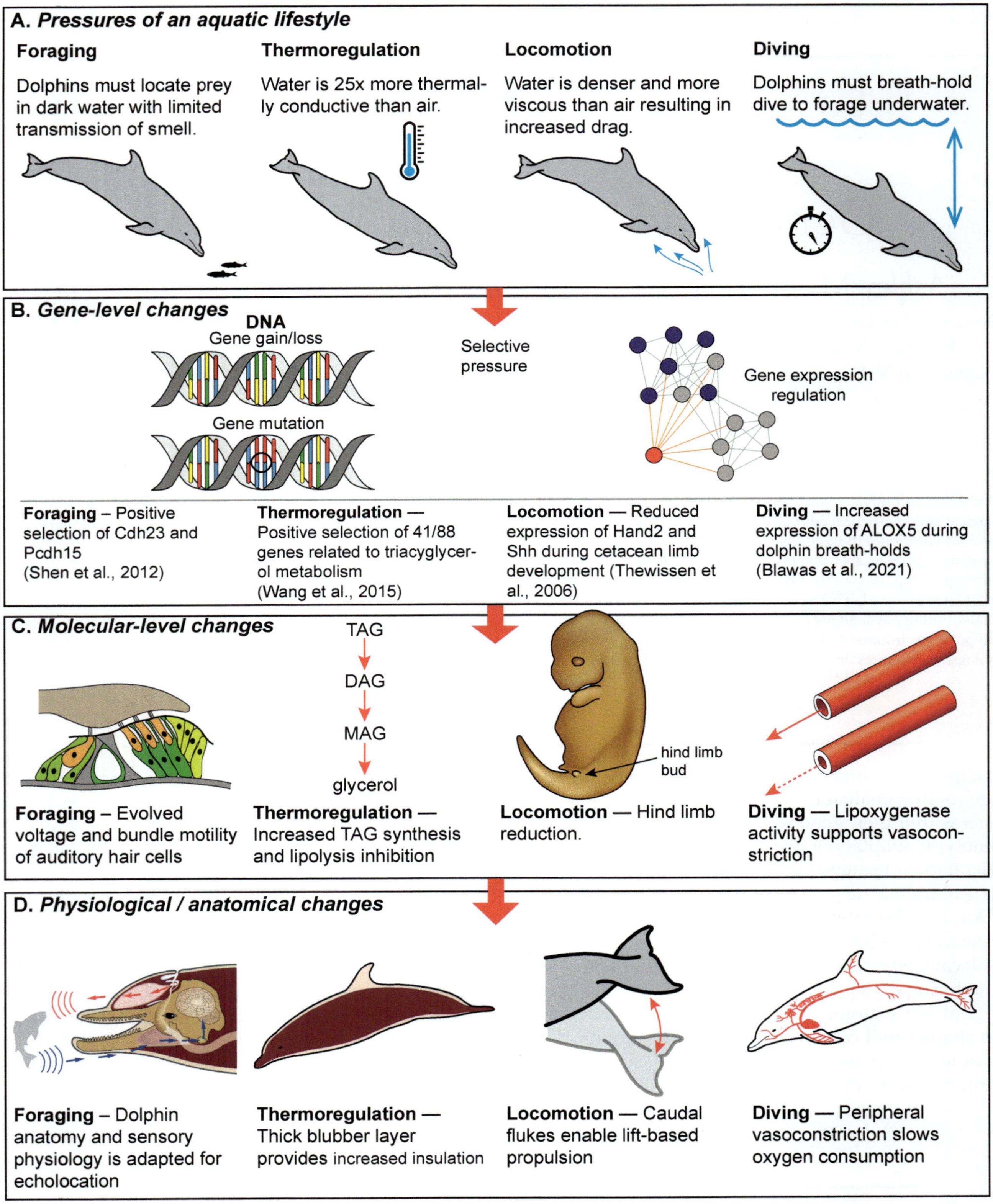

FIG. 1 Relationship among (A) environmental drivers, (B) genes, (C) molecular and (D) physiological/anatomical changes.

classic model organisms have highlighted both the challenges and opportunities in applying genetic and molecular tools to understanding phenotypic traits, ultimately, developing these tools and data sets for dolphins and other marine mammals will lead to an improved understanding of the genetic and molecular adaptations that gave rise to their unique physiological and anatomical phenotypes.

At the genetic level, understanding the phenotypic consequences of single nucleotide mutations remains a substantial challenge. DNA-level changes can have consequences across numerous scales of molecular regulation; for example, encoding differences in transcripts and proteins that affect RNA and protein function, altering RNA or protein binding to the DNA, modifying chromatin architecture and accessibility, and encoding differences in regulatory elements within resulting transcripts or proteins. In addition to these many scales of potential change for just a single point mutation in the DNA, gene-level changes must also be considered within the context of their interaction networks. Genes do not act alone; instead, they produce phenotypes in coordination with other genes within gene regulatory networks. These networks are often composed of tens to hundreds of other genes with multiple feedback loops, redundancies, and mechanisms to buffer signaling noise and tune responses. In the following section, we review mechanisms of genetic and molecular adaptation and the current state of understanding of the genetic and molecular adaptations observed in dolphins, followed by an attempt to connect these genetic and molecular features in dolphins (and other marine mammals) to known physiological adaptations.

DNA-level mechanisms

Gene gain/loss

Protein-coding genes can be gained at the gene level through duplication of genes or entire regions of chromatin, and can be lost via loss-of-function mutations or chromosomal alterations. Resulting changes in gene copy number and/or gene viability can be the result of positive selection, neutral, negative or relaxed selection, and often reflect shifts in phenotypic evolution. For example, the redundancy of function established by gene duplication events provides material on which selection can act without necessitating a loss of the gene's original function (Magadum et al., 2013). By examining gene gain and loss patterns in marine mammals throughout evolutionary time, we can identify the appearance of traits that may have contributed to the adaptation to life in the ocean.

Gene mutations

Mutations in the genetic material can arise spontaneously or may be induced by exposure to known mutagens, such as ionizing radiation and chemical carcinogens. Though not all mutations directly affect phenotype, natural selection can act upon any mutation that affects an organism's fitness. The relative rates of mutation in the genomes of organisms can provide insight into both the direction and magnitude of selection. Genome-wide scans are one method for identifying point mutations across a number of genomes and can help highlight genes that may not have previously been identified as either mutated or under selective pressure in a species of interest. In contrast, examining mutations and selection pressures of specific candidate genes can also reveal important insights related to a phenotype of interest. Both of these methods have been used to identify molecular adaptations in marine mammals that may support their unique physiological function (e.g., Nery et al., 2013a,b; Sun et al., 2013; Feng et al., 2014; Foote et al., 2015; Chikina et al., 2016; Tian et al., 2016; Zhu et al., 2018).

Gene regulatory networks

DNA-level changes can also affect phenotypes by altering the sequences of promoters, binding sites for transcription factors, microRNAs, long noncoding RNAs, splicing factors, other regulatory proteins, and other control elements that determine how genes interact with one another. Gene regulatory networks are often highly interconnected. As a result, it can be incredibly challenging to extricate all of the downstream effects that may be induced by a mutation in a single cis-acting sequence (McAdams and Shapiro, 1995; Shen-Orr et al., 2002). In this sense, the connection between adaptive changes to gene regulatory networks and physiology may be more challenging to discern compared to DNA-level adaptations within regions of protein-coding genes. The development of molecular tools to study variation in regulatory regions within dolphins and other marine mammals could provide new insights into how regulatory elements are associated with a given phenotype.

RNA-level mechanisms

At the cellular level, changes in the environment induce cellular responses that are mediated by gene expression programs. For example, the hypoxia and ischemia induced during a breath-hold can elicit changes in gene expression that enable cell survival in the face of oxygen and nutrient deprivation. Our work in bottlenose dolphins performing trained breath-holds identified upregulation of ALOX5, a gene that induces tolerance to hypoxia in rodent models (Ichinose et al., 2001) and promotes vasoconstriction in humans (Lewis et al., 1990; Kisch et al., 1997). Interestingly, both the mRNA expression of ALOX5 and its enzymatic lipoxygenase activity increase in a time-dependent manner during breath-holding in dolphins (Blawas et al., 2021). We also showed that these changes occurred within physiologically relevant time scales; however, we could not perform functional studies to directly integrate both genetic/molecular and physiological data. Ultimately, functional studies are essential to elucidate the connections between molecular responses and resulting physiological output. Studies from elephant seals that combine physiological measurements with cell culture models and transcriptomics demonstrate the power of these approaches (Pujade Busqueta et al., 2020; Wright et al., 2020; Torres-Velarde et al., 2021).

Relative to gene-level mechanisms, upregulation or downregulation of gene expression in the transcriptome can change rapidly in cells. Because of this, determinations of gene expression only represent cellular transcription during a single snapshot in time. For some genes, gene expression may be relatively constant regardless of external conditions, while expression of other genes can change significantly with tissue type, life stage, and environmental conditions. To better understand how differential gene expression may improve dolphin's and other marine mammals' tolerance to hypoxia and ischemia/reperfusion, for example, we can examine gene expression patterns or changes in gene expression level under varying oxygen and nutrient conditions. These studies may be conducted in vivo, as was done in the aforementioned study during dolphin breath-holds, or in vitro in lab-based studies using cultured cells and tissues.

In vivo studies of gene expression are particularly useful for understanding gene expression and posttranscriptional gene regulation in the context of whole-organism physiology. Logistically, measuring changes in gene expression in a dolphin during a dive is a challenging proposition; however, working with dolphins in human care has allowed for transcriptomic data collection during breath-holds (Blawas et al., 2021). Bio-logging tools that allow for serial blood sampling from a marine mammal during a dive would be extremely useful for expanding our understanding of expression-level changes during diving that may support an increased tolerance for hypoxia and ischemia/reperfusion (Hill, 1986).

Protein-level mechanisms

Differential protein expression and posttranslational modification are two mechanisms by which molecular adaptation may occur at the protein level. Rapid changes in posttranslational modification to protein (phosphorylation, acetylation, etc.) may be able to reflect an animal's changing environment on rapid timescales, making techniques like proteomics an important avenue for future studies of marine mammals.

Adaptations to the marine environment

The marine environment imposes unique selection pressures on mammalian life. Marine mammals have adapted to these unique pressures by altering their strategies for thermoregulation, locomotion, sensing their environment, and breath-hold diving, among other homeostatic challenges. These adaptations can be observed across biological scales, from physiological and anatomical to biochemical, molecular, and genetic. While many of the physiological and anatomical adaptations are reviewed in greater detail in other chapters, here we will briefly review some of these traits, through the lens of physiological genomics, to highlight potential molecular-level drivers that have given rise to the phenotypes of extant marine mammals.

Unique molecular features of marine mammals

As discussed previously, linking a specific genetic alteration to a particular phenotypic trait is often complex and typically requires extensive study to understand the numerous interactions that produce that phenotype. Despite this complexity in linking genotype to phenotype, a number of molecular changes have been identified in dolphins that can be associated with relevant biological processes and may inform physiologic adaptations observed in these animals. For example, a scan of the dolphin genome revealed selection pressures on genes grouped by gene ontology categories, which are based on the gene's function and its products (McGowen et al., 2012). The authors found that genes expressed in the mitochondria were under significantly higher selective pressure in dolphins than in other mammals.

A total of 27 genes that are expressed and/or play a functional role in the nervous system were also positively selected for in dolphins; specifically, many genes related to synaptic plasticity and neuroendocrine function. A similar analysis of the bowhead whale (*Balaena mysticetus*) genome showed that genes related to cancer and aging are under positive selection, even when the bowhead is compared to its close evolutionary neighbor, the minke whale (*Balaenoptera acutorostrata*, Keane et al., 2015). This study also aimed to identify amino acid substitutions to understand potential changes to protein function. In addition to changes in aging and cancer-related genes, multiple genes related to sound perception were found to have unique residues, which were proposed to potentially relate to the species' ability for broadband sound production.

Several additional genetic adaptations related to aging have been identified in marine mammals. Large segmental duplications of DLD and KCNMB1 in the bowhead whale and blue whale (*Balaenoptera musculus*), respectively, were suggested to relate to longevity due to their role in aging (Tollis et al., 2019). Several duplicated tumor suppressor genes were found across the studied cetacean species, including SALL4 in the sei whale (*Balaenoptera borealis*), TGM3 and SEMA3B in the killer whale (*Orcinus orca*), UVRAG in the sperm whale (*Physeter macrocephalus*), North Atlantic right whale (*Eubalaena glacialis*), and bowhead whale, and PDCD5 in all studied species (Tollis et al., 2019). Another study of tumor suppressors in cetaceans identified 71 duplicated tumor suppressors in the cetacean lineage, with 11 of these specifically relating to longevity processes, including senescence promotion (IGFBP3, SIK1, NOTCH3); cell proliferation, apoptosis, and breast cancer (IGFBP3); and arterial blood pressure and sympathetic nervous system activity (SIK1) (Tejada-Martinez et al., 2021). This same study noted that many of the observed duplications occurred in baleen whales, which includes both the largest- and longest-living extant mammals.

While these genome-wide analyses provide insights into groups of genes under selection in dolphins and other marine mammals, additional studies have also begun to associate specific genotypic and molecular alterations with phenotypic and physiologic function in these animals. The following sections are intended to provide an overview of our current understanding of the relationships between genetic/molecular adaptations and how these relate to physiologic function in dolphins.

Thermoregulation

As homeotherms, marine mammals must maintain a constant internal body temperature, often in water that is much cooler than their internal body temperature (discussed in Chapter 3: Thermoregulation). To accomplish this, marine mammals have evolved several physiological and anatomical mechanisms to reduce heat loss to the surrounding cooler water (Favilla and Costa, 2020). One thermoregulatory adaptation marine mammals utilize to increase thermoregulatory efficiency is insulation in the form of thick blubber or dense fur. Multiple genetic alterations have been identified that support increased insulation in marine mammals. Investigation of the genetic basis of thickened blubber in cetaceans revealed positive selection of genes associated with triacylglycerol synthesis, highlighting a potential genetic mechanism for the increased insulation in the cetacean lineage (Wang et al., 2015). In three species of pinnipeds, FASN, a gene that encodes fatty-acid synthase and is related to obesity in humans, was one of 24 rapidly evolving genes in the pinniped clade compared to terrestrial taxa (Park et al., 2018). Other studies have noted that the cetacean homolog of uncoupling protein 1 (UCP1) contains a premature stop codon unique to the lineage (Keane et al., 2015; Gaudry et al., 2017; Zhu et al., 2018). Given that UCP1 plays a role in thermogenesis, this potential loss of function may indicate that the cetaceans studied no longer require mechanisms for elevated levels of heat production due to their other thermoregulatory adaptations (e.g., improved insulation). Furthermore, a study on the dolphin genome revealed positive selection pressures on genes related to lipid metabolism, which may support multiple adaptations, including the development of a thick blubber layer, motor activity, and muscle contraction (Sun et al., 2013).

Locomotion

In contrast to terrestrial locomotion on land, locomotion in water incurs increased drag due to the higher viscosity and density of water compared with air (discussed in Chapter 2: Energetic costs of rest and locomotion in dolphins and Chapter 4: Muscles and movement). To minimize locomotory costs in water, marine mammals have evolved fusiform body shapes and paddle-like appendages (Fish et al., 2008; Davis, 2019). In cetaceans and sirenians, the fully aquatic marine mammal groups, a smooth integument also helps to reduce drag in the water. Several studies have investigated the molecular basis for these locomotory adaptations. For example, the loss of hair-type keratin genes (alpha-keratins, KRT; keratin-associated proteins KAPs or KRTAPs) and other hair-loss genes in cetaceans (hairless, Hr; fibroblast growth factor 5, FGF5; gasdermin A, GSDMA; desmoglein-4, DSG4; desmocollin-1, DSC1) has been suggested to reflect an adaptive response to their aquatic environment (Chen et al., 2013; Nery et al., 2014; Sharma et al., 2018). At least one study has indicated that genetic modifications to sebum-producing genes may underlie changes to skin function in cetaceans (Lopes-Marques et al., 2019). Molecular alterations in the function of cetacean skin may also support improved thermoregulatory capacity and osmoregulation. Themudo et al. detailed 36 gene loss events that support the modern cetacean skin phenotype, including many related to skin keratinization and the skin's inflammatory response (Themudo et al., 2020), which could support expedited skin healing in dolphins (Zasloff, 2011), important for maintaining a smooth integument.

One example of a molecular adaptation of marine mammals that demonstrates the importance of regulatory mechanisms in determining phenotypes is the role that binding site degradation is thought to play in the loss of hindlimbs and development of internal testes in the cetacean lineage (Berger et al., 2018). By examining eroded binding sites and the associated function of those sites, the authors found that the most enriched ontology was that relating to hindlimb loss and implicated ten affected target genes (ADAMTS9, CDH11, FGF5, LIMK1, LMX1B, PITX2, RARG, SCXA, SCXB, and ZEB1). This result was further validated by including the killer whale in the analysis and confirmation of hindlimb loss as the most enriched gene ontology. Further, an analysis of eroded binding sites in bottlenose dolphin, white rhinoceros (*Ceratotherium simum*), and Weddell seal (*Leptonychotes weddellii*) revealed significant enrichment for developmental changes related to internal testes, which are found in all three of these species. Internalizing testes inside the body cavity reduces body drag when swimming in the water and enables a lower testicular temperature to support optimal sperm production in dolphins (Rommel et al., 1994).

Sensory

A critical set of adaptations marine mammals have acquired in their transition to the ocean are those that allow them to sense the underwater environment. Instead of relying on the sensory systems for smell and taste, many mammals in the ocean have become more reliant on their ability to detect prey items using sound and light (Chapter 10: Sensory physiology in delphinids). Although there is evidence that baleen whales have retained the anatomical structures necessary for olfaction, cetaceans typically have minimal olfactory capacity (Thewissen et al., 2011). Marine mammals have fewer olfactory receptor genes than terrestrial mammals, and cetaceans have the least of any marine mammal lineage (Liu et al., 2019). Many olfactory and gustatory genes have been under relaxed evolutionary constraint in marine mammals and have accumulated a high degree of genetic alterations (Chikina et al., 2016). Genetic evidence also suggests that all cetacean species lack the receptors necessary to taste sweet, bitter, and umami flavors (Kishida et al., 2015).

In contrast to olfactory cues, sound can travel vast distances underwater compared to in air which may be particularly useful at depths with minimal or no light penetration where visual cues are absent. Analysis of the Baji genomes revealed hearing genes that exhibit accelerated evolution in both the bottlenose dolphin and little brown bat (*Myotis lucifugus*, Zhou et al., 2013). This study complements several others that have highlighted similar genetic changes in hearing-related genes in species that utilize echolocation, including dolphins and bats (Liu et al., 2010a,b; Davies et al., 2012; Shen et al., 2012). For animals that occupy shallower depths, their vision is adapted to deal with a range of light levels and, in some cases, vision both underwater and in air, like in pinnipeds (Mass and Supin, 2007). Genes that support eye development were found to be under positive selection in greater proportion in dolphins when compared to four terrestrial species (Nery et al., 2013a,b).

Diving

All mammals exhibit the mammalian dive response. There is evidence that some aspects of the response to diving are even conserved across all vertebrates (Davis et al., 2004; Panneton, 2013). This dive response is characterized by apnea, bradycardia, and peripheral vasoconstriction and is suggested to function to (1) slow the rate of oxygen consumption and (2) concentrate oxygenated blood at organs critical for organism function (Scholander, 1940; Andersen, 1966; Butler and Jones, 1997; Foster and Sheel, 2005). Although all mammals undergo these changes in response to diving, dolphins and other marine mammals have adapted a diverse suite of mechanisms to accommodate prolonged dives and many successive dives over a lifetime. While many of the cardiovascular and respiratory adaptations that enable the extended breath-hold dives and high-pressure tolerance of dolphins and other marine mammals are discussed in detail in Chapter 5: Cardiovascular physiology in dolphins and other cetaceans and Chapter 6: Respiratory physiology in the dolphin and other whales, here we review genetic and molecular alterations that could play a role in shaping these structures and functions.

Peripheral vasoconstriction limits blood flow to the peripheral tissues during a dive and is coincident with bradycardia, which reduces cardiac output compared to surface rates (Scholander, 1940; Zapol et al., 1979). Together, these responses help maintain constant blood pressure and enable significant changes in hemodynamics on the tissue and whole-body levels. Loss of SLC6A18 in the cetacean-specific lineage suggests the potential for gene loss that contributes to vasoconstriction during diving (Huelsmann et al., 2019). Indeed, genetic knockout of SLC6A18 in mice induces hypertension, which involves vasoconstriction (Quan et al., 2004). This response is likely secondary to lower arginine levels and nitric oxide, as SLC6A18 encodes an amino acid transporter (Huelsmann et al., 2019). Likewise, an analysis of 17 genes related to hypoxia across 11 cetacean genomes revealed positive selection on 10 genes related to oxygen transport (HBA, HBB, and MB) and vasoconstriction (EDN1, EDN2, EDN3, EDNRA, EDNRB, ADRA1D, and AVP) in cetaceans (Zhu et al., 2018). The strong positive selection and mutations of positively selected sites on HBA, HBB, and MB agree with other studies that provide evidence for modified globins in cetaceans (Dasmeh et al., 2013; Mirceta et al., 2013; Nery et al., 2013a).

One consequence of the dive response is that many peripheral tissues, like the skeletal muscles and viscera, experience dramatic reductions in the supply of oxygen-rich blood during diving. Early studies of forced dives in pinnipeds revealed extreme ischemia in most tissues except the brain (Scholander, 1940; Irving et al., 1942; Zapol et al., 1979). Though studies of ischemia in free-diving animals remain largely absent from the literature due to the technological challenge of measuring blood flow remotely, measurements of skeletal muscle blood flow using laser Doppler flowmetry in juvenile elephant seals during sleep apneas on land demonstrated a 46% decrease in perfusion during apnea compared with eupnea (Ponganis et al., 2008). A recent study using a novel near-infrared spectroscopy device successfully measured changes in blood volume in the blubber of harbor seals (*Phoca vitulina*) and revealed a decrease in blubber blood volume in anticipation of and during a dive (McKnight et al., 2019). In addition to ischemia, the determination of blood gas tensions during free dives of marine mammals indicates a substantial reduction in blood O_2 during diving, indicative of extreme hypoxemia (Qvist et al., 1986; Meir et al., 2008, 2009; McDonald and Ponganis, 2012, 2013; Tift et al., 2017, 2018). This suggests that marine mammals have adapted unique physiologies to tolerate the ischemia/reperfusion and hypoxia associated with diving and enable regular transitions between dives and the surface without subsequent pathological consequences (Fig. 2).

One group of transcription factors that are critically important for O_2-breathing species is the hypoxia-inducible factors (HIFs). Under hypoxic conditions, the HIF-1α subunit associates with the HIF-1β subunit in the nucleus and induces the transcription of genes involved in a wide range of pathways related to hypoxia response, including angiogenesis, erythropoiesis, vasodilation, and glycolysis (Cummins and Taylor, 2005; Schofield and Ratcliffe, 2005). Comparative genomics has demonstrated that five of six genes in the HIF pathway are under positive selection in hypoxia-tolerant species, including high-altitude, subterranean, and diving animals (Zhu et al., 2018). This study also revealed 11 unique substitutions to genes in the HIF pathway in cetaceans, including the oxygen-dependent

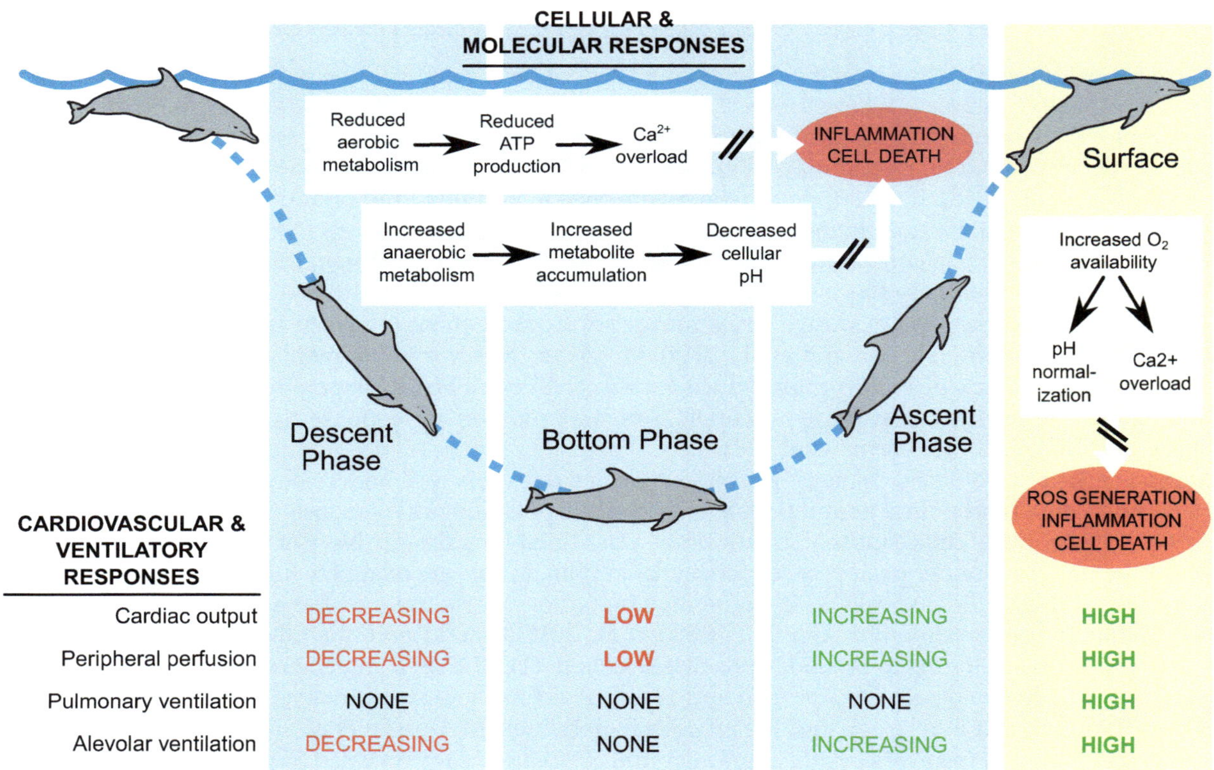

FIG. 2 Dolphins have unique physiologies to tolerate the ischemia/reperfusion and hypoxia that occur in response to cardiovascular and ventilatory changes during diving.

degradation domain and the N-terminal transactivation domain (N-TAD) in HIF-1α and HIF-2α. These findings suggest the HIF pathway genes play an important role in the hypoxia response of cetaceans.

Several studies have explored the transcriptomes, or mRNA transcripts, of marine mammals in vitro to shed light on potential adaptations for hypoxia and ischemia tolerance. In the hooded seal (*Cystophora cristata*), neurons of the visual cortex showed an increase in CLU and S100B expression, compared to the ferret (*Mustela furo*), which may contribute to enhanced hypoxia tolerance in the seal (Fabrizius et al., 2016). The same study noted that many genes related to aerobic metabolism were more highly expressed in the ferret than in the seal, suggesting a lower capacity for aerobic energy production in the visual cortex of the seal. The authors hypothesize that this reduction in aerobic capacity may represent a mechanism to reduce energetic costs. Transcriptomics of tissue from the visual cortex of several cetaceans and cattle revealed that cetaceans display elevated gene expression related to terminal phosphorylation and respiratory electron chain pathways (Krüger et al., 2020). It is suggested that a higher oxidative capacity of the brain tissue may support rapid reoxygenation upon surfacing from a dive. Three genes related to the removal of reactive oxygen species (PRDX6, SOD1, and GSTP1) were upregulated in whale transcriptomes compared to cattle, which could help reduce oxidative stress caused by reperfusion following ischemia.

Conducting controlled experiments where cells are exposed to hypoxic and ischemic conditions enables the identification of genes that are turned on and off in response to changing oxygen and nutrient conditions. For example, a study of hooded seal visual cortex slices exposed to hypoxia and subsequent reoxygenation identified upregulated genes related to cytokines and immediate-early genes (Hoff et al., 2017). This response is similar to the responses of other mammals to hypoxic/ischemic conditions in the brain. Yet, many of the specific genes noted in this study have not been identified in other mammalian models as important in the ischemic brain. In this study, there did not appear to be a specific effect of reoxygenation on gene expression, possibly because of a relatively short reoxygenation window.

In addition to examining transcripts, several studies have analyzed mechanisms of posttranscriptional gene regulatory processes that may be related to diving. For example, elevated expression of miR-499, a cardiac microRNA (miRNA) important for ischemic postconditioning, in the heart tissue of Weddell seals may confer increased tolerance to ischemia/reperfusion in this deep-diving pinniped (Penso-Dolfin et al., 2020). The miRNA miR-424-3p, which plays a role in HIF-1⊠ regulation and promotes vascular growth in mice, was detected at low levels in adult and seal pups but has been shown to respond to hypoxia in human endothelium. Increased expression of miR-29a-3p in the cardiac and muscle cells in adult Weddell seals is expected to decrease the expression of negative regulators of the hypoxia response of HIF-1α.

Loss of function of the PON1 gene in the deep-diving Weddell seal, but not in the shallow-diving Antarctic fur seal (*Arctocephalus gazella*) or walrus (*Odobenus rosmarus*), may indicate changes to lipid and/or antioxidant metabolism in extreme divers. PON1 is known to mediate the production of an enzyme that protects against oxidative damage to high- and low-density lipoproteins and may contribute to preventing atheromas or the buildup of plaques in the arterial lumen (Meyer et al., 2018). If we suppose deep divers have a modified lipid or antioxidant metabolism that supports the differing energetic requirements and oxidative stress of extended dives, PON1 may have been rendered unnecessary in extreme divers contributing to its ultimate loss in the Weddell seal. Interestingly, PON1 is also lost in cetaceans and may render these species sensitive to pesticides (Meyer et al., 2018). Inactivated POLM may also contribute to a beneficial response to oxidative stress that arises from frequent events of ischemia/reperfusion that are commonplace during diving (Huelsmann et al., 2019). POLM encodes for a polymerase involved in an error-prone DNA repair process. A lack of POLM may reduce a cetacean's likelihood of experiencing mutagenic DNA repair by increasing the use of other, less error-prone DNA repair processes in response to oxidative stress (Peto et al., 1975; Caulin and Maley, 2011; Sharma et al., 2018).

In addition to the physiological changes that occur during a dive, it is well understood that marine mammals have several anatomical and biochemical adaptations that augment a physiological response. Two studies have highlighted alterations to the genome of cetaceans that may support a modified lung design. Nery et al. (2013b) revealed that multiple developmental genes under positive selection in the dolphin lineage are related to lung morphogenesis and development, such as RSPO2, LEF1, and FOXP2. These may be important for the unusual architecture and the observed collateral ventilation seen in dolphins (Garcia Párraga et al., 2018). In addition, the following three genes were also identified in bottlenose dolphins to be under positive selection: (1) SCGB3A2, a gene relating to lung surfactant; (2) PLUNC, which encodes for a secretory protein and is expressed in the upper airways; and (3) TMPRSS11D, which encodes for a protease that is released into the mucous membranes (McGowen et al., 2012).

A recent examination of gene loss events in the ancestral cetacean lineage revealed 85 genes containing inactivating mutations that are shared between the baleen whales and toothed whales, but not observed in the closely related hippopotamus (*Hippopotamus amphibius*, Huelsmann et al., 2019). Of these 85 genes, 62 of the gene loss events have not been reported previously, and 11 were identified to have potential functional significance in the transition to life in water. In addition to the previously mentioned SLC6A18—which the authors hypothesize may support enhanced vasoconstriction—several other gene losses identified in this study are expected to benefit diving. For example, the loss of MAP3K19 and SEC14L3 may reduce oxidative stress in the lungs and alter lung surfactant composition, respectively, both of which would benefit a diving mammal that regularly experiences lung collapse and reinflation under the hypoxic/ischemic conditions of diving. Two genes related to blood clotting, F12 and KLKB1, are functionally inactivated in cetaceans. This may limit the risk of a foreign body response to nitrogen microbubbles, which has been observed in humans (Barak and Katz, 2005), that can form during dive ascents and potential activation of a thrombosis cascade. Finally, unique amino acid substitutions have also been identified on the N-methyl-D-aspartate receptor in diving mammals compared to terrestrial mammals, which is hypothesized to improve their tolerance to high pressure (Bliznyuk et al., 2018).

Another critical biochemical adaptation for diving observed in marine mammals is a higher concentration of oxygen-transporting hemoglobin in the blood and oxygen-storing myoglobin in skeletal muscle. Significant attention has been paid to this adaptation and our understanding of its molecular underpinnings. Recent work has revealed positive selection on genes related to myoglobin and an elevated net charge on the molecular surface of marine mammal myoglobin. The elevated net charge of the surface of myoglobin is thought to increase its oxygen storage capacity (Dasmeh et al., 2013; Mirceta et al., 2013; Nery et al., 2013a; Zhu et al., 2018). In addition, in sperm whales, the loss of AMPD3, a protein-encoding gene specific to red blood cells, is believed to reduce the O_2 affinity of hemoglobin, therefore improving O_2 delivery to hypoxic tissues (Sharma et al., 2018).

Marine mammals also have significant modifications to their vasculature, which are thought to help in the modulation of blood flow to and from the heart during diving and to fill air spaces as the lungs collapse (Cozzi et al., 2005; Costidis and Rommel, 2012; Castellini and Mellish, 2015) (Chapter 7: Diving physiology in dolphins and human). For example, an enlarged, elastic aortic bulb aids in maintaining blood pressure during the long diastolic pauses associated with bradycardia during diving (Shadwick and Gosline, 1994; Ponganis, 2018). Some studies have suggested that the size of the aortic bulb is positively correlated with diving capacity (i.e., deeper dives have larger aortic bulbs). For controlling venous return to the heart, a muscular sphincter is located around the inferior vena cava in phocids and some cetaceans (Lillie et al., 2018). Little is known about the genetic alterations that support vascular features that differ from terrestrial mammals.

Future directions

Leveraging natural experiments

Marine mammals entered the ocean environment on four separate occasions, giving rise to the four groups we observe today: cetaceans, pinnipeds, sirenians, and otters. Because each taxon re-entered the water at different points in evolutionary time, their degree of adaptation reflects the extent of their occupation of the aquatic environment. Consistent with this, many traits that afford an organism a high degree of fitness in an aquatic environment are most evident in cetaceans and sirenians and less so in the semiaquatic pinnipeds and otters. This variation in phenotypes across groups based on the timing of their return to the water can also be found within subgroups of marine mammals such as families, genera, and species. At more recent evolutionary scales, ecotypes of a single species that have expanded into unique niches are observed, such as bottlenose dolphins off the east coast of North America and off the coast of California (Hoelzel et al., 1998; Perrin et al., 2011). Capitalizing on this natural variation across the spectrum of phenotypes and adaptation to a particular environment may be a powerful tool for the precise investigation of specific molecular features or traits. For example, comparing gene loss and gain of inshore and offshore Atlantic bottlenose dolphins may help identify gene-level mechanisms that support hypoxia and ischemia tolerance, as these ecotypes are genetically much more similar than two distinct species.

Functional studies: Physiology meets genomics

Functional studies are required to fully elucidate the role of any genetic or molecular mechanisms in determining the physiological, anatomical, and biochemical traits of a dolphin or any other marine mammal. By measuring specific physiological parameters and the concurrent changes in molecular features, we can better understand a gene's role by correlating expression changes with physiological function changes. As the technology that enables genomic analysis has become more accessible, the ease of preparing tissue samples for genomics and performing associated analyses in various experimental settings has increased exponentially. Therefore, whenever possible, samples should also be taken for genetic and molecular studies in parallel with physiological measurements. Though procuring simultaneous physiological and genetic samples can be particularly challenging in marine mammals, even samples collected with low-temporal resolution can still contribute substantially to the understanding of genotype-phenotype relationships in these species. At the cellular level, several methodologies, including the use of CRISPR/Cas-mediated knockouts, mutant lineages of mice or other model organisms, pharmacological blockades, or antisense technology, allow a gene to be directly linked to a particular function (Glueck and Dzau, 2002). Bringing these tools to bear on studies of cetacean cells in culture may provide new insights into how marine mammals have adapted to aquatic life. Identifying the underlying genetic and molecular adaptations that dolphins and other marine mammals have accumulated as a result of the physiological challenges of an aquatic lifestyle will improve our understanding of these unique organisms across multiple physiological scales. As a result, scientists can draw from nature's lessons to understand how dolphins will respond molecularly to a changing ocean environment and even create novel human health treatments.

References

Andersen, H.T., 1966. Physiological adaptations in diving vertebrates. Physiol. Rev. 46, 212–243.

Barak, M., Katz, Y., 2005. Microbubbles: pathophysiology and clinical implications. Chest 128 (4), 2918–2932.

Berger, M.J., Wenger, A.M., Guturu, H., Bejerano, G., 2018. Independent erosion of conserved transcription factor binding sites points to shared hindlimb, vision and external testes loss in different mammals. Nucleic Acids Res. 46 (18), 9299–9308.

Blawas, A.M., Ware, K.E., Schmaltz, E., Zheng, L., Spruance, J., Allen, A.S., West, N., Devos, N., Corcoran, D.L., Nowacek, D.P., Eward, W.C., Fahlman, A., Somarelli, J.A., 2021. An integrated comparative physiology and molecular approach pinpoints mediators of breath-hold capacity in dolphins. Evol. Med. Pub. Health 9 (1), 420–430.

Bliznyuk, A., Golan, H., Grossman, Y., 2018. Marine mammals' NMDA receptor structure: possible adaptation to high pressure environment. Front. Physiol. 9, 1633.

Bult, C.J., Eppig, J.T., Blake, J.A., Kadin, J.A., Richardson, J.E., Mouse Genome Database Group, 2013. The mouse genome database: genotypes, phenotypes, and models of human disease. Nucleic Acids Res. 41, D885–D91.

Butler, P.J., Jones, D.R., 1997. Physiology of diving of birds and mammals. Physiol. Rev. 77, 837–899.

Castellini, M.A., Mellish, J.-A., 2015. Marine Mammal Physiology: Requisites for Ocean Living. CRC Press.

Caulin, A.F., Maley, C.C., 2011. Peto's paradox: evolution's prescription for cancer prevention. Trends Ecol. Evol. 26, 175–182.

Chen, Z., Wang, Z., Shixia, X., Zhou, K., Yang, G., 2013. Characterization of hairless (Hr) and FGF5 genes provides insights into the molecular basis of hair loss in cetaceans. BMC Evol. Biol. 13, 34.

Chikina, M., Robinson, J.D., Clark, N.L., 2016. Hundreds of genes experienced convergent shifts in selective pressure in marine mammals. Mol. Biol. Evol. 33 (9), 2182–2192.

Costidis, A., Rommel, S.A., 2012. Vascularization of air sinuses and fat bodies in the head of the bottlenose dolphin (*Tursiops truncatus*): morphological implications on physiology. Front. Physiol. 3, 243.

Cozzi, B., Bagnoli, P., Acocella, F., Costantino, M.L., 2005. Structure and biomechanical properties of the trachea of the striped dolphin *Stenella coeruleoalba*: evidence for evolutionary adaptations to diving. Anat. Rec. A Discov. Mol. Cell. Evol. Biol. 284 (1), 500–510.

Cummins, E.P., Taylor, C.T., 2005. Hypoxia-responsive transcription factors. Pflugers Archiv: Eur. J. Physiol. 450 (6), 363–371.

Dasmeh, P., Serohijos, A.W.R., Kepp, K.P., Shakhnovich, E.I., 2013. Positively selected sites in cetacean myoglobins contribute to protein stability. PLoS Comput. Biol. 9 (3), e1002929.

Davies, K.T.J., Cotton, J.A., Kirwan, J.D., Teeling, E.C., Rossiter, S.J., 2012. Parallel signatures of sequence evolution among hearing genes in echolocating mammals: an emerging model of genetic convergence. Heredity 108 (5), 480–489.

Davis, R.W., 2019. Marine Mammals: Adaptations for an Aquatic Life. Springer.

Davis, R.W., Polasek, L., Watson, R., Fuson, A., Williams, T.M., Kanatous, S.B., 2004. The diving paradox: new insights into the role of the dive response in air-breathing vertebrates. Comp. Biochem. Physiol. A: Physiol. 138 (3), 263–268.

Fabrizius, A., Hoff, M.L.M., Engler, G., Folkow, L.P., Burmester, T., 2016. When the brain goes diving: transcriptome analysis reveals a reduced aerobic energy metabolism and increased stress proteins in the seal brain. BMC Genomics 17, 583.

Favilla, A.B., Costa, D.P., 2020. Thermoregulatory strategies of diving air-breathing marine vertebrates: a review. Front. Ecol. Evol. 8, 555509.

Feng, P., Zheng, J., Rossiter, S.J., Wang, D., Zhao, H., 2014. Massive losses of taste receptor genes in toothed and baleen whales. Genome Biol. Evol. 6 (6), 1254–1265.

Fish, F.E., Howle, L.E., Murray, M.M., 2008. Hydrodynamic flow control in marine mammals. Integr. Comp. Biol. 48 (6), 788–800.

Foote, A.D., Liu, Y., Thomas, G.W.C., Vinař, T., Alföldi, J., Deng, J., Dugan, S., van Elk, C.E., Hunter, M.E., Joshi, V., Khan, Z., Kovar, C., Lee, S.L., Lindblad-Toh, K., Mancia, A., Nielsen, R., Qin, X., Qu, J., Raney, B.J., Vijay, N., Wolf, J.B.W., Hahn, M.W., Muzny, D.M., Worley, K.C., Gilbert, M.-T.P., Gibbs, R.A., 2015. Convergent evolution of the genomes of marine mammals. Nat. Genet. 47 (3), 272–275.

Foster, G.E., Sheel, A.W., 2005. The human diving response, its function, and its control. Scand. J. Med. Sci. Sports 15 (1), 3–12.

Garcia Párraga, D., Moore, M., Fahlman, A., 2018. Pulmonary ventilation-perfusion mismatch: a novel hypothesis for how diving vertebrates may avoid the bends. Proc. R. Soc. Lond. B 285 (1877), 20180482.

Gaudry, M.J., Jastroch, M., Treberg, J.R., Hofreiter, M., Paijmans, J.L.A., Starrett, J., Wales, N., Signore, A.V., Springer, M.S., Campbell, K.L., 2017. Inactivation of thermogenic UCP1 as a historical contingency in multiple placental mammal clades. Sci. Adv. 3 (7), e1602878.

Glueck, S.B., Dzau, V.J., 2002. Physiological genomics: implications in hypertension research. Hypertension 39 (2 Pt 2), 310–315.

Han, S.K., Kim, I., Hwang, J., Kim, S., 2015. Network modules of the cross-species genotype-phenotype map reflect the clinical severity of human diseases. PLoS One 10 (8), e0136300.

Hill, R.D., 1986. Microcomputer monitor and blood sampler for free-diving weddell seals. J. Appl. Physiol. 61 (4), 1570–1576.

Ho, W.-C., Zhang, J., 2014. The genotype-phenotype map of yeast complex traits: basic parameters and the role of natural selection. Mol. Biol. Evol. 31 (6), 1568–1580.

Hoelzel, A.R., Potter, C.W., Best, P.B., 1998. Genetic differentiation between parapatric 'nearshore' and 'offshore' populations of the bottlenose dolphin. Proc. R. Soc. Lond. B 265 (1402), 1177–1183.

Hoff, M.L., Müller, A.F., Czech-Damal, N.U., Folkow, L.P., Burmester, T., 2017. Transcriptome analysis identifies key metabolic changes in the hooded seal (*Cystophora cristata*) brain in response to hypoxia and reoxygenation. PLoS One 12 (1), e0169366.

Huelsmann, M., Hecker, N., Springer, M.S., Gatesy, J., Sharma, V., Hiller, M., 2019. Genes lost during the transition from land to water in cetaceans highlight genomic changes associated with aquatic adaptations. Sci. Adv. 5 (9), eaaw6671.

Ichinose, F., Zapol, W.M., Sapirstein, A., Ullrich, R., Tager, A.M., Coggins, K., Jones, R., Bloch, K.D., 2001. Attenuation of hypoxic pulmonary vasoconstriction by endotoxemia requires 5-lipoxygenase in mice. Circ. Res. 88 (8), 832–838.

Irving, L., Scholander, P.F., Grinnell, S.W., 1942. The regulation of arterial blood pressure in the seal during diving. Am. J. Physiol. 135 (3), 557–566.

Keane, M., Semeiks, J., Webb, A.E., Li, Y.I., Quesada, V., Craig, T., Madsen, L.B., van Dam, S., Brawand, D., Marques, P.I., Michalak, P., Kang, L., Bhak, J., Yim, H.S., Grishin, N.V., Nielsen, N.H., Heide-Jørgensen, M.P., Oziolor, E.M., Matson, C.W., Church, G.M., Stuart, G.W., Patton, J.C., George, J.C., Suydam, R., Larsen, K., López-Otín, C., O'Connell, M.J., Bickham, J.W., Thomsen, B., de Magalhães, J.P., 2015. Insights into the evolution of longevity from the bowhead whale genome. Cell Rep. 10 (1), 112–122.

Kisch, E.S., Jaffe, A., Knoll, E., Stern, N., 1997. Role of the lipoxygenase pathway in angiotensin II–induced vasoconstriction in the human placenta. Hypertension 29, 796–801.

Kishida, T., Thewissen, J., Hayakawa, T., Imai, H., Agata, K., 2015. Aquatic adaptation and the evolution of smell and taste in whales. Zool. Lett. 1, 9.

Krüger, A., Fabrizius, A., Mikkelsen, B., Siebert, U., Folkow, L.P., Burmester, T., 2020. Transcriptome analysis reveals a high aerobic capacity in the whale brain. Comp. Biochem. Physiol. A: Physiol. 240: 110593.

Lewis, R.A., Austen, K.F., Soberman, R.J., 1990. Leukotrienes and other products of the 5-lipoxygenase pathway. biochemistry and relation to pathobiology in human diseases. N. Engl. J. Med. 323 (10), 645–655.

Lillie, M.A., Wayne Vogl, A., Raverty, S., Haulena, M., McLellan, W.A., Stenson, G.B., Shadwick, R.E., 2018. The caval sphincter in cetaceans and its predicted role in controlling venous flow during a dive. J. Exp. Biol. 221 (11), jeb177212.

Liu, Y., Cotton, J.A., Shen, B., Han, X., Rossiter, S.J., Zhang, S., 2010a. Convergent sequence evolution between echolocating bats and dolphins. Curr. Biol. 20 (2), R53–R54.

Liu, Y., Rossiter, S.J., Han, X., Cotton, J.A., Zhang, S., 2010b. Cetaceans on a molecular fast track to ultrasonic hearing. Curr. Biol. 20 (20), 1834–1839.

Liu, A., He, F., Shen, L., Liu, R., Wang, Z., Zhou, J., 2019. Convergent degeneration of olfactory receptor gene repertoires in marine mammals. BMC Genomics 20 (1), 977.

Lopes-Marques, M., Machado, A.M., Alves, L.Q., Fonseca, M.M., Barbosa, S., Sinding, M.-H.S., Rasmussen, M.H., Alves, L.Q., Fonseca, M.M., Barbosa, S., Sinding, M.-H.S., Rasmussen, M.H., Iversen, M.R., Frost Bertelsen, M., Campos, P.F., da Fonseca, R., Ruivo, R., Castro, L.F.C., 2019. Complete inactivation of sebum-producing genes parallels the loss of sebaceous glands in cetacea. Mol. Biol. Evol. 36 (6), 1270–1280.

Mackay, T.F.C., Huang, W., 2018. Charting the genotype-phenotype map: lessons from the drosophila melanogaster genetic reference panel. Wiley Interdiscip. Rev. Dev. Biol. 7 (1), 10.1002/wdev.289.

Magadum, S., Banerjee, U., Murugan, P., Gangapur, D., Ravikesavan, R., 2013. Gene duplication as a major force in evolution. J. Genet. 92 (1), 155–161.

Mass, A.M., Supin, A.Y., 2007. Adaptive features of aquatic mammals' eye. Anat. Rec. 290 (6), 701–715.

McAdams, H.H., Shapiro, L., 1995. Circuit simulation of genetic networks. Science 269 (5224), 650–656.

McDonald, B.I., Ponganis, P.J., 2012. Lung collapse in the diving sea lion: hold the nitrogen and save the oxygen. Biol. Lett. 8 (6), 1047–1049.

McDonald, B.I., Ponganis, P.J., 2013. Insights from venous oxygen profiles: oxygen utilization and management in diving California sea lions. J. Exp. Biol. 216 (Pt 17), 3332–3341.

McGowen, M.R., Grossman, L.I., Wildman, D.E., 2012. Dolphin genome provides evidence for adaptive evolution of nervous system genes and a molecular rate slowdown. Proc. R. Soc. Lond. B 279 (1743), 0869.

McKnight, J.C., Bennett, K.A., Bronkhorst, M., Russell, D.J.F., Balfour, S., Milne, R., Bivins, M., Moss, S.E.W., Colier, W., Hall, A.J., Thompson, D., 2019. Shining new light on mammalian diving physiology using wearable near-infrared spectroscopy. PLoS Biol. 17 (6), e3000306.

Meir, J.U., Champagne, C.D., Costa, D.P., Ponganis, P.J., 2008. Extreme blood oxygen depletion in diving elephant seals. FASEB J. 22 (S1), 757.7.

Meir, J.U., Champagne, C.D., Costa, D.P., Williams, C.L., Ponganis, P.J., 2009. Extreme hypoxemic tolerance and blood oxygen depletion in diving elephant seals. Am. J. Physiol. Regul. Integr. Comp. Physiol. 297 (4), R927–R939.

Meyer, W.K., Jamison, J., Richter, R., Woods, S.E., Partha, R., Kowalczyk, A., Kronk, C., Chikina, M., Bonde, R.K., Crocker, D.E., Gaspard, J., Lanyon, J.M., Marsillach, J., Furlong, C.E., Clark, N.L., 2018. Ancient convergent losses of yield potential risks for modern marine mammals. Science 361 (6402), 591–594.

Mirceta, S., Signore, A.V., Burns, J.M., Cossins, A.R., Campbell, K.L., Berenbrink, M., 2013. Evolution of mammalian diving capacity traced by myoglobin net surface charge. Science 340 (6138), 1234192.

Nery, M.F., Arroyo, J.I., Opazo, J.C., 2013a. Genomic organization and differential signature of positive selection in the alpha and beta globin gene clusters in two cetacean species. Genome Biol. Evol. 5 (12), 2359–2367.

Nery, M.F., González, D.J., Opazo, J.C., 2013b. How to make a dolphin: molecular signature of positive selection in cetacean genome. PLoS One 8 (6), e65491.

Nery, M.F., Arroyo, J.I., Opazo, J.C., 2014. Increased rate of hair keratin gene loss in the cetacean lineage. BMC Genomics 15, 869.

Panneton, W.M., 2013. The mammalian diving response: an enigmatic reflex to preserve life? Physiology 28 (5), 284–297.

Park, J.Y., Kim, K., Sohn, H., Kim, H.W., An, Y.-R., Kang, J.-H., Kim, E.-M., Kwak, W., Lee, C., Yoo, D., Jung, J., Sung, S., Yoon, J., Kim, H., 2018. Deciphering the evolutionary signatures of pinnipeds using novel genome sequences: the first genomes of *Phoca largha*, *Callorhinus ursinus*, and *Eumetopias jubatus*. Sci. Rep. 8 (1), 16877.

Penso-Dolfin, L., Haerty, W., Hindle, A., Di Palma, F., 2020. microRNA profiling in the weddell seal suggests novel regulatory mechanisms contributing to diving adaptation. BMC Genomics 21 (1), 303.

Perrin, W.F., Thieleking, J.L., Walker, W.A., Archer, F.I., Robertson, K.M., 2011. Common bottlenose dolphins (*Tursiops truncatus*) in California waters: Cranial differentiation of coastal and offshore ecotypes. Mar. Mamm. Sci. 27, 769–792.

Peto, R., Roe, F.J., Lee, P.N., Levy, L., Clack, J., 1975. Cancer and ageing in mice and men. Br. J. Cancer, 32 (4), 411–426.

Ponganis, P.J., 2018. Circulatory system. In: Würsig, B., Thewissen, J.G.M., Kovacs, K.M. (Eds.), Encyclopedia of Marine Mammals, third ed. Academic Press, pp. 191–194.

Ponganis, P.J., Kreutzer, U., Stockard, T.K., Lin, P.-C., Sailasuta, N., Tran, T.-K., Hurd, R., Jue, T., 2008. Blood flow and metabolic regulation in seal muscle during Apnea. J. Exp. Biol. 211 (Pt 20), 3323–3332.

Pujade Busqueta, L., Crocker, D.E., Champagne, C.D., McCormley, M.C., Deyarmin, J.S., Houser, D.S., Khudyakov, J.I., 2020. A blubber gene expression index for evaluating stress in marine mammals. Conserv. Physiol. 8 (1), coaa082.

Quan, H., Athirakul, K., Wetsel, W.C., Torres, G.E., Stevens, R., Chen, Y.T., Coffman, T.M., Caron, M.G., 2004. Hypertension and impaired glycine handling in mice lacking the orphan transporter XT2. Mol. Cell. Biol. 24 (10), 4166–4173.

Qvist, J., Hill, R.D., Schneider, R.C., Falke, K.J., Liggins, G.C., Guppy, M., Elliot, R.L., Hochachka, P.W., Zapol, W.M., 1986. Hemoglobin concentrations and blood gas tensions of free-diving weddell seals. J. Appl. Physiol. 61 (4), 1560–1569.

Rommel, S.A., Pabst, D.A., McLellan, W.A., Williams, T.M., Friedl, W.A., 1994. Temperature regulation of the testes of the bottlenose dolphin (*Tursiops truncatus*): evidence from colonic temperatures. J. Compar. Physiol. B 164 (2), 130–134.

Schofield, C.J., Ratcliffe, P.J., 2005. Signalling hypoxia by HIF hydroxylases. Biochem. Biophys. Res. Commun. 338 (1), 617–626.

Scholander, P.F., 1940. Experimental investigations on the respiratory function in diving mammals and birds. Hvalrådets skr. 22, 1–131.

Shadwick, R.E., Gosline, J.M., 1994. Arterial mechanics in the fin whale suggest a unique hemodynamic design. Am. J. Physiol. 267 (3 Pt 2), R805–R818.

Sharma, V., Hecker, N., Roscito, J.G., Foerster, L., Langer, B.E., Hiller, M., 2018. A genomics approach reveals insights into the importance of gene losses for mammalian adaptations. Nat. Commun. 9 (1), 1215.

Shen, Y.-Y., Liang, L., Li, G.-S., Murphy, R.W., Zhang, Y.-P., 2012. Parallel evolution of auditory genes for echolocation in bats and toothed whales. PLoS Genet. 8 (6), e1002788.

Shen-Orr, S.S., Milo, R., Mangan, S., Alon, U., 2002. Network motifs in the transcriptional regulation network of *Escherichia coli*. Nat. Genet. 31, 64–68.

Sun, Y.-B., Zhou, W.-P., Liu, H.-Q., Irwin, D.M., Shen, Y.-Y., Zhang, Y.-P., 2013. Genome-wide scans for candidate genes involved in the aquatic adaptation of dolphins. Genome Biol. Evol. 5 (1), 130–139.

Tejada-Martinez, D., de Magalhães, J.P., Opazo, J.C., 2021. Positive selection and gene duplications in tumour suppressor genes reveal clues about how cetaceans resist cancer. Proc. R. Soc. Lond. B 288 (1945), 20202592.

Themudo, G.E., Alves, L.Q., Machado, A.M., Lopes-Marques, M., da Fonseca, R.R., Fonseca, M., Ruivo, R., Filipe, L., Castro, C., 2020. Losing genes: the evolutionary remodeling of cetacea skin. Front. Mar. Sci. 7, 592375.

Thewissen, J.G.M., George, J., Rosa, C., Kishida, T., 2011. Olfaction and brain size in the bowhead whale (*Balaena mysticetus*). Mar. Mamm. Sci. 27 (2), 282–294.

Tian, R., Wang, Z., Niu, X., Zhou, K., Xu, S., Yang, G., 2016. Evolutionary genetics of hypoxia tolerance in cetaceans during diving. Genome Biol. Evol. 8 (3), 827–839.

Tift, M.S., Hückstädt, L.A., McDonald, B.I., Thorson, P.H., Ponganis, P.J., 2017. Flipper stroke rate and venous oxygen levels in free-ranging california sea lions. J. Exp. Biol. 220, 1533–1540.

Tift, M.S., Hückstädt, L.A., Ponganis, P.J., 2018. Anterior vena caval oxygen profiles in a deep-diving california sea lion: arteriovenous shunts, a central venous oxygen store and oxygenation during lung collapse. J. Exp. Biol. 221, jeb163428.

Tollis, M., Robbins, J., Webb, A.E., Kuderna, L.F.K., Caulin, A.F., Garcia, J.D., Bèrubè, M., Pourmand, N., Marques-Bonet, T., O'Connell, M.J., Palsbøll, P.J., Maley, C.C., 2019. Return to the sea, get huge, beat cancer: an analysis of cetacean genomes including an assembly for the humpback whale (*Megaptera novaeangliae*). Mol. Biol. Evol. 36 (8), 1746–1763.

Torres-Velarde, J.M., Kolora, S.R.R., Khudyakov, J.I., Crocker, D.E., Sudmant, P.H., Vázquez-Medina, J.P., 2021. Elephant seal muscle cells adapt to sustained glucocorticoid exposure by shifting their metabolic phenotype. Am. J. Physiol. 321 (3), R413–R428.

Wang, Z., Chen, Z., Shixia, X., Ren, W., Zhou, K., Yang, G., 2015. 'Obesity' is healthy for cetaceans? Evidence from pervasive positive selection in genes related to triacylglycerol metabolism. Sci. Rep. 5, 14187.

Wright, T.J., Davis, R.W., Holser, R.R., Hückstädt, L.A., Danesi, C.P., Porter, C., Widen, S.G., Williams, T.M., Costa, D.P., Sheffield-Moore, M., 2020. Changes in northern elephant seal skeletal muscle following thirty days of fasting and reduced activity. Front. Physiol. 11, 564555.

Zapol, W.M., Liggins, G.C., Schneider, R.C., Qvist, J., Snider, M.T., Creasy, R.K., Hochachka, P.W., 1979. Regional blood flow during simulated diving in the conscious weddell seal. J. Appl. Physiol. 47 (5), 968–973.

Zasloff, M., 2011. Observations on the remarkable (and mysterious) wound-healing process of the bottlenose dolphin. J. Invest. Dermatol. 131 (12), 2503–2505.

Zhou, X., Sun, F., Shixia, X., Fan, G., Zhu, K., Liu, X., Chen, Y., Shi, C., Yang, Y., Huang, Z., Chen, J., Hou, H., Guo, X., Chen, W., Chen, Y., Wang, X., Lv, T., Yang, D., Zhou, J., Huang, B., Wang, Z., Zhao, W., Tian, R., Xiong, Z., Xu, J., Liang, X., Chen, B., Liu, W., Wang, J., Pan, S., Fang, X., Li, M., Wei, F., Xu, X., Zhou, K., Wang, J., Yang, G., 2013. Baiji genomes reveal low genetic variability and new insights into secondary aquatic adaptations. Nat. Commun. 4, 2708.

Zhu, K., Ge, D., Wen, Z., Xia, L., Yang, Q., 2018. Evolutionary genetics of hypoxia and cold tolerance in mammals. J. Mol. Evol. 86 (9), 618–634.

CHAPTER

9

Neurophysiology

Peter F. Cook[a], *Stefan Huggenberger*[b], *and Bruno Cozzi*[c]

[a]Department of Psychology, New College of Florida, Sarasota, FL, United States [b]Institute of Anatomy and Clinical Morphology, Witten/Herdecke University, Witten, Germany [c]Department of Comparative Biomedicine and Food Science, University of Padova, Padova, Italy

Abstract

Since the early days of neuroanatomy, the brains of dolphins and whales have attracted attention for their large volume, unusual shape, and intense gyrification. Progress toward describing the central nervous system of cetaceans has been aptly summarized in several seminal articles and reviews. While the authors of these papers have been largely in agreement regarding the key features of dolphin neural organization, there is ongoing and unsettled controversy concerning the relationship between dolphin brains and empathy, social organization, and ultimately the intelligence of these fascinating species. The functional characteristics of the cetacean brain, which are central to these sorts of cognitive and behavioral questions, rely not just on settled anatomical understanding but also on ongoing and incomplete exploration into cetacean neurophysiology. In this chapter, we will briefly summarize what uncontroversial facts are known regarding dolphin (and whale) brains and highlight unsolved issues, with a particular attention to neurophysiology.

Since the early days of neuroanatomy, the brains of dolphins and whales have attracted attention for their large volume, unusual shape, and intense gyrification. Progress toward describing the central nervous system of cetaceans has been aptly summarized in several seminal articles and reviews (Kükenthal and Ziehen, 1893; Langworthy, 1931a, b; Jelgersma, 1934; Breathnach, 1960; Jansen and Jansen, 1969; Pilleri and Gihr, 1970; Morgane and Jacobs, 1972; Jacobs et al., 1979; Morgane et al., 1986; Ridgway, 1990; Cozzi et al., 2017). While the authors of these papers have been largely in agreement regarding the key features of dolphin neural organization, there is ongoing and unsettled controversy concerning the relationship between dolphin brains and empathy, social organization, and ultimately the intelligence of these fascinating species. The functional characteristics of the cetacean brain, which are central to these sorts of cognitive and behavioral questions, rely not just on settled anatomical understanding but also on ongoing and incomplete exploration into cetacean neurophysiology (Fig. 1).

In this chapter, we will briefly summarize what uncontroversial facts are known regarding dolphin (and whale) brains and highlight unsolved issues, with a particular attention to neurophysiology. Where our observations directly relate to further questions into behavior and cognition, we will provide conservative interpretations in line with best current evidence. Due to their overrepresentation in anatomical and physiological analyses of whales, most of our descriptions will be based on the bottlenose dolphins (*Tursiops truncatus*) and other members of the dolphin species, but we will note key facts about other toothed and baleen whales.

Histology

Cell types, their shape, and dimensions

We begin with the small-scale, with a few lines dedicated to the histology of the central nervous system. Neurons of the cetacean brain are similar to those contained in the brains of terrestrial mammals (Kraus and Pilleri, 1969, 1972; Garey et al., 1985, 1989; Ferrer and Perera, 1988; Glezer and Morgane, 1990; Glezer et al., 1993; Hof et al., 1999; Glezer, 2002; Poth et al., 2005; Butti et al., 2015). On a general plane, the shape of neurons is either that of a sphere (and this is the case of smaller neurons such as granules, with many dendrites) or a cone (as for pyramidal cells, with a long axon and a forest of dendrites). Stellate, basket, pigmented, and other neurons are present in whales in similar

The Physiology of Dolphins
https://doi.org/10.1016/B978-0-323-90516-9.00008-7

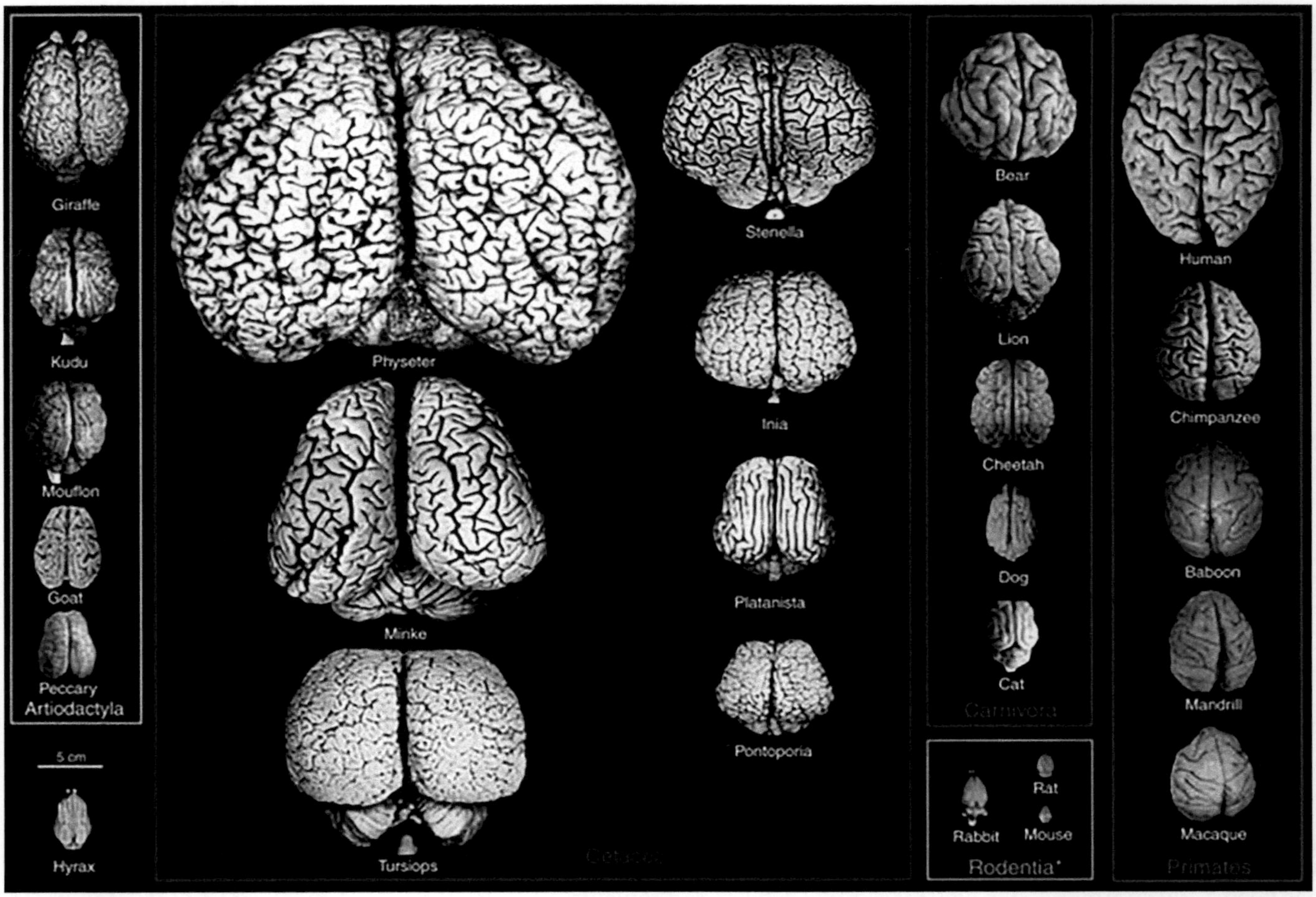

FIG. 1 Photographs of the brains of different species and their masses in grams. Hyracoidea: Rock hyrax (*Procavia capensis*, 12 g). Artiodactyls: giraffe (*Giraffa camelopardalis*, 700 g), kudu (*Strepsiceros strepsiceros*), 166 g, mouflon (*Ovis orientalis*, 118 g), ibex (goat, *Capra pyrenaica*, 115 g), peccary (*Tayassu pecari*, 41 g). Lagomorph: rabbit (*Oryctolagus cuniculus*, 5 g). Rodents: rat (*Rattus rattus*, 2.6 g), mouse (*Mus musculus*, 0.5 g). Odontoceti: sperm whale (*Physeter macrocephalus*, adult), bottlenose dolphin (*Tursiops truncatus*, ca. 1500 g), striped dolphin (*Stenella coeruleoalba*, 1230 g), Amazon River dolphin (*Inia geoffrensis*, 460 g), South Asian River dolphin (*Platanista gangetica*, 292 g), Franciscana (*Pontoporia blainvillei*, 230 g). Mysticeti: minke whale (*Balaenoptera acutorostrata*, 2590 g). Carnivores: bear (*Ursus arctos*, 289 g), lion (*Panthera leo*, 165 g), cheetah (*Acinonyx jubatus*, 119 g), dog (*Canis familiaris*, 95 g), cat (*Felis catus*, 32 g). Primates: human (*Homo sapiens*, 1176 g), chimpanzee (*Pan troglodytes*, 273 g), baboon (*Papio cynocephalus*, 151 g), mandrill (*Mandrillus sphinx*, 123 g), macaque (*Macaca tonkeana*, 110 g). *Modified after Brauer and Schaber (1976), de Felipe et al. (2007). From Cozzi et al. (2017) [originally Fig. 6.1, p. 198].*

positions in the brain as have been observed in other mammals. However, there are a few noteworthy exceptions to this general homology. Butti et al. (2009) noted that von Economo neurons (originally described by Cajal), considered typical of humans and, suggested by some, to be related to our advanced cognitive capability, were present also in the brain of dolphins. This interesting discovery was later limited by the subsequent identification of von Economo cells in many other mammals, thus curtailing their interpretative value in relation to higher brain functions (Butti et al., 2013). Another consideration is that in some cetacean species, pyramidal neurons of (equivalent) area 4 (motor cortex) are true giants, with sizes over 100 μm. Yet, one must note that although the monosynaptic pyramidal tract is not the prevalent motor connection in cetaceans (see below), bottlenose dolphins, sperm whales (*Physeter macrocephalus*), and most baleen whales possess an extremely long spinal cord, and longer axons always possess a larger exit cone. No terrestrial mammal, however big, reaches the size (and especially the length of the spine) of the large whales, and thus a proper comparison of neuronal size is difficult.

Neural density

According to neuron counting isotropic fractionation approaches, the sheer number of cortical neurons in the brain of killer whales (*Orcinus orca*, 40 billion, Ridgway et al., 2019) and pilot whales (*Globicephala* spp., 30 billion, Mortensen et al., 2014) far exceed the estimate for human brain (20 billion, Herculano-Houzel et al., 2015). However, a common general feature of the cetacean brain is that a cortical section from most cetaceans can be easily distinguished from the equivalent section in a macaque, a chimpanzee, or a human by the strikingly lower neuronal density (Kern et al., 2011; Huggenberger et al., 2019; Graïc et al., 2021). Clearly, the issue deserves further study. Older beliefs regarding the lack of variability in cell type of the cetacean cortex have now been proven untrue; as different areas have different cell densities it may be very difficult to obtain estimates of the absolute number of cells. We also add that, contrarily to what has been reported for killer and pilot whales, some species [including beaked and sperm whales (Huggenberger et al., 2019; Ridgway et al., 2019), and at least fin whales (*Balaenoptera physalus*) among Mysticetes (BC and SH personal observations)] appear to have a marked reduction of neurons in most cortical areas if compared to primates. We emphasize that different methodologies may yield different density values. This is especially true when considering very large brains in which a complete sampling of all the areas is practically impossible to perform. Comparisons with primates or other large-brained terrestrial species are also difficult.

Shape and volume of the brain

Shape of the brain

The shape of the brain of dolphins and whales is characterized by intense gyrification (Fig. 2), shortening of the longitudinal axis, and enlargement of the transverse axis, especially in correspondence to the temporal lobes (Langworthy, 1935; Ries and Langworthy, 1936; Kojima, 1951; Morgane et al., 1980; Pilleri, 1989; Duffield et al., 1992; Oelschläger and Oelschläger, 2009). This latter characteristic is present already in the brain of several terrestrial Cetartiodactyls, the closest land relatives of cetaceans. The brain also presents a marked bending, almost a rotation, along the inter-insular axis, so that the frontal part is bent downwards to grossly resemble a boxing glove seen from the side. The general result is a rostro-caudal compression and lateral expansion that give the cranial cavity a special shape. This whole process is part of the evolution of the cetacean skeleton, that modifies the shape of the skull and in toothed whales makes room for the melon and other sound-related structures.

Brain mass

It is well-known that the brains of several species of cetaceans are comparatively very large, sometimes much larger than the human brain (which averages 1350 g). The largest brain ever weighed (9300 g) belonged to a killer whale, followed by one of a sperm whale (9200 g) (Ridgway and Hanson, 2014). Absolute size may be less significant (Ridgway and Brownson, 1984; Huggenberger, 2008) if we consider that the body of many cetaceans is gigantic. A more common way to approach the study of brain mass is to consider the relationship between brain and body weight, and the most accepted equation is Jerison's formula (Jerison, 1973) for the encephalization quotient (EQ) based on calculations that consider both the brain (E) and the body (P) weights ($EQ = Ei/0.12P^{2/3}$). The final value indicates whether a given species possesses a brain mass superior ($EQ > 1$) or inferior ($EQ < 1$) to what would be expected for its body mass.

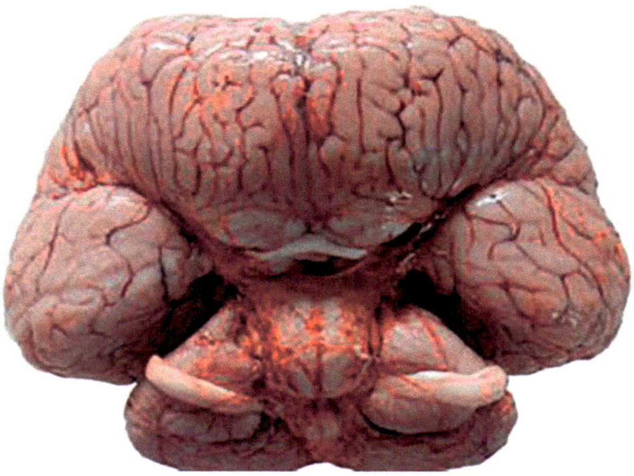

FIG. 2 Ventral view of the brain of the bottlenose dolphin (*T. truncatus*). Note the expansion of the temporal lobes and the depth of the lateral sulcus. Scale bar = 55 cm. *From Huggenberger et al. (2019) [originally Figure at p. 327, top].*

The encephalization quotient and comparison with other mammals

The EQs of many cetacean species, and those of key comparison terrestrial mammals (including terrestrial Cetartiodactyls such as the bovine as well as horses, elephants, and primates), are reported in Huggenberger et al. (2019). Even a superficial comparison immediately suggests that the highest EQs is the human (6.6), but many members of the dolphin family reach considerable EQ values (Ridgway et al., 2016), including rough-toothed dolphin (*Steno bredanensis*) (5.2), false killer whale (*Pseudorca crassidens*) (5.5), white-beaked dolphin (*Lagenorhynchus albirostris*) (5.4), northern right whale dolphin (*Lissodelphis borealis*) (5.7), and Commerson's dolphin (*Cephalorhynchus commersoni*) (5.2) (Huggenberger et al., 2019). One must also consider that some of the available data should be taken with caution, because of possible alterations in the weight of the fixed brain compared with the brain in situ (Ridgway and Van Alstyne, 2017). The bottlenose dolphin's EQ ranges between 2.8 and 3.7, still a remarkable value considering that the closest human relative, the chimpanzee, has an EQ of 2.4. The relation (if any) between EQ and species cognitive capacity and, more specifically intelligence—however defined, remains to be determined (Roth and Dicke, 2005).

Jerrison's equation for EQ is generally biased toward larger mammals, but there is a structural limit to which the gelatinous mass of the brain may be able to maintain its shape even with all the support of a tenacious dura mater and an ossified tentorium. One must consider that the gigantic bodies of the large baleen whales and those of the sperm and beaked whales are possible because of life in the water and the support that the denser medium provides large bodies. Furthermore, some sexual differences in body mass, but not in brain mass, may yield surprising results. For example, the EQ of sperm whales ranges between 0.57 (large males) and 1.01 (females), a difference driven by body size, not brain size (Cozzi et al., 2016). One may also note that female sperm whales lead a social life in communities, whereas adult males do not and are often solitary, which could be hypothesized relevant to brain size in line with ideas about social complexity driving the evolution of bigger brains (Dunbar, 1998). But this must be explored further. No baleen whale has an EQ higher than 0.66 in the minke whale (*Balaenoptera acutorostrata*) (Huggenberger et al., 2019), a fact that once more reflects the expansion of the body mass. River dolphins, being small, on the other hand, show relatively high EQ values (Huggenberger et al., 2019; see also Schwerdtfeger et al., 1984). There is ongoing debate about the relevance of EQ to cognitive capability across species (van Schaik et al., 2021). However, one may safely state that cetaceans have big (sometimes huge) convoluted brains, often larger than expected for the body mass of the species. Members of the dolphin family fare better on measures of EQ than other families, while the large mysticetes and male sperm whales occupy the opposite pole.

The cerebral cortex of whales and dolphins

As we mentioned earlier, the brain of all cetaceans is extremely convoluted, and the cerebral cortex presents many folds. This is not surprising for readers familiar with the basics of cetacean neuroanatomy, but one must remember that intense cortical folding means an increase of cortical surface that is not the general rule for all mammals (i.e., rodents have no cortical folding) (Voogd et al., 1998). Sea cows (Sirenia), the other fully aquatic mammalian group, and even several primate species, including macaques, have an almost flat cortical surface. Of course, many now believe that cortical folding has limited functional relevance, scaling predominantly with the depth and area of the cortex (Mota & Herculano-Houzel, 2015).

The corpus callosum, cerebral ventricles, and blood-brain barrier

The corpus callosum of dolphins and other toothed whales is relatively thin with respect to brain size in comparison with terrestrial mammals of comparable body size (Tarpley and Ridgway, 1994; Keogh and Ridgway, 2008; Cozzi et al., 2017; for baleen whales see Ratner et al., 2011). This data must be considered together with neuron density, as fibers in the corpus callosum reflect the organization as well as the connection pattern of the neocortex.

The ventricular system of the brain has been studied anatomically in the bottlenose dolphin (McFarland et al., 1969), and is of course clearly evident in all studies involving magnetic resonance (see below). The blood-brain barrier has been characterized by Glezer et al. (1987).

Organization of the cetacean cortical column and analogies with other mammals

A general feature of the mammalian cortex is the subdivision in layers (Figs. 3 and 4), whose thickness and composition reflect the specialization of the area; for example, in motor areas pyramidal cells prevail, whereas in

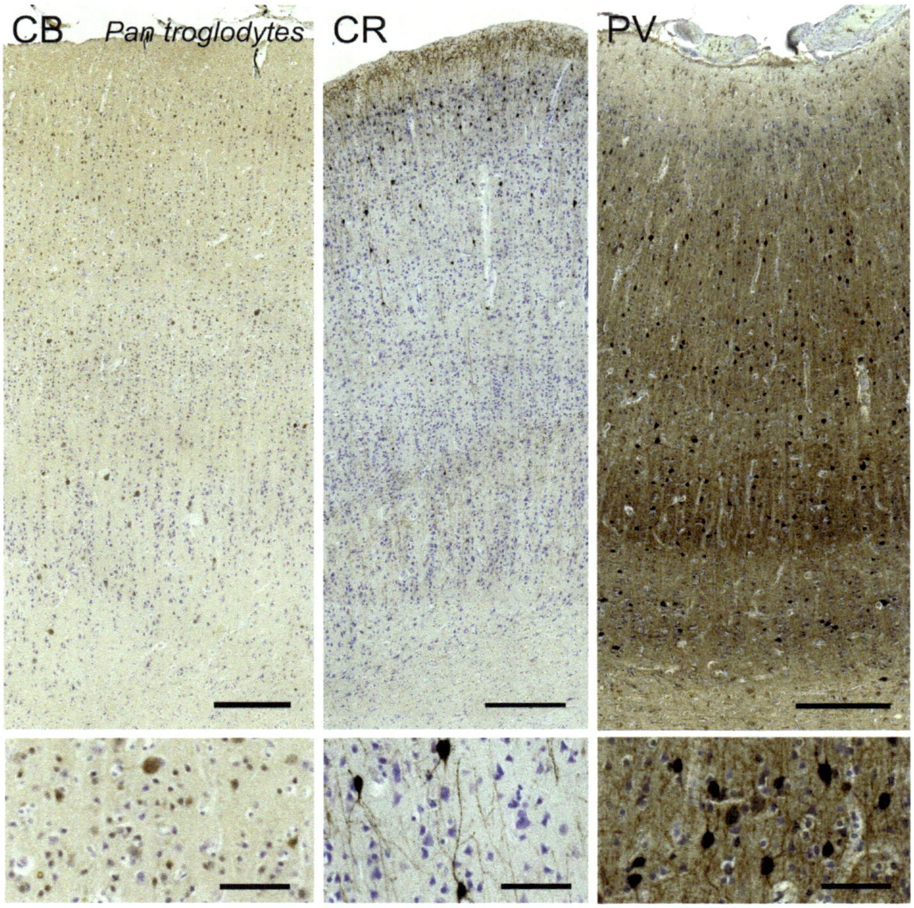

FIG. 3 Microphotographs of Calbindin-immunoreactive (CB-ir, left), Calretinin-immunoreactive (CR-ir, middle), and Parvalbumin (PV-ir, right) neurons in the primary visual cortex of the chimpanzee (*P. troglodytes*), with magnified inserts below. Bars are 300 μm, 100 μm in inserts. *From Graïc et al. (2022) Springer [originally Fig. 6 at p. 1204].*

associative areas granule cells are most abundant (Lodato and Arlotta, 2015). Changes in the relative thickness and composition of the layers are crucial to subdivide the cortex into areas, still commonly identified and named after Korbinian Brodmann's original description (Brodmann, 1909). Most of the human and rodent cortex is characterized by the presence of six layers. However, other mammalian orders including Perissodactyls (horses and rhinos) and Cetartiodactyls (including whales and dolphins) show a reduction in the number of cortical layers (Jacobs et al., 1984; Morgane et al., 1985; Revishchin and Garey, 1991; Hof et al., 1999, 2000; Hof and Van der Gucht, 2007; Furutani, 2008; Kern et al., 2011; Mortensen et al., 2014; Knopf et al., 2016; van Kann et al., 2017; for review see Cozzi et al., 2017). In the cerebral cortex of whales and dolphins, the cortex generally contains five layers, thus missing layer IV also known as the inner granular layer, which in other mammals receives the afferent outputs from the thalamus. The potential functional relevance of the missing layer IV is discussed below.

The visual (Garey and Leuba, 1986; Morgane et al., 1988, 1990; Glezer and Morgane, 1990; Glezer et al., 1992, 1995, 1998; Graïc et al., 2021) and auditory (Ogawa and Arifuku, 1948; Glezer et al., 1995, 1998; Fuzessery et al., 2004) cortices of dolphins and whales have received special attention (Fig. 5). The nuclei and tracts involved in the visual or auditory pathways are the same as in terrestrial mammals, although only a limited number of studies were specifically dedicated to the topic (Zvorykin, 1976).

Projections from the thalamus and different organization of the cortical column

The organization of the cetacean cerebral cortex has puzzled scientists of the second half of the 20th century, mostly because of two key facts. The first is that layer IV, which as stated above is missing in cetaceans, is the thalamo-recipient

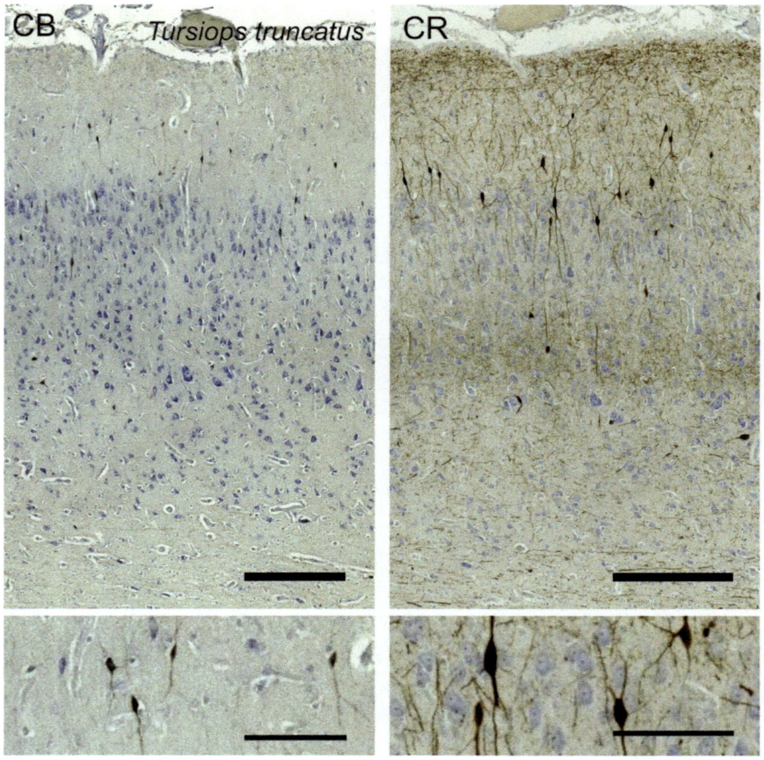

FIG. 4 Microphotographs of CB-ir (left) and CR-ir (right) neurons in the primary visual cortex of the bottlenose dolphin (*T. truncatus*), with magnified inserts below. Bars are 300 μm, 100 μm in inserts. *From Graïc et al. (2022) Springer [originally Fig. 18 at p. 1211].*

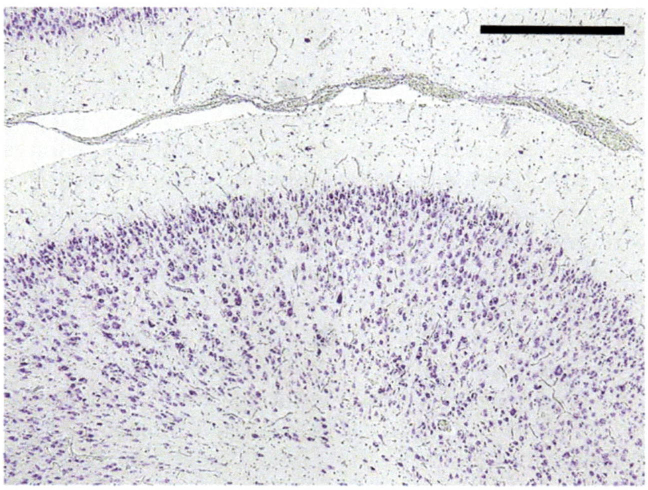

FIG. 5 Whole section (Nissl stain) through the primary acoustic area of a long-finned pilot whale (*G. melas*). Scale bar = 5500 μm. *From Huggenberger et al. (2019) [originally Figure at p. 355, top].*

layer in the generally accepted translational rodent model of the neocortex (Kruger, 1959; Revishchin and Garey, 1990). Cortical layer IV is also thick and characteristic of the human associative areas, considered the cortical region most responsible for higher brain functions. Hence neuroscientists studying whales and dolphins have had to explain two considerable facts: which layer is the target of thalamic projections in the cortex of whales and dolphins, and how to explain species lacking layer IV showing signs of considerable intelligence. Subsequent studies have revealed that in the five-layered cerebral cortex layer II (external granular layer) receives thalamic outputs, with a modification of the circuitry. A functional circuitry has never been demonstrated by electrophysiology in dolphins (for obvious reasons, see also below). However, there is sufficient indirect evidence to consider layer II as the target for thalamic fibers instead of layer IV in dolphins (Morgane et al., 1988; Graïc et al., 2021). Neurochemical analogies and differences between the cetacean cerebral cortex with those of other related or non-related terrestrial mammals have been analyzed and described in many studies (for references and further reading see Hof et al., 1999; Dell et al., 2015).

The question of the different structures of the associative cortex is far more complicated, and is addressed below.

The claustrum, a structure that has received increased attention in recent years for its potential role in attention and possibly consciousness, has been studied only in the bottlenose dolphin (Cozzi et al., 2014). The amygdala, a key structure for perceptual learning, has been seldom described in specific articles, and appears to be relatively small, as with other medial temporal structures (Jansen and Jansen, 1953; Rambaldi et al., 2017).

Cerebral topography

The definition of a given cortical area is based on the composition of the cortical columns, as just outlined above. Extremely detailed maps of cortical areas, including their functions, are available for the human brain, the rat and mouse brain, and the macaque brain (e.g., Roland et al., 1994; Reveley et al., 2017; Chon et al., 2019). Slightly less detailed stereotaxic atlases are present for the chimpanzee, the dog, additional rodent species, and the rabbit (DeLucchi et al., 1965; Girgis and Wang, 1981; Johnson et al., 2020). For most of these species, the function of the areas has been validated by electrophysiology (including human). Atlases for terrestrial or marine Cetartiodactyls are absent or incomplete (for Delphinidae see Morgane et al., 1980; Pilleri et al., 1980; Parolisi et al., 2015; Cozzi et al., 2017; Huggenberger et al., 2019), and electrophysiology has been applied rarely to identify brain areas (Lende and Akdikmen, 1968; Kesarev and Malofeeva, 1969a; for the early context of research see also Lilly, 1961). Consequently, analogies of cortical areas are based on identification of corresponding external landmarks, that is gyri and sulci. But the position, shape, and direction of the gyri in all cetaceans are different from what is observed in primates (rodents have no external gyri) or carnivores (Fig. 6). The orbitofrontal cortex, a key associative area in the human brain, is extremely reduced in sheep, with whom cetaceans share a systematic group, and its projections match only in part those of the human equivalent area (Gerussi et al., 2022). Based on homologies of the gyri, the topography and cellular organization of the orbitofrontal cortex of the dolphin closely resembles that of the sheep. The diminished expansion and the lack of layer IV make it difficult to consider the function of the orbitofrontal cortex of the bottlenose dolphin as even vaguely equivalent to that of primates for computational power. Preliminary results from Berns et al. (2015) did identify typical mammalian patterns of frontal projections from different basal ganglia regions, but the relative density of these pathways was not examined.

We must also note that since no electrophysiology supports the role of the orbitofrontal cortex as the associative area in the dolphin and whale brain, it is perfectly possible that adjacent gyri could be recruited in the expression of higher brain functions linked with frontal cortex in other mammals, such as attention, inhibition, and working memory.

The cerebellum, the brainstem, and the spinal cord

The general organization of the derivates of the mesencephalic and myelencephalic vesicles in the adult cetacean brain follows the same patterns of terrestrial mammals (Hatschek and Schlesinger, 1902; Jansen, 1953; De Graaf, 1967; Jansen and Osen, 1984; Schulmeyer, 1992). However, some noteworthy differences suggest potential differential functional pathways.

Relevant differences with terrestrial mammals

The cerebellum of dolphins and whales (Marino et al., 2000; Kalinichenko and Pushchin, 2008; Hanson et al., 2013; Ridgway and Hanson, 2014) is essentially constituted by two large lateral lobes and a central vermis (Fig. 7). The two lateral lobes (belonging to the neocerebellum) are so well developed that their height and expansion far surpasses that of the vermis, contrary to what happens in the cetaceans' terrestrial relatives Cetartiodactyls and Perissodactyls. This specific organization is found only in apes (and some monkeys) and in the elephant, a fact often considered a sign of cerebral development given the fact that the lateral cerebellar lobes have also been involved in complex neural

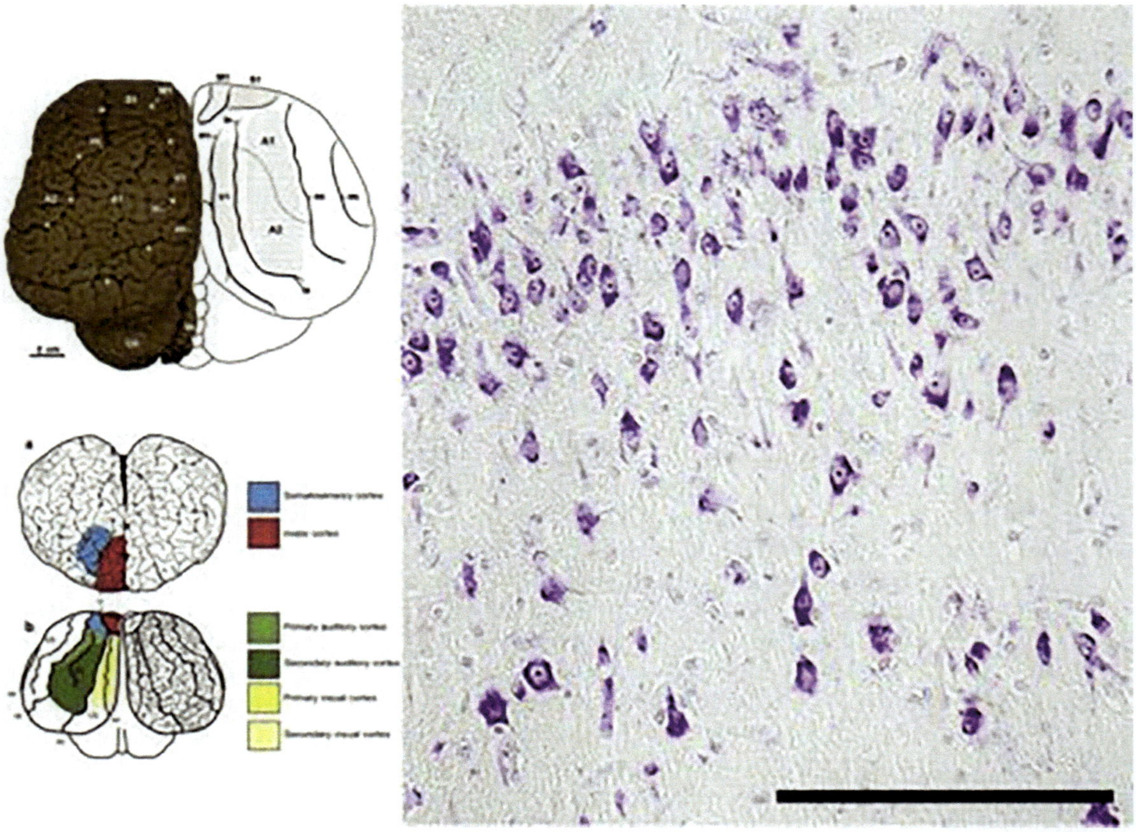

FIG. 6 Top left: localization of functional areas in the brain of the bottlenose dolphin (*T. truncatus*) (arrowheads mark the course of major neocortical sulci. A1, A2, primary, secondary auditory neocortex; c, cruciate sulcus; Ch, cerebellar hemisphere; en, entolateral sulcus; es, ectosylvian sulcus; la, lateral sulcus; M1, motor neocortex; S1, somatosensory neocortex; ss, suprasylvian sulcus; V1, visual neocortex; Ve, vermis). Bottom left: a, rostral view of the dolphin brain showing the natural sulcal pattern with the motor neocortex field (red) and the somatomotor cortex (blue). b, dorsal aspect of the dolphin brain, more schematic. On the right hemisphere, the pattern of sulci and gyri is replicated. On the left hemisphere, only first-order sulci and gyri are shown. cr, cruciate sulcus; ec, lateral sulcus; en, entolateral sulcus; es, ectosylvian sulcus; ES, ectosylvian gyrus; LG, lateral gyrus; ss, suprasylvian sulcus; SS, suprasylvian gyrus. On the right: Nissl stain of motor cortex of the same species. Scale bar = 5200 μm. *The drawings on the left originally appeared as Figs. 6.14 and 6.15 of Cozzi et al. (2017). Top left: modified by Jutta Oelschläger from Morgane et al. (1986), pp. 529 and Kern et al. (2011). Bottom a: modified and redrawn by Jutta Oelschläger; a, from Morgane et al. (1980); bottom b, from Supin et al. (2001). From Huggenberger et al. (2019) [originally Figure at p. 354].*

activities (including language), as demonstrated in humans by the presence of connections with the frontal cortex and by PET studies as well as a growing body of functional imaging results (Buckner et al., 2011; Tirapu-Ustarroz et al., 2011; Wagner and Luo, 2020).

The vermis of the cerebellum of beaked whales is rather indistinct (Oelschläger et al., 2010), almost mingled with the later lobes. Whether this latter morphological characteristic has any bearing with function remains undetermined.

The brainstem is rather bent because of the shape of the cerebral fossa (Fig. 8), and is characterized by the enormous development of acoustic structures, including the medial geniculate nuclei (in fact belonging to the thalamus) and the inferior (= caudal) colliculi. As expected, the cochlear nuclei complex is relevant and functional to the role played by sound in daily life (Osen and Jansen, 1965; Schulmeyer et al., 2000; Malkemper et al., 2012). The mesencephalic *nucleus ellipticus* (present in elephants too, and considered a modified nucleus of Darkschewitsch), is also related to sound as it controls the muscles of the nasal sacs (Dailly, 1972; Cozzi et al., 2017). Several articles described individual nuclei of the cetacean brainstem (Seki, 1973, 1984, 1985; Popov and Supin, 1990; Manger et al., 2003, 2004; Kern et al., 2009; Sacchini et al., 2018). The inferior (= posterior) olivary complex is huge and is the origin of the olivospinal tract (potentially related to the motor control of rotatory movements of the body induced by vertebral muscles) (Cozzi et al., 2017).

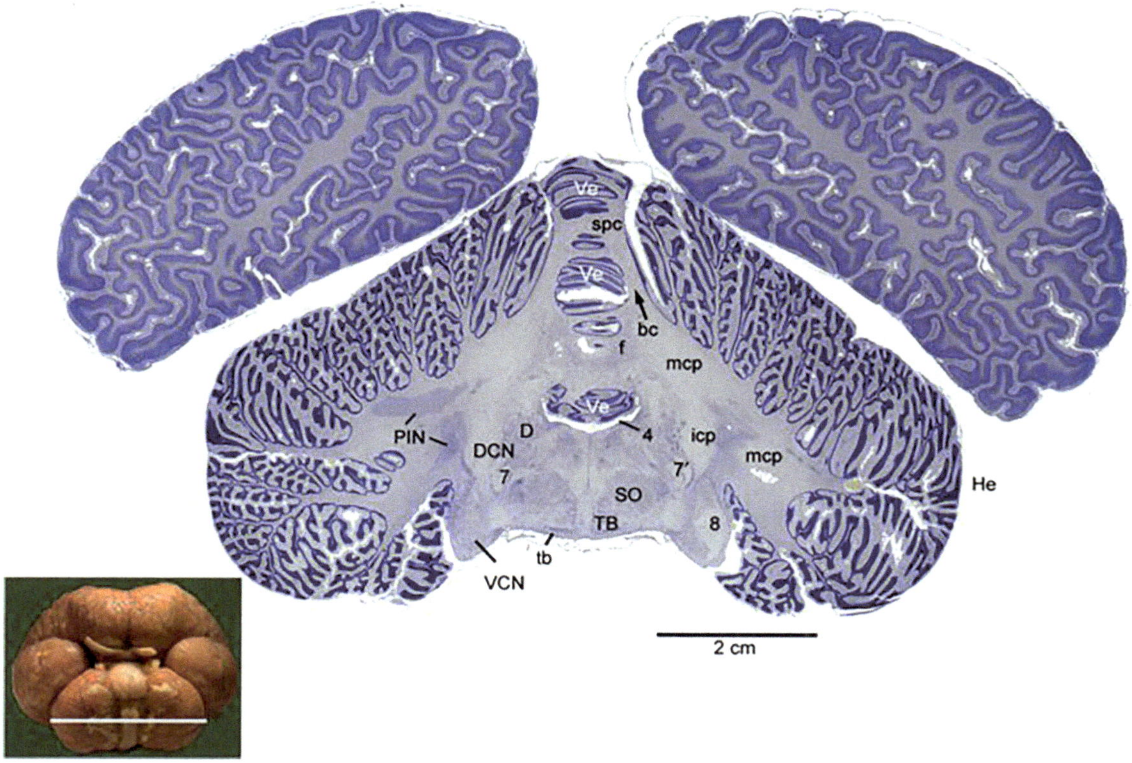

FIG. 7 Common dolphin (*Delphinus delphis*). Coronal section (inset) through the brainstem, cerebellum, and the rear part of the temporal lobes (see pilot). The cerebellum is cut at its maximal diameter, with the cortex of the vermis interrupted by bands of white substance. Note the inverse correlation between the thickness of the cortical plate and its folding density in the cerebral and cerebellar cortices. The flat part of the fourth ventricle (4) extends between the lower contour of the vermis and the upper contour of the myelencephalon. The cerebellar nuclei (PIN, f) are embedded in the white substance of the cerebellum. Abbreviations: 7, facial nerve; 7′, facial nucleus; 8, cochlear nerve; bc, brachium conjunctivum; D, Deiters' nucleus (lateral vestibular nucleus); DCN, dorsal cochlear nucleus; f, nucleus fastigii; icp, inferior (caudal) cerebellar peduncle (restiform body); He, cerebellar hemisphere; mcp, middle cerebellar peduncle; PIN, posterior interposed nucleus; scp, superior (rostral) cerebellar peduncle; SO, superior olive; tb, trapezoid bod; TB, nucleus of the trapezoid body; VCN, ventral cochlear nucleus; Ve, vermis. *From Cozzi et al. (2017) [originally Fig. 6.35, p. 254].*

Several indications suggest the existence of a central loop in the dolphin brainstem (Figs. 9–11), involving the elliptical nucleus, the (medial accessory) inferior olive, the cerebellar paraflocculus, and posterior interposed nucleus. All these structures, rather hypertrophic if compared to terrestrial mammals, are involved in feedback reciprocal loops that, together with the medial tegmental tract, are related to premotor control and integration of sensory perception. Apparently, this circuit is the *equivalent* of the circuit involving the (parvocellular) red nucleus and central tegmental pathways in primates (Cozzi et al., 2017). We must note that most of our current concepts of physiology of the cetacean brainstem derive from anatomical observations and the logical interpretation of the development (or reduction) of structures whose functions are clear in terrestrial mammals, and especially laboratory rodents, primates, and domestic herbivores. There is no verification of this connectivity by modern morphological methods and no electrophysiological demonstration of their actions in cetaceans.

The spinal cord

The spinal cord of whales and dolphins is characterized by the absence of the central medullary canal (sometimes substituted by a blood vessel), and the lack of a lumbar intumescence, this latter being a logical consequence of the absence of the pelvic limbs and the relative nerves (Seki, 1958; Ogden et al., 1981; Cozzi et al., 2017). The ventral horn is much larger than the dorsal, thus suggesting a prevalence of motor control over sensory inputs in the composition of peripheral nerves. A series of articles described the anatomy and neurochemistry of the spinal cord and spinal ganglia in dolphins (Bombardi et al., 2010, 2011, 2013).

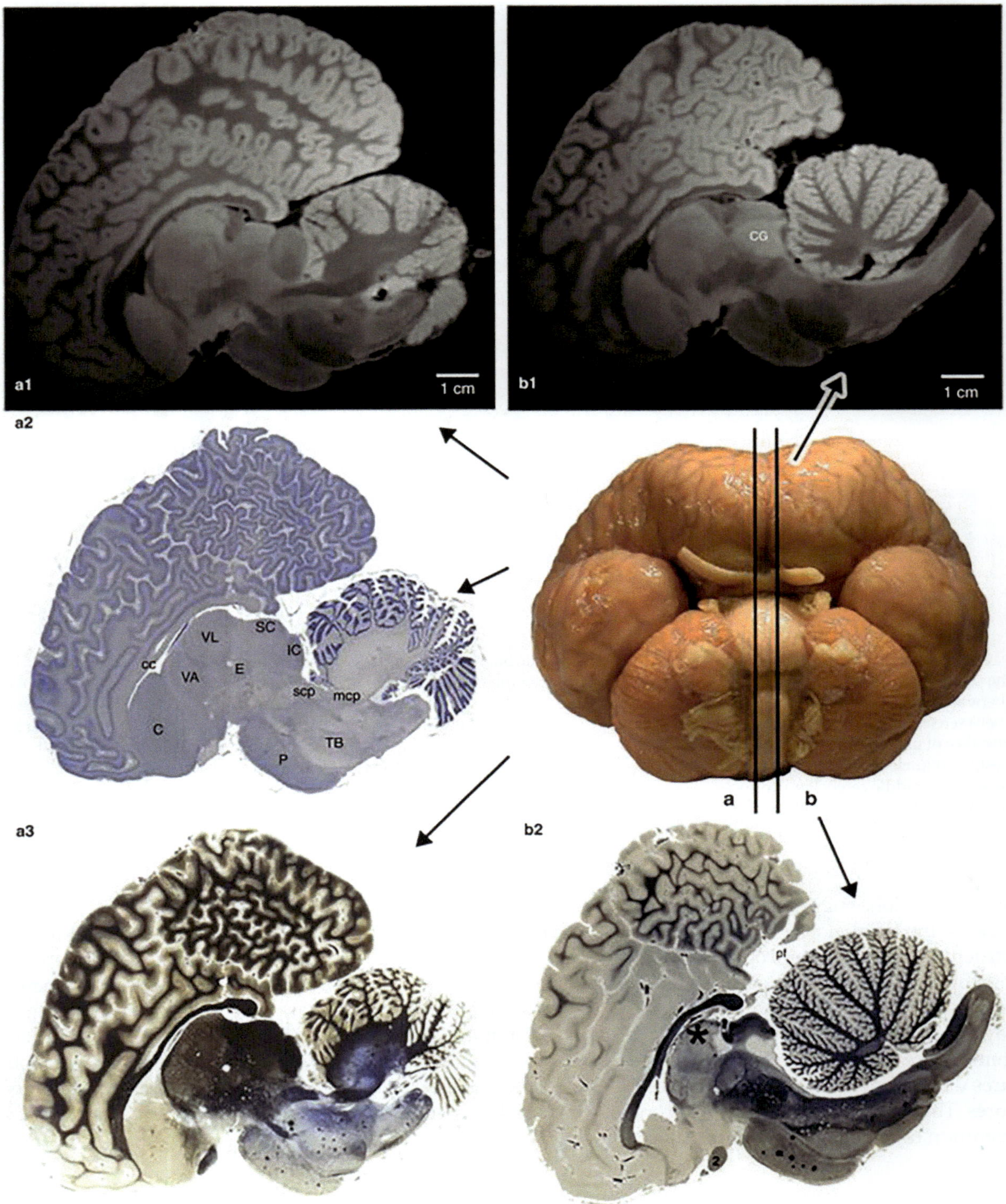

FIG. 8 Near-midsagittal scans and sections through the medial wall of the telencephalic hemisphere and the brainstem. (a1–a3) A more lateral parasagittal plane against (b1–b3) a more medial parasagittal plane. a1 and b1 are MR scans; a2, Nissl section, a3 and b2 are sections stained for fibers (Heidenhain-Woelcke). Important details are the caudate nucleus (C), subnuclei of the thalamus (VA, VL), elliptic nucleus (E), pons (P) as well as nuclei of the auditory system: trapezoid body and nucleus (TB), nucleus of inferior (posterior) colliculus (IC). White asterisk in b2: commissural complex. The cerebellum (Cer) is seen together with its superior (rostral) and middle peduncle (scp, mcp). cc, corpus callosum; SC, superior (rostral) colliculus. Black dots in histologic sections: staining artifacts. *Modified from Oelschläger et al. (2008). From Cozzi et al. (2017) [originally Fig. 6.30, p. 248].*

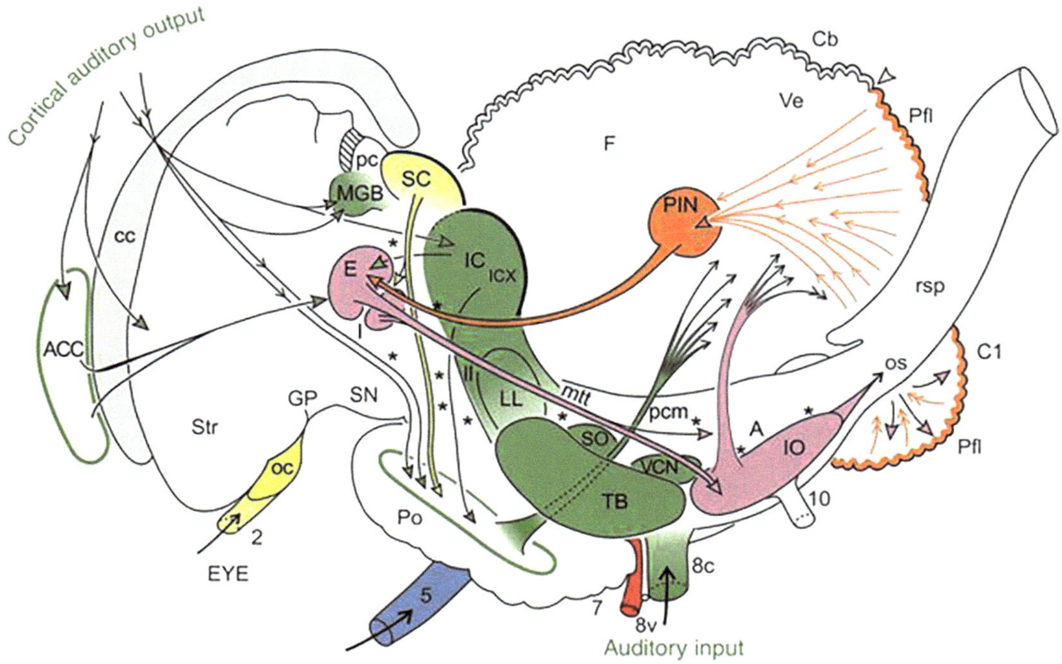

FIG. 9 Simplified scheme of the feedback loops in the dolphin brainstem most important for premotor information processing. The auditory system (green) as the dominant sensory system is the didactic center here. Other sources of sensory information are the visual system (yellow) and the trigeminal (blue), which, for the most part, are left out here. The auditory brainstem (cochlear nuclei, trapezoid body, and nucleus, superior olive, lateral lemniscus, and nucleus) gets information from the vestibulocochlear nerve (8) as well as from the auditory cortex in the telencephalic hemisphere. This extended cortex feeds into the pontine nuclei, the anterior cingulate cortex (ACC), an integrative part of the limbic system as well as in the elliptic nucleus, which is associated with Cajal's interstitial nucleus, a center involved in rotational axial body movements (turns) in terrestrial mammals. The auditory input into the pontine nuclei is projected into the cerebellar cortex (paraflocculus) and to cerebellar nuclei (PIN, posterior interposed nucleus) and via the superior cerebellar peduncle to the elliptic nucleus. From here, multisensory input from various sources is projected to the inferior olivary complex and further to the cerebellar cortex, and returned to the elliptic nucleus via the cerebellar nuclei. A, nucleus ambiguus; ACC, anterior cingulate cortex; C1, first cervical nerve; Cb, cerebellum; cc, corpus callosum; E, elliptic nucleus (Darkschewitsch); F, nucleus fastigii; GP, globus pallidus; I, interstitial nucleus of Cajal; IC, inferior (caudal) colliculus; ICX, external cortex of IC; IO, inferior olive; ll, lateral lemniscus; LL, nucleus of the lateral lemniscus; MGB, medial geniculate body; mtt, medial tegmental tract; oc, optic chiasm; os, olivospinal tract; pc, posterior commissure; pcm, pedunculus cerebellaris medius; Pfl, paraflocculus; PIN, posterior interposed nucleus; Po, pons; rsp, reticulospinal tract; SC, superior (rostral) colliculus; SN, substantia nigra; So, superior olive; Str, striatum; TB, trapezoid body; VCN, ventral cochlear nucleus; Ve, vermis; 2, optic nerve; 5, trigeminal nerve; 7, facial nerve; 8c, cochlear nerve; 8v, ventricular nerve; 10, vagus nerve; ***, periaqueductal gray and reticular formation. *Modified by Jutta Oelschläger. From Cozzi et al. (2017) [originally Fig. 6.60, p. 296].*

The topography of the white matter and therefore of the descending and ascending pathways will be discussed with the motor and sensory systems (see below).

The spinal cord is surrounded by the *retia mirabilia*, that receives blood by the thoracic arteries and supply the brain (Viamonte et al., 1968). The *retia mirabilia* are blood sponges contained within a fixed space (the vertebral canal) and may be essential for extended breath-holding dives, as the repeated dorso-flexion of the flukes induces changes in the abdominal and thoracic pressure that, in turn, moves blood from the thorax into the *retia* and consequently into the brain circulation (Bonato et al., 2019).

Motor systems

Motor cortex and control of body movement

Mimic muscles are virtually absent, and the predominant muscles of the head are those related to the control of the nasal complex including the air sacs and jaw opening and closing (but not chewing) (Mead, 1975; Green et al., 1980; Cozzi et al., 2016). Locomotion in the water is based on dorsiflexion of the fluke induced by the powerful vertebral muscles, that are also capable, together with thoracic and abdominal musculature, to induce rotation and spinning

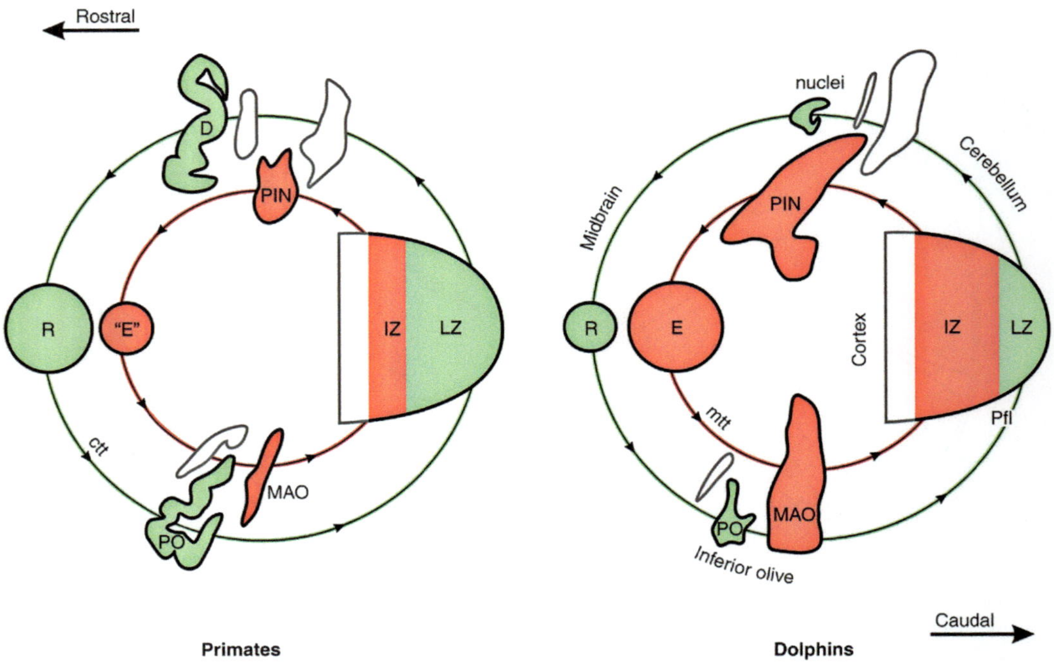

FIG. 10 The central premotor feedback loops in the brainstem of primates and dolphins. In mammals, there are two loops that run in parallel in both groups of animals: the outer ring (in green) consists of the red nucleus (R), principal olive (PO), lateral cortical zone of the cerebellum (LZ), as well as its dentate nucleus (D). The inner ring here (in red) consists of the elliptic nucleus ("E" and E, respectively; nucleus of Darkschewitsch), median accessory olivary nucleus (MAO), intermediate C2 zone of the cerebellar cortex (IZ), and the posterior interposed nucleus (PIN). In both groups of animals, the components of one loop show an alternative quantitative correlation to those of the other loop; whereas primates have the red nucleus and the conjoined elements large, dolphins show hypertrophy in the elliptic nucleus and its associated elements. This principal deviation in the two groups of animals may be related to a different organization of the extrapyramidal system with respect to the motor innervation of the body stem (see also Fig. 6.52). *Redrawn and modified by Jutta Oelschläger after Striedter (2005). From Cozzi et al. (2017) [originally Fig. 6.42, p. 262].*

of the body. All of these changes bring a transformation and a reduction of the nervous control of the musculature of the limbs, and a shift toward fine control of the nasal musculature for breathing and vocalization.

Prevalence of the extrapyramidal system

The need for fine control of the movements of the hand and generally of the limbs is one of the main evolutionary reasons for the existence of monosynaptic pyramidal pathways of primates (Fig. 12). Consequently, the current opinion is that multisynaptic extrapyramidal pathways constitute the prevailing motor control system in dolphins, and possibly also in larger toothed and baleen whales (Cozzi et al., 2017). Old electrophysiological experiments identified the motor cortex in a very frontal position (Lende and Akdikmen, 1968).

Terrestrial Perissodactyls and Cetartiodactyls (the closest relatives to whales and dolphins) require control of complex repetitive motor schemes, generally organized by pattern generator centers in the brainstem. Multisynaptic extrapyramidal pathways are the prevailing motor systems in cetaceans, a logical consequence of the absence of the pelvic limbs and the consistent reduction of movements of the distal segments of thoracic limbs. However, recent studies suggest that the presence and importance of pyramidal pathways in terrestrial Cetartiodactyls may have been underestimated (Peruffo et al., 2019). We emphasize that the precise extent of the ipsi- and contralateral pyramidal tracts in the brainstem and spinal cord of dolphins and whales is yet to be determined.

Blood supply

The blood supply to the mammalian brain derives from a series of arteries that constitute a (more or less complete) circle on its ventral surface (Purves et al., 2001). In many species, the cerebral circle receives blood from the left and

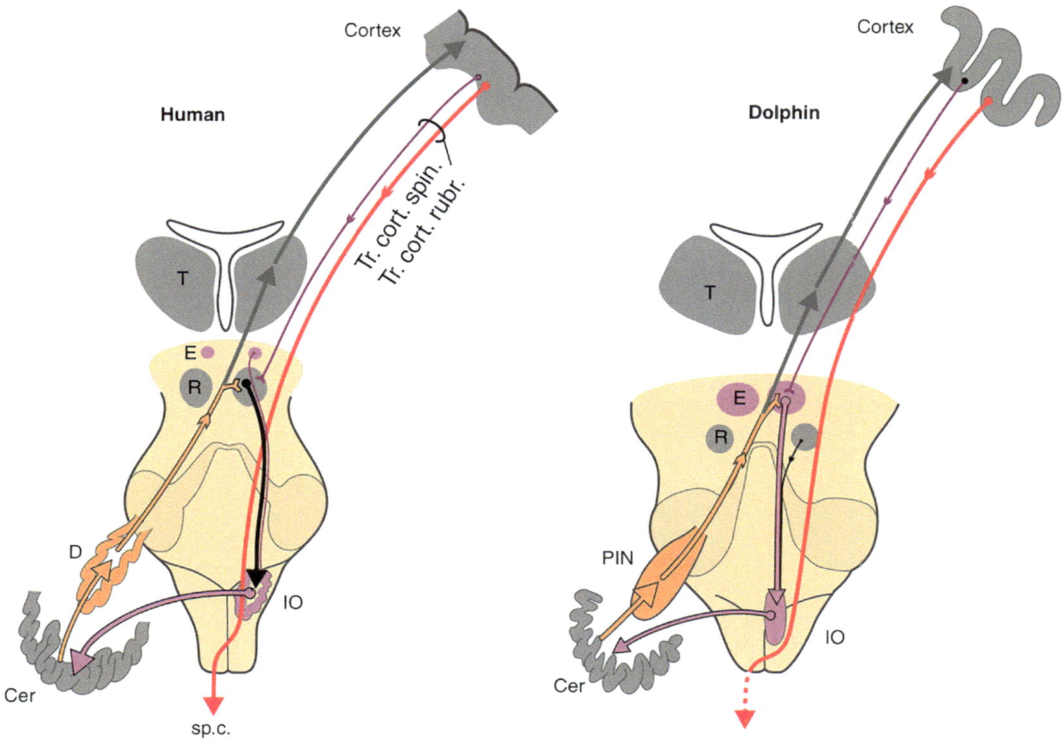

FIG. 11 Simplified scheme of the two central premotor feedback loops in the brainstem of primates and dolphins (cf. Figs. 6.41 and 6.42). In the human, the red nucleus (R) is the main gate for extrapyramidal (premotor) information processing. The large size of this nucleus is correlated with a similar dimension of the brain structures following in this loop, i.e., the inferior olivary complex (IO, principal nucleus), the cerebellar cortex (Cer, lateral zone of hemisphere), dentate nucleus (D), and back to the red nucleus (R). In the dolphin, a sister loop running in parallel is engaged in the extrapyramidal innervation projecting to the motor neurons in the ventral horn of the spinal gray matter. Here, the elliptic nucleus (E) is the main gate for extrapyramidal (premotor) information processing; and the other members in this sister loop are the medial accessory subnucleus of the inferior olivary complex, the intermediate C2 zone of the cerebellar cortex (paraflocculus), which projects to the posterior intermediate nucleus (PIN, syn. Globose nucleus of primates) and back to the elliptic nucleus. However, whereas in primates the size of the two "rivaling" gateways is extremely different (very large size of red nucleus, very small nucleus Darkschewitsch [ED]), the situation is different in dolphins. Here the elliptic nucleus (E; syn. Nucleus Darkschewitsch) is about as large as the red nucleus in primates, but their own red nucleus is comparatively well developed. It is possible that dolphins (like hoofed animals) use both main gates in extrapyramidal (premotor) information processing. They rely heavily on these systems with respect to mass movements of their body stem while their pyramidal system is underdeveloped to lacking in the spinal cord. A similar feedback loop in mammals is known using the nuclei pontis instead of the red nucleus or elliptic nucleus for the additional integration of sensory (particularly auditory) information. *Modified by Jutta Oelschläger from Duus (1976). From Cozzi et al. (2017) [originally Fig. 6.43, p. 263].*

right internal carotid arteries and from the basal artery. Three cerebral arteries (anterior, middle, and posterior) and one cerebellar (anterior) arise from each side of the circle and go deep into the cerebral tissue. Moreover, one or two cerebellar arteries (for each side), that stem directly from the basilar artery, complete the cerebellar circulation. However, the blood supply system in cetaceans is different than in most mammals (McFarland et al., 1979; Cozzi et al., 2017) and follows a pattern more similar to the terrestrial ruminants (Fig. 13).

The internal carotid system

The common carotid artery splits into the (left and right) external and internal carotid, this latter being one of the vessels essential for the cerebral circle. In dolphins, the internal carotid artery obliterates soon after birth and does not vascularize the brain. Most of the blood of the cerebral circle derives from meningeal arteries that originate from the cervical part of the spinal *rete mirabilis* (Nagel et al., 1968). The absence of a pervious internal carotid does not compromise the blood supply to the brain. When cerebral blood supply derives from a *rete* and not from carotid artery, any effect induced by changes in pressure is dampened by the nature of the arterial mesh, a fact that may have an important bearing on species that move rapidly up and down the water column with continuous variations of external environmental pressure.

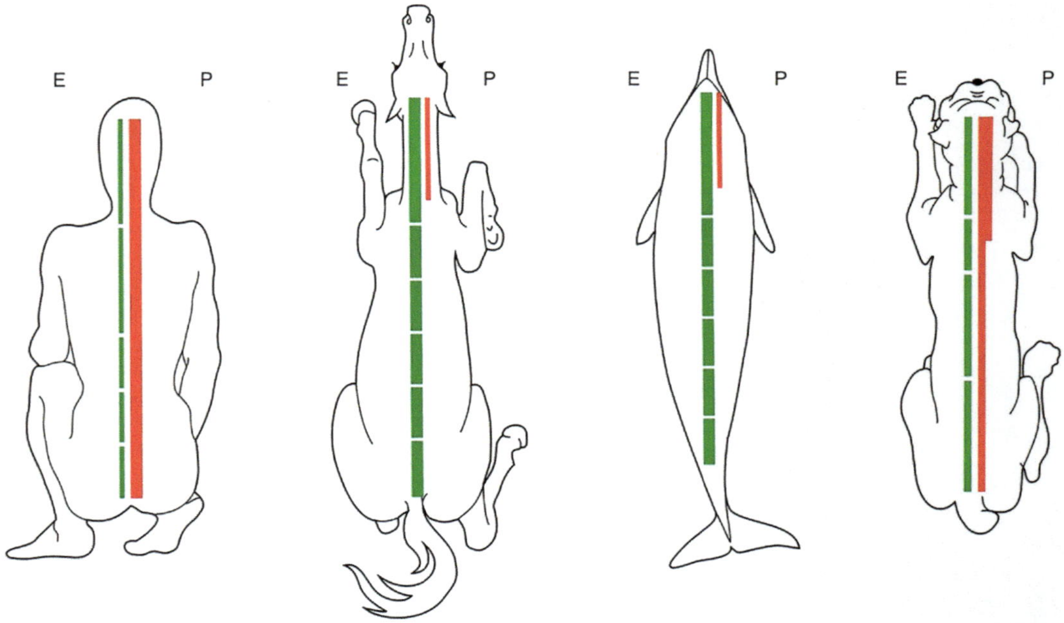

FIG. 12 Comparison of the pyramidal system (P, red) and extrapyramidal (E, green) systems in different mammals. From left to right: human, horse, dolphin, dog. Spinal apparatus brought to about the same length. Note that these spinal systems are very similar in the horse (hoofed animal) and the dolphin (cetacean), in contrast to the situation in the human and the dog. The latter two mammals are rather similar to each other, although the pyramidal tract is weaker in the dog and the extrapyramidal system somewhat more pronounced in the human. *Redrawn and modified by Jutta Oelschläger from Dyce et al. (1991). From Cozzi et al. (2017) [originally Fig. 6.52, p. 276].*

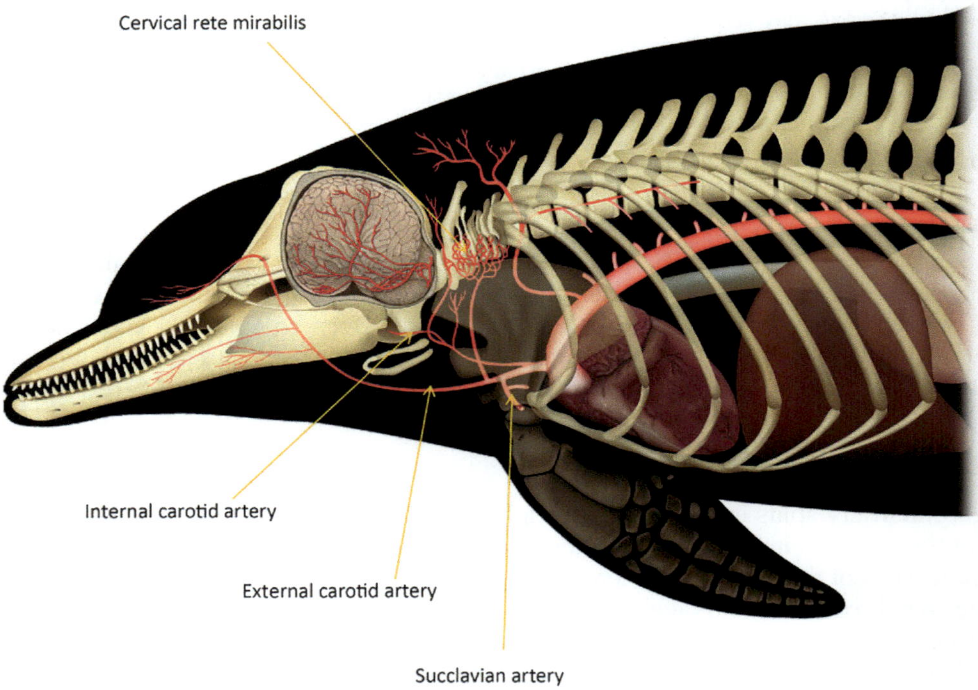

FIG. 13 Vascularization of the brain of the bottlenose dolphin (*Tursiops truncatus*). Note that in the adult the internal carotid artery is obliterated, and the brain receives most of its blood supply from the cervical retia mirabilia. *Illustration by Uko Gorter.*

The spinal and thoracic retia mirabilia

A *rete mirabilis* (plural *retia mirabilia*) is a complex mesh of small arteries surrounded by veins. The spinal *rete mirabilis* (or *retia mirabilia*, if the vertebral canal does not contain a unique rete but a sequence of consecutive interconnected ones) receives blood from the thoracic *rete mirabilis*, supplied by intercostal arteries (Bonato et al., 2019). The spinal *rete mirabilis* is placed within the closed walls of the vertebral canal. Downward movements of the flukes, mostly caused by contraction of the *rectus abdominis* muscle and the other muscles of the abdominal wall, induce an increase in abdominal pressure, and ultimately push the diaphragm forward. This latter movement, in turn, increases the intrathoracic pressure. When a dolphin or a whale is swimming at depth, external pressure constricts the thoracic walls, and the ribs and adjacent musculature are pushed inwards. Since the increased intrathoracic pressure caused by movements of the diaphragm cannot be absorbed by the expansion of the thorax because of the environmental pressure, the final effect is to squeeze more blood into intercostal arteries and then into the *retia mirabilia*. Considering that the spinal *retia* are housed within the vertebral canal, which cannot expand, blood is driven into the meningeal arteries that originate from the cervical part of the *retia*. Therefore, a diving animal faces bradycardia, but a certain rhythmicity or pulsatile supply of blood to the brain is guaranteed by the chain of mechanical actions that start with movements of the fluke and terminate with circulation in the meningeal arteries that supply the cerebral circle (Lillie et al., 2022).

Vascular shunts and the theory of the half-brain

A series of experiments demonstrated that, during diving, dolphins are able to shift part of the arterial blood supply only to active parts of the brain (Houser et al., 2010). The visceromotor control of said mechanism has not been clarified, but the advantages of this process are obvious. Since extended breath-holding induces a series of negative consequences, including the potential onset of mild to severe brain hypoxia, the capacity to maximize the available circulating oxygen would constitute a major evolutionary adaptive factor.

Functional organization of the cetacean central nervous system

Most of the present knowledge on the regions and connections of the brain is based on tracing and electrophysiological studies on rodents and monkeys, and pathological observations in humans. Experimental methods have been seldom employed in cetaceans.

Limits to research

The use of invasive methodologies on dolphins is nowadays prevented by broadly prohibitive ethical and legal restrictions. Some pioneering studies gave us some insights into the electrophysiological characteristics of the dolphin brains but cannot be verified or replicated (Fig. 6). There are very few studies on non-infective pathologies of the cetacean brain, so in this case, a correlation between a certain lesion and the consequent loss of function is difficult to make.

Furthermore, the brain of cetacean species significantly larger than adult humans is (and was) simply impossible to reach for measurements in live individuals, thus limiting the existing knowledge to smaller species.

Anatomical techniques

Studies based on anatomical dissections, serial sections, and tracing

Early knowledge on the cetacean brains derived from anatomical dissections, and studies on thorough series of consecutive sections (Jansen and Jansen, 1969; Morgane et al., 1980; Pilleri et al., 1980). As briefly discussed earlier in this chapter, these studies quantified the size of the brain and some sub-structures and described gyrification and cell distribution patterns. They did not offer great insight into functional specialization of the dolphin brain in specific sensory and cognitive domains. The advent of new, promising non-invasive techniques (see below) may greatly benefit from traditional anatomical dissection for precise identification of structures and connections. Therefore, the use of tractography may help us understand the structure of the main pathways in the cetacean brain, and, consequently, define a more detailed localization of functional territories in the cortex.

Role of magnetic resonance imaging studies

The advent of magnetic resonance imaging allowed some steps forward from purely anatomical description on the cadaver or isolated brains. Comparisons were made among cetacean species and with terrestrial mammals and allowed the basis for a far more general knowledge of the brain of marine mammals (Marino et al., 2003; Oelschläger et al., 2008, 2010; Alonso-Farre et al., 2015; Wright et al., 2017), as no dissection was required and topographical relationships with other body structures during the course of their ontogenesis (including skull bones, bony parts of the ear, and nasal sacs) were easier to understand (Rauschmann et al., 2006; Haddad et al., 2012).

Studies based on diffusion tensor imaging

The use of computerized diffusion tensor imaging (DTI) applied to high-quality MRI allowed identification of tracts and connections in the live brain, or, with new imaging protocols, in perfused brains. This technique is now being applied and is producing novel findings. Berns et al. (2015) demonstrated a putative pathway from inferior colliculus through medial geniculate nucleus to dorsal temporal lobe, which neatly matches evidence from terrestrial mammals, as opposed to prior chemical tracer findings in dolphins indicating a much more dorsal/occipital location for primary auditory cortex (Revishchin and Garey, 1990). Prior mapping of the sensory cortex of cetaceans has shown intense changes in the composition of the cortical column (Morgane and Jacobs, 1986a, b). More work will be required to determine the location of primary auditory cortex and its relation to auditory projections. Wright et al. (2018) examined auditory motor projections in the dolphin cortex, and found a striking asymmetry that may bear relevance for understanding auditory production and hearing in dolphins. Even more recently, Orekhova et al. (2022) have used DTI coupled with histology to examine auditory nuclei projections and dolphin's neurochemical response to hypoxia. In the coming years, we will surely see some improvement in our knowledge of the functionality of the dolphin and whale brain.

Presently, DTI studies on the cetacean brain are limited either because of the dimensions of the magnet (cetacean heads or even some isolated brains are too large for the bores of stronger magnets) or because the body weight of live dolphins (or whole cadavers) is generally close to the upper limit of tolerance of the bed of most MR scanners. Postmortem brains of all but the biggest species can now be successfully imaged in 3 T scanners. Examining the connectivity of smaller subsections of brain, including brain stem nuclei, is likely also technically possible with post-mortem material, but such studies have yet to be carried out.

The role of cranial nerves

A number of studies have been conducted on the projections and size of cetacean cranial nerves. The relevant size of these nerves may be useful for driving inferences and hypotheses about the relative importance of different sensory modalities. Classically, the nerves of the brain are numbered I to XII from rostral to caudal. Some of these cranial nerves (CNs) are purely or largely somatomotor, such as the nerves of the eye muscles (oculomotor (CNIII), trochlear (CNIV), abducens (CNVI)), and the hypoglossal nerve (CNXII). Others represent the mixed innervation (trigeminal (CNV), facial (CNVII), glossopharyngeal (CNIX), vagus (CNX), accessory (CNXI)) of the pharyngeal arches, which constitute the visceral part of the head. Of particular interest are the pure sensory CNs innervating the three main cranial sensory organs of mammals, the nose, eyes, and ears: the olfactory (CNI), optic (CNII), and vestibulocochlear (CNVIII) nerves (marked in gray in Table 1; Cozzi et al., 2017).

Comparative studies of vertebrates demonstrate that the relative size of the CNs, expressed in diameters or fiber numbers, are related to the relevance of the corresponding system regarding the adaptation to the environment (Romer and Parsons, 1986). Table 1 shows that in humans the optic nerve (CNII) is by far the largest of the three purely

TABLE 1 Relative fiber counts of CNs in dolphin (*Tursiops truncatus*) and human (*Homo sapiens*) (Glezer, 2002, b).

	CNI	CNII	CNIII	CNIV	CNV	CNVI	CNVII	CNVIII	CNIX	CNX	CNXI	CNXII
T.truncatus	0	29.6	1.8	0.6	29.6	0.4	9.9	19.8	4.0	1.5	1.5	1.4
H.sapiens	0.8	75.5	1.8	0.3	10.6	0.5	1.0	3.8	0.8	2.6	1.9	0.6

Values are given as percentages of the total fiber count of all cranial nerves in the species.

sensory nerves while in dolphins it is the vestibulocochlear nerve (CNVIII). Since the CNVIII consists mainly of auditory fibers—the vestibular system should be of minor importance in dolphins (Kern et al., 2009)—the hearing systems may then be the most important sensory system.

As mentioned earlier, dolphins do not have a CNI, an olfactory bulb, and tract. Thus, the sense of smell is not existent in most if not all toothed whales which is in contrast to many other marine vertebrate carnivores which tend to rely heavily on olfaction to locate prey—sharks may be a famous example (Tester, 1963).

Medial to the CNI and the olfactory bulbs, there are fibers of the terminal nerve (CN0) in many vertebrates; thus associated with the chemoreceptors in the nose (Sonne et al., 2017). The terminal system with its neurons producing gonadotropin releasing hormones (GnRH) is the initial part of the hypothalamo-hypophyseal-gonadal axis which is a critical part in the regulation of the reproductive system (among others) (Peña-Melián et al., 2019). Fibers of the terminal nerve were verified in the bottlenose dolphin (Ridgway et al., 1987) but their function is enigmatic. An extended overview of this system can be found in Cozzi et al. (2017).

As stated above, CNII is the largest cranial nerve (more than 3/4 of all CN fibers) in humans and many other primates. This fact demonstrates that humans rely heavily on vision in nearly all our routines. In contrast, dolphins have a thick CNII although it represents only 1/3 of all their CN fibers (Table 1). This may be related to some species of dolphins relying less on vision than audition in dark and turbid waters. Dolphins use sight for orientation at day time but they rely more on echolocation in murky waters or at night. This explains the thickness of CNVIII (see above). In this context, it is noteworthy that the trigeminal (CNV) and the facial (CNVII) nerves should have relevant functions in dolphins since they also contain high numbers of fibers (Table 1). These high numbers of the trigeminal nerve may correlate with a sensory (CNV; 29.6% of CN fibers) and the motor (CNVII; 9.9% of CN fibers) control of the hypertrophied nasal sound generator and transmitter used for communication and echolocation.

The relative thickness of the trigeminal nerve (CNV) may be explained by two additional senses not known from primates. First, the electric sense (see above), which receptors are known to be the vibrissal crypts at the rostrum of dolphins, must be mediated by this nerve since the maxillaris nerve of CNV has the only sensory nerve fibers in this region. Second, the receptor of the magnetic sense is thought to be part of the dura mater near the occipital pole of the brain (Zoeger et al., 1981). Since most of the dura mater if not all is innervated by the CNV magnetosensation also may be a part of the trigeminal system. However, as mentioned above, so far nothing is known about the representation within the cetacean brain of both systems: electrosensation and magnetosensation.

Physiological techniques

Studies based on electrophysiology

Electrophysiologic studies performed by Russian scientists allowed a clear comprehension of the mechanism of sleep in dolphins, showing that cerebral activity can be synchronized within half-brain at a time for a few minutes (Supin and Mukhametov, 1986). This feature, which is known only from few mammalian species, allows dolphins to maintain vigilance while simultaneously letting part of the brain rest.

Stimulation studies performed in the 1960s on live anesthetized bottlenose dolphins yielded the only available maps of the motor control areas of the brain and confirm that boundaries set in terrestrial species by sulci apply also to the equivalent gyri of the dolphin brain (Kesarev and Malofeeva, 1969b).

Electroencephalogram and related techniques

Dolphins have proved susceptible to non-invasive electroencephalographic recording of multiple brain regions. The majority of this work has focused on hemispheric symmetry and asymmetry in processing and evoked auditory responses. EEG work has replicated previous findings from invasive electrophysiology demonstrating unihemispheric sleep, specifically, that during phase 3 sleep, when delta waves make up 50% or more of the signal, only one hemisphere at a time enters this state (Ridgway, 1986; Howard et al., 2006). Conversely, activity appears relatively symmetrical between hemispheres during waking (Ridgway, 2002).

Pioneering studies by a group of Russian scientists (Mukhametov et al., 1977) discovered that dolphins can sleep with only half their brain at a time, as briefly outlined above. The finding was subsequently confirmed (for details and references see Lyamin et al., 2008). Later studies confirmed the presence of electrical asymmetries in the neural activities of an anesthetized bottlenose dolphin (Ridgway, 2002). Other subsequent studies clarified the organization of the sleep-related neural system in small and large cetaceans nature of sleep (Dell et al., 2016a, b). There is currently no published study on sleep activities on larger species, including, beaked, sperm, and baleen whales.

As discussed in Chapter 10: Sensory physiology in delphinids, dolphins and other whales have also undergone hearing tests using the auditory evoked potential technique, in which auditory brainstem responses are recorded from surface electrodes placed by the base of a subject's skull (Picton and Durieux-Smith, 1988). Results from these methods applied to captive and even wild, free-ranging cetaceans have produced similar results to carefully controlled laboratory psychophysics methods, suggesting they are an accessible and easy way to examine hearing in whales, including wild whales (Ridgway and Carder, 2001; Nachtigall et al., 2007).

fNIRS studies

The use of functional near-infrared spectroscopy (fNIRS) methodology, a technique proven extremely useful in human newborn infants, has been applied in vivo to the sheep and a few other terrestrial mammals to get an insight into the functions of the living brain (Wilcox and Biondi, 2015; Kim et al., 2017). fNIRS is based on the capacity of the instrument, applied to the surface of the head, to discriminate between oxygenated and nonoxygenated hemoglobin carried by blood vessels a few centimeters below. An increase in oxyhemoglobin in a certain brain area corresponds to an intensification in neural activity. The initial enthusiasm for the technique has been partially dampened by the difficulty to precisely identify the area whose activity is recorded by the instrument at the surface. The position of the brain in several mammals is such that the fNIRS has difficulty in reaching cerebral areas not immediately connected to motor activities, not to mention the potential disturbing signal given by the intensely vascularized skin and heavy musculature.

To date, a single published study has applied fNIRS to a living pinniped (McKnight et al., 2021). Preliminary tests performed in our laboratories indicated that the instruments currently available do not penetrate deep enough to record signal from the cerebral vessels of dolphins. However, improvements of the technique may lead to some interesting discoveries in cetaceans.

Modern studies based on behavioral experiments

While physiological studies of whale neural processing have been limited, there are numerous studies of dolphin behavior and cognition that may provide insight into the type of neural processing dolphins undertake. An important caveat: there is no broad consensus in the comparative cognition research community for how to assess intelligence and how to definitively compare capabilities of different species. While attention-grabbing findings regarding certain cognitive abilities in select species abound, it is very difficult to prove a negative—we cannot be certain that other species (either not tested, or not tested appropriately) do not share the capabilities in question. To illustrate this point, emerging research in insects and other invertebrates has identified many of the behaviors long believed to be hallmarks of vertebrate intelligence, including memory, communication, and decision-making (Simons and Tibbetts, 2019). Some comparative researchers have gone so far as to suggest that there is no substantive difference in cognitive capability across large swathes of animals (Macphail, 1985). While we believe that different ecological and social factors will necessarily drive different cognitive capabilities, including in cetaceans (e.g., Bauer et al., 2020) it is nonetheless important to acknowledge that, while behavioral work may tell us what dolphins can do, it seldom tells us what they are uniquely capable of.

While confidently ranking dolphin intelligence in relation to other species is difficult, they show behavioral capabilities in sound production and hearing that appear fairly rare in the animal kingdom, and that do bear directly on their ability for mental representation. Dolphins are robust vocal production learners, able to learn new vocal sequences and patterns of both whistle contour and buzzes from hearing (Janik, 2014). While the extent to which their vocalizations are used in communication remains uncertain, a number of dolphin species show evidence of individually identifying contact calls ("signature" whistles) and whales apparently possess regional "dialects" (Winn et al., 1981; Filatova et al., 2012).

Their capability for echolocation has been well studied, and, in part due to the speed of sound underwater, successful identification of stimuli (and, impressively, stimulus characteristics such as shape, size, and texture) requires extremely rapid auditory integration and processing. This likely bears on features of dolphin auditory processing early in the ascending auditory stream (as noted above, brainstem and midbrain auditory nuclei are clearly hypertrophic in dolphins). The use of complex auditory processing for communication signals and echolocation may also be relevant to segregation of auditory processing later in the neural hierarchy, for which there is some behavioral evidence matching preliminary diffusion connectivity findings (Ridgway et al., 2015).

Notably, dolphins are also capable of cross-modal representational referencing, correctly identifying objects with subtle feature differences experienced only with vision or echolocation when subsequently presented to only the other sense (Harley et al., 1996). Dolphins are certainly not unique in this type of cross-modal recognition (humans do this with touch/sight, even at birth, e.g., Streri and Gentaz, 2003), but it does emphasize the high resolution of their three-dimensional auditory representation.

In addition to complex auditory processing, dolphins also show apparently high capability for certain types of social cognition. They are robust mimics (Bauer and Harley, 2001), both in auditory and visuomotor domains, and many species live in large, gregarious, fission-fusion groups with complex, long-term social relationships, and complex social foraging behaviors (Connor and Mann, 2006; Connor, 2007). Again, none of these capabilities are apparently unique, but they are notable. There is some evidence that dolphin encephalization correlates with group size and social complexity (Fox et al., 2017).

Much attention has fallen on the apparent ability of dolphins to demonstrate mirror self-recognition (Reiss and Marino, 2001). While the original study did not control for the possibility that dolphins treated the mirror image as a conspecific, a clever new study does show a clear distinction between how dolphins treat their reflection as opposed to how they interact with conspecifics (Loth et al., 2022) While self-recognition likely relates to complex abstract representational capability, it has been noted that mirror recognition tests (which have also been "passed" by animals such as goats (Hals, 2016) and involve behavioral alterations that can be observed in insects (Riojas-Schnier and Toth, 2022)) may more directly relate to visuomotor contingency testing than robust and abstract self-concept (Brandl, 2018).

Although impressive in the experimental design and elegance, some of the dolphin mirror studies may also contain an involuntary hidden bias. Dolphins rely heavily on their sight when outside the water, and their image is something they may be relatively familiar with, as they can mirror themselves in the water surface while swimming on the side, although the mirror image has no meaning for their echolocation system. They therefore may have extensive experience interacting with mirrored self-images. A definite conclusion is hard to reach.

While dolphins and whales demonstrate many other notable cognitive capacities, including fairly robust memory (despite an apparently atrophic limbic lobe) and decision-making (for review see: Herman, 2010; Janik, 2013), they do not clearly distinguish themselves cognitively outside the area of auditory and vocal processing and social interaction.

The opinion of the authors of the present chapter is that dolphins and whales evolved the best possible brain to live in the water within their specific trophic niche (or vast habitat), and that comparisons with species that live on land may be impossible, or simply conceptually wrong to propose. Until comparative psychologists can alight on a series of tests that definitively show general intelligence in nonverbal animals, gross comparisons of cognitive capability have little value beyond exciting the public.

The diving brain

The central nervous system and diving

Breath-holding may affect the human central nervous system in many different ways. The absence of oxygen intake stimulates an increase of blood pressure and changes in the heart rate. The subsequent progressive increase of carbon physiological mechanism currently dioxide levels in the blood may have serious consequences if anaerobic metabolism is not interrupted. Breath-holding *diving* further adds external pressure to the picture, and changes due to descent and then ascent in the water column increase the risks of drowning and even decompression sickness in very deep and particularly long dives. However, dolphins and whales have adapted to living in the water and developed some mechanisms to stand repeated dives and their potential lethal effects on the CNS and other organs.

The effects of pressure on the brain

Pressure does not directly affect the brain, protected by the brain cage, because of the liquid content represented by blood. Theoretically, gas pressure should not constitute a risk factor, as air breathed at the surface cannot expand during ascent to a volume higher than the original one. However, a too rapid ascent may induce the formation of bubbles since the velocity of bubble formation and expansion may surpass the capacity of the surrounding tissues, including nervous tissue, to absorb them.

Decompression sickness, nitrogen narcosis, and oxygen toxicity

Apparently, the body of dolphins and whales can balance absorption of gases during descent and readjust for their expansion during ascent. The instances of decompression sickness registered in cetaceans are rare (Dennison et al.,

2012; Fernández et al., 2017), and generally occurred when external factors intervene. Military sonar beams may intercept deep-diving whales (as described for beaked whales) and convey so much energy as to induce a change of status in gases (= air) dissolved in the blood. The bubbles that ensue may affect diverse organs and systems, including the brain, where bubbles may form in the brain stem and white matter, leading to tissue damage.

Long-term consequences of deep dives are visible in the skeleton, mostly on the head of the homers and the lumbar and/or early caudal parts of the spine. However, long-term consequences of repetitive dives on the brain of whales and dolphins are not known. Beaked whales could be vulnerable to sustained and repetitive brain hypoxia since these animals showed more extensive pathological changes such as β-amyloid plaques and antineurofibrillary tangles, hallmarks of Alzheimer's disease in humans, than species diving less deep (Sacchini et al., 2020). Nevertheless, these pathologies (plaques and tangles) were described also as spontaneous changes in the brains of odontocetes recently (Vacher et al., 2022).

Thermoregulation and the development of the cetacean brain

One diffuse theory states that one of the key factors responsible for the evolution of the human mind is the need to adequately control a hand endowed with independently movable fingers and a thumb capable of opposition. The situation clearly does not apply to cetaceans since their fingers are incapable of independent movement. However, the degree of fine-grained motor control they can implement for vocal production is notable, and there is evidence of dolphin tool use (Wild et al., 2019).

The unknowns of the cetacean nervous system

The reader has surely realized that we have left several blank spots and unsolved questions in our paper. In fact, the many unknowns of cetacean research have an intrinsic value, because they force us to steer our attention toward uncharted venues and perhaps uncomfortable paths. As discussed above, new techniques and new applications of techniques in brain imaging and behavioral research may clarify some of these areas.

Here we tentatively point out some examples of these unsolved issues, as a further stimulus toward research and objectivity.

The development of the cetacean brain

The embryonic and fetal development of dolphins has been well described (for references see Štěrba et al., 2000; Cozzi et al., 2017), but studies dedicated to the development of the cetacean brain are relatively few (i.e., see Buhl and Oelschläger, 1988; Kükenthal and Ziehen, 1893; Oelschläger and Kemp, 1998). It would be useful to understand when some of the key features of the brain start to appear and the cortical column takes shape in different cetacean species. This is in line with the question of how the cetacean brain develops shortly after birth. Whales and dolphins are extremely precocial mammals but, as mentioned earlier, they have to learn significant capabilities during their life span. Knowledge of their post-natal brain development would thus explain the timing of the species' learning curves and thus contribute to potent protective measures.

The question of body size

The brains of the large whales (beaked, sperm, baleen) decay very rapidly after death, because heat accumulates within the brain cavity and thus accelerates tissue deterioration (Kruger, 1966). Therefore, there are relatively few studies on the structure of the brain of these species (see references in Oelschläger and Oelschläger, 2009; Cozzi et al., 2017; Huggenberger et al., 2019), and virtually nothing on neurophysiology. However, some of the large cetaceans are deep divers, and their nervous system is capable of features hard to explain because of the prolonged hypoxia: terrestrial mammals could not sustain prolonged neural activity under these conditions. Some cetacean species may hold their breath for more than one hour, and it is difficult to explain how they are still able to perform the higher brain functions required for orientation and hunting during the long dives. Potential explanations (discussed elsewhere in this book) include high affinity for oxygen of the cetacean myoglobin, the existence of the retia mirabilia, and so forth. However further studies are required. Even the vascular system of the large whales is poorly described, and this lack of knowledge hinders further discoveries and insights.

One way out of the situation, even considering the current lack of innovative techniques adapted to study such large species, lies in cooperation and sharing of specimens. There are very few available brains of sperm whales, bottlenose whales (*Hyperoodon* spp.), right whales, blue (*Balaenoptera musculus*) and fin whales, and even humpback whales (*Megaptera novaeangliae*). If newly recovered specimens could be treated according to an internationally agreed protocol (i.e., see Knudsen et al., 2002), better grade tissue specimens could be made available to the community of scientists worldwide.

Opinionated studies

No other group of animals has attracted the attention and passion dolphins and whales do. The numerous unknowns of their nervous system have led to different opinions on their neural abilities. Since electrophysiologic studies are scarce—a few were discussed above—we base our current identification of functional areas of the cetacean brain on older studies (see references in Cozzi et al., 2017). New techniques and new paradigms of studies are required if we want to move from Galean theory to Vesalian demonstration. For example, recent fMRI studies (see references above) are leading in this direction. Nevertheless, some scientists (Lilly, 1961, 1967; Bearzi and Stanford, 2008) have claimed the higher brain functions of some cetacean species are comparable to those of apes, if not humans. In some instances, independent, provocative but well-documented opinion papers that related the development of a large brain of cetaceans to thermoregulation (Manger, 2006) have been countered by manifestos signed by scientists convinced of the contrary (Marino et al., 2008). Polarization of opinion without engagement of testable hypotheses is detrimental to science.

Chemoreception: Olfaction

For obvious reasons, recent and older papers focus on the primary senses of toothed whales to interpret their adaptations to the environment (see Chapter 10). However, structures missing from the mammalian brain bauplan may be of equal value to explain these adaptations. Toothed whales have no olfactory nerve, no olfactory bulbs, and to all effects show no awareness of odors. Residual parts of the olfactory systems have been described in several baleen whales which may play a role in prey location (Thewissen et al., 2011; Bouchard et al., 2019). The changes in the olfactory apparatus are accompanied by a reduction in the development of the whole limbic lobe (Filimonoff, 1966; Jacobs et al., 1971; Morgane et al., 1982; Morgane and Jacobs, 1986a,b; Patzke et al., 2015). Notably the hippocampus, although it follows the mammalian bauplan, in a cetacean brain requires attention and is by no means an easy feature to identify for those familiar only with the equivalent found in terrestrial mammals. Recently, the subventricular zone and the hippocampus of dolphins have received specific attention because of the debated neurogenesis in adult mammals (Parolisi et al., 2017, 2018).

Here we also mention that there are old, but apparently reliable, drawings of the brain of the bottlenose whale clearly showing the presence of olfactory bulbs (Kükenthal and Ziehen, 1893). Was it the attempt to complete the drawing of a damaged brain by superimposing the anatomy of the ventral surface of a human brain, or is the bottlenose whale endowed with olfactory bulbs? And if so, why?

Although toothed whales do not have a functional olfactory system, they possess remnants of the terminal nerve, a structure potentially related to the nasal organ (Buhl and Oelschläger, 1986; Oelschläger et al., 1987; Ridgway et al., 1987).

Perception of environmental pressure and deep diving

Terrestrial mammals are seldom exposed to changes in environmental pressure, except when living at high altitudes. Humans feel variations in pressure mostly on the eardrum, because of its position intermediate between the external environment and the cavity of the middle ear. Equalization is operated by swallowing and consequent opening of the pharyngeal end of the auditory tube connected to the middle ear. The structures of the external and middle ear are extremely modified in cetaceans, and although there are indications that the lining of external auditory canal possesses lamellar corpuscles capable of feeling pressure (De Vreese et al., 2020), the location and mechanisms that convey awareness of increased external pressure remain undetected. This kind of sensation should be extremely important for beaked and sperm whales that reach great depths, with external pressure of the order of hundreds of atmospheres (up to 20 or even 30 MPa).

Neuro-hormonal control of circadian and circannual rhythmicity

Contrarily to most mammals (with the notable exception of elephants), cetaceans (with very few individual exceptions) do not possess a pineal gland. The absence of a pineal body, where the circadian hormone melatonin is produced, is difficult to explain in species that in many cases rely on long seasonal migrations. Pinnipeds have large pineal glands (Cuello and Tramezzani, 1969; Montie et al., 2009) and live in the same environment. However, melatonin is present in the blood of bottlenose dolphins (Panin et al., 2012), therefore some alternative source of melatonin must be present.

The organization of the five (or four or even three) layered cortex

Some parts of the cortex of large whales show marked reduction of neural layers. Models of the organization of a five-layered cortex are based on the agranular motor cortex of rodents and primates. However, what may apply to motor areas is more difficult to understand for sensory (including acoustic) areas. Cortical columns with very low cell density, a further reduction of the layers, or where layers are poorly defined and difficult to assess, probably require new functional models.

The question of the magnetic sense

Some preliminary indications suggest that dolphins may be sensible to magnetic fields, presumably relevant to navigation, as in birds (Zoeger et al., 1981; Bauer et al., 1985; Klinowska, 1986; Begall et al., 2014; Kremers et al., 2014) although there is no knowledge about the nature of this sense in cetaceans. The possibility that cetaceans use magnetic fields to migrate implies the existence of magnetic receptors and corresponding neural centers, not yet identified.

Differential vascularization of selected brain areas

Pioneering studies showed that, under specific circumstances, dolphins may direct most of blood perfusion to a specific lobe (Houser et al., 2010), thus getting the maximum possible oxygenation of active parts while simultaneously saving on oxygen stores. Although further studies are required to better describe and understand this amazing physiological feature, we also must understand what kind of visceromotor control is operated by the nervous system, possibly simultaneously involving the parasympathetic system to vasodilate parts of the brains and the sympathetic system to vasoconstrict others.

Brain metabolism and the long and deep dives

We already pointed out the paradox of the neural metabolism in long, deep dives. A recent seminal article linked neuron density and size of the cerebellum to dive duration in some cetacean species (Ridgway et al., 2019). To simplify the concept, the human nervous system requires approx. 750/mL/min of oxygenated blood for every 100 g of brain tissue in a resting individual (Fox, 2016). The values are based on the human brain, and therefore we have no proof that the same may apply to the brain of other species, depending also on the density of neurons and glia (more cells require more oxygen, and vice versa). But let's imagine, as a working hypothesis, that the same may be true for the deep-diving whales. We also know that neurons require a constant supply of oxygen, and in humans a period longer than a few minutes (4–10, depending on exercise, training, etc.) without oxygen may lead to serious consequences, including permanent brain damage and death. We also know that breath-holding athletes have apnea records that vary between 3 or 4 min (for the deep dives involving active swimming) and 15 or 16 min (for stating apnea). Therefore, the dives performed by beaked and sperm whales, that may last from 40 to 90 min, are completely beyond any physiological mechanism currently understood (cf. *The question of body size*).

Delphinid language

Cetaceans learn socially and their vocalizations in groups and subgroups remain stable over long periods (Rendell and Whitehead, 2010). In the 1960s, John Cunningham Lilly's work was probably the first modern research to note a tendency among dolphins to communicate with each other using complex vocalizations (Fox, 2020). At his time, Lilly was convinced that dolphins possessed a discernible and comprehensible language (e.g., Lilly, 1961, 1967). Today, it is known that trained dolphins are able to correctly follow symbolic instructions on a screen and that they easily understand the meaning of a human's pointing gesture (Herman, 2010). Only humans and dolphins appear capable of imitation in multiple modalities (Hauser et al., 2002). However, there is no clear evidence of productive syntax, the primary hallmark of human language.

The complexity of the sound generation structures in dolphins, mainly their nasal complex, are the prerequisite for multifaceted vocalizations. The strong trigeminal and facial nerves serving the nasal structures sensory and motorically, respectively, build the neural background for this multifaceted vocalization and the ear system, which is extremely adapted to the aquatic environment in toothed whales, represent the appropriate receiver. These anatomical requirements as well as the dolphin's intricate behavioral repertoire and the splendid anatomy of their large brains may represent the sensory-motor system and the conceptual-intentional system needed as prerequisite for language use. However, so far there is no evidence that cetaceans have the faculty of language in the narrow sense, i.e., an open-ended and limitless system of productive syntactic combination of units of meaning (Hauser et al., 2002).

Conclusion

The many unusual features and the evolutionary history of the cetacean nervous system continue to perplex and motivate scientists in equal measure. New methodology promises to unravel some of these mysteries in the coming decade, although the size and inaccessibility of cetacean brain samples are a serious hurdle. The work will go on. As suggested by Oelschläger (2008), the many unique features of the dolphin brain make it an excellent test bench for theories of the relationship between brain anatomy, physiology, and cognitive capability across species, including humans.

References

Alonso-Farre, J.M., Gonzalo-Orden, M., Barreiro-Vazquez, J.D., Barreiro-Lois, A., Andre, M., Morell, M., Llarena-Reino, M., Monreal-Pawlowsky, T., Degollada, E., 2015. Cross-sectional anatomy, computed tomography, and magnetic resonance imaging of the head of common dolphin (*Delphinus delphis*) and striped dolphin (*Stenella coeruleoalba*). Anat. Histol. Embryol. 44, 13–21.

Bauer, G.B., Cook, P.F., Harley, H.E., 2020. The relevance of ecological transitions to intelligence in marine mammals. Front. Psychol. 11, 2053.

Bauer, G.B., Fuller, M., Perry, A., Dunn, J.R., Zoeger, J., 1985. Magnetoreception and biomineralization of magnetite in cetaceans. In: Kirschvink, J.L., Jones, D.S., MacFadden, B.J. (Eds.), Magnetite Biomineralization and Magnetoreception in Organisms. Springer US, Boston, MA, pp. 489–507.

Bauer, G.B., Harley, H.E., 2001. The mimetic dolphin. Behav. Brain Sci. 24 (2), 326.

Bearzi, M., Stanford, C.B., 2008. Beautiful Minds: The Parallel Lives of Great Apes and Dolphins. Harvard University Press, pp. 1–351.

Begall, S., Burda, H., Malkemper, E.P., 2014. Magnetoreception in mammals. In: Advances in the Study of Behavior. Elsevier, pp. 45–88.

Berns, G.S., Cook, P.F., Foxley, S., Jbabdi, S., Miller, K.L., Marino, L., 2015. Diffusion tensor imaging of dolphin brains reveals direct auditory pathway to temporal lobe. Proc. R. Soc. Lond. B 282, 20151203.

Bombardi, C., Cozzi, B., Nenzi, A., Mazzariol, S., Grandis, A., 2011. Distribution of nitrergic neurons in the dorsal root ganglia of the bottlenose dolphin (*Tursiops truncatus*). Anat. Rec. (Hoboken) 294, 1066–1073.

Bombardi, C., Grandis, A., Gardini, A., Cozzi, B., 2013. Nitrergic neurons in the spinal cord of the bottlenose dolphin (*Tursiops truncatus*). Anat. Rec. (Hoboken) 296, 1603–1614.

Bombardi, C., Grandis, A., Nenzi, A., Giurisato, M., Cozzi, B., 2010. Immunohistochemical localization of substance P and cholecystokinin in the dorsal root ganglia and spinal cord of the bottlenose dolphin (*Tursiops truncatus*). Anat. Rec. (Hoboken) 293, 477–484.

Bonato, M., Bagnoli, P., Centelleghe, C., Maric, M., Brocca, G., Mazzariol, S., Cozzi, B., 2019. Dynamics of blood circulation during diving in the bottlenose dolphin (*Tursiops truncatus*): the role of the retia mirabilia. J. Exp. Biol. 222, jeb198457.

Bouchard, B., Barnagaud, J.-Y., Poupard, M., Glotin, H., Gauffier, P., Torres Ortiz, S., Lisney, T.J., Campagna, S., Rasmussen, M., Célérier, A., 2019. Behavioural responses of humpback whales to food-related chemical stimuli. PLoS One 14 (2), e0212515.

Brandl, J.L., 2018. The puzzle of mirror self-recognition. Phenomenol. Cogn. Sci. 17 (2), 279–304.

Brauer, K., Schaber, W., 1976. Catalog of Mammalian Brains. VEB Gustav Fischer Verlag, Jena.

Breathnach, A.S., 1960. The cetacean central nervous system. Biol. Rev. 35, 187–230.

Brodmann, K., 1909. Vergleichende lokalisationslehre der grosshirnrinde in ihren prinzipien dargestellt auf grund des zellenbaues. Verlag von Johann Ambrosius Barth, Leipzig.

Buckner, R.L., Krienen, F.M., Castellanos, A., Diaz, J.C., Yeo, B.T., 2011. The organization of the human cerebellum estimated by intrinsic functional connectivity. J. Neurophysiol. 106 (5), 2322–2345.

Buhl, E.H., Oelschläger, H.A., 1986. Ontogenetic development of the nervus terminalis in toothed whales. Evidence for its non-olfactory nature. Anat. Embryol. (Berl.) 173, 285–294.

Buhl, E.H., Oelschläger, H.A., 1988. Morphogenesis of the brain in the harbour porpoise. J. Comp. Neurol. 277, 109–125.

Butti, C., Janeway, C.M., Townshend, C., Wicinski, B.A., Reidenberg, J.S., Ridgway, S.H., Sherwood, C.C., Hof, P.R., Jacobs, B., 2015. The neocortex of cetartiodactyls: I. A comparative Golgi analysis of neuronal morphology in the bottlenose dolphin (*Tursiops truncatus*), the minke whale (*Balaenoptera acutorostrata*), and the humpback whale (*Megaptera novaeangliae*). Brain Struct. Funct. 220, 3339–3368.

Butti, C., Santos, M., Uppal, N., Hof, P.R., 2013. Von Economo neurons: clinical and evolutionary perspectives. Cortex 49 (1), 312–326.

Butti, C., Sherwood, C.C., Hakeem, A.Y., Allman, J.M., Hof, P.R., 2009. Total number and volume of Von Economo neurons in the cerebral cortex of cetaceans. J. Comp. Neurol. 515, 243–259.

Chon, U., Vanselow, D.J., Cheng, K.C., Kim, Y., 2019. Enhanced and unified anatomical labeling for a common mouse brain atlas. Nat. Commun. 10 (1), 1–12.

Connor, R.C., 2007. Dolphin social intelligence: complex alliance relationships in bottlenose dolphins and a consideration of selective environments for extreme brain size evolution in mammals. Philos. Trans. R. Soc. Lond. B. 362 (1480), 587–602.

Connor, R., Mann, J., 2006. Social cognition in the wild: Machiavellian dolphins? In: Hurley, S., Nudds, M. (Eds.), Rational Animals? Oxford University Press, pp. 329–368.

Cozzi, B., Huggenberger, S., Oelschläger, H.H.A., 2017. Anatomy of Dolphins: Insights into Body Structure and Function. Academic Press, London, pp. 1–438.

Cozzi, B., Mazzariol, S., Podestà, M., Zotti, A., Huggenberger, S., 2016. An unparalleled sexual dimorphism of sperm whale encephalization. Int. J. Comp. Psychol. 29, 1–9.

Cozzi, B., Roncon, G., Granato, A., Giurisato, M., Castagna, M., Peruffo, A., Panin, M., Ballarin, C., Montelli, S., Pirone, A., 2014. The claustrum of the bottlenose dolphin *Tursiops truncatus* (Montagu 1821). Front. Syst. Neurosci. 8, 42.

Cuello, A.C., Tramezzani, J.H., 1969. The epiphysis cerebri of the Weddell seal: its remarkable size and glandular pattern. Gen. Comp. Endocrinol. 12 (1), 154–164.

Dailly, M., 1972. Contribution on the study of the VIIIth vestibular nerve in cetaceans: the elliptical nucleus. In: Investigations on Cetacea. vol. IV, pp. 231–250.

de Felipe, J., et al., 2007. Specializations of the cortical microstructure of humans. In: Kaas, J.H., Preuss, T.M. (Eds.), Evolution of the Nervous Systems. A Comprehensive Reference. Elsevier Academic Press, Amsterdam, pp. 167–190.

De Graaf, A.S., 1967. Anatomical Aspects of the Cetacean Brainstem. Van Gorcum, pp. 1–170.

De Vreese, S., André, M., Cozzi, B., Centelleghe, C., van der Schaar, M., Mazzariol, S., 2020. Morphological evidence for the sensitivity of the ear canal of odontocetes as shown by immunohistochemistry and transmission electron microscopy. Sci. Rep. 10, 4191.

Dell, L.A., Karlsson, K.A., Patzke, N., Spocter, M.A., Siegel, J.M., Manger, P.R., 2016a. Organization of the sleep-related neural systems in the brain of the minke whale (*Balaenoptera acutorostrata*). J. Comp. Neurol. 524, 2018–2035.

Dell, L.A., Patzke, N., Spocter, M.A., Siegel, J.M., Manger, P.R., 2016b. Organization of the sleep-related neural systems in the brain of the harbour porpoise (*Phocoena phocoena*). J. Comp. Neurol. 524, 1999–2017.

Dell, L.A., Spocter, M.A., Patzke, N., Karlson, K.A.E., Alagaili, A.N., Bennett, N.C., Muhammed, O.B., Bertelsen, M.F., Siegel, J.M., Manger, P.R., 2015. Orexinergic bouton density is lower in the cerebral cortex of cetaceans compared to artiodactyls. J. Chem. Neuroanat. 68, 61–76.

DeLucchi, M.R., Dennis, B.J., Adey, W.R., 1965. A Stereotaxic Atlas of the Chimpanzee Brain (*Pan Satyrus*). California Univ Los Angeles Brain Research Inst.

Dennison, S., Moore, M.J., Fahlman, A., Moore, K., Sharp, S., Harry, C.T., Wells, R.S., 2012. Bubbles in live-stranded dolphins. Proc. R. Soc. Lond. B 279 (1732), 1396–1404.

Duffield, D.W., Haldiman, J.T., Henk, W.G., 1992. Surface morphology of the forebrain of the bowhead whale, *Balaena mysticetus*. Mar. Mamm. Sci. 8 (4), 354–378.

Dunbar, R.I., 1998. The social brain hypothesis. In: Evolutionary Anthropology: Issues, News, and Reviews. vol. 6, pp. 178–190.

Duus, P., 1976. Neurologisch-Topische Diagnostik. Anatomie, Physiologie, Klinik. Georg Thieme Verlag, Stuttgart.

Dyce, K.M., et al., 1991. Anatomie der Haustiere. Ferdinand Enke Verlag Stuttgart.

Fernández, A., Sierra, E., Díaz-Delgado, J., Sacchini, S., Sánchez-Paz, Y., Suárez-Santana, C., Bernaldo de Quirós, Y., 2017. Deadly acute decompression sickness in Risso's dolphins. Sci. Rep. 7 (1), 1–9.

Ferrer, I., Perera, M., 1988. Structure and nerve cell organisation in the cerebral cortex of dolphin *Stenella coeruleoalba*, A Golgi study with special attention to the primary auditory area. Anat. Embryol. (Berl) 178 (2), 161–173.

Filatova, O.A., Deecke, V.B., Ford, J.K., Matkin, C.O., Barrett-Lennard, L.G., Guzeev, M.A., Hoyt, E., 2012. Call diversity in the North Pacific killer whale populations: implications for dialect evolution and population history. Anim. Behav. 83 (3), 595–603.

Filimonoff, I.N., 1966. On the so-called rhinencephalon in the dolphin. J. Hirnforsch. 8, 1–23.

Fox, S., 2016. Human Physiology, fourteenth ed. McGraw Hill, p. 471.

Fox, K., 2020. The social cetaceans. Inference 5, 1–6.

Fox, K.C., Muthukrishna, M., Shultz, S., 2017. The social and cultural roots of whale and dolphin brains. Nat. Ecol. Evol. 1 (11), 1699–1705.

Furutani, R., 2008. Laminar and cytoarchitectonic features of the cerebral cortex in the Risso's dolphin (*Grampus griseus*), striped dolphin (*Stenella coeruleoalba*), and bottlenose dolphin (*Tursiops truncatus*). J. Anat. 213, 241–248.

Fuzessery, Z.M., Feng, A.S., Supin, A.Y., 2004. Central auditory processing of temporal information in bats and dolphins. In: Thomas, J.A., Moss, C.F., Vater, M. (Eds.), Echolocation in Bats and Dolphins. University of Chicago Press, Chicago, IL, pp. 115–122.

Garey, L.J., Leuba, G., 1986. A quantitative study of neuronal and glial numerical density in the visual cortex of the bottlenose dolphin: evidence for a specialized subarea and changes with age. J. Comp. Neurol. 247, 491–496.

Garey, L.J., Takás, J., Revishchin, A.V., Hámori, J., 1989. Quantitative distribution of GABA-immunoreactive neurons in cetacean visual cortex is similar to that in land mammals. Brain Res. 485, 278–284.

Garey, L.J., Winkelmann, E., Brauer, K., 1985. Golgi and Nissl studies of the visual cortex of the bottlenose dolphin. J. Comp. Neurol. 240, 305–321.

Gerussi, T., Graïc, J.-M., Grandis, A., Peruffo, A., Cozzi, B., 2022. The orbitofrontal cortex of the sheep. Topography, organization, neurochemistry, digital tensor imaging and comparison with the chimpanzee and human. Brain Struct. Funct. 227, 1871–1891.

Girgis, M., Wang, S.C., 1981. New Stereotaxic Atlas of the Rabbit Brain. WH Green.

Glezer, I.I., 2002. Neural morphology. In: Hoelzel, A.R. (Ed.), Marine Mammal Biology—An Evolutionary Approach. Blackwell Science, Oxford, pp. 98–115.

Glezer, I.I., Hof, P.R., Istomin, V.V., Morgane, P.J., 1995. Comparative immunocytochemistry of calcium-binding protein-positive neurons in visual and auditory systems of cetacean and primate brains. In: Kastelein, R.A., Thomas, J.A., Nachtigall, P.E. (Eds.), Sensory Systems of Aquatic Mammals. De Spil Publishers, Woerden, The Netherlands, pp. 477–513.

Glezer, I.I., Hof, P.R., Leranth, C., Morgane, P.J., 1993. Calcium-binding protein-containing neuronal populations in mammalian visual cortex: a comparative study in whales, insectivores, bats, rodents, and primates. Cereb. Cortex 3, 249–272.

Glezer, I.I., Hof, P.R., Morgane, P.J., 1992. Calretinin-immunoreactive neurons in the primary visual cortex of dolphin and human brains. Brain Res. 595, 181–188.

Glezer, I.I., Hof, P.R., Morgane, P.J., 1998. Comparative analysis of calcium-binding protein-immunoreactive neuronal populations in the auditory and visual systems of the bottlenose dolphin (*Tursiops truncatus*) and the macaque monkey (*Macaca fascicularis*). J. Chem. Neuroanat. 15, 203–237.

Glezer, I.I., Jacobs, M.S., Morgane, P.J., 1987. Ultrastructure of the blood-brain barrier in the dolphin (*Stenella coeruleoalba*). Brain Res. 414 (2), 205–218.

Glezer, I.I., Morgane, P.J., 1990. Ultrastructure of synapses and Golgi analysis of neurons in neocortex of the lateral gyrus (visual cortex) of the dolphin and pilot whale. Brain Res. Bull. 24, 401–427.

Graïc, J.-M., Peruffo, A., Grandis, A., Cozzi, B., 2021. Topographical and structural characterization of the V1-V2 transition zone in the visual cortex of the long-finned pilot whale *Globicephala melas* (Traill, 1809). Anat. Rec. (Hoboken) 304, 1105–1118.

Graïc, J.M., et al., 2022. The primary visual cortex of Cetartiodactyls: organization, cytoarchitectonics, and comparison with perissodactyls and primates. Brain Struct. Funct. 227, 1195–1225.

Green, R.F., Ridgway, S.H., Evans, W.E., 1980. Functional and descriptive anatomy of the bottlenosed dolphin nasolaryngeal system with special reference to the musculature associated with sound production. In: Busnel, R.-G., Fish, J.F. (Eds.), Animal Sonar Systems. Springer, Boston, MA, pp. 199–238.

Haddad, D., Huggenberger, S., Haas-Rioth, M., Kossatz, L.S., Oelschläger, H.H.A., Haase, A., 2012. Magnetic resonance microscopy of prenatal dolphins (Mammalia, Odontoceti, Delphinidae)—ontogenetic and phylogenetic implications. Zool. Anz. 251, 115–130.

Hals, S.K., 2016. Responses to Mirrors in Domestic Goats (*Capra aegagrus* Hircus): Assessing Mirror Use to Solve a Problem and in Self-Recognition. Master's thesis, Norwegian University of Life Sciences, Ås.

Hanson, A., Grisham, W., Sheh, C., Annese, J., Ridgway, S., 2013. Quantitative examination of the bottlenose dolphin cerebellum. Anat. Rec. (Hoboken) 296, 1215–1228.

Harley, H.E., Roitblat, H.L., Nachtigall, P.E., 1996. Object representation in the bottlenose dolphin (*Tursiops truncatus*): integration of visual and echoic information. J. Exp. Psychol. Anim. Behav. Process. 22 (2), 164.

Hatschek, R., Schlesinger, H., 1902. Der Hirnstamm des Delphins (*Delphinus delphis*). Arb. Aus Dem Neurol. Inst. Wien. Univ. 9, 1–117.

Hauser, M.D., Chomsky, N., Fitch, W.T., 2002. The faculty of language: what is it, who has it, and how did it evolve? Science 298 (5598), 1569–1579.

Herculano-Houzel, S., Catania, K., Manger, P.R., Kaas, J.H., 2015. Mammalian brains are made of these: a dataset of the numbers and densities of neuronal and nonneuronal cells in the brain of glires, primates, scandentia, eulipotyphlans, afrotherians and artiodactyls, and their relationship with body mass. Brain Behav. Evol. 86 (3–4), 145–163.

Herman, L., 2010. What laboratory research has told us about dolphin cognition. Int. J. Comp. Psychol. 23, 310–330.

Hof, P.R., Glezer, I.I., Condé, F., Flagg, R.A., Rubin, M.B., Nimchinsky, E.A., Vogt Weisenhorn, D.M., 1999. Cellular distribution of the calcium-binding proteins parvalbumin, calbindin, and calretinin in the neocortex of mammals: phylogenetic and developmental patterns. J. Chem. Neuroanat. 16, 77–116.

Hof, P.R., Glezer, I.I., Nimchinsky, E.A., Erwin, J.M., 2000. Neurochemical and cellular specializations in the mammalian neocortex reflect phylogenetic relationships: evidence from primates, cetaceans, and artiodactyls. Brain Behav. Evol. 55, 300–310.

Hof, P.R., Van der Gucht, E., 2007. Structure of the cerebral cortex of the humpback whale, *Megaptera novaeangliae* (Cetacea, Mysticeti, Balaenopteridae). Anat. Rec. (Hoboken) 290 (1), 1–31.

Houser, D.S., Moore, P.W., Johnson, S., Lutmerding, B., Branstetter, B., Ridgway, S.H., Trickey, J., Finneran, J.J., Jensen, E., Hoh, C., 2010. Relationship of blood flow and metabolism to acoustic processing centers of the dolphin brain. J. Acoust. Soc. Am. 128 (3), 1460–1466.

Howard, R.S., Finneran, J.J., Ridgway, S.H., 2006. Bispectral index monitoring of unihemispheric effects in dolphins. Anesth. Analg. 103 (3), 626–632.

Huggenberger, S., 2008. The size and complexity of dolphin brains—a paradox? J. Mar. Biol. Assoc. U. K. 88, 1103–1108.

Huggenberger, S., Oelschläger, H.H.A., Cozzi, B., 2019. Atlas of the Anatomy of Dolphins and Whales. Academic Press, London.

Jacobs, M.S., Galaburda, A.M., McFarland, W.L., Morgane, P.J., 1984. The insular formations of the dolphin brain: quantitative cytoarchitectonic studies of the insular component of the limbic lobe. J. Comp. Neurol. 225, 396–432.

Jacobs, M.S., McFarland, W.L., Morgane, P.J., 1979. The anatomy of the brain of the bottlenose dolphin (*Tursiops truncatus*). Rhinic lobe (rhinencephalon): the archicortex. Brain Res. Bull. 4 (Suppl. 1), 1–108.

Jacobs, M.S., Morgane, P.J., McFarland, W.L., 1971. The anatomy of the brain of the bottlenose dolphin (*Tursiops truncatus*). Rhinic lobe (rhinencephalon). I. The paleocortex. J. Comp. Neurol. 141, 205–271.

Janik, V.M., 2013. Cognitive skills in bottlenose dolphin communication. Trends Cogn. Sci. 17 (4), 157–159.

Janik, V.M., 2014. Cetacean vocal learning and communication. Curr. Opin. Neurobiol. 28, 60–65.

Jansen, J., 1953. Studies on the cetacean brain. The gross anatomy of the rhombencephalon of the fin whale *Balaenoptera physalus* (L.). Hvalrådets skr. 37, 1–36.

Jansen, J., Jansen, J., 1953. A note on the amygdaloid complex in the fin whale [*Balaenoptera physalus* (L.)]. Hvalrådets skr. 39, 1–14.

Jansen, J., Jansen, J.K.S., 1969. The nervous system of Cetacea. In: Andersen, H.T. (Ed.), The Biology of Marine Mammals. Academic Press, New York, pp. 175–252.

Jansen, J., Osen, K.K., 1984. Morphogenesis and morphology of the brain stem nuclei of cetacean. II. The nuclei of the accessory, vagal and glossopharyngeal nerves in baleen whales. J. Hirnforsch. 25 (1), 53–87.

Jelgersma, G., 1934. Das gehirn der wassersäugetiere. Eine anatomische untersuchung. Von Johann Ambrosium Bart Verlag, Leipzig, pp. 82–91.

Jerison, H.J., 1973. Evolution of the Brain and Intelligence. Academic Press, New York.

Johnson, P.J., Luh, W.M., Rivard, B.C., Graham, K.L., White, A., FitzMaurice, M., Barry, E.F., 2020. Stereotactic cortical atlas of the domestic canine brain. Sci. Rep. 10 (1), 1–16.

Kalinichenko, S.G., Pushchin, I.I., 2008. Calcium-binding proteins in the cerebellar cortex of the bottlenose dolphin and harbour porpoise. J. Chem. Neuroanat. 35, 364–370.

van Kann, E., Cozzi, B., Hof, P.R., Oelschläger, H.H.A., 2017. Qualitative and quantitative analysis of primary neocortical areas in selected mammals. Brain Behav. Evol. 90, 193–210.

Keogh, M.J., Ridgway, S.H., 2008. Neuronal fiber composition of the corpus callosum within some odontocetes. Anat. Rec. Adv. Integr. Anat. Evol. Biol. 291, 781–789.

Kern, A., Seidel, K., Oelschläger, H.H.A., 2009. The central vestibular complex in dolphins and humans: functional implications of Deiters' nucleus. Brain Behav. Evol. 73, 102–110.

Kern, A., Siebert, U., Cozzi, B., Hof, P.R., Oelschläger, H.H., 2011. Stereology of the neocortex in Odontocetes: qualitative, quantitative, and functional implications. Brain Behav. Evol. 77, 79–90.

Kesarev, V.S., Malofeeva, L.I., 1969a. Structural organization of the dolphin motor cortex. Arkh. Anat. Gistol. Embriol. 56, 48–55.
Kesarev, V.S., Malofeeva, L.I., 1969b. Structural organization of the dolphin motor cortex. Neurosci. Transl. 3 (4), 33–39.
Kim, H.Y., Seo, K., Jeon, H.J., Lee, U., Lee, H., 2017. Application of functional near-infrared spectroscopy to the study of brain function in humans and animal models. Mol. Cells 40 (8), 523.
Klinowska, M., 1986. The cetacean magnetic sense—evidence from stranding. In: Bryden, M.M., Harrison, R. (Eds.), Research on Dolphins. Oxford University Press, New York, pp. 401–432.
Knopf, J.P., Hof, P.R., Oelschläger, H.H.A., 2016. The neocortex of Indian river dolphins (genus Platanista): comparative, qualitative and quantitative analysis. Brain Behav. Evol. 88, 93–110.
Knudsen, S.K., Mørk, S., Øen, E.O., 2002. A novel method for in situ fixation of whale brains. J. Neurosci. Methods 120, 35–44.
Kojima, T., 1951. On the brain of the sperm whale (*Physeter catodon* L.). Sci. Rep. Whales Res. Inst. Tokyo 6, 49–72.
Kraus, C., Pilleri, G., 1969. Zur Histologie der Groβhirnrinde von Balaenoptera borealis. In: Investigations on Cetacea. vol. I, pp. 151–172.
Kraus, C., Pilleri, G., 1972. On the histology of the large pyramidal cells in cortical layer V of cetaceans (*Delphinus delphis* and *Balaenoptera borealis*). In: Investigations on Cetacea. vol. IV, pp. 207–214.
Kremers, D., Lopez Marulanda, J., Hausberger, M., Lemasson, A., 2014. Behavioural evidence of magnetoreception in dolphins: detection of experimental magnetic fields. Naturwissenschaften 101, 907–911.
Kruger, L., 1959. The thalamus of the dolphin (*Tursiops truncatus*) and comparison with other mammals. J. Comp. Neurol. 111, 133–194.
Kruger, L., 1966. Specialized features of the cetacean brain. In: Norris, K.S. (Ed.), Whales, Dolphins, and Porpoises. University of California Press, pp. 232–254.
Kükenthal, W., Ziehen, T., 1893. Über das Centralnervensystem der Cetaceen nebst Untersuchungen über die vergleichende Anatomie des Gehirns bei Placentaliern. Denkschr. Med. Naturwiss. Ges. Jena 3, 77–198.
Langworthy, O.R., 1931a. Central nervous system of the porpoise *Tursiops truncatus*. J. Mammal. 12, 381–389.
Langworthy, O.R., 1931b. A description of the central nervous system of the porpoise (*Tursiops truncatus*). J. Comp. Neurol. 54, 437–499.
Langworthy, O.R., 1935. The brain of the whalebone whale, *Balaenoptera physalus*. Bull. Johns Hopkins Hosp. 57, 143–147.
Lende, R.A., Akdikmen, S., 1968. Motor field in cerebral cortex of the bottlenose dolphin. J. Neurosurg. 29, 495–499.
Lillie, M.A., Vogl, A.W., Gerard, S.G., Raverty, S., Shadwick, R.E., 2022. Retia mirabilia: protecting the cetacean brain from locomotion-generated blood pressure pulses. Science 377 (6613), 1452–1456.
Lilly, J.C., 1961. Man and Dolphin. Doubleday & Co.
Lilly, J.C., 1967. The Mind of the Dolphin. Avon Books, New York.
Lodato, S., Arlotta, P., 2015. Generating neuronal diversity in the mammalian cerebral cortex. Annu. Rev. Cell Dev. Biol. 31 (1), 699–720.
Loth, A., Güntürkün, O., von Fersen, L., Janik, V.M., 2022. Through the looking glass: how do marked dolphins use mirrors and what does it mean? Anim. Cogn. 25 (5), 1151–1160.
Lyamin, O.I., Manger, P.R., Ridgway, S.H., Mukhametov, L.M., Siegel, J.M., 2008. Cetacean sleep: an unusual form of mammalian sleep. Nuerosci. Biobehav. Rev. 32, 1451–1484.
Macphail, E.M., 1985. Vertebrate intelligence: the null hypothesis. Philos. Trans. R. Soc. Lond. B. 308 (1135), 37–51.
Malkemper, P., Huggenberger, S., Oelschläger, H.H.A., 2012. The dolphin cochlear nucleus—topography, histology and functional implications. J. Morphol. 273, 173–185.
Manger, P.R., 2006. An examination of cetacean brain structure with a novel hypothesis correlating thermogenesis to the evolution of a big brain. Biol. Rev. 81, 293–338.
Manger, P.R., Fuxe, K., Ridgway, S.H., Siegel, J.M., 2004. The distribution and morphological characteristics of catecholaminergic cells in the diencephalon and midbrain of the bottlenose dolphin (*Tursiops truncatus*). Brain Behav. Evol. 64, 42–60.
Manger, P.R., Ruidgway, S.H., Siegel, J.M., 2003. The locus coeruleus complex of the bottlenose dolphin (*Tursiops truncatus*) as revealed by tyrosine hydroxylase immunohistochemistry. J. Sleep Res. 12, 149–155.
Marino, L., Butti, C., Connor, R.C., Fordyce, R.E., Herman, L.M., Hof, P.R., Lefebvre, L., Lusseau, D., McCowan, B., Nimchinsky, E.A., Pack, A.A., Reidenberg, J.S., Reiss, D., Rendell, L., Uhen, M.D., Van der Gucht, E., Whitehead, H., 2008. A claim in search of evidence: reply to Manger's thermogenesis hypothesis of cetacean brain structure. Biol. Rev. Camb. Philos. Soc. 83 (4), 417–440.
Marino, L., Rilling, J.K., Lin, S.K., Ridgway, S.H., 2000. Relative volume of the cerebellum in dolphins and comparison with anthropoid primates. Brain Behav. Evol. 56, 204–211.
Marino, L., Sudheimer, L., Pabst, D.A., McLellan, W.A., Johnson, J.I., 2003. Magnetic resonance images of the brain of a dwarf sperm whale (*Kogia simus*). J. Anat. 203, 57–76.
McFarland, W.L., Jacobs, M.S., Morgane, P.J., 1979. Blood supply to the brain of the dolphin, *Tursiops truncatus*, with comparative observations on special aspects of the cerebrovascular supply of other vertebrates. Neurosci. Biobehav. Rev. 3, 1–93.
McFarland, W.L., Morgane, P.J., Jacobs, M.S., 1969. Ventricular system of the brain of the dolphin, *Tursiops truncatus*, with comparative anatomical observations and relations to brain specializations. J. Comp. Neurol. 135, 275–368.
McKnight, J.C., Ruesch, A., Bennett, K., Bronkhorst, M., Balfour, S., Moss, S.E.W., Milne, R., Tyack, P.L., Kainerstorfer, J.M., Hastie, G.D., 2021. Shining new light on sensory brain activation and physiological measurement in seals using wearable optical technology. Philos. Trans. R. Soc. Lond. B. 376, 20200224.
Mead, J.G., 1975. Anatomy of the external nasal passages and facial complex in the Delphinidae (Mammalia, Cetacea). Smithson. Contrib. Zool. 207, 1–72.
Montie, E.W., Pussini, N., Schneider, G.E., Battey, T.W., Dennison, S., Barakos, J., Gulland, F., 2009. Neuroanatomy and volumes of brain structures of a live California sea lion (*Zalophus californianus*) from magnetic resonance images. Anat. Rec. 292 (10), 1523–1547.
Morgane, P.J., Glezer, I.I., Jacobs, M.S., 1988. Visual cortex of the dolphin: an image analysis study. J. Comp. Neurol. 273, 3–25.
Morgane, P.J., Glezer, I.I., Jacobs, M.S., 1990. Comparative anatomy of the visual cortex of the dolphin. In: Jones, E.G., Peters, A. (Eds.), Cerebral Cortex. vol. 8B. Plenum Publishing Co, New York, pp. 215–262.
Morgane, P.J., Jacobs, M.S., 1972. Comparative anatomy of the cetacean nervous system. In: Harrison, R.J. (Ed.), Functional Anatomy of Marine Mammals. vol. 1. Academic Press, London, pp. 117–244.
Morgane, P.J., Jacobs, M.S., 1986a. A morphometric Golgi and cytoarchitectonic study of the hippocampal formation of the bottlenose dolphin, *Tursiops truncatus*. The Hippocampus 3, 369–432.

Morgane, P.J., Jacobs, M.S., 1986b. A morphometric Golgi and cytoarchitectonic study of the hippocampal formation of the bottlenose dolphin, *Tursiops truncatus*. In: Isaacson, R.L., Pribram, K.H. (Eds.), The Hippocampus. Plenum Press, New York and London, pp. 369–432.

Morgane, P.J., Jacobs, M.S., Galaburda, A., 1985. Conservative features of neocortical evolution in dolphin brain. Brain Behav. Evol. 26 (3–4), 176–184.

Morgane, P.J., Jacobs, M.S., Galaburda, A.M., 1986. Evolutionary morphology of the dolphin brain. In: Schusterman, R.J., Thomas, J.A., Wood, F.G. (Eds.), Dolphin Cognition and Behavior: A Comparative Approach. Lawrence Erlbaum Associates, Hillsdale, pp. 5–29.

Morgane, P.J., Jacobs, M.S., McFarland, W.L., 1980. The anatomy of the brain of the bottlenose dolphin (*Tursiops truncatus*). Surface configuration of the telencephalon of the bottlenose dolphin with comparative anatomical observations in four other cetacean species. Brain Res. Bull. 5 (Suppl. 3), 1–107.

Morgane, P.J., McFarland, W.L., Jacobs, M.S., 1982. The limbic lobe of the dolphin brain: a quantitative cytoarchitectonic study. J. Hirnforsch. 23, 465–552.

Mortensen, H.S., Pakkenberg, B., Dam, M., Dietz, R., Sonne, C., Mikkelsen, B., Eriksen, N., 2014. Quantitative relationships in delphinid neocortex. Front. Neuroanat. 8, 132.

Mota, B., Herculano-Houzel, S., 2015. Cortical folding scales universally with surface area and thickness, not number of neurons. Science 349 (6243), 74–77.

Mukhametov, L.M., Supin, A.Y., Polyakova, I.G., 1977. Interhemispheric asymmetry of the electroencephalographic sleep pattern in dolphins. Brain Res. 134, 581–584.

Nachtigall, P.E., Mooney, T.A., Taylor, K.A., Yuen, M.M., 2007. Hearing and auditory evoked potential methods applied to odontocete cetaceans. Aquat. Mamm. 33 (1), 6.

Nagel, E.L., Morgane, P.J., McFarlane, W.L., Galliano, R.E., 1968. Rete mirabile of dolphin: its pressure-damping effect on cerebral circulation. Science 161, 898–900.

Oelschläger, H.H.A., 2008. The dolphin brain—a challenge for synthetic neurobiology. Brain Res. Bull. 75, 450–459.

Oelschläger, H.A., Buhl, E.H., Dann, J.F., 1987. Development of the nervus terminalis in mammals including toothed whales and humans. Ann. N. Y. Acad. Sci. 519, 447–464.

Oelschläger, H.H.A., Haas-Rioth, M., Fung, C., Ridgway, S.H., Knauth, M., 2008. Morphology and evolutionary biology of the dolphin (*Delphinus* sp.) brain-MR imaging and conventional histology. Brain Behav. Evol. 71, 68–86.

Oelschläger, H.H.A., Kemp, B., 1998. Ontogenesis of the sperm whale brain. J. Comp. Neurol. 399, 210–228.

Oelschläger, H.H.A., Oelschläger, J.S., 2009. Brain. In: Perrin, W.F., Würsig, B., Thewissen, J.G.M. (Eds.), Encyclopedia of Marine Mammals. Academic Press, San Diego, CA, pp. 134–149.

Oelschläger, H.H.A., Ridgway, S.H., Knauth, M., 2010. Cetacean brain evolution: dwarf sperm whale (*Kogia sima*) and common dolphin (*Delphinus delphis*)—an investigation with high-resolution 3D MRI. Brain Behav. Evol. 75, 33–62.

Ogawa, T., Arifuku, S., 1948. On the acoustic system in the cetacean brains. Sci. Rep. Whales Res. Inst. Tokyo 2, 1–20.

Ogden, J.A., Lee, K.E., Conlogue, G.J., Barnett, J.S., 1981. Prenatal and postnatal development of the cervical portion of the spine in the short-finned pilot whale *Globicephala macrorhyncha*. Anat. Rec. 200, 83–94.

Orekhova, K., Selmanovic, E., De Gasperi, R., Gama Sosa, M.A., Wicinski, B., Maloney, B., Hof, P.R., 2022. Multimodal assessment of bottlenose dolphin auditory nuclei using 7-Tesla MRI, immunohistochemistry and stereology. Vet. Sci. 9 (12), 692.

Osen, K.K., Jansen, J., 1965. The cochlear nuclei in the common porpoise, *Phocaena phocaena*. J. Comp. Neurol. 125, 223–257.

Panin, M., Gabai, G., Ballarin, C., Peruffo, A., Cozzi, B., 2012. Evidence of melatonin secretion in cetaceans: plasma concentration and extrapineal HIOMT-like presence in the bottlenose dolphin *Tursiops truncatus*. Gen. Comp. Endocrinol. 177, 238–245.

Parolisi, R., Cozzi, B., Bonfanti, L., 2017. Non-neurogenic SVZ-like niche in dolphins, mammals devoid of olfaction. Brain Struct. Funct. 222, 2625–2639.

Parolisi, R., Cozzi, B., Bonfanti, L., 2018. Humans and dolphins: decline and fall of adult neurogenesis. Front. Neurosci. 12, 497.

Parolisi, R., Peruffo, A., Messina, S., Panin, M., Montelli, S., Giurisato, M., Cozzi, B., Bonfanti, L., 2015. Forebrain neuroanatomy of the neonatal and juvenile dolphin (*T. truncatus* and *S. coeruloalba*). Front. Neuroanat. 9, 140.

Patzke, N., Spocter, M.A., Karlsson, K.A.E., Bertelsen, M.F., Haagensen, M., Chawana, R., Streicher, S., Kaswera, C., Gilissen, E., Alagaili, A.N., Mohammed, O.B., Reep, R.L., Bennett, N.C., Siegel, J.M., Ihunwo, A.O., Manger, P.R., 2015. In contrast to many other mammals, cetaceans have relatively small hippocampi that appear to lack adult neurogenesis. Brain Struct. Funct. 220, 361–383.

Peña-Melián, Á., Cabello-de la Rosa, J.P., Gallardo-Alcañiz, M.J., Vaamonde-Gamo, J., Relea-Calatayud, F., González-López, L., Villanueva-Anguita, P., Flores-Cuadrado, A., Saiz-Sánchez, D., Martínez-Marcos, A., 2019. Cranial pair 0: the nervus terminalis. Anat. Rec. (Hoboken) 302 (3), 394–404.

Peruffo, A., Corain, L., Bombardi, C., Centelleghe, C., Grisan, E., Graïc, J.-M., Bontempi, P., Grandis, A., Cozzi, B., 2019. The motor cortex of the sheep: laminar organization, projections and diffusion tensor imaging of the intracranial pyramidal and extrapyramidal tracts. Brain Struct. Funct. 224, 1933–1946.

Picton, T.W., Durieux-Smith, A., 1988. Auditory evoked potentials in the assessment of hearing. Neurol. Clin. 6 (4), 791–808.

Pilleri, G., 1989. Further material on the brain anatomy of the Sei whale (*Balaenoptera borealis*). In: Investigations on Cetacea. vol. XXII, pp. 342–377.

Pilleri, G., Gihr, M., 1970. The central nervous system of the mysticete and odontocete whales. In: Investigations on Cetacea. vol. II, pp. 89–128.

Pilleri, G., Peixun, C., Zuohua, S., 1980. Concise Macroscopical Atlas of the Brain of the Common Dolphin (*Delphinus delphis* Linnaeus, 1758). University of Berbe, pp. 1–26.

Popov, V.V., Supin, A.Y., 1990. Auditory brain stem responses in characterization of dolphin hearing. J. Comp. Physiol. A 166 (3), 385–393.

Poth, C., Fung, C., Güntürkün, O., Ridgway, S.H., Oelschläger, H.H.A., 2005. Neuron numbers in sensory cortices of five delphinids compared to a physeterid, the pygmy sperm whale. Brain Res. Bull. 66, 357–360.

Purves, D., Augustine, G.J., Fitzpatrick, D., Katz, L.C., LaMantia, A.S., McNamara, J.O., Williams, S.M., 2001. The organization of the nervous system. In: Purves, D., Augustine, G.J., Fitzpatrick, D., Katz, L.C., LaMantia, A.S., McNamara, J.O., et al. (Eds.), Neuroscience. Sinauer Associates, Sunderland, MA.

Rambaldi, A.M., Cozzi, B., Grandis, A., Canova, M., Mazzoni, M., Bombardi, C., 2017. Distribution of calretinin immunoreactivity in the lateral nucleus of the bottlenose dolphin (*Tursiops truncatus*) amygdala. Anat. Rec. 300 (11), 2008–2016.

Ratner, C., Riise, J., Eriksen, N., Pakkenberg, B., 2011. A post-mortem study of the corpus callosum in the common minke whale (*Balaenoptera acutorostrata*). Mar. Mamm. Sci. 17 (4), 688–700.
Rauschmann, M.A., Huggenberger, S., Kossatz, L.S., Oelschläger, H.H.A., 2006. Head morphology in perinatal dolphins: a window into phylogeny and ontogeny. J. Morphol. 267, 1295–1315.
Reiss, D., Marino, L., 2001. Mirror self-recognition in the bottlenose dolphin: a case of cognitive convergence. Proc. Natl. Acad. Sci. U. S. A. 98 (10), 5937–5942.
Rendell, L., Whitehead, H., 2010. Culture in whales and dolphins. Behav. Brain Sci. 24, 309–324.
Reveley, C., Gruslys, A., Ye, F.Q., Glen, D., Samaha, J., Russ, B.E., Saleem, K.S., 2017. Three-dimensional digital template atlas of the macaque brain. Cereb. Cortex 27 (9), 4463–4477.
Revishchin, A.V., Garey, L.J., 1990. The thalamic projection to the sensory neocortex of porpoise, *Phocoena phocoena*. J. Anat. 169, 85–102.
Revishchin, A.V., Garey, L.J., 1991. Laminar distribution of cytochrome oxidase staining in cetacean isocortex. Brain Behav. Evol. 37, 355–367.
Ridgway, S.H., 1986. Physiological observations on dolphin brains. In: Dolphin Cognition and Behavior: A Comparative Approach, pp. 31–59.
Ridgway, S.H., 1990. The central nervous system of the bottlenose dolphin. In: Leatherwood, S., Reeves, R.R. (Eds.), The Bottlenose Dolphin. Academic Press, San Diego, CA, pp. 69–97.
Ridgway, S.H., 2002. Asymmetry and symmetry in brain waves from dolphin left and right hemispheres: some observations after anesthesia, during quiescent hanging behavior, and during visual obstruction. Brain Behav. Evol. 60 (5), 265–274.
Ridgway, S.H., Brownson, R.H., 1984. Relative brain size and cortical surface areas in odontocetes. Acta Zool. Fenn. 172, 149–152.
Ridgway, S.H., Brownson, R.H., Van Alstyne, K.R., Hauser, R.A., 2019. Higher neuron densities in the cerebral cortex and larger cerebellums may limit dive times of delphinids compared to deep-diving toothed whales. PLoS One 14 (12), e0226206.
Ridgway, S.H., Carder, D.A., 2001. Assessing hearing and sound production in cetaceans not available for behavioral audiograms: experiences with sperm, pygmy sperm, and gray whales. Aquat. Mamm. 27 (3), 267–276.
Ridgway, S.H., Carlin, K.P., Van Alstyne, K.R., Hanson, A.C., Tarpley, R.J., 2016. Comparison of dolphins' body and brain measurements with four other groups of cetaceans reveals great diversity. Brain Behav. Evol. 88, 235–257.
Ridgway, S.H., Demski, L.S., Bullock, T.H., Schwanzel-Fukuda, M., 1987. The terminal nerve in odontocete cetaceans. Ann. N. Y. Acad. Sci. 519, 201–212.
Ridgway, S.H., Hanson, A.C., 2014. Sperm whales and killer whales with the largest brains of all toothed whales show extreme differences in cerebellum. Brain Behav. Evol. 83 (4), 266–274.
Ridgway, S., Samuelson Dibble, D., Van Alstyne, K., Price, D., 2015. On doing two things at once: dolphin brain and nose coordinate sonar clicks, buzzes and emotional squeals with social sounds during fish capture. J. Exp. Biol. 218 (24), 3987–3995.
Ridgway, S.H., Van Alstyne, K.R., 2017. The blue whale brain misrepresented by an alcohol dehydrated brain of 3,636 grams. Mar. Mamm. Sci. 33, 386–388.
Ries, F.A., Langworthy, O.R., 1936. A study of the surface structure of the brain of the whale (*Balaenoptera physalus* and *Physeter catodon*). Brain 68, 1–47.
Riojas-Schnier, J.L., Toth, A.L., 2022. Insights into insect cognition through mirror-induced behaviour in paper wasps. Behaviour 1, 1–24.
Roland, P.E., Graufelds, C.J., Wăhlin, J., Ingelman, L., Andersson, M., Ledberg, A., Zilles, K., 1994. Human brain atlas: for high-resolution functional and anatomical mapping. Hum. Brain Mapp. 1 (3), 173–184.
Romer, A.S., Parsons, T.S., 1986. The Vertebrate Body, sixth ed. Saunders College Publishing, Philadelphia.
Roth, G., Dicke, U., 2005. Evolution of the brain and intelligence. Trends Cogn. Sci. 9 (5), 250–257.
Sacchini, S., Arbelo, M., Bombardi, C., Fernández, A., Cozzi, B., Bernaldo de Quiròs, Y., Herráez, P., 2018. Locus coeruleus complex of the family *Delphinidae*. Sci. Rep. 8, 5486.
Sacchini, S., Díaz-Delgado, J., Espinosa de los Monteros, A., Paz, Y., Bernaldo de Quirós, Y., Sierra, E., Arbelo, M., Herráez, P., Fernández, A., 2020. Amyloid-beta peptide and phosphorylated tau in the frontopolar cerebral cortex and in the cerebellum of toothed whales: aging versus hypoxia. Biol. Open 9, bio054734.
van Schaik, C.P., Triki, Z., Bshary, R., Heldstab, S.A., 2021. A farewell to the encephalization quotient: a new brain size measure for comparative primate cognition. Brain Behav. Evol. 96 (1), 1–12.
Schulmeyer, F.J., 1992. Morphologische Untersuchungen am Hirnstamm der Delphine unter besonderer Berücksichtigung des La Plata-Delphins, *Pontoporia blainvillei* (Inaugural-Dissertation, Fachbereich Medizin). Johann Wolfgang Goethe-Universität, Frankfurt am Main.
Schulmeyer, F.J., Adams, J., Oelschläger, H.H.A., 2000. Specialized sound reception in dolphins—a hint for the function of the dorsal cochlear nucleus in mammals. Hist. Biol. 14, 53–56.
Schwerdtfeger, W.K., Oelschläger, H.H.A., Stephan, H., 1984. Quantitative neuroanatomy of the brain of the La Plata dolphin, *Pontoporia blainvillei*. Anat. Embryol. (Berl.) 170, 11–19.
Seki, Y., 1958. Observations on the spinal cord of the right whale. Sci. Rep. Whales Res. Inst. Tokyo 13, 231–251.
Seki, Y., 1973. An anatomical study on the lower extension of the dorsal vagal nucleus to the upper cervical cord in the sperm whale. Sci. Rep. Whales Res. Inst. Tokyo 25, 241–249.
Seki, Y., 1984. Anatomical study on the cell column located closely medial to the nucleus of the spinal root of the trigeminal nerve in the sperm and the pygmy sperm whale. Sci. Rep. Whales Res. Inst. Tokyo 35, 47–56.
Seki, Y., 1985. Anatomical observations on the lower brain stem of the right whale. Sci. Rep. Whales Res. Inst. Tokyo 36, 49–87.
Simons, M., Tibbetts, E., 2019. Insects as models for studying the evolution of animal cognition. Curr. Opin. Insect Sci. 34, 117–122.
Sonne, J., Reddy, V., Lopez-Ojeda, W., 2017. Neuroanatomy, Cranial Nerve 0 (Terminal Nerve).
Štěrba, O., Klima, M., Schildger, B., 2000. Embryology of dolphins—staging and ageing of embryos and fetuses of some cetaceans. Adv. Anat. Embryol. Cell Biol. 157, 1–133.
Streri, A., Gentaz, E., 2003. Cross-modal recognition of shape from hand to eyes in human newborns. Somatosens. Mot. Res. 20 (1), 13–18.
Striedter, G.F., 2005. Principles of Brain Evolution. Sinauer Associates, Sunderland, MA.
Supin, A.Y., Mukhametov, L.M., 1986. Some mechanisms of the unihemispheric slow wave sleep in dolphins. In: Sokolov, V.E. (Ed.), Electrophysiology of sensory systems in marine mammals. Nauka, Moscow, pp. 188–207 (in Russian).
Supin, A.Y., Popov, V.V., Mass, A.M., 2001. The Sensory Physiology of Aquatic Mammals. Kluwer Academic Publishers, Boston.
Tarpley, R.J., Ridgway, S.H., 1994. Corpus callosum size in delphinid cetaceans. Brain Behav. Evol. 44, 156–165.

Tester, A.L., 1963. The role of olfaction in shark predation. Pac. Sci. 17 (2), 145–170.
Thewissen, J.G.M., George, J., Rosa, C., Kishida, T., 2011. Olfaction and brain size in bowhead whale (*Balaena mysticetus*). Mar. Mamm. Sci. 27, 282–294.
Tirapu-Ustarroz, J., Luna-Lario, P., Iglesias-Fernandez, M.D., Hernaez-Goni, P., 2011. Cerebellar contribution to cognitive process: current advances. Rev. Neurol. 53 (5), 301–315.
Vacher, M., Durrant, C., Rose, J., Hall, A., Spires-Jones, T., Gunn-Moore, F., Dagleish, M., 2022. Alzheimer's disease-like neuropathology in three species of oceanic dolphin. Eur. J. Neurosci. 57 (7), 1161–1179.
Viamonte, M., Morgane, P.J., Galliano, R.E., Nagel, E.L., McFarland, W.L., 1968. Angiography in living dolphin and observations on blood supply to the brain. Am. J. Phys. 214, 1225–1249.
Voogd, J., Nieuwenhuyis, R., Van Dongen, P.A.M., ten Donkelaar, H.J., 1998. Mammals. In: Nieuwenhuyis. R., ten Donkelaar, H.J., Nicholson, C. (Eds.), The Central Nervous System of Vertebrates. Springer, Berlin, pp. 1637–2097.
Wagner, M.J., Luo, L., 2020. Neocortex–cerebellum circuits for cognitive processing. Trends Neurosci. 43 (1), 42–54.
Wilcox, T., Biondi, M., 2015. fNIRS in the developmental sciences. Wiley Interdiscip. Rev. Cogn. Sci. 6 (3), 263–283.
Wild, S., Allen, S.J., Krützen, M., King, S.L., Gerber, L., Hoppitt, W.J., 2019. Multi-network-based diffusion analysis reveals vertical cultural transmission of sponge tool use within dolphin matrilines. Biol. Lett. 15 (7), 20190227.
Winn, H.E., Thompson, T.J., Cummings, W.C., Hain, J., Hudnall, J., Hays, H., Steiner, W.W., 1981. Song of the humpback whale—population comparisons. Behav. Ecol. Sociobiol. 8 (1), 41–46.
Wright, A., Scadeng, M., Stec, D., Dubowitz, R., Ridgway, S., Leger, J.S., 2017. Neuroanatomy of the killer whale (*Orcinus orca*): a magnetic resonance imaging investigation of structure with insights on function and evolution. Brain Struct. Funct. 222, 417–436.
Wright, A.K., Theilmann, R.J., Ridgway, S.H., Scadeng, M., 2018. Diffusion tractography reveals pervasive asymmetry of cerebral white matter tracts in the bottlenose dolphin (*Tursiops truncatus*). Brain Struct. Funct. 223, 1697–1711.
Zoeger, J., Dunn, J.R., Fuller, M., 1981. Magnetic material in the head of the common Pacific dolphin. Science 213, 892–894.
Zvorykin, V.P., 1976. Specific features of the structural organization of the external geniculate body in the dolphin *Phocaena phocaena* as compared with other dolphins. Arkh. Anat. Gistol. Embriol. 70, 58–66.

C H A P T E R

10

Sensory physiology in delphinids

Frederike D. Hanke[a], *T. Aran Mooney*[b], *and Vincent M. Janik*[c]

[a]Neuroethology, Institute for Biosciences, University of Rostock, Rostock, Germany [b]The Sensory Ecology and Bioacoustics Lab, Biology Department, Woods Hole Oceanographic Institution, Woods Hole, MA, United States [c]Scottish Oceans Institute, School of Biology, University of St. Andrews, St. Andrews, United Kingdom

Abstract

Perception of sensory information is a crucial part of an animal's life. Delphinids provide an intriguing example of how mammalian sensory systems can adapt to the aquatic environment. Yet, our knowledge on how delphinids perceive the world is patchy. For some sensory systems, extensive research efforts have provided detailed information as is the case for audition, while others, such as chemo- or electroreception, are hardly understood or have only recently been discovered. Here, we summarize delphinid sensory physiology, highlight gaps in our understanding of delphinid sensory systems, and provide a tentative assessment of the role each modality plays in their lives. This review not only addresses the individual sensory modalities but also underscores the combination of sensory information available to delphinids, which provides a complex, rich experience of their environment.

Introduction

There is a multitude of information that an organism perceives in every second of its life. Its senses provide a constant stream of inputs from which it needs to extract what is relevant for maintaining and increasing its inclusive fitness. Animal senses, from periphery to neuroanatomy, are adapted to not only gather information on essential environmental stimuli but also have limitations in what physical parameters they are sensitive to. Each sense focuses and filters information that reaches the brain. Much of the ever-present input is, however, inconsequential and therefore can be considered "noise." Masking effects caused by such noise make the extraction of relevant information challenging. Examining sensory systems and their physiology is crucial to understand what information animals can rely on to make decisions affecting their behavior and to assess how anthropogenic activities affect their sensory world.

In this chapter, we summarize the current knowledge on delphinid sensory physiology focusing on exteroception, the perception of information from the environment. Delphinids are mammals that secondarily adapted to a fully aquatic lifestyle over 10–12 million years ago. This transition from air to water had a dramatic effect on their sensory systems due to changes in the physical properties of components of the sensory systems, the surrounding media, and the environmental stimuli. Thus, new structures evolved, existing systems were modified, and some were reduced or even lost.

Out of necessity, much of our review is focused on the bottlenose dolphin (*Tursiops truncatus*), since it is the best-studied delphinid species by far. This is one of the main limitations in this field since a closer look at the over 35 extant species in this family makes it obvious that they are highly diverse. Significant differences can be found in morphology, distribution and habitat, foraging, reproduction, communication, and even echolocation capabilities (Jeffersen and LeDuc, 2018). For some species, we have only just scratched the surface when studying their sensory physiology, and for many, information is absent altogether. Moreover, even for bottlenose dolphins, our understanding of some sensory systems is far more advanced than others. The extent of these imbalances will become clear in the following sections in which we highlight the numerous experimental findings from bottlenose dolphins and try to contrast them with data available for other species in an attempt to assess how sensory physiology differs between species.

The Physiology of Dolphins
https://doi.org/10.1016/B978-0-323-90516-9.00010-5

Vision

Vision is often considered a sense of minor importance for cetaceans, in part because of limitations of light-transmission underwater. However, good viewing conditions can be encountered for example when at or close to the water surface. Even at depth, bioluminescence might provide light, allowing the visual system to provide valuable information. Consequently, vision is suggested to play a role in numerous contexts (for review, see Kinne, 1975; Madsen and Herman, 1980; Mobley and Helweg, 1990; Würsig et al., 1990; Kremers et al., 2016a), including (i) orientation and navigation such as when using the sun for orientation or when orienting to underwater or surface structures, (ii) for prey detection and capture, (iii) avoidance of predators, (iv) mediating the coordination of activity within a group of dolphins, and (v) for species or individual recognition, including the communication of behavioral states.

A multitude of studies since the late 19th century have described the cetacean eyes to be highly derived and well-adapted to the environment and lifestyle of their possessors. The main optical challenge of vision is that the cornea is mostly ineffective in water leaving the lens as the main refracting, light-bending, element. Although adapted to aquatic vision, the cetacean eye retained some ability to see in air, in which the cornea may add to the refractive power of the eye. Thus, delphinids potentially encounter the challenge of constant changes in refractive power of the eye when transiting between air and water. Moreover, their ability to dive to great depth, as well as to travel over large distances, requires the eye to cope with changes in light intensity and spectrum, density, or temperature of the surrounding medium.

Specifics of the eye globe, its position, and mobility

The eyes of delphinids are oblong (Fig. 1A; Pütter, 1903). In the bottlenose dolphin, the eye measures approximately 25mm axially but has an equatorial diameter of 35mm (Dawson et al., 1972; Dral, 1975; Cartee et al., 1995; Harvey, 2019). The eyes of the common dolphin (*Delphinus delphis*) are comparable in size to the eyes of the bottlenose dolphin (Rochon-Duvigneaud, 1940; but much larger diameters were indicated by Pütter, 1903), whereas the eyes of the tucuxi (*Sotalia fluviatilis*) and the killer whale (*Orcinus orca*) are slightly smaller (Mass and Supin, 1999; Mass et al., 2012). When compared to other mammals, all cetaceans have smaller eyes than expected by body mass (Harvey, 2019), but their eyes are comparatively large in absolute terms. Having a large eye comes with the benefit of an increased size of the image on the retina. Although increasing a small eye in size has a large effect on vision, increasing the size of a large eye by the same amount does not result in the same improvement (law of diminishing returns; Nilson et al., 2012). Moreover, the benefits from having a large eye need to be balanced with the additional costs of maintaining it.

All delphinids have laterally placed eyes (Fig. 1B). Consequently, they often watch objects with a single eye, i.e., monocularly, and show lateralization of eye use—and of various behaviors in general (see summary in Ridgway, 1986; Blois-Heulin et al., 2012). Despite the lateral eye position, delphinids might have binocular vision of 20–30 deg to the front at some distance to the rostrum and in the downward gaze direction (Slijper, 1962; Yablokov et al., 1972; Pryor, 1973). Downward gaze direction is facilitated by a narrow lower lid. The extent of the visual fields

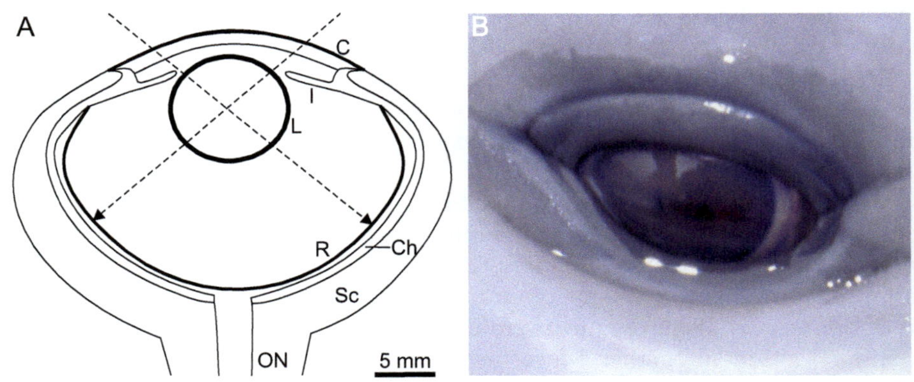

FIG. 1 (A) Schematic drawing of an eye of a bottlenose dolphin (*Tursiops truncatus*). The slightly oblong eye has the typical mammalian lens eye bauplan. Specific characteristics are its spherical lens (L) and thick sclera (Sc). The two arrows indicate the two viewing axes, which are set by the two pupillary pinholes present in bright light conditions. C, cornea; Ch, choroid; I, iris; L, lens; ON, optic nerve; R, retina; Sc, sclera. Scale 5mm. *Modified from Mass and Supin (1995).* (B) Eye of a bottlenose dolphin in which the pupil, which is circular in darkness, is closed to two small slits (appearing dark on the photo). These slits ultimately close to two pinholes in bright light conditions. *Photo: Tim Hüttner.*

seems to fill perception gaps of the echolocation system providing a view on objects like prey when it is below the mandible (Madsen and Herman, 1980).

When monocularly viewing, the delphinids can oversee approximately 130 deg of visual space in the horizontal and 100 deg in the vertical plane. The visual field can be extended when the eyes are protruded (Yablokov et al., 1972; Dawson et al., 1987a; Supin et al., 2001) or by eye movements mediated by six ocular muscles, which are well developed in delphinids. A strong muscle that retracts the eye into the orbit (musculus retractor bulbi; Hosokawa, 1951) allows the bottlenose dolphin to protrude its eyes by 10–15 mm and retract its eyes by 20 mm or more as the orbit has a depth of 80 mm with only the upper part being ossified (Pütter, 1903; Dawson et al., 1972; Yablokov et al., 1972). Eye movements and large visual fields are essential in an animal that (i) has fused vertebrae and thus very limited possibilities to move its head and (ii) most likely benefits from staying streamlined even when scanning the environment to avoid increased locomotor costs caused by reorientating its whole body.

Unique features of the cetacean eye globe include a thick sclera (Pütter, 1903; Dawson, 1980) and well-developed vascularization. Although the sclera of delphinids is generally thick, it is not as thick as that of, for example, some mysticetes (see, for example, Zhu et al., 2001). Cartee et al. (1995) described the sclera of the bottlenose dolphin to increase in thickness toward the optic nerve (Fig. 1A). The thick sclera and cornea probably evolved to protect the eye during swimming, breaching, and diving (Walls, 1942; Dawson, 1980; Mass and Supin, 2007), to withstand the high eye pressure (Dawson, 1980) or the force of the retractor muscle as suggested for the sperm whale (*Physeter macrocephalus*) eye (Bjerager et al., 2003), or to reduce heat loss in water (Walls, 1942; Mengual et al., 2014).

Delphinids display extensive vascularization of the eye, which allows blood access to supply nutrients and oxygen, as well as heat to maintain the eye's functionality even during deep dives (Ninomiya and Yoshida, 2007). Delphinids have a well-developed ophthalmic rete surrounding the optic nerve (Schoenfuss, 1997), likely to maintain temperature at the optic nerve and the retina (Dawson, 1980) and to dampen blood flow to minimize changes in blood pressure and oxygen supply during dives (Nagel et al., 1968; Ninomiya and Yoshida, 2007). According to Ninomiya and Yoshida (2007), the rete lacks the anatomical characteristics that would allow it to play a role during retraction/protrusion of the eye as speculated for the sperm whale (Bjerager et al., 2003).

The dioptric apparatus and the iris

Although the dioptric or light-bending apparatus consists of a cornea and a lens, the lens is the main refractive element in the eyes of aquatic animals underwater. The cornea is most often optically ineffective, when the eye is submersed, as its refractive index (its light-bending abilities) equals that of sea water. In the delphinid eye, the slightly oval-shaped cornea (Pütter, 1903) is thicker in the periphery than in the center (Pütter, 1903; Yablokov et al., 1972). This difference in thickness might render the cornea a weak diverging lens underwater, as was speculated for the harbor porpoise (*Phocoena phocoena*; Kröger and Kirschfeld, 1992, 1993, 1994). Examination of the cornea of the bottlenose dolphin provides inconclusive evidence whether the cornea is generally or locally astigmatic or not or whether local flattenings occur and have to be considered the "normal" or a pathological corneal status in dolphins (see, for example, Dawson et al., 1987b; and discussion in Dral, 1974). In general, corneal examinations are complicated by a viscous secretion found on the cornea of the bottlenose dolphin with a refractive index similar to the cornea itself and to the surrounding sea water (Dawson et al., 1987b). This secretion most likely serves to protect the eye.

After passing the cornea, light enters the eye via the pupil, which operates over a broad range of ambient luminance (Fig. 1B). When the pupil is dilated in darkness or under dim light conditions, it is of circular shape with a diameter of 10–15 mm in the bottlenose dolphin (see photo in Dawson et al., 1979; Dawson, 1980). With increasing light intensity, the pupil constricts adopting the shape of a "u" (Fig. 1B) as the central pupil gets covered by the dorsal operculum of the iris. When fully constricted in bright light conditions, only two pinholes remain open (photo of the closed pupil of a bottlenose dolphin in Madsen and Herman, 1980; Rivamonte, 2009); these two pinholes define two main viewing axes, which might differentially be used for aerial/underwater vision (see Section "The retina and visual function"). In this final stage of constriction, the depth of focus of the eye is large, and putative deleterious effects of the dioptric apparatus are minimized. The pupil response characteristics of two delphinids proved to be comparable to humans (Dawson et al., 1979). However, in contrast to humans, the dolphin pupil is only partially consensual. If one eye is stimulated with light, the pupil of this eye constricts; however, the pupil constriction of the unstimulated eye is only 60% of that of the stimulated eye (Dawson, 1980). This reaction contrasts with human eyes in which the pupil reflex in the unstimulated eye equals the pupil reflex of the stimulated eye. This independent reaction of the pupils of the two eyes seems adaptive for delphinids with laterally placed, independently moving eyes (McCormick, 1969; Dawson et al., 1981) most likely experiencing different light inputs on a regular basis.

Behind the iris, a transparent lens is the main refracting element for the dolphin eye underwater. It is spherical to provide the refractive power required for high-resolution vision (Walls, 1942; Dawson, 1980; Mass and Supin, 2007). The diameter of the bottlenose dolphin's lens is 9.3 mm axially and 9.9 mm equatorially, whereas the lenses of other delphinids range from approximately 5 mm in Risso's dolphins (*Grampus griseus*) to 12 mm in the short-finned pilot whale (*Globicephala macrorhynchus*) and the common dolphin (Pütter, 1903; Rochon-Duvigneaud, 1940; Mass and Supin, 1995, 1999; Harvey, 2019). Spherical lenses might suffer from spherical aberration (i.e., the refractive power of the lens differs from periphery to center), but lens optics have not been measured for delphinids. It is not resolved whether the lenses of cetaceans can change their refractive power or, more generally, whether the eyes of cetaceans can accommodate; maybe dolphins do not need to accommodate as their pinhole pupil provides a large depth of focus. Among delphinids, a ciliary muscle was evident in the long-finned pilot whale (*Globicephala melas*) and most likely in the bottlenose dolphin (Dral, 1972; Waller, 1992). However, even with a ciliary muscle present, it is highly unlikely that the shape of the lens can be changed as it is spherical. An alternative accommodative mechanism could include a change in the axial position of the lens by the involvement of the retractor muscle, but this was questioned due to the rigidity of the eye (Rivamonte, 2009). So far, refractive measurements did not reveal accommodative abilities in the bottlenose dolphin (see, for example, Dawson et al., 1987a; Litwiler and Cronin, 2001). This aligns with the finding that visual acuity varies with distance (Herman et al., 1975; data of Peacock et al., 1974, reported in Rivamonte, 1976).

The retina and visual function

In the back of the eye, the light is falling onto the retina showing the typical mammalian laminar bauplan. However, the cetacean retina is approximately twice as thick as the retina of terrestrial mammals (Dawson et al., 1983; Mass and Supin, 2007; Mengual et al., 2014). In the photoreceptor layer, a photoreceptor density of approximately 290,000 cells/mm^2 can be found in the peripheral retina and a density of approximately 400,000 cells/mm^2 in the central retina (Dral, 1977; Dawson, 1980). The delphinid retina is equipped with two different photoreceptors, rods and cones. Light-sensitive rods outnumber the cones by far, and a rod-dominated retina is considered an adaptation to dim light conditions. Rod density, assessed at unspecified locations within the retina, varies from a density of 200,000 rods/mm^2 in bottlenose dolphin to 400,000 rods/mm^2 in the short-finned pilot whale (Peichl et al., 2001). Cone density was assessed as 3700–4000 cones/mm^2 in the bottlenose dolphin and ranged between 3000 and 6500 cones/mm^2 in delphinids; thus, approximately 1.5%–2% of the photoreceptors were cones (Peichl et al., 2001). For comparison, the human retina has approximately 5% cones (Jonas et al., 1992), whereas some mysticetes lack cones entirely (Meredith et al., 2013).

As most cetaceans, delphinids possess only one type of cone, a long-wavelength sensitive cone (L cone; Perez, 1972; Dawson and Perez, 1973; Dawson, 1980). Although the gene for a short-wavelength sensitive cone (S cone) is present, it is not expressed (Fasick et al., 1998). The L cone visual pigment maximally absorbs light with a wavelength of 524 nm (Fasick and Robinson, 1998; Fasick et al., 1998). The rod pigment of delphinids has a wavelength of maximum absorbance of 484–489 nm (McFarland, 1971; Fasick and Robinson, 1998; Fasick et al., 1998; Fasick and Robinson, 2000; Bischoff et al., 2012; for review, see Fasick and Robinson, 2016), and thus, its maximum absorbance is clearly shifted to shorter wavelengths in comparison with the rod pigment of terrestrial mammals (Jacobs, 1993). Fasick and Robinson (1998) showed that this blue-shift was caused by the substitution of only three amino acids when compared to the bovine pigment gene sequence. Overall, the wavelength of maximum absorbance reflects the light conditions encountered by pelagically hunting delphinids (McFarland, 1971).

Several hypotheses have been put forward to explain why marine mammals have lost the S cone (Peichl et al., 2001) such as (i) the S cone loss might have been adaptive to reduce sensitivity to high degrees of scattering of short wavelengths of light under water (Levenson and Dizon, 2003); (ii) offset pigments are adaptive as the figure background contrast is increased at short viewing distances (Madsen and Herman, 1980); (iii) the low density of S cones results in S cones being unable to produce contrasts essential in the underwater light environment, which is also dominated by a small range of wavelengths (Newman and Robinson, 2005); (iv) with only one cone, animals do not need to cope with chromatic aberration, as, especially in the dim light encountered underwater, when the pupil is fully dilated and chromatic aberration would have maximal effect, they do not need to focus light of short and long wavelengths (Newman and Robinson, 2005).

As delphinids are monochromats, they should not have cone-based color vision. Accordingly, researchers trying to train bottlenose dolphins to discriminate between monochromatic lights had no success (Madsen, 1976; Simons and Huigen, 1977). Griebel and Schmid (2002) concluded from their training results that bottlenose dolphins have color vision. However, their results might have been confounded by brightness cues. Some color information could potentially be gained if the dolphins made use of mesopic color vision on the basis of rods and cones both being active under mesopic light conditions. This possibility needs addressing in future studies.

The receptor signal is ultimately conveyed via the inner retinal layers to the ganglion cells. The photoreceptor to ganglion cell ratio amounts to 80:1 in the periphery and 55:1 in the center in bottlenose dolphins (Dawson, 1980). This large degree of spatial summation suggests again that vision is adapted to function under low light conditions (Walls, 1942; Kröger and Katzir, 2008). For several delphinid species, the ganglion cell layer of the retina was analyzed in detail (for summary, see Lisney and Collin, 2018; Mass and Supin, 2018). One remarkable feature of the ganglion cell layer of cetaceans is the presence of giant ganglion cells with diameters of up to 100 μm (Dawson and Perez, 1973; Dawson et al., 1982; Dral, 1983; Mass and Supin, 1995; Murayama and Somiya, 1998; Mass et al., 2012). In delphinids, these cells reach diameters of up to 60–88 μm (Dral, 1983; Mass and Supin, 1995, 1999; Mengual et al., 2014).

The axons of the ganglion cells form the optic nerve. Within the optic nerve, axons with diameters larger than 10 μm can be found (Dawson et al., 1983), which may improve conduction velocities in large animals with long optic nerves (Lisney and Collin, 2018). However, such giant axons are not present in all large animals. Alternatively, large ganglion cells might improve the detection of low frequency and moving stimuli (Dawson, 1980).

In delphinids, the total number of ganglion cells ranges from 45,400 cells in the tucuxi (Mass and Supin, 1999) to 219,900 cells in the bottlenose dolphin (Mass and Supin, 1995). Ganglion cell density is not equal across the retina; instead, it peaks in two areas in the nasal and temporal retina. The two areas of increased ganglion cell density align with the two viewing axes through the two pupillary openings in bright light conditions (Fig. 1B; Rivamonte, 2009). In the bottlenose dolphin, the nasal spot seems to have a slightly higher ganglion cell density than the temporal spot; densities vary from 460 and 510 cells/mm^2 reported by Dral (1975) to approximately 670 cells/mm^2 reported by Mass and Supin (1995). Maximum ganglion cell density reaches values ranging from approximately 190 cells/mm^2 in the tucuxi (Mass and Supin, 1999) to 500–550 cells/mm^2 in the common dolphin and in the Pacific white-sided dolphin (*Lagenorhynchus obliquidens*; Dral, 1983; Murayama and Somiya, 1998; Mass and Supin, 1999). In most delphinids, except for the long-finned pilot whale (Mengual et al., 2014) and Pacific white-sided dolphin (Murayama and Somiya, 1998), the two areas of increased ganglion cell density are linked by a visual streak, a horizontal band stretching from nasal to temporal retina in which ganglion cell density is substantially higher than that in the rest of the retina but lower than that in areas with peak ganglion cell densities. This ganglion cell topography allows acute vision over a large part of the visual field, thereby also compensating for the restricted mobility of the head (Mass and Supin, 1995).

From the peak ganglion cell densities, the visual resolution in air and under water was calculated as 8–12 min of arc in various delphinid species (Dral, 1975, 1983; Mass and Supin, 1995; Murayama and Somiya, 1998; Mass et al., 2012) in comparison with 1 min of arc or even below in humans. The tucuxi, however, has a two to three times inferior resolution (Mass and Supin, 1999), which correlates with the highly murky water of their riverine habitats and thus the decreased utility of vision therein. Early behavioral experiments that attempted to assess visual acuity in the Pacific white-sided dolphin and the killer whale had overestimated visual acuity as brightness cues and variable viewing distance during experiments had confounded the results (Spong and White, 1971; White et al., 1971a). In bottlenose dolphins, behavioral assessments revealed an underwater visual acuity of 8–15 min of arc (but see Hall et al., 1972; Noordenbos and Boogh, 1974), whereas aerial visual acuity was determined as 12–22 min of arc (Madsen, 1972 cited after Pepper and Simmons, 1973; Herman, 1975; Nachtigall, 1986; data reported in Rivamonte, 1976; for overview see Mass and Supin, 2018). Interestingly, some behavioral visual acuity studies revealed equal visual acuity, at a particular viewing distance, in air and underwater (Herman et al., 1975; data reported in Rivamonte, 1976).

Many explanations have been put forward to explain this air–water acuity similarity. However, unless more experimental evidence (on the properties of the cornea and lens, particularly) is collected, these explanations will remain speculative. Many researchers seem to favor the idea that the dolphin uses different viewing axes in air and underwater with (putative) different properties of cornea, lens, and retina. Indeed, dolphins were reported to use different viewing axes in the two media. When viewing in air, dolphins look at objects with both eyes, thus use ventronasal viewing involving their temporal retina. In contrast, underwater objects are mainly looked at monocularly using posterolateral viewing involving the nasal retina, although ventronasal viewing might occasionally happen, too. Due to the existence of the two viewing axes and putative differences along the different axes, refractive measurements that have so far revealed very puzzling results (see, for example, Dral, 1972; Litwiler and Cronin, 2001) seem to be quite challenging and only informative if measurements are perfectly controlled. Even if refractive errors were present, the pinhole pupil of delphinids might minimize their deleterious effects in the end.

The ganglion cells give rise to the optic nerve that has a diameter of approximately 5 mm in the bottlenose dolphin (Dawson et al., 1972; Dawson, 1980) and 3 mm in the common dolphin (Rochon-Duvigneaud, 1943). The number of axons within the optic nerve of the bottlenose dolphin has been repeatedly assessed, and the values vary drastically [between a minimum of 147,000 axons and a maximum of 390,000 axons (Morgane and Jacobs, 1972; Dawson, 1980; Dawson et al., 1982, 1983; Gao and Zhou, 1992)]. In contrast, there are slightly fewer axons, 111,000 and 20,000, respectively, in the optic nerves of the striped dolphin (*Stenella coeruleoalba*) and the Guiana dolphin (*Sotalia guianensis*; Morgane and Jacobs, 1972; Mazzatenta et al., 2001). The low number of optic nerve axons within the Guiana dolphin again illustrates that vision is less effective in species living in highly turbid rivers. The optic nerves of the left and right eye cross entirely at the optic chiasm (Longworthy, 1931; Morgane and Jacobs, 1972; Jacobs et al., 1975; Tarpley et al., 1994). Thus, delphinids might not have stenopic vision (i.e., spatial vision that relies on the input from both eyes), an aspect that is, however, still unresolved.

In the back of the eye, in the eye fundus, a thick tapetum fibrosum can be found (Walls, 1942; Young et al., 1988; Mass and Supin, 2007). Under low light conditions, a tapetum is an effective means to assure that photons that are not initially absorbed at first passage through the retina may be reflected back to allow absorbance at second passage. In the bottlenose dolphin or the melon-headed whale (*Peponocephala electra*), it covers the entire fundus (Dawson, 1980; Young et al., 1988; fundus photo of the melon-headed whale in Ninomiya and Yoshida, 2007); it usually only covers the ventral fundus in terrestrial mammals (Dawson, 1980; Cartee et al., 1995). In the tucuxi (Mass and Supin, 1999), it did not cover the fundus entirely. Waller (1984) speculated that differences in diving habit might cause the tapetum to cover the entire fundus or only some part, but in all species examined it covers at least the upper two-thirds of the fundus. With a fully tapetalized fundus, delphinids can maximize absorbance of photons entering the eye from any direction, which is beneficial in an animal that has all axes of freedom regarding rotational movements in the three-dimensional underwater environment.

In first behavioral studies, in which delphinids, mainly bottlenose dolphins, were asked to discriminate between 2D visual stimuli, they did not learn the tasks for unknown reasons (see, for example, Forestell and Herman, 1988; White et al., 1971b; reviewed in Herman, 1980; Nachtigall, 1986). The dolphins, however, learned visual tasks when presented with 3D stimuli that separated clearly from background (Herman et al., 1989). After this positive experience, the animals also performed successfully with 2D stimuli (Herman, 1990). Numerous studies thereafter showed bottlenose dolphins are able to classify visual objects (Herman et al., 1989; Mercado III et al., 2000; Kilian et al., 2003) or to imitate visually observed behaviors (Bauer and Johnson, 1994). Moreover, bottlenose dolphins are able to recognize objects cross-modally integrating information from echolocation and vision (Pack and Herman, 1995; Harley et al., 2003). More behavioral experiments need to be conducted to assess whether the statement of Herman (1980) that visual learning is slower and less stable over time than, for example, auditory learning in bottlenose dolphins is indeed correct.

Mechanoreception

Hearing

Hearing is central to odontocete sensory ecology, enabling complex communication, navigation, and foraging. Not only do delphinids use the acoustic domain as their main way of communication (Janik, 2009), they also are capable of actively echolocating to explore their environment. For example, experimental studies have shown that they can detect a 2.5-cm solid steel sphere at a distance of 72 m by echolocation (Au and Snyder, 1980). Correspondingly, odontocete audition has long been a focus of study with researchers as early as the 18th century noting the parallels of marine mammal to terrestrial mammal ears, including an external opening, a tympanic membrane, the Eustachian tube, ossicles, cochlea, and semicircular canals (Hunter, 1787). With World War II and the advancement of underwater acoustic measurements, it was realized that cetaceans could make (and thus likely perceive) ultrasound (Schevill and Lawrence, 1949; Kellogg and Kohler, 1952), a result that was soon supported by histological examination of their cochlea (Yamada and Yoshizaki, 1959). Yet, it was not until Ken Norris that we were presented with the "jaw-hearing" hypothesis, which surmised that sound enters the odontocete head through the mandibular fat bodies termed the "acoustic window." These fats lead directly to the tympanic bulla, and Norris hypothesized that they may provide a low-impedance pathway guiding sound to the ears (Norris, 1964, 1968). A number of behavioral and anatomical studies supported this idea, somewhat culminating in an application of auditory evoked potential (AEP) methods applied to a bottlenose dolphin that helped to noninvasively map sound sensitivity around the odontocete's head and lower jaw and showing that the dolphin detected lower frequencies better around the mandible, compared to sounds presented near the meatus (Møhl et al., 1999). More recently, some diversity in the jaw-hearing pathway hypothesis has been noted. Odontocete hearing thresholds are generally low, but not necessarily lowest at the lower jaw's pan bone region. Rather, hearing pathway variations appear to exist among odontocete species and are at least partially dependent on head morphology (Mooney et al., 2008). For example, with a blunter rostrum, porpoises may have relatively less acoustic shadowing from their rostrum and greater sensitivity along the ventral midline of their bodies when listening to sounds (Mooney et al., 2014). Thus, while there is some diversity, together, these studies helped establish the validity of Norris's unconventional theory, leading to a paradigm shift by uncovering a whole new mechanism for mammalian hearing.

Auditory anatomy

The outer ear of delphinids is very different from those of terrestrial animals. Delphinids have no pinnae or air-filled canals leading to the middle ear (Fig. 2). An air-filled space would in fact be detrimental to hearing underwater since it would reflect sounds at the air–water interface. An opening to what would be an outer ear canal in terrestrial animals can be seen on a delphinid's head from outside, but this canal is filled with wax and does not appear to be used in sound conduction. Instead, sound travels via fatty acid structures in the lower jaw, one of which belonging to the lateral channel resembles a pinna embedded in tissue (Ketten, 2000). Sound can enter dolphin tissues without the help of specific sound-conducting structures since the density between water and tissue is very similar, so that very little

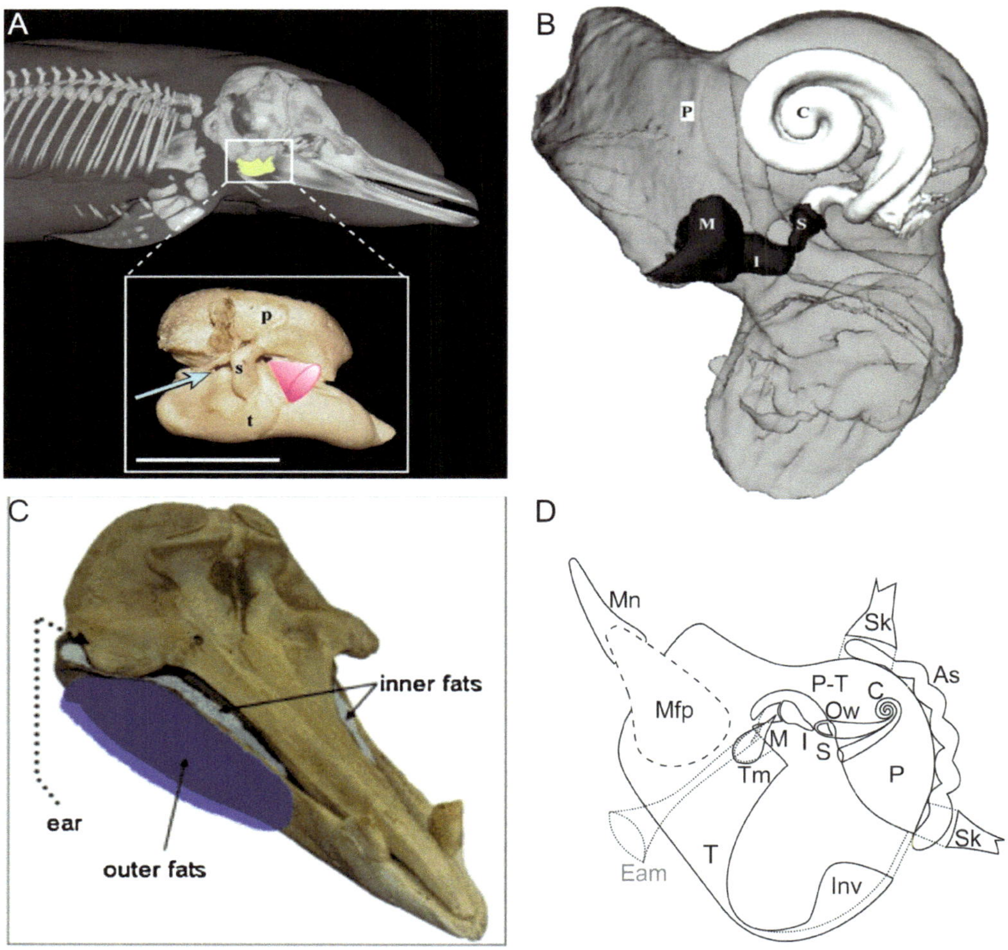

FIG. 2 The auditory anatomy of odontocetes. (A) CT-based three-dimensional reconstructions of a spotted dolphin (*Stenella attenuata*; USNM 504048). The tympanoperiotic complex houses the middle and inner ear structures, and is highlighted in yellow. The insert image is a photograph of the tympanoperiotic complex also from a spotted dolphin (USNM 487880). The tympanic (T) and periodic (P) bones housing the inner and middle ear showing location and placement with respect to the body of a spotted dolphin. The pink cone shows the likely sound input channel. The tympanic aperture is indicated by the blue arrow, and the acoustic funnel is illustrated by the pink cone. Scale bar 2 cm. *Modified from Yamato and Pyenson (2015).* (B) The odontocete "outer" ear (here a Gervais' beaked whale, *Mesoplodon europaeus*) highlighting the inner and outer mandibular fats, an acoustic guide to the ear, relative to the skull. *Modified from Koopman et al. (2006).* (C) CT reconstruction of the periotic (P) bone, malleus (M), incus (I), and stapes (S), with the cochlea (C) overlaid. *Modified from Miller et al. (2006).* (D) Schematic of the odontocete middle and inner ear showing parts already noted, as well as the external auditory meatus (Eam and dotted line for clarity), tympanic membrane (Tm), oval window (Ow), air sinuses (As), skull (Sk), and involucrum (Inv), mandible (Mn), and mandibular fat pad (Mfp). The dashed line indicates the mandibular fat pad, a feature which is absent in the ears of terrestrial mammals. *Modified from Nummela et al. (2004).*

energy is lost in reflections at the interface. Channels of fatty acids conduct sound that enters the lower jaw and guide it toward the oval window of the inner ear. To shield the inner ear from bone conduction and sound inputs from outside fatty channel pathways, the temporal bones that house the cochlea are extremely dense and are not directly connected to the skull (Cozzi et al., 2017). This isolates the ear from other inputs, which is thought to help with directional hearing.

Middle ear structures of odontocetes are not wholly unlike those of terrestrial mammals, with a malleus, incus, and stapes, which attach directly to the oval window of the inner ear and presumably set a basilar membrane in motion when sound is detected. However, empirical evidence of the middle ear functioning underwater remains elusive leading to some debate as to whether these bones are actually used to conduct and amplify sound, as in terrestrial mammals (although invasive studies at the water's surface suggest they can operate similar to other mammals; McCormick et al., 1970). It is still unclear whether these ossicles contribute to sound conduction from the lower jaw or whether fatty channels can transmit sound to the oval window without the involvement of the ossicles.

In the inner ear, sound sets a portion of the basilar membrane within the cochlea in motion, the location of which is dependent upon the acoustic frequency of the received sound, and this in turn will deflect hair cells and generate an auditory response. Generally, the inner ear is similar to that of terrestrial mammals although it has some substantial adaptations enabling impressive auditory capabilities described as follows. For example, auditory ganglion cell densities for odontocetes range from 2500 (in delphinids such as bottlenose dolphins) to around 3500 cells/mm (in harbor porpoises and belugas (*Delphinapterus leucas*)) (Wever et al., 1971; Ketten, 2000), more than doubling the density of humans, and provide a ganglion-to-hair cell density of 7.3–6.5, dwarfing that of humans, cats, and bats (2.4, 3.7, and 4,1, respectively; Gacek and Rasmussen, 1961; Bruns and Schmieszek, 1980; Nadol Jr, 1988). Similarly, auditory fiber diameters and spiral cell bodies are substantially larger than in other mammals. Such properties have long been speculated to support superior acoustic and echolocation signal analyses and rapid processing (Wever et al., 1971; Ketten, 2000). Within cetaceans, there is interspecies variation in the inner ear fine structure in relation to the kind of signals that are most important to each species. This affects mostly the size and thickness of the basilar membrane, the nature of supporting structures that modify membrane stiffness (bony spiral laminae), and the number of revolutions of the cochlea. Among echolocating odontocetes, two types of cochlea have been identified, a type I and a type II version (Ketten, 2000). Type I can be found in species that produce narrow-band, high-frequency echolocation clicks such as porpoises and presumably dolphins belonging to the genus *Cephalorhynchus*, while type II cochlea is typical of species with broadband, lower-frequency echolocation clicks such as the bottlenose dolphin or common dolphin. Type I species have a shorter and stiffer basilar membrane with more bony spiral lamina support than type II species as a specialization to perceive the higher frequencies that these species use in their echolocation compared to type II species.

Hearing abilities

Behavioral studies have been instrumental in determining hearing thresholds in cetaceans. Early work by Johnson (1967) defined the hearing thresholds for the bottlenose dolphin, and this work sets the standard for many odontocetes in terms of level of detail of the audiogram and overall sensitivity of the animal. These psychophysical experiments have been expanded to examine a range of auditory capabilities. For example, Au and colleagues used two directional sound sources and a masking (background noise) paradigm, to address the receiving beam of the bottlenose dolphin (Au and Moore, 1984). They showed that the receiving beam (i.e., the auditory directivity index) was narrower at higher frequencies and generally less-directional and more symmetrical in the horizontal plane of the animal compared with its vertical midline axis. Another study revealed that while hearing thresholds are often measured using pure tones, dolphins are more sensitive to broadband pulses; noting that echolocation and most natural environmental signals are typically broadband, it is not surprising that odontocete auditory systems are probably better tuned to broadband sounds (Au et al., 2002). Furthermore, classic target detection studies have utilized behavioral conditioning to evaluate how well dolphins hear and use echolocation pulses, for example, showing echo integration time to be as fast at 264 μs (Au et al., 1988).

Many of the recent advancements in our understanding of auditory sensory physiology have, however, utilized the aforementioned AEP methods. Leveraged from terrestrial mammal science, this technique involves recording neurological responses of the auditory brainstem typically with surface electrodes placed on (or suction-cupped to) the surface of the odontocete's skin. Stimuli are often tone pips, sinusoidally amplitude-modulated (SAM) tones or a series of clicks, generating a series of 5–7 neurophysiological "waves" in the case of tone pips or envelope following responses when a series of waves are following SAM tones or clicks (Fig. 3). The method is noninvasive, typically using electrodes embedded in suction cups to record the evoked neurophysiological responses from the surface of the skin. The method produces an audiogram similar to those from behavioral studies, although AEP thresholds can be elevated by up to 18 dB when compared to behavioral audiograms (Yuen et al., 2005; Finneran and Houser, 2006), and the nature of the short tones used in most odontocete AEP studies limits the frequency range of hearing that can be measured to signals

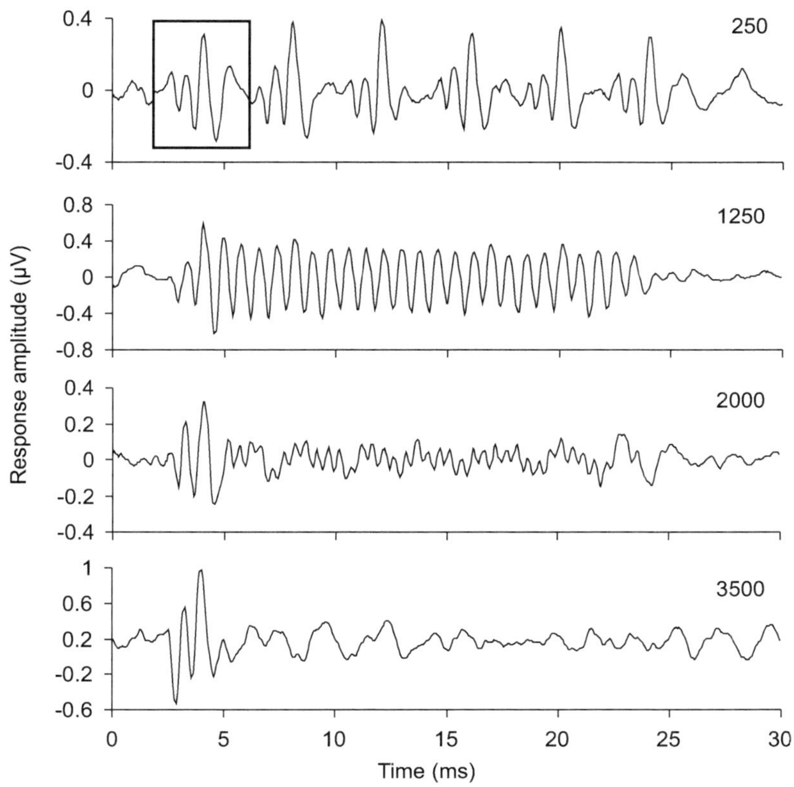

FIG. 3 Auditory evoked potential waveforms of a finless porpoise (*Neophocaena phocaenoides asiaeorientalis*) in response to increasing rates of broadband pulses. Rates shown on the right in Hz (pulses per second). There is a clear set of waveforms to the 250 Hz rate (boxed). This waveform becomes an envelope following response (EFR) at higher click rates (1250 Hz), but this EFR diminishes at 2000 Hz and we only see an onset AEP response at 3500 Hz, as if the animal is responding to a continuous tone, not a series of clicks, suggesting that it cannot follow this rate of pulses. *Modified from Mooney et al. (2011).*

above 4–8 kHz (Supin et al., 2001; Finneran et al., 2016). Yet, there are clear advantages to the method and it has now been used to measure hearing thresholds and audiograms in relatively short amounts of time (tens of minutes) allowing researchers to quickly assess hearing thresholds and determine audiograms in wild animals, a key advancement as many cetaceans cannot be kept in aquaria. These methods are now even being employed in free-swimming animals via custom-built tag sensors (Fig. 4; Smith et al., 2021).

Consequently, we now know more about hearing sensitivity in wild odontocetes than perhaps in any other animal family. Generally, these animals have a wide hearing range, from 200 Hz to 150 kHz (Fig. 5; Johnson, 1967; Mooney et al., 2012). Odontocete thresholds can be low compared to many other marine taxa, near 40–50 dB re 1 µPa at their most sensitive, with good hearing (<80 dB) ranging from 16 to 100 kHz (Mooney et al., 2012, 2018b). Visualization of hearing sensitivity results in a U-shaped audiogram (Fig. 5), with lower-frequency thresholds steadily improving (decreasing in sound level) to the middle-frequency range. Thresholds typically increase rapidly beyond 100 kHz as animals approach their high-frequency hearing limit. Some groups such as porpoises routinely show sound detection up to 180 kHz (Kastelein et al., 2002) while hearing ranges of killer whales and beaked whales [Blainville's beaked whale (*Mesoplodon densirostris*); Gervais' beaked whale (*Mesoplodon europaeus*)] appear to be slightly narrower (Finneran et al., 2009; Pacini et al., 2011; Branstetter et al., 2017). For the killer whale, sensitive hearing (i.e., within 20 dB of best sensitivity) ranges from 5 to 81 kHz and from 22 to 64 kHz for the Blainville's beaked whale.

Some hearing curves however deviate from this general pattern. Bottlenose dolphins routinely exposed to noise show age-related hearing loss similar to other mammals (Houser and Finneran, 2006). Although intriguingly, for some wild belugas ($n = 26$) that live in quieter environments, such age-related hearing loss is not as apparent (Fig. 5A and B). However, since ages of these individuals were only estimated, it is hard to say if age-related hearing loss was not noted, or if the sample of 26 wild individuals did not include older animals (Fig. 5A and B; Mooney et al., 2018b). Hearing data from stranded animals show frequent hearing loss occurrence (Mann et al., 2010). The authors suggested that hearing

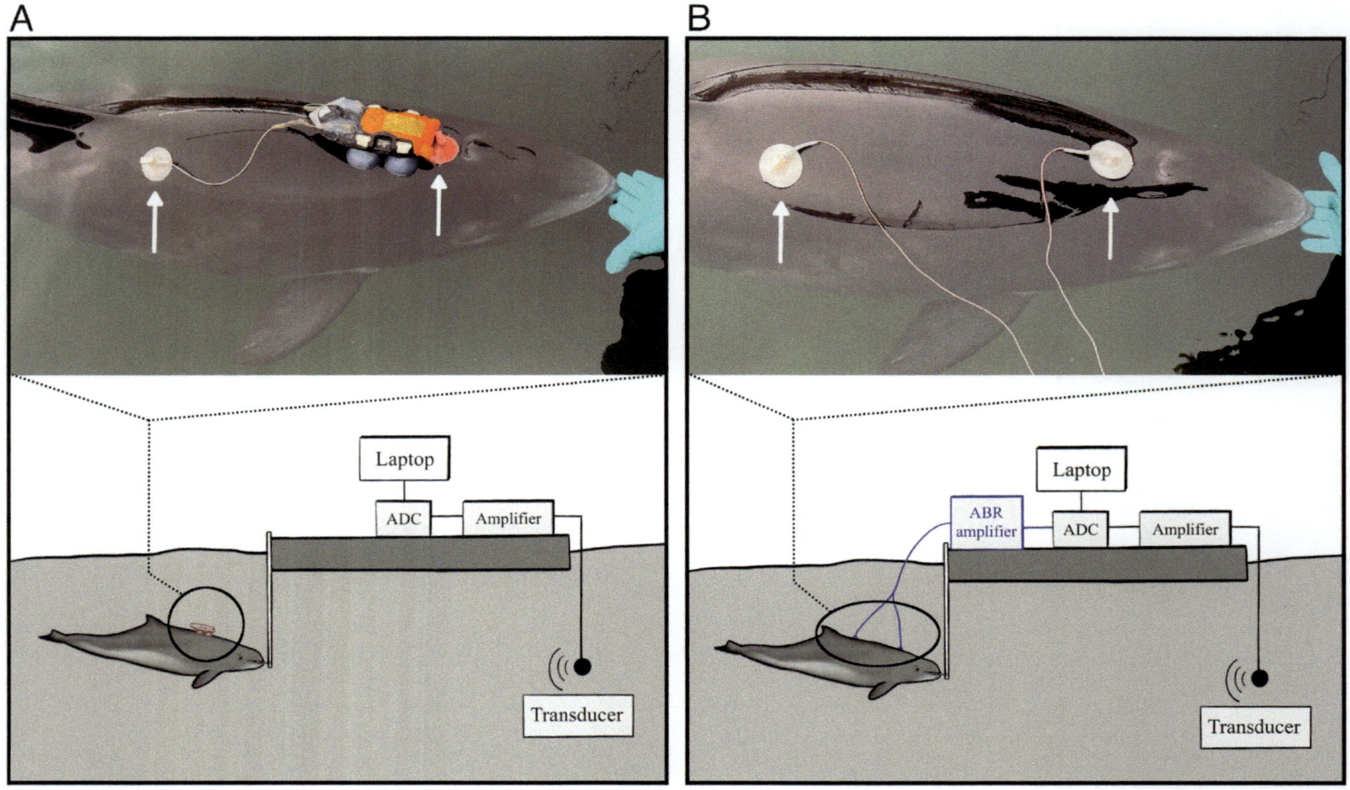

FIG. 4 New tools in auditory physiology. Pictures (top panels) and diagrams (bottom panels) illustrating the on-animal placements and equipment setup of the AEP recording system. (A) Shows the DTAG3-ECG, and AEP tag, and (B) shows a "standard" surface-based ("tethered") data acquisition system. White arrows indicate the placements of the suction cup electrodes. ADC analog-to-digital converter. *Modified from Smith et al. (2021).*

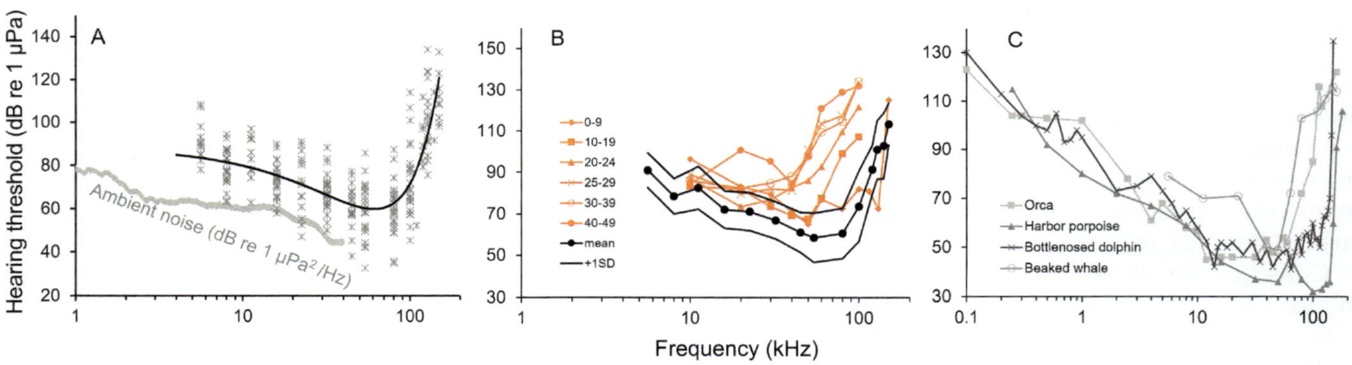

FIG. 5 Evoked potential audiograms of 26 belugas (*Delphinapterus leucas*). (A) Hearing thresholds across all frequencies measured and the modeled audiogram fitted to those data ($y = 0.0077x^2 - 0.9383x + 88.41$; $R^2 = 0.56$) of those 26 wild belugas. Animals were caught, studied, and released in Bristol Bay, Alaska, USA. The modeled data reduce the influence of any one particular audiogram or individual, and such methods are now used to address noise impacts to a species. *Modified from Mooney et al. (2018b).* (B) Comparing the mean beluga audiogram (± quartiles) to age-based, mean audiograms of bottlenose dolphins (*Tursiops truncatus*). *Modified from Houser and Finneran (2006) and Mooney et al. (2018b).* (C) Audiograms of 4 odontocete species, showing differences between taxa and some hearing groups. Hearing curves include the killer whale (*Orcinus orca*), harbor porpoise (*Phocoena phocoena*), bottlenose dolphin (*Tursiops. truncatus*), and a Blainville's beaked whale (*Mesoplodon densirostris*). *Adapted from Pacini et al. (2011), Branstetter et al. (2017), Kastelein et al. (2002), and Johnson (1967).*

loss can potentially play a role in strandings since stranded animals are more likely to suffer from hearing loss than others (measured when caught in fishing gear or in captivity) in the same population (Mann et al., 2010).

The AEP method has also been used to address a variety of questions regarding sensory physiology. For example, auditory processing measurements of pulsed sounds (similar to echolocation clicks) suggest that odontocetes can follow pulses well at modulation rates up to 1250 Hz, after which response amplitudes declined and even extinguished beyond 2500 Hz (Fig. 3; Mooney et al., 2006; Smith et al., 2020). Furthermore, response latencies of a delphinid's auditory system are extremely short; the delay of the first wave, called PI, nears 2.76 ms in the Risso's dolphin. These results predict that species such as the Risso's dolphin should have the ability to follow clicks and echoes while foraging at close range.

Similarly, Popov and Supin (2009) used AEP methods to study directional hearing by mapping the receiving beam patterns (e.g., angle of best sensitivity) of both ears. As expected, they found that beam widths decrease with increasing frequencies (3-dB beam width of beluga: ±33.5° at 8 kHz and ± 14.3° at 128 kHz; 3 dB beam width bottlenose dolphins: ± 19.9° at 8 kHz and ±6.3° at 128 kHz) and that receiving beams were always wider for the beluga than the bottlenose dolphin, possibly because of different rostrum morphologies.

Accomando et al. (2020) investigated directionality of hearing at frequencies that dolphins use in communication (2–30 kHz) psychophysically using a "go/no-go" detection task, in which the dolphin reported the detection of the acoustic stimulus presented at multiple angles around it. They found a surprising directionality with hearing thresholds raised by 18.6 dB when sound was projected from behind the animal in comparison with frontal sound presentation and an 8 dB less sensitive receiving pattern when sound came from above the head rather than from below. Belugas show a similar drop in sensitivity from clicks presented from behind (lowest thresholds), compared to those from the side and in front of the animal (highest thresholds; Mooney et al., 2008), underscoring that shadowing effects of the body could serve to protect animals from overexposure of noise by decreasing received levels at the ears.

Older studies on auditory capabilities of delphinids have demonstrated that frequency difference limens (i.e., the difference in Hz needed to notice a difference between a standard and a comparison frequency, usually expressed as Weber fraction defined as the quotient of difference limen and standard stimulus) for bottlenose dolphins fall between 0.002 and 0.004 for frequencies between 2 and 53 kHz, rising to 0.008 when considering higher frequencies up to 130 kHz (Thompson and Herman, 1975). Thus, their frequency discrimination abilities are comparable to those of humans at lower frequencies but superior at higher frequencies (Thompson and Herman, 1975). Similarly, dolphins have sound intensity discrimination abilities comparable to that of humans since they can discriminate between clicks that differ by only 1–2 dB in received level (Bullock et al., 1968; Dubrovskiy, 1990).

Automatic gain control

Studies addressing echolocation processing have shown that odontocetes have an active auditory "automatic gain control" mechanism by which they can dampen or increase hearing sensitivity to reduce forward masking (a preceding sound influencing hearing abilities) and enable hearing both the high amplitudes of outgoing pulses and the substantially lower levels of returning echoes, all while adjusting for the time-varying delay between the clicks and echoes (Fig. 6; Nachtigall and Supin, 2008). Essentially, these tests have shown that (i) odontocetes can hear high-amplitude outgoing clicks and much lower amplitude returning echoes at comparable physiological (AEP) response levels (i.e., similar hearing levels); (ii) they have protective mechanisms that dampen the response to high-intensity outgoing echolocation pulses, reducing these sensation levels by about 100-fold, thus generating an evoked response similar to those of the returning echoes; and (iii) when echo return levels are lowered, either by reducing target strength or moving it farther away, the auditory evoked responses from these echoes remain at approximately the same amplitude (almost regardless of the intensity of the outgoing click). Yet, if targets are made substantially smaller and thus harder to echolocate, odontocetes will amplify sensation responses of the outgoing echolocation pulse – seemingly to heighten overall hearing sensitivity to keep the echo at a detectable level. Together, these AEP responses underscore that echolocation is an active auditory process.

Such gain control, and protection from high-amplitude sounds, can also be learned or conditioned. In a series of experiments, Nachtigall and colleagues have measured AEP responses to low-amplitude warning stimuli (test tones) during the time an animal might expect to hear a (following) loud sound (Nachtigall et al., 2018). The odontocetes tested [false killer whale (*Pseudorca crassidens*), bottlenose dolphin, harbor porpoise, and beluga whale] showed a dampening of hearing sensitivity by approximately 15 dB prior to the loud sound, suggesting a preconditioning to reduce the impacts of the high-amplitude sound. The extent of this "dampening" is frequency-dependent; for example, in the harbor porpoise, dampening was 17 dB at 45 kHz and 11 dB at 32 kHz (Nachtigall et al., 2016), suggesting the protection effect is neural (i.e., not simply mechanical as in the stapedial reflex in the middle ear in terrestrial mammals). The authors suggest such protection may be afforded by efferent innervation of the hair cells of the basilar membrane.

Noise impacts

The impact of anthropogenic noise has been one of the largest areas of marine mammal research in the past two decades. One particular area of focus has been noise-induced hearing loss or auditory threshold shifts that are typically

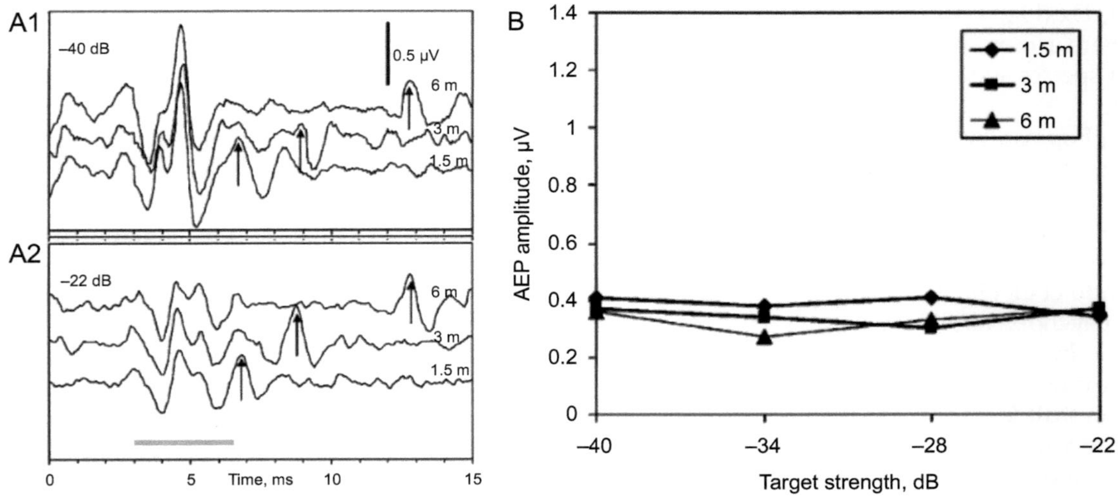

FIG. 6 Changes in hearing sensitivity during echolocation. (A1, A2) AEPs recorded from a false killer whale (*Pseudorca crassidens*) during echolocation of −40 dB (A1) and −22 dB (A2) strength targets at distances of 1.5, 3, and 6 m, as indicated near the traces. The emission-related click AEPs (above the grey bar) are aligned and notably similar. Arrows mark the echo-related responses of different latencies, depending on the target distance. Note that all echo-related AEPs are of approximately equal amplitudes. (B) Echo-related AEP amplitude dependence on target and distance. Note that the AEP level stays constant. Click output did vary but to a lesser extent than needed to compensate by output level alone (Supin et al., 2005).

temporary in experimental settings. Such temporary threshold shifts (TTS) have been induced with sources such as broadband noise, tones, sonar, and simulated explosions (Finneran et al., 2002, 2005; Nachtigall et al., 2004; Mooney et al., 2009b). When the noise is confined to an octave-band (i.e., a frequency range in which the highest frequency is twice the lowest), TTS is maximal at the upper portion of that band or a half-octave above the noise (similar to other mammals; Fig. 7; Mooney et al., 2009a; Popov et al., 2011). Given the place-based model of hearing, whereby frequency discrimination can be enabled by auditory filters, TTS has also been measured using multiple simultaneous tone pips and thus complex AEP responses (Finneran et al., 2007).

Predicting the sound levels and durations of noise, which may induce these impacts, is important for practical conservation, as well as management and regulation of anthropogenic activities to determine when noise might cause harm. When sound exposure levels (dB re: 1 uPa^{2}s) are kept constant, threshold shifts are induced by longer duration exposures, but higher sound energy levels are required for shorter-duration exposures, suggesting that an increasing

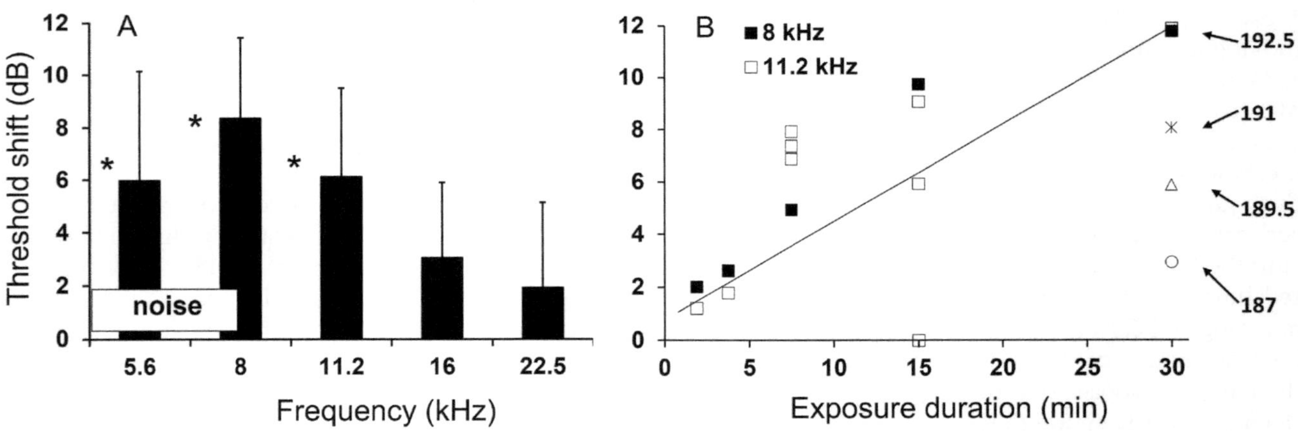

FIG. 7 Temporary threshold shift (TTS) measured physiologically in a bottlenose dolphin (*Tursiops truncatus*). (A) TTS as a function of frequency with the greatest shifts often notable at the center frequency of the noise (5.6 kHz), as well as a higher shift at the limit of the octave band-noise (8 kHz) and a half-octave above the noise (11.2 kHz). The noise here is an octave band (4–8 kHz); its frequency range is depicted by the white "noise" bar overlaying the hearing frequencies tested. (B) The amount of TTS is also dependent on exposure duration, as well as sound exposure level (SEL; dB re 1 uPa^{2}s). Numbers on the right indicate an increasing SEL showing a correspondingly increasing amount of shift. TTS also increased with a constant 192.5 dB SEL (squares) but increasing exposure durations. Two hearing-test frequencies were measured for those data, 8 and 11.2 kHz. *Modified from Mooney et al. (2009a).*

energy model (as duration decreases) is a good predictor of TTS (Mooney et al., 2009a). Thus, for short-duration signals such as sonar pings, exposures would need to be repeated at a very high level (near 214 dB re: 1 uPa^2 s) to induce TTS (Mooney et al., 2009b).

Somewhat alarmingly, there are several physiological hearing studies showing potential noise-induced permanent hearing loss in odontocetes. One study noted widespread hearing loss (10–100 kHz) across the most sensitive hearing range of a stranded melon-headed whale and suggested the loss was from transient, intense anthropogenic sonar or chronic shipping noise exposures (Wang et al., 2021). Another study addressed the hearing of several stranded spinner dolphins (*Stenella longirostris*) and rough-toothed dolphins (*Steno bredanensis*) and noted three of the four animals showed profound hearing loss, and no hearing response beyond 22.5 kHz. Given the timing of some nearshore explosions around the time of the stranding, the authors suggest hearing loss was induced by the acoustic impacts of dynamite fishing (Pacini et al., 2016).

Beyond hearing loss, noise may mask (Branstetter and Sills, 2022) or saturate the auditory system and limit the detection of biologically important sounds. First masking studies were published shortly after the first hearing studies using behavioral methods to evaluate detection abilities in the presence of masking noise (e.g., Johnson, 1968, 1971). Generally, it is thought that the auditory system acts as a series of overlapping band-pass filters and each filter processes frequency energy within a certain range (a critical band) while attenuating peripheral frequency energy (Fig. 8). Interestingly, at least for porpoises, auditory filter bandwidths are somewhat constant as frequency increases, providing them with improved frequency and temporal resolution (Fig. 8; Popov et al., 2006). The relationship between the center frequency of a filter and the bandwidth of a filter can be described as a quality factor, *Q*. While porpoises (and some bats) may narrow their Q (and thus filter bandwidth at higher frequencies), other odontocetes seem to have a more constant-quality filter (and thus perhaps lower-frequency resolution at higher frequency bands). Frequency is not the only component of masking. Direction of the sound and essentially shadowing by the body and head (and the receiving beam of the animal) can provide spatial release from masking because a sound arriving from one direction is usually not masked by noise from other directions (Mooney et al., 2008; Popov et al., 2020). Temporal fluctuations in a noise may also improve signal detection, despite frequency overlapping (Branstetter and Finneran, 2008; Trickey et al., 2010). Thus, it is critical to evaluate a range of parameters when considering if masking is likely to occur.

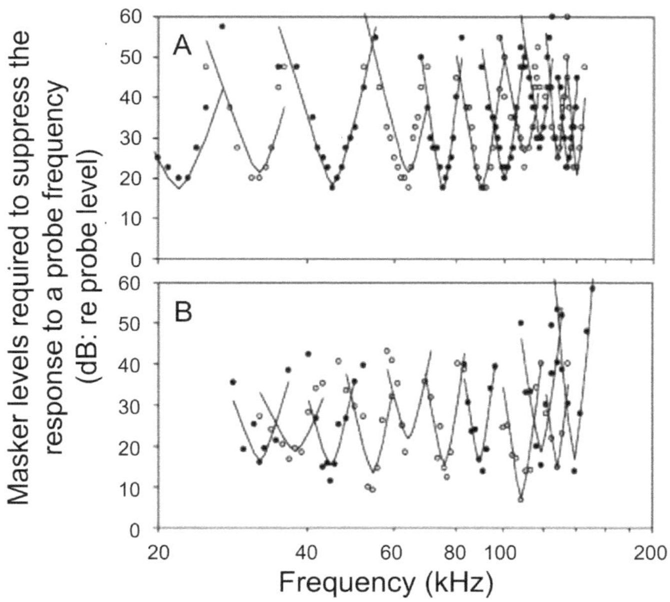

FIG. 8 Auditory filter shapes obtained in (A) the harbor porpoise (*Phocoena phocoena*) and (B) the finless porpoise (*Neophocaena phocaenoides*). The curves are masked thresholds as functions of masker frequency. Dots represent experimental data, curves—roex approximations. Note that the thresholds shown here are ones that indicate at what masker level above the probe level an amplitude envelope following response (EFR) to a probe frequency disappears depending on the probe-to-masker frequency difference. Thus, the masker is most efficient when the threshold is lowest. At this minimum, the EFR can be suppressed by the masker at low levels because the masker frequency at that point is the same as the probe signal that the animal needs to detect. Dots to the left and right of the probe frequency (the minimum of each threshold) are measurements when the masker frequency was shifted off the probe frequency. Each line is the threshold at one probe frequency. 12 probe frequencies were tested in (A) and 11 were tested in (B). Generally, for odontocetes, the bandwidths of auditory filters widen as frequency increases giving them improved temporal resolution. Yet, for porpoises (and some bats), the filter bandwidth narrows at higher frequencies (particularly notable in A) providing both high-temporal and high frequency resolution, especially at the frequencies of echolocation, near 120 kHz. *Modified from Popov et al. (2006).*

Touch and hydrodynamics

From an observational viewpoint, touch and hydrodynamics play a clear role in odontocete behavior and ecology. Many delphinids contact each other via the flukes, flippers, and rostrum during social and sexual interactions (Herzing, 1996; Dudzinski et al., 2010). It has been suggested that acoustic signals such as burst pulses have been directed at conspecifics' genital areas (Herzing, 1996). Belugas (Smith et al., 1992) and false killer whales (T. A. Mooney, personal obs.) have been noted to self-rub against structures for both play and hygiene, supported by tactile sensation. Navigationally, riverine odontocetes have specific swimming behaviors during which the animal makes contact with the benthos (e.g., side-swimming by the Indus River dolphins (*Plantanista minor*); Pilleri, 1974), or swim very near to the benthos (Bristol Bay belugas), likely for orientation to the bottom in very shallow, dynamic river environments (Mooney et al., 2018a; Castellote et al., 2021). Such sensing may provide key cues in these turbid environments where vision is limited and echolocation or passive listening may be acoustically challenging.

Mechanosensory stimuli can be detected by the skin or by vibrissae. To date, there are relatively few physiological measurements of the sensitivity of odontocete skin to mechanosensory stimulation. Ridgway and Carder (1990, 1993) measured somatic evoked potentials to an array of stimulus types, including finger taps and water drops, finding areas of greatest sensitivity around the head, including the gape of the jaws and around the eyes, with less but noticeable sensitivity around the blowhole. These results are comparable to a previous study, which found lowest thresholds below the eye and near the blowhole (Kolchin and Belkovich, 1973 described in Bauer et al., 2018). These pressure thresholds were $10\,\text{mg/mm}^2$ for the head and $40\,\text{mg/mm}^2$ at the body trunk; responses that are comparable to those of human hands and suggest the ability to detect water turbulence (Ridgway and Carder, 1990, 1993). Moreover, substantial innervation and cutaneous ridges have been observed around the body of many odontocetes [although the ridges are notably faint or absent on most of the head and ventral region in some species (Ridgway and Carder, 1990, 1993; Shoemaker and Ridgway, 1991)]. The ridges tend to run circumferential around the body trunk. The function of the cutaneous ridges and innervation has not been established, but it has been speculated that they could support tactile sensing, pressure detection, and/or sensing hydrodynamic characteristics around an animal (Ridgway and Carder, 1990, 1993; Shoemaker and Ridgway, 1991). A pronounced, serrated leading edge on Commerson's dolphin (*Cephalorhynchus commersonii*) and other dolphin pectoral fins is used to make physical contact with conspecifics, suggesting that mechanoreception plays an important role in their social behavior (Johnson and Moewe, 1999).

Vibrissae can be found in odontocete fetuses and neonates. These hairs are lost shortly before or after birth in most odontocetes. In contrast, mysticetes have numerous sensory hairs in the facial region, as well as close to the blowhole even as adults (see, for example, Drake et al., 2015; Reichmuth et al., 2022). In young animals, vibrissae may help maintaining close contact with their mother during nursing. This could include providing information about the position of the neonate's mouth relative to the mother, and the mother's nipple, as well as aiding in stimulating milk ejection via touching the papilla mammae (Gerussi et al., 2021).

There are also reports that some adult odontocetes have small hairs within the follicle [i.e., the bottlenose dolphin (Gerussi et al., 2021) and two species of river dolphins (*Inia geoffrensis* and *Platanista* spp; Ling, 1977)]. These hair shafts could aide in proprioception, provide information about water movements around the head, or serve mechanoreception of prey (Ling, 1977; Gerussi et al., 2021). Given the limitation of vision in cetaceans and their habitats, such detection could prove important, but experimental tests would aid in supporting this hypothesis. For bottlenose dolphins, Gerussi et al. (2021) described a hair shaft lying entirely within the vibrissal follicle, resembling the follicle–sinus complex (FSC) of other marine mammal vibrissae. However, Hüttner et al. (2022) found the adult vibrissal follicles ("vibrissal crypts") to be different from a classic FSC and could not find a hair. In behavioral experiments, Hüttner et al. (2022) described the vibrissal crypts to be functional electroreceptors (see Section "Electroreception").

Electroreception

Electroreception is the ability of an animal to sense or perceive electric fields such as those generated by prey items or conspecifics. Two delphinids, the bottlenose dolphin and the Guiana dolphin, have recently been described being able to sense weak electric fields with most likely the empty vibrissal crypts on the rostrum of the dolphins serving as electroreceptors (Fig. 9; Czech-Damal et al., 2012; Hüttner et al., 2022; for review, see Dehnhardt et al., 2020). Electroreception might assist benthic feeding behavior. In bottlenose dolphin, for example, the dolphins could use electroreception to detect their prey hidden in the substrate during crater-feeding (Rossbach and Herzing, 1997).

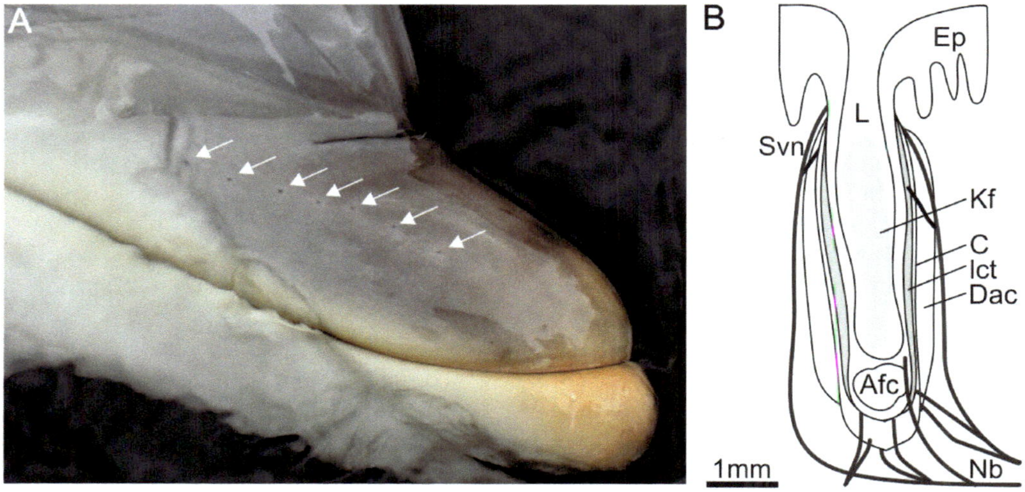

FIG. 9 (A) Rostrum of a bottlenose dolphin (*Tursiops. truncatus*) showing seven vibrissal crypts marked with white arrows. (*Photo: Tim Hüttner*). (B) Schematic drawing of a vibrissal crypt of a dolphin. The crypt is an invagination of the epidermis (Ep). Its lumen is filled with a meshwork of keratinous fibers (Kf). The crypt is densely innervated. Afc, agglomeration of fat cells; C, capsule; Dac, dermal layer with adipose cells; Ep, epidermis; Ict, loose connective tissue; Kf, keratinous fibers; L, lumen; Nb, nerve bundles; Svn, superficial vibrissal nerve. Scale 1 mm. *Modified from Hüttner et al. (2022).*

Originally, these crypts were described as vestigial structures (Yablokov and Klevezal, 1969; Ling, 1977), as they are lacking most characteristics of FSCs. First evidence that these crypts are functional sensory units was obtained when examining the crypts with an infrared thermographic camera revealing an intense thermal radiation at the crypts (Mauck et al., 2000). Subsequent histological examination (Czech-Damal et al., 2012; Hüttner et al., 2022) described the crypts to be well supplied with blood, highly innervated, and of ampullary shape reminiscent of the ampullae of Lorenzini of sharks (Fig. 9B). With this similarity in mind, researchers initiated a successful behavioral study training a Guiana dolphin on weak electric fields. In this first experiment, the Guiana dolphin's sensory threshold was determined at 4.6 μV/cm (Czech-Damal et al., 2012), thus showing a sensitivity that is comparable to an "electro-specialist," the platypus (*Ornithorhynchus anatinus*) (for review, see Czech-Damal et al., 2013). Moreover, strong evidence for the vibrissal crypts acting as electroreceptors was obtained during the training with electrosensory stimuli: the crypts of the Guiana dolphin increased in diameter, and when the crypts were covered by a plastic shell, the dolphin did not respond to even high electric fields (Czech-Damal et al., 2012). The transduction process, however, still needs to be addressed in future studies.

From histological work, including vibrissal crypts of numerous odontocetes (Czech, 2007), an electroreceptive function of the follicle crypts of the bottlenose dolphin besides the Guiana dolphin was deduced. Indeed, recently Hüttner et al. (2022) showed that four bottlenose dolphins spontaneously reacted to electric stimuli during the first trial of presentation and three of the four dolphins successfully transferred a go/no-go paradigm, in which the animal needed to give a response ("go") every time it perceived a stimulus, but needed to stay ("no-go") when it did not perceive a stimulus, from acoustic, optic, and tactile to electrosensory stimuli of 1.5 mV/cm^2 and 0.5 mv/cm^2 electric field strength throughout testing. Thresholds for electrosensory stimuli could be determined for two dolphins at 2.4 μV/cm and 5.5 μV/cm, respectively (Hüttner, 2022).

Chemosensation

Olfaction

Chemosensitivity for odors is a key component of the mammalian perception of the environment. By using receptor combinations to encode different odors, humans and other mammals are estimated to distinguish over 1 trillion olfactory stimuli (Busdid et al., 2014). With the transition to a fully aquatic lifestyle, these inputs have reduced their relevance to delphinids since they hold their breath most of the time and only inhale air from outside their main environment. Furthermore, it has been hypothesized that the substantial re-organization of the nasal passages for complex sound production may not have been compatible with the sensory function of the anterior olfactory system with mechanical stress leading to the reduction of the olfactory mucous membrane (Cozzi et al., 2017). Over time, the main olfactory structures in delphinids have either disappeared or been greatly reduced. On the genetic level, this is reflected in a high proportion of pseudogenes in the olfactory receptor genome (Kishida et al., 2007), a very low number of functional olfactory receptor genes (Kishida et al., 2015), and the inactivation of the olfactory marker protein gene in odontocetes (Springer and Gatesy, 2017). In terms of sensory cells and neural pathways, the vomeronasal system, the olfactory nerve, and the olfactory bulb are not present in adult dolphins (Oelschläger, 1992). Thus, it appears toothed whales have lost the sense of smell.

Nevertheless, behavioral tests were specifically employed to study olfaction in dolphins (Kremers et al., 2016b; Bouchard et al., 2022). A barrel with rotting fish and an empty control barrel was left by a dolphin pool, and the researchers compared the frequency of surfacings next to each barrel over several days. The authors found that dolphins surfaced more often next to the fish barrel than the control barrel on later days when it had developed a strong, rotting smell. This is a surprising result given the drastic reduction of olfactory structures mentioned previously. Unfortunately, reactions of all animals were pooled and not much detail was given about the placing of barrels over the days so that confounding effects cannot be excluded. Bouchard et al. (2022) tested reactions of bottlenose dolphins and pilot whales in the wild to in air and underwater release of dimethyl sulfide, which is released by phytoplankton and indicates high-productivity areas. The authors found no reactions except for a close-range approach by bottlenose dolphins once they were within the underwater plume of the release suggesting gustatory but not olfactory detection. Altogether, more research on olfactory perception needs to be conducted to characterize olfaction further.

Gustation

Chemosensitivity for tastes has also been reduced during evolution. Delphinids lost four of the five basic taste receptor genes with only the salty taste receptor genes still intact (Feng et al., 2014), and taste buds are either absent or greatly reduced on the tongue of delphinids (for review, see Cozzi et al., 2017). However, the main neural pathways for taste reception are preserved or even expanded as is the case for the olfactory tubercle, epithalamus, and the mediodorsal nucleus and the main cranial nerves carrying taste information in mammals (VIIth, IXth, and Xth cranial nerves) (Cozzi et al., 2017). While these structures also have other functions besides carrying taste information, behavioral research suggests that delphinids can distinguish more gustatory stimuli than their anatomy and molecular biology would suggest. Kuznetzov (1990; summarized in Nachtigall, 1986) reported that bottlenose dolphins could detect and discriminate a variety of substances in water by taste alone. A more controlled detection test using solutions of sour, bitter, salty, and sweet tastes indicated that a trained bottlenose dolphin detected all of these when compared to distilled water but at thresholds much lower than those found in humans (Friedl et al., 1990). Several studies also found food preferences by presenting dolphins with different fish taste samples (Dill et al., 2003; Kremers et al., 2016b; Bouchard et al., 2017), but often confounding effects of how stimuli were presented could not be excluded. Taste discrimination on an advanced level has been demonstrated by Bruck et al. (2022) by pouring dolphin urine samples in front of untrained dolphins and measuring the time they spent investigating the plume. Bottlenose dolphins spent more time investigating urine than water samples and investigating urine from familiar individuals than samples from unfamiliar dolphins. When urine samples were paired with individually distinctive signature whistle playbacks of conspecifics, dolphins spent longer investigating the area if the urine sample came from the signature whistle owner than if the samples came from different individuals. This demonstrates that dolphins can recognize individuals by the taste of their urine, which is a fairly sophisticated form of taste discrimination making it likely that other information such as sex, diet, and hormonal status can also be decoded by taste.

Anatomical evidence for additional gustatory sensors comes from cells that resemble gustatory von Ebner's glands that help with lipid hydrolysis on the tongue (Ferrando et al., 2010) and some reports of modified taste receptors at the base of the tongue at least in young individuals (Suchowskaja, 1972; Yamasaki et al., 1978). Given the behavioral evidence, it is likely that other biochemical routes for taste perception are still undiscovered. For example, dolphins possess positively selected genes for CD36 proteins (Wang et al., 2015), which enable lipid taste perception in other mammals (Laugerette et al., 2005). Furthermore, studies on gustatory receptors and possible neural pathways are still needed to understand how dolphins perceive taste.

Conclusions

Our review has revealed a rich literature on a multitude of aspects of delphinid sensory physiology. However, the information is unbalanced in at least two ways. First, most sensory research has centered on the bottlenose dolphin. However, delphinids vary tremendously in almost all aspects of their biology leading Jeffersen and LeDuc (2018), to

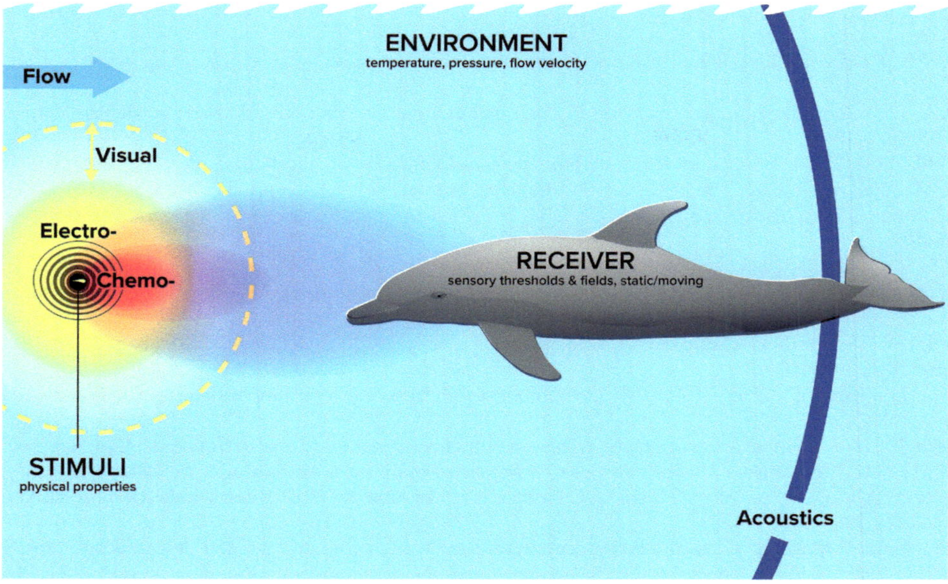

FIG. 10 Overview of delphinid sensory physiology. To fully understand how delphinids perceive their world, we need to consider sensory stimuli and their physical properties, the environment (e.g., temperature, pressure, and flow velocity), sensory fields and thresholds of the receiver, and receiver movement.

call it a "taxonomic trashbox." Thus, results obtained in one species cannot necessarily be generalized to another. While systematic research in one species such as the bottlenose dolphin might advance the field by resolving long-lasting mysteries, such as amphibious vision, or by documenting population or individual differences, future research needs to focus on sensory physiology in other delphinids to assess variability. Second, audition and vision have been studied a lot more than any of the other senses. For example, until very recently we knew little about their skills in electroreception or taste recognition of other individuals. Even in the better-studied modalities, studies are often not building on each other. For vision, a multitude of studies has been conducted but without a clear framework so that key aspects have not been addressed or contradictory results do not allow us to draw clear conclusions. Adding new sensory skills to our picture of dolphin perception raises the question how much they rely on each sense and what information they extract from where. New findings reveal a diversity of inputs that we have not fully acknowledged in the past. Finally, most experiments are carried out on stationary animals, which is not the most common scenario in the wild. Additional flow noise caused by movement can lead to a more difficult noise field, but movement also allows animals to sample stimuli from different positions and integrate information across these, adding another facet to the complex perception of the environment. Thus, we need a more inclusive research approach that does not focus on primary senses but on the integration of information across senses in the marine environment (Fig. 10).

Observing this imbalance in sensory research regarding the sensory modalities, we should not be misled to believe that a large amount of studies suggests priority of one modality over another. Audition, about which much has been published in delphinids, is clearly an important sense, but the low transmission loss for sounds also creates challenges such as substantial background noise that receivers have to cope with. Low transmission loss also makes it difficult to limit interactions to selected receivers (Janik, 2005) . Thus, other modalities may be more important than we have previously recognized.

This review has shown that the senses of delphinids show numerous adaptations that will most likely allow different senses to take on different roles as animals interact dynamically with the environment. Cognitive studies have demonstrated that dolphins represent their environment in a multimodal fashion (Harley et al., 2003; Bruck et al., 2022). A key aspect of future research should be to understand this interplay of the senses and their representation to gain a more accurate picture of the dolphin's perception of the world (DeVreese et al., 2023).

References

Accomando, A.W., Mulsow, J., Branstetter, B.K., Schlundt, C.E., Finneran, J.J., 2020. Directional hearing sensitivity for 2–30 kHz sounds in the bottlenose dolphin (*Tursiops truncatus*). J. Acoust. Soc. Am. 147 (1), 388–398.

Au, W.W.L., Lemonds, D.W., Vlachos, S., Nachtigall, P.E., Roitblat, H.L., 2002. Atlantic bottlenose dolphin (*Tursiops truncatus*) hearing thresholds for brief broadband signals. J. Comp. Psychol. 116 (2), 151–157.

Au, W.W.L., Moore, P.W.B., 1984. Receiving beam patterns and directivity indices of the Atlantic bottlenose dolphin, *Tursiops truncatus*. J. Acoust. Soc. Am. 75 (1), 255–262.

Au, W.W.L., Moore, P.W.B., Pawloski, D.A., 1988. Detection of complex echoes in noise by an echolocating dolphin. J. Acoust. Soc. Am. 83 (2), 662–668.

Au, W.W.L., Snyder, K.J., 1980. Long-range target detection in open waters by an echolocating Atlantic bottlenose dolphin (*Tursiops truncatus*). J. Acoust. Soc. Am. 68, 1077–1084.

Bauer, G.B., Johnson, C.M., 1994. Trained motor imitation by bottlenose dolphins (*Tursiops truncatus*). Percept. Mot. Skills 79, 1307–1315.

Bauer, G.B., Reep, R.L., Marshall, C.D., 2018. The tactile senses of marine mammals. Int. J. Comp. Psychol. 31, 1–28.

Bischoff, N., Nickle, B., Cronin, T.W., Velasquez, S., Fasick, J.I., 2012. Deep-sea and pelagic rod visual pigments identified in the mysticete whales. Vis. Neurosci. 29, 95–103.

Bjerager, P., Heegaard, S., Tougaard, J., 2003. Anatomy of the eye of the sperm whale (*Physeter macrocephalus* L.). Aquat. Mamm. 29 (1), 31–36.

Blois-Heulin, C., Crevel, M., Böye, M., Lemasson, A., 2012. Visual laterality in dolphins: importance of the familiarity of stimuli. BMC Neurosci. 13, 9–16.

Bouchard, B., Barnagaud, J.-Y., Verborgh, P., Gauffier, P., Campagna, C., Celerier, A., 2022. A field study of chemical senses in bottlenose dolphins and pilot whales. Anat. Rec. 305, 668–679.

Bouchard, B., Lisney, T.J., Campagna, C., Celerier, A., 2017. Do bottlenose dolphins display behavioural response to fish taste? Appl. Anim. Behav. Sci. 194, 120–126.

Branstetter, B.K., Finneran, J.J., 2008. Comodulation masking release in bottlenose dolphins (*Tursiops truncatus*). J. Acoust. Soc. Am. 124 (1), 625–633.

Branstetter, B.K., Sills, J.M., 2022. Mechanisms of auditory masking in marine mammals. Anim. Cogn. 25, 1029–1047.

Branstetter, B.K., St. Leger, J., Acton, D., Stewart, J., Houser, D., Finneran, J.J., Jenkins, K., 2017. Killer whale (*Orcinus orca*) behavioral audiograms. J. Acoust. Soc. Am. 141 (4), 2387–2398.

Bruck, J.N., Walmsley, L., Janik, V.M., 2022. Cross-modal perception of identity by sound and tate in bottlenose dolphins. Sci. Adv. 8, eabm7684.

Bruns, V., Schmieszek, E.T., 1980. Cochlear innervation in the greater horseshoe bat: demonstration of an acoustic fovea. Hear. Res. 3, 27–43.

Bullock, T.H., Grinnell, A.D., Ikezono, F., Kameda, K., Katsuki, Y., Nomoto, M., Sato, O., Suga, N., Yanagisava, K., 1968. Electrophysiological studies of the central auditory mechanisms in cetaceans. Z. Vergl. Physiol. 59, 117–156.

Busdid, C., Magnasco, M.O., Vosshall, L.B., Keller, A., 2014. Humans can discriminate more than 1 trillion olfactory stimuli. Science 343, 1370–1372.

Cartee, R.E., Brosemer, K., Ridgway, S.H., 1995. The eye of the bottlenose dolphin (*Tursiops truncatus*) evaluated by B mode ultrasonography. J. Zoo Wildl. Med. 26 (3), 414–421.

Castellote, M., Mooney, A., Andrews, R., Deruiter, S., Lee, W.-J., Ferguson, M., Wade, P., 2021. Beluga whale (*Delphinapterus leucas*) acoustic foraging behavior and applications for long term monitoring. PLoS One 16 (11), e0260485.

Czech-Damal, N., Liebschner, A., Miersch, L., Klauer, G., Hanke, F.D., Marshall, C.D., Dehnhardt, G., Hanke, W., 2012. Electroreception in the Guiana dolphin (*Sotalia guianensis*). Proc. R. Soc. Lond. B 279 (1729), 663–668.

Czech-Damal, N.U., Dehnhard, G., Manger, P., Hanke, W., 2013. Passive electroreception in aquatic mammals. J. Comp. Physiol. A 199, 555–563.

Cozzi, B., Huggenberger, S., Oelschläger, H., 2017. Anatomy of Dolphins: Insights into Body Structure and Function. Academic Press, London.

Czech, N., 2007. Functional Morphology and Postnatal Transformation of Vibryssal Crypts in Toothed Whales (Odontoceti). Ruhr-University (PhD thesis), Bochum.

Dawson, W., Birndorf, L., Perez, J., 1972. Gross anatomy and optics of the dolphin eye. Cetology 10, 1–12.

Dawson, W.W., 1980. The cetacean eye. In: Herman, L.M. (Ed.), Cetacean Behavior: Mechanisms and Processes. Academic Press, New York, pp. 53–100.

Dawson, W.W., Adams, C.K., Barris, M.C., Litzkow, C.A., 1979. Static and kinetic properties of the dolphin pupil. Am. J. Physiol. 237 (5), R301–R305.

Dawson, W.W., Carder, D.A., Ridgway, S.H., Schmeisser, E.T., 1981. Synchrony of dolphin eye movements and their power density spectra. Comp. Biochem. Physiol. A: Physiol. 68, 443–449.

Dawson, W.W., Hawthorne, M.N., Jenkins, R.L., Goldston, R.T., 1982. Giant neural systems in the inner retina and optic nerve of small whales. J Comp Neurol 205, 1–7.

Dawson, W.W., Hope, G.M., Ulshafer, R.J., Hawthrone, M.N., Jenkins, R.L., 1983. Contents of the optic nerve of a small cetacean. Aquat. Mamm. 10 (2), 45–56.

Dawson, W.W., Perez, J.M., 1973. Unusual retinal cells in the dolphin eye. Science 181 (4101), 747–749.

Dawson, W.W., Schroeder, J.P., Dawson, J.F., 1987a. The ocular fundus of two cetaceans. Mar. Mamm. Sci. 3 (1), 1–13.

Dawson, W.W., Schroeder, J.P., Sharpe, S.N., 1987b. Corneal surface properties of two marine mammal species. Mar. Mamm. Sci. 3 (2), 186–197.

Dehnhardt, G., Miersch, L., Marshall, C.D., Von Fersen, L., Hüttner, T., 2020. Passive electroreception in mammals. In: Fritzsch, B. (Ed.), The Senses: A Comprehensive Reference. Elsevier, pp. 385–392.

DeVreese, S., Sörensen, K., Biolsi, K., Fasick, J.I., Reidenberg, J.S., Hanke, F.D., 2023. Open questions in marine mammal sensory research. Biol. Open 12, bio059904.

Dill, L.M., Dill, E.S., Charles, D., 2003. Feeding preferences of the Monkey Mia dolphins: results from a simultaneous choice protocoll. Mar. Mamm. Sci. 19, 650–660.

Drake, S.E., Crish, S.D., George, J.C., Stimmelmayr, R., Thewissen, J.G.M., 2015. Sensory hairs in the bowhead whale, *Balaena mysticetus* (Cetacea, Mammalia). Anat. Rec. 298, 1327–1335.

Dral, A.D.G., 1972. Aquatic and aerial vision in the bottle-nosed dolphin. Neth. J. Sea Res. 5, 510–513.

Dral, A.D.G., 1974. Problems of image-focusing and astigmatism in cetacea—a state of affair. J. Anatomy Mamm. 2, 22–28.

Dral, A.D.G., 1975. Some quantitative aspects of the retina of *Trusiops truncatus*. Aquat. Mamm. 2, 28–31.

Dral, A.D.G., 1977. On the retinal anatomy of cetaceans. In: Harrison, R.J. (Ed.), Functional Anatomy of Marine Mammals. Academic, London, pp. 81–134.

Dral, A.D.G., 1983. The retinal ganglion cells of *Delphinus delphis* and their distribution. Aquat. Mamm. 10 (2), 57–68.

Dubrovskiy, N.A., 1990. On the two auditory subsystems in dolphins. In: Thomas, J.A., Kastelein, R.A. (Eds.), Sensory Abilities of Cetaceans: Laboratory and Field Evidence. Springer, pp. 233–254.

Dudzinski, K.M., Gregg, J.D., Paulos, R.D., Kuczaj II, S.A., 2010. A comparison of pectoral fin contact behaviour for three distinct dolphin populations. Behav. Processes 84 (2), 559–567.

Fasick, J.I., Cronin, T.W., Hunt, D.M., Robinson, P.R., 1998. The visual pigments of the bottlenose dolphin (*Tursiops trunatus*). Vis. Neurosci. 15, 643–651.

Fasick, J.I., Robinson, B.H., 1998. Mechanism of spectral tuning in the dolphin visual pigments. Biochemistry 37 (2), 433–438.

Fasick, J.I., Robinson, P.R., 2000. Spectral-tuning mechanisms of marine mammal rhodopsins and correlations with foraging depth. Vis. Neurosci. 17, 781–788.

Fasick, J.I., Robinson, P.R., 2016. Adaptations of cetacean retinal pigments to aquatic environments. Front. Ecol. Evol. 4 (70).

Feng, P., Zheng, J., Rossiter, S.J., Wang, D., Zhao, H., 2014. Massive losses of taste receptor genes in toothed and baleen whales. Genome Biol. Evol. 6, 1254–1265.

Ferrando, T., Caresano, F., Ferrando, S., Gallus, L., Wurtz, M., Tagliafierro, G., 2010. The tongue morphology and lingual gand histochemistry of Ligurian Sea odontocetes. Mar. Mamm. Sci. 26, 588–601.

Finneran, J.J., Carder, D.A., Schlundt, C.E., Ridgeway, S.H., 2005. Temporary threshold shift in bottlenose dolphins (*Tursiops truncatus*) exposed to mid-frequency tones. J. Acoust. Soc. Am. 118 (4), 2696–2705.

Finneran, J.J., Houser, D.S., 2006. Comparison of in-air evoked potential and underwater behavioral hearing thresholds in four bottlenose dolphins (*Tursiops truncatus*). J. Acoust. Soc. Am. 119 (5), 3181–3192.

Finneran, J.J., Houser, D.S., Mase-Guthrie, B., Ewing, R.Y., Lingenfelser, R.G., 2009. Auditory evoked potentials in a stranded Gervais' beaked whale (*Mesoplodon europaeus*). J. Acoust. Soc. Am. 126 (1), 484–490.

Finneran, J.J., Mulsow, J., Houser, D.S., Burkard, R.F., 2016. Place specificity of the click-evoked auditory brainstem response in the bottlenose dolphin *(Tursiops truncatus)*. J. Acoust. Soc. Am. 140 (4), 2593–2602.

Finneran, J.J., Schlundt, C.E., Branstetter, B.K., Dear, R., 2007. Assessing temporary threshold shift in a bottlenose dolphin (*Tursiops truncatus*) using multiple simultaneous auditory evoked potentials. J. Acoust. Soc. Am. 122 (2), 1249–1264.

Finneran, J.J., Schlundt, C.E., Dear, R., Carder, D.A., Ridgway, S.H., 2002. Temporary shift in masked hearing thresholds in odontocetes after exposure to single underwater impulses from a seismic watergun. J. Acoust. Soc. Am. 111 (6), 2929–2940.

Forestell, P.H., Herman, L.M., 1988. Delayed matching of visual materials by a bottlenosed dolphin aided by auditory symbols. Anim. Learn. Behav. 16 (2), 137–146.

Friedl, W.A., Nachtigall, P.E., Moore, P.W.B., Chun, N.K.W., 1990. Taste reception in the bottlenose dolphin (*Tursiops truncatus gilli*) and the California sea lion (*Zalophus californianus*). In: Thomas, J.A., Kastelein, R.A. (Eds.), Sensory Abilities of Cetaceans. Plenum Press, New York, pp. 447–451.

Gacek, R.R., Rasmussen, G.L., 1961. Fiber analysis of the statoacoustic nerve of guinea pig, cat, and monkey. Anat. Rec. 139 (4), 455–463.

Gao, G., Zhou, K., 1992. Fiber analysis of the optic and cochlear nerves of small cetaceans. In: Thomas, J., Kastelein, R.A., Supin, A.Y. (Eds.), Marine Mammal Sensory Systems. Plenum Press, New York, pp. 39–52.

Gerussi, T., Graïc, J.M., De Vreese, S., Grandis, A., Tagliavia, C., De Silva, M., Huggenberger, S., Cozzi, B., 2021. The follicle-sinus complex of the bottlenose dolphin (*Tursiops truncatus*). Functional anatomy and possible evolutional significance of its somato-sensory innervation. J. Anat. 238 (4), 942–955.

Griebel, U., Schmid, A., 2002. Spectral sensitivity and color vision in the bottlenose dolphin (*Tursiops truncatus*). Mar. Freshw. Behav. Physiol. 35 (3), 129–137.

Hall, N.R., Hall, H.I., Caldwell, M.C., Caldwell, D.K., 1972. Visual acuity under conditions of near darkness and discrimination learning in the Atlantic bottle-nosed dolphin (*Tursiops truncatus*). Cetology 12, 1–7.

Harley, H.E., Putman, E.A., Roitblat, H.L., 2003. Bottlenose dolphins perceive object features through echolocation. Nature 424, 667–669.

Harvey, R., 2019. Comparative analysis of ceteacean eye morpholoy using micro-computed tomography. Duke University (Honors thesis), Surham, North Carolina.

Herman, L.M., 1975. Interference and auditory short-term memory in the bottlenosed dolphin. Anim. Learn. Behav. 3 (1), 43–48.

Herman, L.M., 1980. Cognitive characteristics of dolphins. In: Herman, L.M. (Ed.), Cetacean Behavior: Mechanisms and Processes. Academic Press, New York, pp. 363–429.

Herman, L.M., 1990. Cognitive performance of dolphins in visually-guided tasks. In: Thomas, J.A., Kastelein, R.A. (Eds.), Sensory Abilities of Cetaceans—Laboratory and Field Evidence. Plenum Press, New York, pp. 455–462.

Herman, L.M., Hovancik, J.R., Gory, J.D., Bradshaw, G.L., 1989. Generalization of visual matching by a bottlenosed dolphin (*Tursiops truncatus*): evidence for invariance of cognitive performance with visual and auditory materials. J. Exp. Psychol. 15 (2), 124–136.

Herman, L.M., Peacock, M.F., Yunker, M.P., Madsen, C.J., 1975. Bottlenosed dolphin: double-slit pupil yields equivalent aerial and underwater diurnal acuity. Science 189, 651–652.

Herzing, D.L., 1996. Vocalizations and associated underwater behavior of free-ranging Atlantic spotted dolphins, *Stenella frontalis* and bottlenose dolphins, *Tursiops truncatus*. Aquat. Mamm. 22, 61–80.

Hosokawa, H., 1951. On the extrinsic eye muscles of the whale, with special remarks upon the innervation and function of the musculus retractor bulbi. Scient. Rep. Whale Res. Inst. 6, 1–33.

Houser, D.S., Finneran, J.J., 2006. Variation in the hearing sensitivity of a dolphin population determined through the use of evoked potential audiometry. J. Acoust. Soc. Am. 120 (6), 4090–4099.

Hunter, J., 1787. Observations on the structure and oeconomy of whales. Philos. Trans. R. Soc. Lond. B. 77, 371–450.

Hüttner, T., 2022. Experimental evidence for electroreception in the bottlenose dolphin (*Tursiops truncatus*). University of Rostock (PhD thesis), Rostock.

Hüttner, T., Von Fersen, L., Miersch, L., Czech, N.U., Dehnhardt, G., 2022. Behavioral and anatomical evidence for electroreception in the bottlenos dolphin (*Tursiops truncatus*). Anat. Rec. 305 (3), 592–608.

Jacobs, G.H., 1993. The distribution and nature of colour vision among the mammals. Biol. Rev. 68 (3), 413–471.

Jacobs, M.S., Morgane, P.J., McFarland, W.L., 1975. Degeneration of visual pathways in the bottlenose dolphin. Brain Res. 88, 346–352.

Janik, V.M., 2005. Acoustic communication networks in marine mammals. In: McGregor, P.K. (Ed.), Animal Communication Networks. Cambridge University Press, Cambridge, pp. 390–415.

Janik, V.M., 2009. Acoustic communication in delphinids. Adv. Study Behav. 40, 123–157.

Jeffersen, T.A., LeDuc, R., 2018. Delphinids, Overview. In: Würsig, B., Thewissen, J.G.M., Kovacs, K.M. (Eds.), Encyclopedia of Marine Mammals, third ed. Academic Press, London, pp. 242–246.

Johnson, C.S., 1967. Sound detection thresholds in marine mammals. In: Tavolga, W.N. (Ed.), Marine Bioacoustics. 2. Pergamon Press, New York, pp. 247–260.
Johnson, C.S., 1968. Masked tonal thresholds in the bottlenosed porpoise. J. Acoust. Soc. Am. 44, 965–967.
Johnson, C.S., 1971. Auditory masking of one pure tone by another in the bottlenosed porpoise. J. Acoust. Soc. Am. 49, 1317–1318.
Johnson, C.M.J., Moewe, K., 1999. Pectoral fin preference during contact in Commerson's dolphins *(Cephalorhynchus commersonii)*. Aquat. Mamm. 25 (2), 73–77.
Jonas, J.B., Schneider, U., Naumann, G.O., 1992. Count and density of human retinal photoreceptors. Graefes Arch. Clin. Exp. Ophthalmol. 230 (6), 505–510.
Kastelein, R.A., Bunskoek, P., Hagedoorn, M., Au, W.W.L., de Haan, D., 2002. Audiogram of a harbor porpoise (*Phocoena phocoena*) measured with narrow-band frequency-modulated signals. J. Acoust. Soc. Am. 112, 334–344.
Kellogg, W.N., Kohler, R., 1952. Reactions of the porpoise to ultrasonic frequencies. Science 116, 250–252.
Ketten, D.R., 2000. Cetacean ears. In: Au, W.W.L., Fay, R.J., Popper, A.N. (Eds.), Hearing by Whales and Dolphins. Springer-Verlag, New York, pp. 43–108.
Kilian, A., Yaman, S., von Fersen, L., Güntürkün, O., 2003. A bottlenose dolphin discriminates visual stimuli differing in numerosity. Learn. Behav. 31 (2), 133–142.
Kinne, O., 1975. Marine Ecology - a comprehensive, integrated treatise on life in oceans and coastal waters. Volume II Physiological mechanisms Part 2, John Wiley & Sons, London New York Sydney Toronto.
Kishida, T., Kubota, S., Shirayama, Y., Fukami, H., 2007. The olfactory receptor gene repertoires in secondary-adapted marine vertebrates: evidence for reduction of the functional proportions in cetaceans. Biol. Lett. 3, 428–430.
Kishida, T., Thewissen, J.G.M., Hayakawa, T., Imai, H., Agata, K., 2015. Aquatic adaptation and the evolution of smell and taste in whales. Zoolog. Lett. 1, 9.
Kolchin, S., Belkovich, V., 1973. Tactile sensitivity in *Delphinus delphis*. Zool. Zhurnal 52, 620–622.
Koopman, H.N., Budge, S.M., Ketten, D.R., Iverson, S., 2006. Topographic distribution of lipids inside the mandibular fat bodies of odontocetes: remarkable complexity and consistency. IEEE J. Ocean. Eng. 31 (1), 95–106.
Kremers, D., Célérier, A., Schaal, B., Campagna, S., Trabalon, M., Böye, M., Hausberger, M., Lemasson, A., 2016a. Sensory perception in cetaceans: Part I—Current knowledge about dolphin senses as a representative species. Front. Ecol. Evol. 4, 49.
Kremers, D., Célérier, A., Schaal, B., Campagna, S., Trabalon, M., Böye, M., Hausberger, M., Lemasson, A., 2016b. Sensory perception in cetaceans: Part II - Promising experimental approaches to study chemoreception in dolphins. Front. Ecol. Evol. 4, 50.
Kröger, R.H.H., Katzir, G., 2008. Comparative anatomy and physiology of vision in aquatic tetrapods. In: Thewissen, J.G.M., Nummela, S. (Eds.), Sensory Evolution on the Threshold - Adaptations in Secondarily Aquatic Vertebrates. University of California Press, Berkeley, pp. 121–147.
Kröger, R.H.H., Kirschfeld, K., 1992. The cornea as an optical element in the cetacean eye. In: Thomas, J., et al. (Eds.), Marine Mammal Sensory Systems. Plenum Press, New York, pp. 97–106.
Kröger, R.H.H., Kirschfeld, K., 1993. Optics of the harbor porpoise eye in water. J. Opt. Soc. Am. A 10 (7), 1481–1489.
Kröger, R.H.H., Kirschfeld, K., 1994. Refractive index in the cornea of a harbour porpoise (*Phocoena phocoena*) measured by two-wavelengths laser-interferometry. Aquat. Mamm. 20, 99–107.
Kuznetzov, V.B., 1990. Chemical sense of dolphins. In: Thomas, J.A., Kastelein, R.A. (Eds.), Sensory Abilities of Cetaceans. Plenum Press, New York, pp. 481–503.
Laugerette, F., Passilly-Degrace, P., Patris, B., Niot, I., Febbraio, M., Montmayeur, J.P., Besnard, P., 2005. CD36 involvement in orosensory detection of dietary ipids, sponteneous fat preference, and digestive secretions. J. Clin. Investig. 115, 3177–3184.
Levenson, D.H., Dizon, A., 2003. Genetic evidence for the ancestral loss of short-wavelength-sensitive COE pigments in mysticete and odontocete cetaceans. Proc. R. Soc. Lond. B 270, 673–679.
Ling, J.K., 1977. Vibrissae of marine mammals. In: Harrison, R.J. (Ed.), Functional Anatomy of Marine Mammals. Academic Press, London, pp. 387–415.
Lisney, T.J., Collin, S.P., 2018. Retinal topography in two species of baleen whale (Cetacea: Mysticeti). Brain Behav. Evol. 92, 97–116.
Litwiler, T.L., Cronin, T.W., 2001. No evidence of accommodation in the eyes of the bottlenose dolphin, *Tursiops truncatus*. Mar. Mamm. Sci. 17 (3), 508–525.
Longworthy, O.R., 1931. Central nervous system of the porpoise *Tursiops truncatus*. J. Mammology 12 (4), 381–389.
Madsen, C.J., 1972. Visual Acuity in the Bottlenose Dolphin, *Tursiops truncatus* (Montagu, 1821), McGill University (MSc thesis), Montreal.
Madsen, C.J., 1976. Test for color discrimination and spectral sensitivity in the bottlenose dolphin (*Tursiops truncatus*). University of Hawaii (PhD thesis).
Madsen, C.J., Herman, L.M., 1980. Social and ecological correlates of cetacean vision and visual appearance. In: Herman, L.M. (Ed.), Cetacean Behavior: Mechanisms and Functions. Wiley Interscience, New York, pp. 101–147.
Mann, D., Hill-Cook, M., Manire, C., Greenhow, D., Montie, E.W., Powell, J., Wells, R., Bauer, G., Cunningham-Smith, P., Lingenfelser Jr., R.R.D., Stone, A., Brodsky, M., Stevens, R., Kieffer, G., Hoetjes, P., 2010. Hearing loss in stranded odontocete dolphins and whales. PLoS One 5 (11), e13824.
Mass, A.M., Supin, A.Y., 1995. Ganglion cell topography of the retina in the Bottlenose dolphin, *Tursiops truncatus*. Brain Behav. Evol. 45, 257–265.
Mass, A.M., Supin, A.Y., 1999. Retinal topography and visual acuity in the riverine tucuxi (*Sotalia fluviatilis*). Mar. Mamm. Sci. 15 (2), 351–365.
Mass, A.M., Supin, A.Y., 2007. Adaptive features of aquatic mammals' eyes. Anat. Rec. 290, 701–715.
Mass, A.M., Supin, A.Y., 2018. Vision. In: Würsig, B., Thewissen, J.G.M., Kovacs, K.M. (Eds.), Encyclopedia of Marine Mammals, third ed. Academic Press, London, pp. 1035–1044.
Mass, A.M., Supin, A.Y., Mukhametov, L.M., Rozanova, E.I., Arbramov, A.V., 2012. Morphological estimation of retinal resolution of a killer whale. Dokl. Biol. Sci. 442, 34–37.
Mauck, B., Eysel, U., Dehnhardt, G., 2000. Selective heating of vibrissal follicles in seals (*Phoca vitulina*) and dolphins (*Sotalia fluviatilis guianensis*). J. Exp. Biol. 203, 2125–2131.
Mazzatenta, A., Caleo, M., Baldaccini, N.E., Maffei, L., 2001. A comparative morphometric analysis of the optic nerve in two cetacean species, the striped dolphin (*Stenella coeruleoalba*) and fin whale (*Balaenoptera physalus*). Vis. Neurosci. 18, 319–325.

McCormick, J.G., 1969. Relationship of sleep, respiration, and anesthesia in the porpoise: a preliminary report. Proc. Natl. Acad. Sci. U. S. A. 62, 697–703.

McCormick, J.G., Wever, E.G., Palin, J., Ridgeway, S.H., 1970. Sound conduction in the dolphin ear. J. Acoust. Soc. Am. 48, 1418–1428.

McFarland, W.N., 1971. Cetacean visual pigments. Vision Res. 11, 1065–1076.

Mengual, R., Segovia, Y., Garcia, M., 2014. Morphological characteristics of pilot whales retin a (*Globicephala melas*; Traill, 1809) and their relationship to habitat. Int. J. Morphol. 32 (4), 1399–1406.

Mercado III, E., Killebrew, D.A., Pack, A.A., Macha, I.V.B., Herman, L.M., 2000. Generalization of 'same-different' classification abilities in bottle-nosed dolphins. Behav. Processes 50, 79–94.

Meredith, R.W., Gatesby, J., Emerling, C.A., York, V.M., Springer, M.S., 2013. Rod monochromacy and the coevolution of cetacean retinal opsins. PLoS Genet. 9 (4), e1003432.

Miller, B.S., Zosuls, A.L., Ketten, D.R., Mountain, D.C., 2006. Middle-ear stiffness of the bottlenose dolphin *Tursiops truncatus*. IEEE J. Ocean. Eng. 31 (1), 87–94.

Mobley, J.R., Helweg, D.A., 1990. Visual ecology and cognition in cetaceans. In: Thomas, J., Kastelein, R. (Eds.), Sensory Abilities of Cetaceans. Plenum Press, New York, pp. 519–536.

Møhl, B., Au, W.W.L., Pawloski, J.L., Nachtigall, P.E., 1999. Dolphin hearing: relative sensitivity as a function of point of application of a contact sound source in the jaw and head region. J. Acoust. Soc. Am. 105 (6), 3421–3424.

Mooney, T.A., Castellote, M., Jones, I., Quakenbush, L., Hobbs, R.C., Gaglione, E., Goertz, C.E., 2018a. Local environmental soundscapes relative to hearing sensitivity in beluga whales (*Delphinapterus leucas*). J. Ecoacous. 2, #QZD9Z5.

Mooney, T.A., Castellote, M., Quakenbush, L., Hobbs, R., Gaglione, E., Goertz, C., 2018b. Variation in hearing within a wild population of beluga whales (*Delphinapterus leucas*). J. Exp. Biol. 221, jeb171959.

Mooney, T.A., Li, S., Ketten, D.R., Wang, K., Wang, D., 2011. Auditory temporal resolution and evoked responses to pulsed sounds for the Yangtze finless porpoises (*Neophocaena phocaenoides asiaeorientalis*). J. Comp. Physiol. A 197, 1149–1158.

Mooney, T.A., Li, S., Ketten, D.R., Wang, K., Wang, D., 2014. Hearing pathways in the Yangtze finless porpoise, *Neophocaena asiaeorientalis asiaeorientalis*. J. Exp. Biol. 217, 444–452.

Mooney, T.A., Nachtigall, P.E., Breese, M., Vlachos, S., Au, W.W.L., 2009a. Predicting temporary threshold shifts in a bottlenose dolphin (*Tursiops truncatus*): the effects of noise level and duration. J. Acoust. Soc. Am. 125 (3), 1816–1826.

Mooney, T.A., Nachtigall, P.E., Castellote, M., Taylor, K.A., Pacini, A.F., Esteban, J.-A., 2008. Hearing pathways and directional sensitivity of the beluga whale, *Delphinapterus leucas*. J. Exp. Mar. Biol. Ecol. 362, 108–116.

Mooney, T.A., Nachtigall, P.E., Vlachos, S., 2009b. Sonar-induced temporary hearing loss in dolphins. Biol. Lett. 5, 565–567.

Mooney, T.A., Nachtigall, P.E., Yuen, M.M.L., 2006. Temporal resolution of the Risso's dolphin, *Grampus griseus*, auditory system. J. Comp. Physiol. A 192, 373–380.

Mooney, T.A., Yamato, M., Branstetter, B.K., 2012. Hearing in cetaceans: from natural history to experimental biology. Adv. Mar. Biol. 63, 197–246.

Morgane, P.J., Jacobs, M., 1972. Comparative anatomy of the cetacean nervous system. In: Harrison, R.J. (Ed.), Functional Anatomy of Marine Mammals. Academic Press, New York, pp. 117–244.

Murayama, T., Somiya, H., 1998. Distribution of ganglion cells and object localizing ability in the retina of three cetaceans. Fisheries Sci. 64 (1), 27–30.

Nachtigall, P.E., 1986. Vision, audition and chemoreception in dolphins and other marine mammals. In: Schusterman, R.J., Thomas, J.A., Wood, F.G. (Eds.), Dolphin Cognition and Behavior. Lawrence Erlbaum Associates, Hillsdale, New Jersey, pp. 79–113.

Nachtigall, P.E., Supin, A.Y., 2008. A false killer whale adjusts its hearing when it echolocates. J. Exp. Biol. 211 (11), 1714–1718.

Nachtigall, P.E., Supin, A.Y., Pacini, A.F., Kastelein, R.A., 2016. Conditioned hearing sensitivity change in the harbor porpoise (*Phocoena phocoena*). J. Acoust. Soc. Am. 140 (2), 960–967.

Nachtigall, P.E., Supin, A.Y., Pacini, A.F., Kastelein, R.A., 2018. Four odontocete species change hearing levels when warned of impending loud sound. Integr. Zool. 13 (2), 160–165.

Nachtigall, P.E., Supin, A.Y., Pawloski, J.L., Au, W.W.L., 2004. Temporary threshold shifts after noise exposure in the bottlenose dolphin (*Tursiops truncatus*) measured using evoked auditory potentials. Mar. Mamm. Sci. 20 (4), 673–687.

Nadol Jr., J.B., 1988. Quantification of human spiral ganglion cells by serial section reconstruction and segmental density estimates. Am. J. Otolaryngol. 9 (2), 47–51.

Nagel, E.L., Morgane, P.J., McFarland, W.L., Gallano, R.E., 1968. Rete mirabile of dolphin: its pressure-damping effect on cerebral circulation. Science 161, 898–900.

Newman, L.A., Robinson, P.R., 2005. Cone visual pigments of aquatic mammals. Vis. Neurosci. 22, 873–879.

Nilson, D.-E., Warrant, E.J., Johnson, S., Hanlon, R., Shashar, N., 2012. A unique advantage for giant eyes in giant squid. Curr. Biol. 22, 683–688.

Ninomiya, H., Yoshida, E., 2007. Functional anatomy of the ocular circulatory system: vascular corrosion casts of the cetacean eye. Vet. Ophthalmol. 10 (4), 231–238.

Noordenbos, J.W., Boogh, C.J., 1974. Underwater visual acuity in the bottle-nosed dolphin. Aquat. Mamm. 2 (3), 15–24.

Norris, K.S., 1964. Some problems of echolocation in cetaceans. In: Tavolga, W.N. (Ed.), Marine Bioacoustics. Pergamon, New York, pp. 316–336.

Norris, K.S., 1968. The evolution of acoustic mechanisms in odontocete cetaceans. In: Drake, E.T. (Ed.), Evolution and Environment. Yale University Press, New York, pp. 297–324.

Nummela, S., Thewissen, J.G.W., Rajpai, S., Hussain, T., Kumar, K., 2004. Eocene evolution of whale hearing. Nature 430, 776–778.

Oelschläger, H., 1992. Development of the olfactory and terminalis systems in whales and dolphins. In: Doty, R.L., Müller-Schwarze, D. (Eds.), Chemical Signals in Vertebrates. Plenum Press, New York, pp. 141–147.

Pacini, A.F., Nachtigall, P.E., Quintos, C.T., Schofield, T.D., Look, D.A., Levine, G.A., Turner, J.P., 2011. Audiogram of a stranded Blainville's beaked whale (*Mesoplodon densirostris*) measured using auditory evoked potentials. J. Exp. Biol. 214, 2409–2415.

Pacini, A.F., Nachtigall, P.E., Smith, A.B., Suarez, L.J., Magno, C., Laule, G.E., Aragones, L.V., Braun, R., 2016. Evidence of hearing loss due to dynamite fishing in two species of odontocetes. In: Proceedings of Meetings on Acoustics 4ENAL. Acoustical Society of America, vol. 1, p. 010043.

Pack, A.A., Herman, L.M., 1995. Sensory integration in the bottlenosed dolphin: immediate recognition of complex shapes across the senses of echolocation and vision. J. Acoust. Soc. Am. 98 (2), 722–733.

Peacock, M.F., Yunker, M.P., Madsen, C.J., Herman, L.M., 1974. In-air and in-water visual acuities in the bottlenose dolphin. In: 54th Annual Convention Western Psychological Association, pp. 1–10.

Peichl, L., Behrmann, G., Kröger, R.H.H., 2001. For whales and seals the ocean is not blue: a visual pigment loss in marine mammals. Eur. J. Neurosci. 13, 1520–1528.

Pepper, R.L., Simmons, J.V., 1973. In-air visual acuity of the bottlenose dolphin. Exp. Neurol. 41, 271–276.

Perez, J.M., 1972. Retinal anatomy of the bottlenose dolphin. Cetology 11, 1–11.

Pilleri, G., 1974. Side-swimming, vision and sense of touch in *Platanista indi* (Cetacea, Platanistidae). Experientia 30 (1), 100–104.

Popov, V.V., Supin, A.Y., 2009. Comparison of directional selectivity of hearing in a beluga whale and a bottlenose dolphin. J. Acoust. Soc. Am. 126 (3), 1581–1587.

Popov, V.V., Supin, A.Y., Gvozdeva, A.P., Nechaev, D.I., Tarakanov, M.B., Sysueva, E.V., 2020. Spatial release from masking in a bottlenose dolphin *Tursiops truncatus*. J. Acoust. Soc. Am. 147 (3), 1719–1726.

Popov, V.V., Supin, A.Y., Wang, D., Wang, K., 2006. Nonconstant quality of auditory filters in the porpoises, *Phocoena phocoena*, and *Neophocaena phocaenoides* (Cetacea, Phocoenidae). J. Acoust. Soc. Am. 119 (5), 3173–3180.

Popov, V.V., Supin, A.Y., Wang, D., Wang, K., Dong, L., Wang, S., 2011. Noise-induced temporary threshold shift and recovery in Yangtze finless porpoises *Neophocaena phocaenoides asiaeorientalis*. J. Acoust. Soc. Am. 130 (1), 574–584.

Pryor, K., 1973. Behavior and learning in porpoises and whales. Naturwissenschaften 60, 413–430.

Pütter, A., 1903. Die Augen der Wassersäugethiere. Zoolog. Jahrbuech. Abteil. Anat. Ontogen. Tiere 17, 99–402.

Reichmuth, C., Casey, C., Friedlaender, A., 2022. In-situ observations of the sensory hairs of Antarctic minke whales (*Balaenoptera bonaerensis*). Anat. Rec. 305 (3), 568–576.

Ridgway, S.H., 1986. Physiological observations on the dolphin brain. In: Schusterman, R.J., Thomas, J.A., Wood, F.G. (Eds.), Dolphin Cognition and Behavior. Lawrence Erlbaum Associates, Hillsdale, New Jersey, pp. 31–59.

Ridgway, S.H., Carder, D.A., 1990. Tactile sensitivity, somatosensory responses, skin vibrations, and the skin surface ridges of the bottle-nose dolphin, *Tursiops truncatus*. In: Sensory Abilities of Cetaceans. Springer, pp. 163–179.

Ridgway, S.H., Carder, D.A., 1993. Features of dolphin skin with potential hydrodynamic importance. IEEE Eng. Med. Biol. Mag. 12 (3), 83–88.

Rivamonte, L.A., 1976. Eye model to account for comparable aerial and underwater acuities of the bottlenose dolphin. Neth. J. Sea Res. 10 (4), 491–498.

Rivamonte, L.A., 2009. Bottlenose dolphin (*Tursiops truncatus*) double-slit pupil asymmetries enhance vision. Aquat. Mamm. 35 (2), 269–280.

Rochon-Duvigneaud, A., 1940. L'oeil des cétacés. In: Archives du Müséum National d'Histoire naturelle de Paris, pp. 57–90.

Rochon-Duvigneaud, A., 1943. Les yeux et la vision des vertébrés. Libraires de l'académie de médicine, Paris, France.

Rossbach, K.A., Herzing, D.L., 1997. Underwater observations of benthic-feeding bottlenose dolphins (*Tursiops truncatus*) near Grand Bahama Island, Bahamas. Mar. Mamm. Sci. 13 (3), 498–504.

Schevill, W.E., Lawrence, B., 1949. Underwater listening to the white porpoise (*Delphinapterus leucas*). Science 109, 143–144.

Schoenfuss, H.L., 1997. Anatomy of the eye of the cetaceans, an annotated translation. In: Occasional Papers of the Museum of Natural Science. Louisiana State University, Baton Rouge, pp. 1–38.

Shoemaker, P., Ridgway, S., 1991. Cutaneous ridges in odontocetes. Mar. Mamm. Sci. 7 (1), 66–74.

Simons, D., Huigen, M., 1977. Analysis of an experiment on colour vision in dolphins. Aquat. Mamm. 5, 27–33.

Slijper, E.J., 1962. Whales. Basic Books, New York.

Smith, T.G., St Aubin, D.J., Hammill, M.O., 1992. Rubbing behaviour of belugas, *Delphinapterus leucas*, in a high Arctic estuary. Can. J. Zool. 70, 2405–2409.

Smith, A., Pacini, A.F., Nachtigall, P.E., 2020. Modulation rate transfer functions from four species of stranded odontocete (*Stenella longirostris, Feresa attenuata, Globicephala melas*, and *Mesoplodon densirostris*). J. Comp. Physiol. A. 204, 377–389.

Smith, A.B., Madsen, P.T., Johnson, M., Tyack, P., Wahlberg, M., 2021. Toothed whale auditory brainstem responses measured with a non-invasive, on-animal tag. JASA Expr. Lett. 1 (9), 091201.

Spong, P., White, D., 1971. Visual acuity and discrimination learning in the dolphin (*Lagenorhynchus obliquidens*). Exp. Neurol. 31, 431–436.

Springer, M.S., Gatesy, J., 2017. Inactivation of the olfactory marker protein (OMP) gene in river dolphins and other odontocete cetaceans. Mol. Phylogenet. Evol. 109, 375–387.

Suchowskaja, L.I., 1972. The morphology of taste organs in dolphins. Invest. Cetacea 4, 201–207.

Supin, A.Y., Nachtigall, P.E., Au, W.W.L., Breese, M., 2005. Invariance of echo-responses of the target strength and distance in an echolocation false killer whale: evoked-potential study. J. Acoust. Soc. Am. 117 (6), 3928–3935.

Supin, A.Y., Popov, V.V., Mass, A.M., 2001. The Sensory Physiology of Aquatic Mammals. Kluwer Academic Publishers, Boston, Dordrecht, London.

Tarpley, R.J., Gelderd, J.B., Bauserman, S., Ridgway, S.H., 1994. Dolphin peripheral visual pathway in chronic unilateral ocular atrophy: complete decussation apparent. J. Morphol. 222, 91–102.

Thompson, R.K., Herman, L.M., 1975. Underwater frequency discrimination in the bottlenosed dolphin (1–140 kHz) and the human (1–8 kHz). J. Acoust. Soc. Am. 57 (4), 943–948.

Trickey, J.S., Branstetter, B.K., Finneran, J.J., 2010. Auditory masking of a 10 kHz tone with environmental, comodulated, and Gaussian noise in bottlenose dolphins (*Tursiops truncatus*). J. Acoust. Soc. Am. 128 (6), 3799–3804.

Waller, G.N.H., 1984. The ocular anatomy of cetacea—an historical perspective. Invest. Cetacea 16, 138–148.

Waller, G.N.H., 1992. Ciliary muscles in the eye of the long-finned pilot whale *Globicephala melaena* (Traill, 1809). Aquat. Mamm. 18 (2), 36–39.

Walls, G.L., 1942. The Vertebrate Eye and its Adaptive Radiation. Hafner Press, New York.

Wang, Z.-T., Supin, A.Y., Akamatsu, T., Duan, P.-X., Yang, Y.-N., Wang, K.-X., Wang, D., 2021. Auditory evoked potential in stranded melon-headed whales (*Peponocephala electra*): With severe hearing loss and possibly caused by anthropogenic noise pollution. Ecotoxicol. Environ. Saf. 228, 113047.

Wang, Z., Chen, Z., Yu, S., Ren, W., Zhou, K., Yang, G., 2015. 'Obesity' is healthy in cetaceans? Evidence from pervasive positive selection in genes related to triacylglycerol metabolism. Sci. Rep. 5 (14187).

Wever, E.G., Mccormick, J.G., Palin, J., Ridgway, S.H., 1971. The cochlea of the dolphin, *Tursiops truncatus*: hair cells and ganglion cells. Proc. Natl. Acad. Sci. U. S. A. 68 (12), 2908–2912.

White, D., Cameron, N., Spng, P., Bradford, J., 1971a. Visual acuity of the killer whale (*Orcinus orca*). Exp. Neurol. 32, 230–236.

White, D., Spong, P., Cameron, N., Bradford, J., 1971b. Visual discrimination learning in the killer whale (*Orcinus orca*). Behav. Res. Meth. Instrum. 3 (4), 187–188.

Würsig, B., Kieckhefer, T.R., Jeffersen, T.A., 1990. Visual displays for communication in cetaceans. In: Thomas, J., Kastelein, R. (Eds.), Sensory Abilities of Cetaceans. Plenum Press, New York, pp. 545–559.

Yablokov, A.V., Bel'kovich, Y.M., Borisov, V.I., 1972. Whales and Dolphins (Translation from Russian). Israel Program for Scientific Translations, Jerusalem.

Yablokov, A.V., Klevezal, G.A., 1969. Whiskers of whales and seals and their distribution structure, and significance. In: Kleinenberg, S.E. (Ed.), Morphological Characteristics of Aquatic Mammals. Isdatel'stvo "Nauka", Moscow, pp. 48–81.

Yamada, M., Yoshizaki, F.U., 1959. Osseous labyrinth of Cetacea. Sci. Rep. Whale Res. Inst. (Tokyo) 14, 291–304.

Yamasaki, F., Komatusu, S., Kmaiya, T., 1978. Taste buds in the pits at the posterior dorsum of the tongue of *Stenella coeruleoalba*. Sci. Rep. Whales Res. Inst. 20, 285–290.

Yamato, M., Pyenson, N.D., 2015. Early development and orientation of the acoustic funnel provides insight into the evolution of sound reception pathways in cetaceans. PLoS One 10 (3), e0118582.

Young, N.M., Hope, G.M., Dawson, W.W., Jenkins, R.L., 1988. The tapetum fibrosum in the eyes of two small whales. Mar. Mamm. Sci. 4 (4), 281–290.

Yuen, M.M.L., Nachtigall, P.E., Breese, M., Supin, A.Y., 2005. Behavioral and auditory evoked potential audiograms of a false killer whale (*Pseudorca crassidens*). J. Acoust. Soc. Am. 118 (4), 2688–2695.

Zhu, Q., Hillmann, D.J., Henk, W.G., 2001. Morphology of the eye and surrounding structures of the bowhead whale, *Balaena mysticetus*. Mar. Mamm. Sci. 17 (4), 729–750.

CHAPTER

11

Kidneys and osmoregulation

Rudy M. Ortiz[a] *and Miwa Suzuki*[b]

[a]Department of Molecular & Cell Biology, School of Natural Sciences, University of California, Merced, CA, United States
[b]Department of Marine Science and Resources, College of Bioresource Sciences, Nihon University, Fujisawa, Kanagawa, Japan

Abstract

The reniculate kidneys of dolphins are a remarkable feature of the cetacean anatomy and confer the requisite evolutionary adaptations necessary to thrive in an aquatic environment regardless of salinity. While the renal anatomy is similar among all marine mammals, it is unique within the class Mammalia. Despite the anatomical similarity among cetaceans, the maximum urine concentrating ability varies among the different species that have been examined, including the common bottlenose dolphin. Nonetheless, most marine mammals that are strictly marine possess only a slightly greater ability to concentrate urine than humans and many other terrestrial mammals. Many dolphins routinely travel through aquatic environments of varying salinity (from salt to fresh water), and studies demonstrate delphinids possess the requisite physiological mechanisms to accommodate shifts between environments without detriment. While a vast majority of research on dolphin kidneys has focused on its osmoregulatory functions, several studies have demonstrated that they are susceptible to accumulation of environmental pollutants and toxins, as well as formation of kidney stones, which could ultimately compromise function. This chapter will focus on data summarizing kidney function and other data around renal pathologies.

Introduction

The challenges of living in an aquatic environment regardless of salinity present unique and exciting opportunities to study the regulation of water and electrolytes in marine mammals. Studies have spanned over the last two centuries (see references in Fetcher Jr., 1939). Marine mammals have evolved robust mechanisms that have allowed for their adaptation to both hypo- and hyperosmotic environments, including routine transitions between them. To properly osmoregulate in a marine habitat, physiological mechanisms intended to conserve fresh water and thus avoid dehydration are required. However, as challenging as enduring life in a high salt environment may be, aquatic mammals also diverged and inhabited freshwater niches as well. Although adapting to freshwater habitats may appear beneficial because conserving fresh water no longer poses a problem, animals adopting these habitats are confronted with a different osmotic challenge. Living in a salt-"free" environment requires the appropriate physiological mechanisms to conserve primarily Na^+. However, some marine mammals have adapted to yet a third, and potentially greater osmotic challenge, which is extended periods of food deprivation (fasting) (i.e., Ortiz et al., 1978). The ability to maintain water balance and electrolyte homeostasis during prolonged periods of complete abstinence from food and water requires even more robust mechanisms, which have evolved to conserve both salts (Na^+) and water to maintain internal homeostasis. Occupying either extreme in environmental salinity, or adapting to both, as well as adapting to protracted fasts provide an indication of the dynamic scope in osmoregulatory capacity of aquatic-adapted mammals such as dolphins, which has made them an intriguing model for the study of renal function over the last two centuries.

Kidney structure

The kidneys of aquatic mammals are structurally different from those of their terrestrial predecessors, suggesting that these "specialized," or reniculate, kidneys allow them to occupy habitats with a broad range of salinity. These differences in renal architecture may be important to understanding how marine mammals in general are able to tolerate expansion into aquatic environments because kidneys are the principal organs of water and electrolyte regulation. Members of the order Cetacea (dolphins and whales) (Hedges et al., 1979; Pfeiffer, 1997) and Pinnipedia (seals, sea lions, and walruses) (Bester, 1975; Vardy and Bryden, 1981) possess reniculate kidneys (Figs. 1 and 2) similar to bears

The Physiology of Dolphins
https://doi.org/10.1016/B978-0-323-90516-9.00006-3

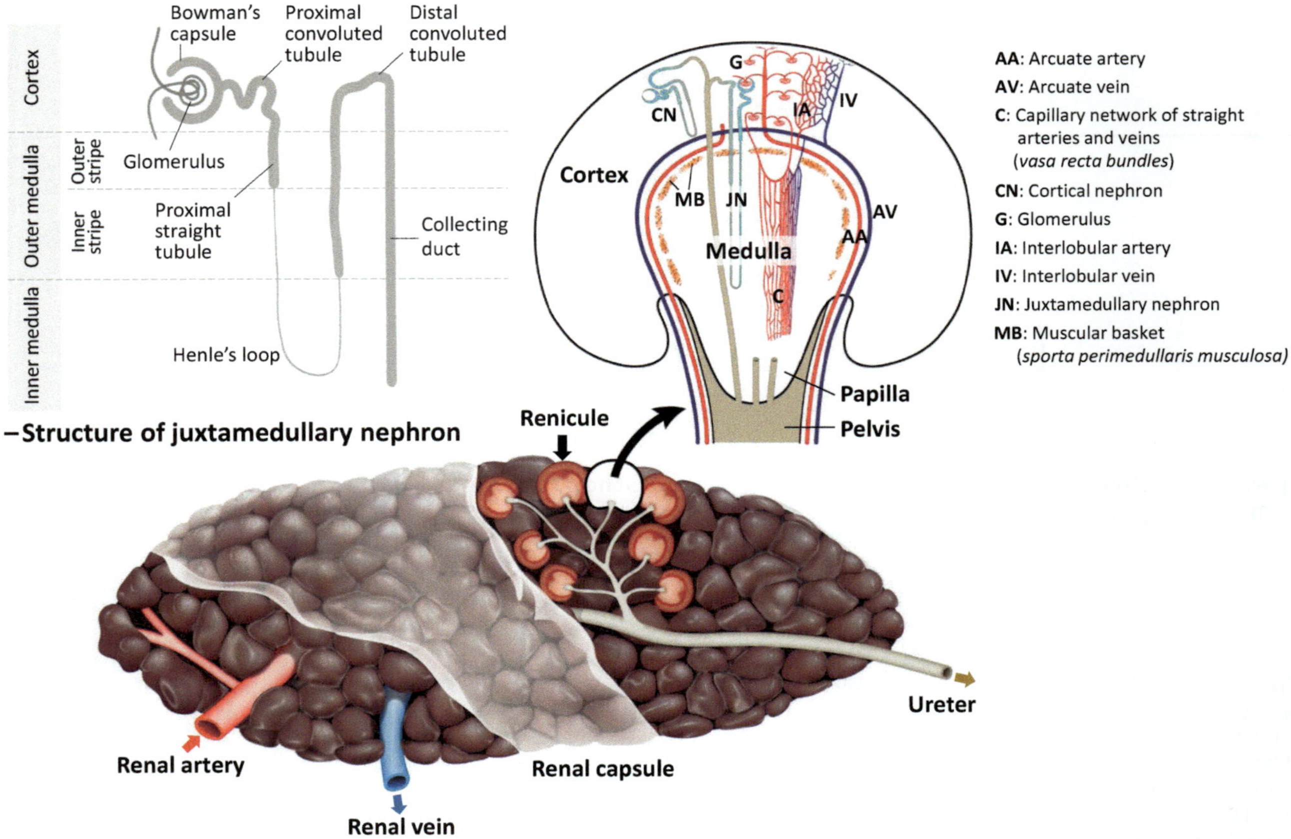

FIG. 1 A drawing of dolphin kidney and renicule (*right, illustration by Uko Gorter*) and the structure of the juxtamedullary nephron (*left*). A renicule is a completely functional renal unit and its structure is similar to kidneys of other mammals, except for the presence of the muscular basket and quite well-developed capillary network of straight arteries and veins.

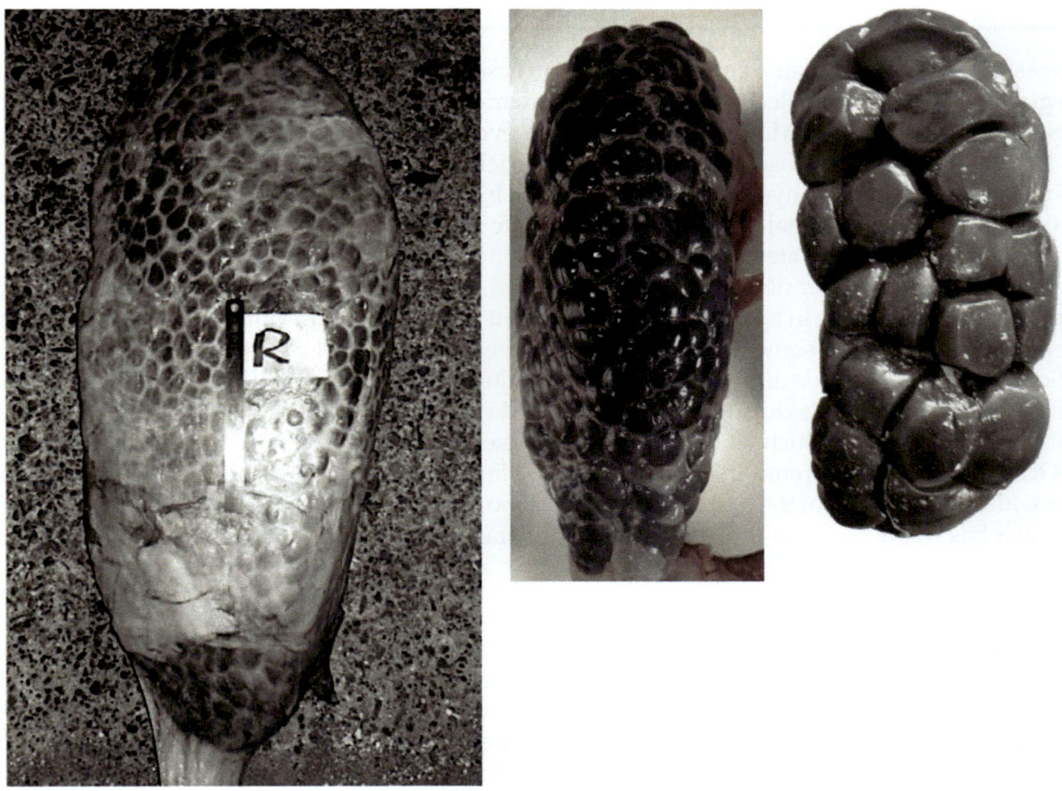

FIG. 2 The kidney of a stranded gray whale (*Eschrichtius robustus, left*), a spotted dolphin (*Stenella frontalis, middle*), and a cow (*Bos taurus, right*). The kidneys of marine mammals, especially in cetaceans, are more lobulated than in their close related groups and composed of hundreds and thousands of renicules. *Photo by M. Suzuki and K. Ueda.*

and otters. These kidneys can be composed of hundreds of individual lobes, or reniculi, each of which contains discrete cortical tissue and a single medullary pyramid inserted in a single calyx (Bester, 1975; Vardy and Bryden, 1981) (Fig. 1).

A histological dissection of the cetacean kidney reveals structures most likely adapted to accommodate diving behavior (Pfeiffer, 1997). A comparison between the marine and terrestrial mammal kidney has revealed some unique features in the cetacean renal architecture, including (1) specialized glycogen stores in the proximal convoluted tubule epithelial cells, (2) highly concentrated bundles of medullary blood vessels (vasa recta bundles), and (3) the sporta perimedullaris musculosa, a layer of collagen, elastic fibers, and smooth muscle separating the cortex from the medulla (Pfeiffer, 1997).

Dolphin kidneys possess anatomical features such as increased medullary thickness, which is necessary to produce a highly concentrated urine and is especially important for mammals in a hyperosmotic environment. Despite the unique anatomical architecture, the reniculate kidney does not achieve the concentrating ability that could be predicted (Bester, 1975). A close correlation between relative medullary thickness and maximum recorded urine osmolality exists, with pinnipeds exhibiting ratios between 1.1 and 1.7 (Vardy and Bryden, 1981). However, these low values suggest that the kidneys of marine mammals are relatively poor concentrators of urine, given their anatomy (Vardy and Bryden, 1981). The urine-to-plasma ratio in osmolality serves as a robust index of concentrating ability. As such, the concentrating ability of marine mammals is only slightly better than humans, similar to domestic dogs and the camel, and much lower than desert rodents (kangaroo rat and hopping mouse) (Ortiz, 2001). Thus the large body size and exceptional diving abilities of whales and dolphins more than likely served as the primary selection pressure for the evolution of the reniculate kidney; more so than the osmotic challenges induced by hypo- or hyperosmotic environments (Bester, 1975; Vardy and Bryden, 1981). Therefore the unique kidney morphology present in cetaceans does not appear to provide any greater benefit when compared to the urine concentrating ability of terrestrial mammals, suggesting that the adaptation to variable aquatic habitats was accomplished via more conventional mechanisms such as hormonal regulation of urine concentration and volume and/or regulation of renal hemodynamics. However, considering that urinary concentrating ability is negatively correlated with body size (Beuchat, 1990), it is likely that cetacean kidneys are concentrating urine commensurate with their physiological requirements.

At the cellular level, the kidneys of the rough-toothed dolphin (*Steno bredanensis*) and melon-headed whale (*Peponocephala electra*) presented with relatively higher concentrations of sulfoglycolipids, which are a type of acidic glycosphingolipids that are ubiquitously expressed in all vertebrate cells (Nagai et al., 2008). It is suspected that these glycolipids are important in facilitating cellular osmotic balance; however, the mechanisms of action remain unresolved. In addition to the evolution of a unique renal anatomy present in marine mammals, delphinids also present with unique biochemical composition of their cell membranes to contribute to their osmoregulatory capabilities.

Urine concentration and kidney function

Because robust studies of the kidney require the collection of urine to accurately assess function, the collection of urine from dolphins can be extremely challenging, especially in wild populations. However, innovative practices have been employed to facilitate the collection of urine such as urethral catheterization in a suspended animal (see Ridgway and Venn-Watson, 2010) and by using trained animals in professional human care (see Jiménez-Zucchet et al., 2019). These methods have already been shown to be important in helping advance the study of kidney function (Ridgway and Venn-Watson, 2010) as well as to facilitate the monitoring of animal health (Jiménez-Zucchet et al., 2019).

Although sea water consumption is not a common behavior in marine mammals in general (with a few exceptions), these animals can produce urine concentrations greater than that of sea water (Bester, 1975; Costa, 1982; Maluf, 1989). The highest urine osmolality measured in any marine mammal was 2658 $mOsm\,L^{-1}$ (Ridgway, 1972) (range: 1150–2658, Ortiz, 2001). Even mammals found in freshwater habitats can concentrate their urine in response to a hyperosmotic stimulus (Irvine et al., 1980; Hong et al., 1982), suggesting that marine mammals that are found in fresh water rather than marine environments maintained their urine concentrating abilities (Ortiz, 2001). Originally, it was thought that marine mammals commonly and purposefully consumed seawater, a behavior known as *mariposia*, because their urine osmolality was greater than that of sea water based on the assumption that the composition of their urine was primarily Na^+ and Cl^- (Portier, 1910). However, for marine mammals to obtain a net positive gain in solute-free water after consuming sea water, they would have to excrete Na^+ and Cl^- in concentrations greater than that of the water they drink (Albrecht, 1950). With few exceptions, marine mammals do not regularly concentrate Na^+ and Cl^- above that of sea water, further suggesting that they do not rely on sea water consumption to maintain fluid balance (Ortiz, 2001). Conversely, most of the data on the osmoregulatory capacity of dolphins comes from studies in marine-adapted dolphins, including the bottlenose dolphin (*Tursiops truncatus*, which is also frequently observed in brackish and freshwater environments). However, a few freshwater-adapted dolphins (and porpoises) exist and are not sufficiently studied. The Yangtze finless porpoise (*Neophocaena asiaeorientalis*) resides solely in freshwater rivers of China. The plasma and urine osmolality and urinary urea concentration of this species is relatively lower than its marine counterparts while plasma Na^+ is equivalent, suggesting that these cetacean species have the appropriate counterregulatory mechanisms to maintain osmotic homeostasis and avoid hyponatremia (Guo et al., 2014).

Urea transporters (UT-A and UT-B) are important for reabsorbing urea from the filtrate and contribute to the concentrating capacity of the kidney. UTs have been cloned from multiple cetacean species (but not delphinids) (Janech et al., 2002; Birukawa et al., 2008; Wang et al., 2015). The data demonstrated that two putative glycosylation consensus sites are conserved between whales and terrestrial mammals; however, protein kinase (A/C) consensus sites demonstrate much more variability, which may contribute to the differences in the urine concentrating ability among cetaceans (Birukawa et al., 2008). Nonetheless, renal UTs are highly conserved among mammals (Janech et al., 2002). A study of the molecular evolution of the UT genes (SLC14A1/2) suggested that these genes were subjected to positive selection, which likely contributed to the ability of terrestrial mammals to radiate into the ocean and provide enhanced urine concentrating ability (Xu et al., 2013).

The assessment of kidney function requires an estimate of renal hemodynamics to robustly interpret urinary excretion. Logistically, this can be challenging because of the aquatic nature of delphinids; however, employing validated biomarkers such as inulin (renal blood flow, RBF) and creatinine (glomerular filtration rate, GFR) has been useful. Similar to pinnipeds, the secretion of creatinine from the renal tubules into the filtrate was not observed in dolphins (Malvin and Rayner, 1968), suggesting that the use of creatinine clearance provides a valid estimate of GFR in dolphins as well. More recently, Venn-Watson et al. (2008) developed an algorithm to predict GRF ($mL/min/m^2$) based on the equation validated in the Modification of Diet in Renal Disease Study (Hunsicker et al., 1997) in humans using serum creatinine and a correction factor for use in dolphins. This is a significant advancement in renal physiology of marine mammals because it alleviates the need for simultaneous 24-h urine collections and matched plasma for creatinine measurements in both, which may be very challenging to obtain. While the application of the equation has been demonstrated to be useful in dolphins to closely monitor renal health using only serum creatinine concentration, it may require more robust validation for use in other marine mammals during other conditions expected to alter GFR. Nonetheless, this equation is promising and greatly simplifies the incorporation of predicted GFR measurements to allow for more comprehensive calculations of kidney function in dolphins and likely other marine mammals as well. Complementing urinary electrolyte excretion measurements and their plasma concentrations with estimated GFR provides an opportunity to calculate the fractional excretion of electrolytes, enhancing the interpretation of the results.

Do dolphins drink salt water?

Early studies of osmoregulation in any cetacean consisted of an oral administration of hypertonic NaCl in two bottlenose dolphins (Fetcher Jr. and Fetcher, 1942). The calculated changes in serum, fecal, and urinary Cl^- led the authors to conclude that dolphins do not drink salt water (Fetcher Jr. and Fetcher, 1942). However, more recently, isotopic dilution techniques were incorporated to address the question of water flux and consumption in four species of delphinids: common dolphin (*Delphinus delphis*), Pacific white-sided dolphin (*Lagenorhynchus obliquidens*), Pacific bottlenose dolphin (*Tursiops truncatus gilli*), and short-finned pilot whale (*Globicephala macrorhynchus*). These studies demonstrated that under fasting conditions, these delphinids consumed sea water at rates ranging between 4.5 and 13 mL/kg/d (Telfer et al., 1970; Hui, 1981). While Telfer et al. (1970) concluded that absorption of water through the skin was unlikely, later studies reported that the skin may be a significant avenue of water flux, accounting for as much as 70% of the total flux in fasting dolphins (Hui, 1981). Water flux across the skin was also reported to be dependent on the osmotic gradient of the environment, suggesting that delphinids with access to a hypo-osmotic habitat could experience a net gain in fresh water (Andersen and Nielsen, 1983).

Feeding and fasting studies

Fasted dolphins have been shown to produce a dilute urine, suggesting that at least at the time of the measurements, renal mechanisms had not been initiated to stimulate free water reabsorption, which would result in a concentrated urine (Bentley, 1963; Telfer et al., 1970; Hui, 1981). Telfer et al. (1970) suggested that perhaps the animals become acutely dehydrated, which serves as a stimulus to increase metabolic water production primarily from the oxidation of fat, resulting in the increased excretion of free water and a diluted urine. Hui (1981) demonstrated that metabolic water was sufficient to preclude the need for mariposia and of excreting a concentrated urine to maintain water balance. Therefore the studies demonstrating increased water influx (drinking and/or transdermal absorption) in fasting dolphins (Telfer et al., 1970; Hui, 1981) is intriguing and raises questions as to the need and mechanisms for drinking. Acute (24-h) fasting in bottlenose dolphins in professional human care decreased the ratio of plasma urea:creatinine (U:C), suggesting that GFR was reduced (Ortiz et al., 2010); however, because fasting can increase urea simultaneously despite the increase in creatinine, the use of urea:creatinine ratio during fasting to assess changes in GFR has its limitations. This decrease in GFR is likely a mechanism to decrease urinary loss of water and electrolytes and to maintain osmotic homeostasis, which was demonstrated by static measurements of plasma osmolality over the course of the study (Ortiz et al., 2010) and consistent with a similar study in dolphins (Ridgway and Venn-Watson, 2010). After refeeding, U:C ratio increased to prefasting levels, suggesting that GFR was normalized (Ortiz et al., 2010).

Like pinnipeds, GFR, renal plasma flow (RPF), and urine output in dolphins increased after feeding, increasing urinary electrolyte excretion (Malvin and Rayner, 1968; Ridgway, 1972), which likely reflected the increased electrolyte intake and the mechanisms to maintain osmotic homeostasis. A high-protein meal of raw fish fed to dolphins increased GFR and RPF, which were likely induced by an increased nitrogen load on the kidney similar to that observed in the harbor seal (Hiatt and Hiatt, 1942).

Loading experiments

Exogenous administration of fluids (hypertonic saline or freshwater) are useful experiments to assess the adaptive responses to these stimuli. Such experiments have been performed in dolphins by inserting a stomach tube to acutely manipulate the osmotic status of the animal (Fetcher and Fetcher, 1942; Malvin and Rayner, 1968; Ridgway, 1972; Ridgway and Venn-Watson, 2010). Interestingly, hypertonic saline increased urine flow (Fetcher and Fetcher, 1942; Ridgway, 1972; Ridgway and Venn-Watson, 2010); however, a freshwater load (2.5% of body mass) did not induce a diuresis but rather increased urine osmolality 2 h post-administration (Malvin and Rayner, 1968). Malvin and Rayner (1968) concluded, based on theirs and Fetcher and Fetcher's (1942) data, that the increased diuresis following the hypertonic infusion resulted from the expansion of extracellular fluid consistent with similar studies performed in northern elephant seal (*Mirounga angustirostris*) pups (Ortiz et al., 2002). However, Ridgway (1972) repeated the tests of Malvin and Rayner (1968) using bottlenose dolphins and demonstrated a diuresis following the infusion of freshwater. Although Ridgway (1972) mentioned that a species difference may have accounted for the discrepancy in the results, both animals were of similar body mass (187 vs. 160 kg) and received the same dose. Probably the most glaring discrepancy between the two studies was the urine collection time: 2 h in one study (Malvin and Rayner, 1968) and 24 h in the other (Ridgway, 1972). Therefore a time-delayed response may have contributed more to the difference between the two studies, providing important insights to the potential adaptability of the dolphin kidney to freshwater, especially if freshwater-adapted dolphins actively drink.

Probably, the most elaborate set of experiments designed to exogenously manipulate osmotic balance were performed by Ridgway and Venn-Watson (2010) in common bottlenose dolphins in professional human care. Results from a series of experiments that included fasting, freshwater loading, seawater loading, and hypertonic saline + high-protein fish meal demonstrated that these animals either maintain or regain osmotic homeostasis within 24 h following the perturbations. The changes in urine flow rate and osmolality reflect the expected changes following the prescribed manipulations. Interestingly, as demonstrated previously, the infusion of hypertonic saline (seawater) initially stimulated a diuresis that waned after the first 5 h associated with an increase in urine osmolality and urine:plasma ratio of osmolality, which is suggestive of an increase in osmotic clearance and reduced free water clearance (calculated estimates), which would implicate the stimulation of vasopressinergic activity (although not directly measured) (Ridgway and Venn-Watson, 2010).

Renal endocrinology of dolphins

The first published account of a bioassay using extracted cetacean tissue was Eichelberger et al. (1940). Extracted renin from the kidney of a bottlenose dolphin was injected into dogs at two doses, 4 and 2 mL. In the initial trial, 4 mL of extract immediately increased blood pressure (BP) by 48% and remained 38% higher than control pressure for 10 min. In a separate trial, an initial 2 mL dose of extract acutely increased BP by 21% but was not sustained. A second 2 mL dose increased BP by 27%, which also was not sustained. However, the two doses collectively were able to induce a 9%

increase in BP that was sustained. A third dose of 2 mL produced a 20% increase in BP that sustained an additional 4% increase in BP. These experiments demonstrated that: (1) renin in dolphins (and probably all cetaceans) causes a vascular pressor effect as in other mammals, most likely through the increased generation of angiotensin II, (2) the pressor effect is dose-dependent, and (3) the response is graded until a threshold is reached, which results in a sustained elevation in BP. Unfortunately, the concentrations of the doses were not reported. However, sufficient information on the extract was provided to enable repetition using known concentrations to shed light on the sensitivity and kinetics of the renin-induced increase in BP, and thus, the dynamic effects of the renin–angiotensin system in dolphins.

Concentrations of circulating renin were reported 27 years later (Malvin and Vander, 1967). Eleven years after that, Malvin et al. (1978) demonstrated a positive correlation between plasma renin activity (PRA) and aldosterone, suggesting that the renin–angiotensin–aldosterone system (RAAS) is present and active in dolphins (Table 1). The genes for two important components of RAAS, angiotensin-converting enzyme (ACE) and angiotensinogen (Agt; the precursor to angiotensin I), were subjected to positive selection based on the molecular analysis in multiple cetaceans, providing an indication of the importance of RAAS to osmoregulation and endocrine-mediated kidney function (Xu et al., 2013). The sensitivity of RAAS appears to be the same as that of California sea lions, suggesting that the function of this system is conserved among marine mammals (Ortiz, 2001). Several studies have shown that aldosterone, the adrenal steroid primarily responsible for Na + reabsorption in the gut and kidney of mammals, serves as a potential biomarker of physiological stress (Houser et al., 2011; Fair et al., 2014; Hart et al., 2015; Miller et al., 2021); however, its contributions to directly regulating renal Na + reabsorption have not been elucidated in any dolphin but expected to induce the typical mammalian mechanism. Interestingly, bottlenose dolphins exposed to low salinity environments for extended periods (years) presented with lower serum electrolytes and higher aldosterone concentrations, providing some initial evidence that the adrenal glands in dolphins are responsive to the alternations in environmental salinity and change accordingly to try and restore electrolyte balance (McClain et al., 2020). Additionally, a well-controlled, simulation of transport-invoked stress induced an initial (first 2 h) increase in aldosterone (and cortisol) independent of an increase in PRA (Parker and Schroeder, 1987), suggesting that aldosterone secretion was stimulated by adrenocorticotropin (ACTH) and not Ang II. Over the next 22 h, adrenal steroids gradually waned while PRA increased linearly and urinary Na + excretion decreased (along with Na^+:K^+ ratio), suggesting that renal Na + reabsorption was increased and associated with sustained aldosterone (Parker and Schroeder, 1987). Collectively, these associative datasets provide some profound evidence of the antinatriuretic effects of aldosterone, although controlled studies designed to directly evaluate the contributions of aldosterone to renal Na + handling are warranted.

While aldosterone stimulates Na + reabsorption, the atrium-derived peptide, atrial natriuretic peptide (ANP), antagonizes RAAS and directly promotes urinary Na + excretion (natriuresis). The peptide sequence of ANP in dolphins is highly conserved and similar to other mammals, including humans (Naka et al., 2007). Although there is some evidence in elephant seals that aldosterone may regulate urinary Na + excretion (Ortiz et al., 2002) and water retention (Ortiz et al., 2006), its contributions to promoting natriuresis have not been examined in dolphins.

The only other known bioassay of cetacean tissue was that of plasma-extracted arginine vasopressin (AVP) on water retention in water-loaded rats by measuring the electrical conductivity of the urine (Malvin et al., 1971). In most of the

TABLE 1 The principal components of the renin–angiotensin–aldosterone system. Listed in sequential order for promoting the cascade.

Protein	Primary site of secretion/synthesis	Primary function[a]
Angiotensinogen (Agt)	Liver	Precursor for Ang I
Renin	Kidney—juxtaglomerular cells	Enzyme that catalyzes the formation of Ang I
Angiotensin I (Ang I)	Enzymatically in circulation	Precursor for Ang II
Angiotensin-converting enzyme (ACE)	Lung	Catalyzes formation of Ang II
Angiotensin II (Ang II)	Enzymatically in circulation	Stimulates vasoconstriction and secretion of aldosterone from adrenal gland, and lesser extent, Na + reabsorption in renal tubule
Aldosterone	Adrenal gland—zona glomerulosa	Stimulates Na + reabsorption (and K^+ excretion) in distal tubule of the kidney

[a] *Primary functions as they relate to the kidney; Ang II and aldosterone have other functions in other tissues.*

dolphin samples obtained, the extracted AVP did not significantly change urine conductivity, suggesting that AVP concentration was not altered as determined by a bioassay. For those samples that showed detectable amounts of AVP, concentrations were very low, as were the amounts from pituitary extracts. Furthermore, a correlation between urine flow rate and plasma AVP concentrations in fasted dolphins was not observed, suggesting that AVP does not significantly regulate urine volume and thus water retention in these animals (Malvin et al., 1971). However, significantly greater concentrations of AVP have been measured recently in free-ranging dolphins, which may be the result of advances in assay techniques for hormone measurements (Ortiz and Worthy, 2000). Because aquaporin-2 (AQP2), the AVP-sensitive water channel, has been identified in the collecting duct of bottlenose dolphins relatively high density, the potential for AVP to stimulate free water reabsorption from the collecting duct is present in dolphins (and other cetaceans as well) (Suzuki et al., 2008). Additionally, based on the molecular analysis of the AQP2 gene in cetaceans, it has been suggested that AQP2 was subjected to positive selection, suggesting that its contribution to regulating urinary water loss is relevant to the ability of cetaceans, including delphinids to adapt to hyperosmotic habitats (Xu et al., 2013).

Renal pathologies and toxicology

Some renal pathologies have been identified in dolphins with most of these cases involving nephrolithiasis of unresolved etiology. Interestingly, urate nephrolithiasis seems to be more prevalent in dolphins in professional human care than in wild populations (Venn-Watson et al., 2010a, 2010b; Smith et al., 2013; Ardente et al., 2017). Differences in diet between managed and wild populations associated with the dietary cation–anion difference and/or dietary purine content likely contribute to the development of nephroliths observed in dolphins (Ardente et al., 2017; Le-Bert et al., 2018). Polycystic kidney disease has also been observed in rough-toothed dolphins, presenting with a similar pathology as in humans, suggesting that the genetic factors contributing to the disease may be similar to humans, but the condition is considered relatively rare in dolphins (Pires et al., 2021).

Evidence of acute, renal tubule necrosis likely resulting from stress-induced ischemia of the renal vasculature was identified in various dolphin species captured in the eastern tropical Pacific during operations associated with the tuna fishery (Cowan and Curry, 2002). The authors cautioned that the injuries described cannot be ascribed to the fishery activities and may be residual from stressors the animals naturally encountered in their environment. Nonetheless, these reports are valuable because they provide evidence that the kidneys of these animals are susceptible to potential stressors, especially if these perturbations induce renal ischemia. This is important because apnea studies (to emulate ischemia–reperfusion) in elephant seals demonstrated that acute apneustic bouts did not induce oxidative injury (Vázquez-Medina et al., 2011), suggesting that the reported renal, ischemic injury was likely the result of chronic and/or frequent stress.

Chemical analyses of kidneys from rough-toothed dolphins, which died while stranded in Florida in 1997, demonstrated the accumulation of selenium (Se) and mercury (Hg) with levels higher in older animals, suggesting that dolphins accumulate certain metals over time, which can potentially lead to renotoxicity and impaired renal function (Mackey et al., 2003). Some of these dead-stranded animals also presented with fibrotic nodules in the kidneys, which can also potentially reduce function. Collectively, while the accumulation of metals and presence of renal fibroids may not be the cause of death, the observations suggest that the kidneys of dolphins and likely all cetaceans are susceptible to heavy metals in the environment. However, the accumulation of other heavy metals (cadmium: Cd, chromium: Cr, copper: Cu, zinc: Zn) in the kidney of bottlenose dolphins caught in the wild in the Indian Ocean were not exceptional and were considered relatively low (Mapunda et al., 2017). Except for Cr, none of the other metals demonstrated an age-dependent correlation.

Metallothioneins (MTs) are cysteine-rich, low molecular weight, metal proteins that protect against metal toxicity by trapping metals on the cysteine residues and serve as an important bank of sequestered essential metals in tissues. Many studies have reported on the tissue levels of MTs across a large range of cetaceans and primarily delphinids (for full list of studies to date, see Cáceres-Saez et al., 2016). The lack of a difference in MTs levels in liver and kidney in Commerson's dolphin (*Cephalorhynchus commersonii*) between immature and mature animals suggested that the sequestration of metals (environmental exposure) was not sufficient to induce an increase in older animals (Cáceres-Saez et al., 2016). The authors suggested that these levels were consistent with other wild delphinids for these tissues and may serve as a good baseline for evaluating health status in other wild populations.

Summary

For logistical reasons, little information on the osmoregulatory capacity of dolphins exists; however, there have been significant strides over the last 20+ years, which have been facilitated by technical and methodological advancements. Studies prior to 1940 consisted primarily of measurements of freezing point depression (osmolality) and electrolyte concentrations in different biological fluids from various species (Fetcher Jr., 1939). While these measurements remain important for better assessing the function of dolphin kidneys, more studies are warranted. The evolution of the unique renal architecture and biochemical composition of kidney cells have provided the structures to support the radiation of dolphins into aquatic environments of variable salinities. The current data provide robust evidence of the capacity to alter at least GFR accordingly to maintain osmotic homeostasis. Along these lines, these animals appear to possess a viable RAAS to regulate the renal handling of primarily Na^+. Aldosterone likely contributes to Na + regulation and balance in these animals as it typically does in other mammals; however, targeted studies of aldosterone and urinary Na + excretion will help confirm its contribution to electrolyte balance. The manipulative studies of diet and salinity demonstrate the flexibility and adaptability of the dolphin kidney. Regardless of the perturbation, dolphins are able to either maintain osmotic homeostasis throughout the bout, or if altered, homeostasis is quickly regained. Dolphins in managed care appear to be more susceptible to urate nephrolithiasis than wild populations; however, it cannot be ascertained if this pathology impairs kidney function in these animals. Lastly, the levels of metallothioneins do not appear to increase with age as may be expected, given the potential of the kidney to sequester heavy metals. Given that innovative methods of collecting urine samples and of easily estimating GFR from serum creatinine have been developed, the potential to rapidly advance the study of kidney function in dolphins is high. For now, it would appear that the primary physiological mechanisms to regulate kidney function are conserved in dolphins; however, further studies are needed to confirm and to explore potentially unique adaptations that are not readily apparent.

References

Albrecht, C.B., 1950. Toxicity of sea water in mammals. Am. J. Phys. 163, 370–385.

Andersen, S.H., Nielsen, E., 1983. Exchange of water between the harbor porpoise, *Phocoena phocoena*, and the environment. Experientia 39, 52–53.

Ardente, A.J., Wells, R.S., Smith, C.R., Walsh, M.T., Jensen, E.D., Schmitt, T.L., Colee, J., Vagt, B.J., Hill, R.C., 2017. Dietary cation–anion difference may explain why ammonium urate nephrolithiasis occurs more frequently in common bottlenose dolphins (*Tursiops truncatus*) under human care than in free-ranging common bottlenose dolphins. J. Anim. Sci. 95 (3), 1396–1406.

Bentley, P.J., 1963. Composition of the urine of the fasting humpback whale (*Megaptera nodosa*). Comp. Biochem. Physiol. 10, 257–259.

Bester, M.N., 1975. The functional morphology of the kidney of the cape fur seal, *Arctocephalus pusillus* (Schreber). Modoqua Ser. II 4, 69–92.

Beuchat, C.A., 1990. Body size, medullary thickness, and urine concentrating ability in mammals. Am. J. Physiol. Regul. Integr. Comp. Physiol. 258 (2), R298–R308.

Birukawa, N., Ando, H., Goto, M., Kanda, N., Pastene, L.A., Urano, A., 2008. Molecular cloning of urea transporters from the kidneys of baleen and toothed whales. Comp. Biochem. Physiol. B: Biochem. 149 (2), 227–235.

Cáceres-Saez, I., Polizzi, P., Romero, B., Dellabianca, N.A., Guevara, S.R., Goodall, R.N.P., Cappozzo, H.L., Gerpe, M., 2016. Hepatic and renal metallothionein concentrations in Commerson's dolphins (*Cephalorhynchus commersonii*) from Tierra del Fuego, South Atlantic Ocean. Mar. Pollut. Bull. 108 (1–2), 263–267.

Costa, D.P., 1982. Energy, nitrogen, electrolyte flux and sea water drinking in the sea otter *Enhydra lutris*. Physiol. Zool. 55, 35–44.

Cowan, D.F., Curry, B.E., 2002. Histopathological assessment of dolphins necropsied onboard vessels in the eastern tropical Pacific tuna fishery. National Marine Fisheries Service, Southwest Fisheries Science Center. NMFS SWFSC administrative report LJ-02-24C.

Eichelberger, L., Leiter, L., Geiling, E.M.K., 1940. Water and electrolyte content of dolphin kidney and extraction of pressor substance (renin). Proc. Soc. Exp. Biol. Med. 44, 356–359.

Fair, P.A., Schaefer, A.M., Romano, T.A., Bossart, G.D., Lamb, S.V., Reif, J.S., 2014. Stress response of wild bottlenose dolphins (*Tursiops truncatus*) during capture–release health assessment studies. Gen. Comp. Endocrinol. 206, 203–212.

Fetcher Jr., E.S., 1939. The water balance in marine mammals. Q. Rev. Biol. 14, 451–459.

Fetcher Jr., E.S., Fetcher, G.W., 1942. Experiments on the osmotic regulation of dolphins. J. Cell. Comp. Physiol. 19, 123–130.

Guo, A., Hao, Y., Wang, J., Zhao, Q., Wang, D., 2014. Concentrations of osmotically related constituents in plasma and urine of finless porpoise (*Neophocaena asiaeorientalis*): implications for osmoregulatory strategies for marine mammals living in freshwater. Zool. Stud. 53 (1), 1–6.

Hart, L.B., Wells, R.S., Kellar, N., Balmer, B.C., Hohn, A.A., Lamb, S.V., Rowles, T., Zolman, E.S., Schwacke, L.H., 2015. Adrenal hormones in common bottlenose dolphins (*Tursiops truncatus*): influential factors and reference intervals. PLoS One 10 (5), e0127432.

Hedges, N.A., Gaskin, D.E., Smith, G.J.D., 1979. Rencular morphology and renal vascular system of the harbour porpoise *Phocoena phocoena* (L.). Can. J. Zool. 57, 868–875.

Hiatt, E.P., Hiatt, R.B., 1942. The effect of food on the glomerular filtration rate and renal blood flow in the harbor seal (*Phoca vitulina* L.). J. Cell. Comp. Physiol. 19, 221–227.

Hong, S.K., Elsner, R., Claybaugh, J.R., Ronald, K., 1982. Renal functions of the Baikal seal *Pusa sibirica* and ringed seal *Pusa hispida*. Physiol. Zool. 55, 289–299.

Houser, D.S., Yeates, L.C., Crocker, D.E., 2011. Cold stress induces an adrenocortical response in bottlenose dolphins (*Tursiops truncatus*). J. Zoo Wildlife Med., 565–571.

Hui, C., 1981. Seawater consumption and water flux in the common dolphin *Delphinus delphis*. Physiol. Zool. 54, 430–440.

Hunsicker, L.G., Adler, S., Caggiula, A., England, B.K., Greene, T., Kusek, J.W., Rogers, N.L., Teschan, P.E., Beck, G., Modification of Diet in Renal Disease Study Group, 1997. Predictors of the progression of renal disease in the modification of diet in renal disease study. Kidney Int. 51 (6), 1908–1919.

Irvine, A.B., Neal, R.C., Cardeilhac, R.T., Popp, J.A., Whiter, F.H., Jenkins, R.C., 1980. Clinical observations on captive and free-ranging West Indian manatees, *Trichechus manatus*. Aquat. Mamm. 8, 2–10.

Janech, M.G., Chen, R., Klein, J., Nowak, M.W., McFee, W., Paul, R.V., Fitzgibbon, W.R., Ploth, D.W., 2002. Molecular and functional characterization of a urea transporter from the kidney of a short-finned pilot whale. Am. J. Phys. Regul. Integr. Comp. Phys. 282 (5), R1490–R1500.

Jiménez-Zucchet, N., Alejandro-Zayas, T., Alvarado-Macedo, C.A., Arreola-Illescas, M.R., Benítez-Araiza, L., Bustamante-Tello, L., Cruz-Martínes, D., Falcón-Robles, N., Garduño-González, L., López-Romahn, M.C., Martínez-Taylor, A.M., 2019. Baseline urinalysis values in common bottlenose dolphins under human care in the Caribbean. J. Vet. Diagn. Investig. 31 (3), 426–433.

Le-Bert, C.R., Smith, C.R., Poindexter, J., Ardente, A., Meegan, J., Wells, R.S., Venn-Watson, S., Jensen, E.D., Sakhaee, K., 2018. Comparison of potential dietary and urinary risk factors for ammonium urate nephrolithiasis in two bottlenose dolphin (*Tursiops truncatus*) populations. Am. J. Physiol. Ren. Physiol. 315 (2), F231–F237.

Mackey, E.A., Oflaz, R.D., Epstein, M.S., Buehler, B., Porter, B.J., Rowles, T., Wise, S.A., Becker, P.R., 2003. Elemental composition of liver and kidney tissues of rough-toothed dolphins (*Steno bredanensis*). Arch. Environ. Contam. Toxicol. 44 (4), 0523–0532.

Maluf, N.S.R., 1989. Renal anatomy of the manatee, *Trichechus manatus* (Linnaeus). Am. J. Anat. 184, 269–286.

Malvin, R.L., Rayner, M., 1968. Renal function and blood chemistry in Cetacea. Am. J. Phys. 214, 187–191.

Malvin, R.L., Vander, A.J., 1967. Plasma renin activity in marine teleosts and Cetacea. Am. J. Phys. 213, 1582–1584.

Malvin, R.L., Bonjour, J.P., Ridgway, S., 1971. Antidiuretic hormone levels in some cetaceans. Proc. Soc. Exp. Biol. Med. 136, 1203–1205.

Malvin, R.L., Ridgway, S., Cornell, L., 1978. Renin and aldosterone levels in dolphins and sea lions. Proc. Soc. Exp. Biol. Med. 157, 665–668.

Mapunda, E.C., Othman, O.C., Akwilapo, L.D., Bouwman, H., Mwevura, H., 2017. Concentrations of metallic elements in kidney, liver, and lung tissue of Indo-Pacific bottlenose dolphin *Tursiops aduncus* from coastal waters of Zanzibar, Tanzania. Mar. Pollut. Bull. 122 (1–2), 483–487.

McClain, A.M., Daniels, R., Gomez, F.M., Ridgway, S.H., Takeshita, R., Jensen, E.D., Smith, C.R., 2020. Physiological effects of low salinity exposure on bottlenose dolphins (*Tursiops truncatus*). J. Zool. Bot. Gardens 1 (1), 61–75.

Miller, L.J., Lauderdale, L.K., Walsh, M.T., Bryant, J.L., Mitchell, K.A., Granger, D.A., Mellen, J.D., 2021. Reference intervals and values for fecal cortisol, aldosterone, and the ratio of cortisol to dehydroepiandrosterone metabolites in four species of cetaceans. PLoS One 16 (8), e0250331.

Nagai, K.I., Tadano-Aritomi, K., Niimura, Y., Ishizuka, I., 2008. Higher expression of renal sulfoglycolipids in marine mammals. Glycoconj. J. 25 (8), 723–726.

Naka, T., Katsumata, E., Sasaki, K., Minamino, N., Yoshioka, M., Takei, Y., 2007. Natriuretic peptides in cetaceans: identification, molecular characterization and changes in plasma concentration after landing. Zool. Sci. 24 (6), 577–587.

Ortiz, R.M., 2001. Osmoregulation in marine mammals. J. Exp. Biol. 204 (11), 1831–1844.

Ortiz, C.L., Costa, D., Le Boeuf, B.J., 1978. Water and energy flux in elephant seal pups fasting under natural conditions. Physiol. Zool. 51, 166–178.

Ortiz, R.M., Wade, C.E., Costa, D.P., Ortiz, C.L., 2002. Renal responses to plasma volume expansion and hyperosmolality in fasting seal pups. Am. J. Phys. Regul. Integr. Comp. Phys. 282 (3), R805–R817.

Ortiz, R.M., Crocker, D.E., Houser, D.S., Webb, P.M., 2006. Angiotensin II and aldosterone increase with fasting in breeding adult male northern elephant seals (*Mirounga angustirostris*). Physiol. Biochem. Zool. 79 (6), 1106–1112.

Ortiz, R.M., Long, B., Casper, D., Ortiz, C.L., Williams, T.M., 2010. Biochemical and hormonal changes during acute fasting and re-feeding in bottlenose dolphins (*Tursiops truncatus*). Mar. Mamm. Sci. 26 (2), 409–419.

Ortiz, R.M., Worthy, G.A., 2000. Effects of capture on adrenal steroid and vasopressin concentrations in free-ranging bottlenose dolphins (*Tursiops truncatus*). Comp. Biochem. Physiol. A: Physiol. 125 (3), 317–324.

Parker, H.R., Schroeder, J.P., 1987. Renal Function, A Possible Indicator of Stress in Dolphins. Naval Ocean Systems Center, San Diego, CA.

Pfeiffer, C.J., 1997. Renal cellular and tissue specializations in the bottlenose dolphin (*Tursiops truncatus*) and beluga whale (*Delphinapterus leucas*). Aquat. Mamm. 23, 75–84.

Pires, B.G., da Nóbrega, D.F., Domit, C., Bracarense, A.P.F.R.L., 2021. Polycystic kidney disease in rough toothed dolphins (*Steno bredanensis*) founded in the Paraná Coast, Southern Brazil. Acta Sci. Vet. 49 (Suppl. 1), 1–7.

Portier, P., 1910. Pression osmotique des liquides des oiseaux et mammifères marins. J. Physiol. Path. Gener. 12, 202–208.

Ridgway, S.H., 1972. Homeostasis in the aquatic environment. In: Ridgway, S.H. (Ed.), Mammals of the Sea: Biology and Medicine. Charles C. Thomas Publisher, Springfield, IL, pp. 590–747.

Ridgway, S., Venn-Watson, S., 2010. Effects of fresh and seawater ingestion on osmoregulation in Atlantic bottlenose dolphins (*Tursiops truncatus*). J. Comp. Physiol. B. 180 (4), 563–576.

Smith, C.R., Venn-Watson, S., Wells, R.S., Johnson, S.P., Maffeo, N., Balmer, B.C., Jensen, E.D., Townsend, F., Sakhaee, K., 2013. Comparison of nephrolithiasis prevalence in two bottlenose dolphin (*Tursiops truncatus*) populations. Front. Endocrinol. 4, 145.

Suzuki, M., Endo, N., Nakano, Y., Kato, H., Kishiro, T., Asahina, K., 2008. Localization of aquaporin-2, renal morphology and urine composition in the bottlenose dolphin and the Baird's beaked whale. J. Comp. Physiol. B. 178 (2), 149–156.

Telfer, N., Cornell, L.H., Prescott, J.H., 1970. Do dolphins drink water? J. Am. Vet. Med. Assoc. 157, 555–558.

Vardy, P.H., Bryden, M.M., 1981. The kidney of *Leptonychotes weddelli* (Pinnipedia: Phocidae) with some observations on the kidneys of two other southern phocid seals. J. Morphol. 167, 13–34.

Vázquez-Medina, J.P., Zenteno-Savín, T., Tift, M.S., Forman, H.J., Crocker, D.E., Ortiz, R.M., 2011. Apnea stimulates the adaptive response to oxidative stress in elephant seal pups. J. Exp. Biol. 214 (24), 4193–4200.

Venn-Watson, S., Smith, C.R., Johnson, S., Daniels, R., Townsend, F., 2010a. Clinical relevance of urate nephrolithiasis in bottlenose dolphins *Tursiops truncatus*. Dis. Aquat. Org. 89 (2), 167–177.

Venn-Watson, S.K., Townsend, F.I., Daniels, R.L., Sweeney, J.C., McBain, J.W., Klatsky, L.J., Hicks, C.L., Staggs, L.A., Rowles, T.K., Schwacke, L.H., Wells, R.S., 2010b. Hypocitraturia in common bottlenose dolphins (*Tursiops truncatus*): assessing a potential risk factor for urate nephrolithiasis. Comp. Med. 60 (2), 149–153.

Venn-Watson, S., Smith, C.R., Dold, C., Ridgway, S.H., 2008. Use of a serum-based glomerular filtration rate prediction equation to assess renal function by age, sex, fasting, and health status in bottlenose dolphins (*Tursiops truncatus*). Mar. Mamm. Sci. 24 (1), 71–80.

Wang, J., Yu, X., Hu, B., Zheng, J., Xiao, W., Hao, Y., Liu, W., Wang, D., 2015. Physicochemical evolution and molecular adaptation of the cetacean osmoregulation-related gene UT-A2 and implications for functional studies. Sci. Rep. 5 (1), 1–17.

Xu, S., Yang, Y., Zhou, X., Xu, J., Zhou, K., Yang, G., 2013. Adaptive evolution of the osmoregulation-related genes in cetaceans during secondary aquatic adaptation. BMC Evol. Biol. 13 (1), 1–9.

CHAPTER

12

Reproductive physiology of dolphins

Shannon Atkinson

Department of Fisheries, College of Fisheries and Ocean Sciences, University of Alaska Fairbanks, Juneau, AK, United States

Abstract

The reproductive system is a bodily system that is integral to the existence of a species. Bottlenose dolphins represent a key species from which baseline and benchmark physiological information can be used to assist in the perpetuation of other small cetacean populations. This chapter reviews basic reproductive physiology and endocrinology of bottlenose dolphins, including sexual maturation, reproductive cycles, gestation and lactation, as well as social and environmental impacts on reproduction in both sexes. The work includes studies conducted on animals in the wild as well as those performed under managed care settings, where bottlenose dolphin studies have a long history. In addition, the breadth of the studies includes assisted reproductive and contraceptive technologies, as well as use of ex-situ conservation efforts for small cetaceans, of which several populations are at extreme risks of extirpation.

Introduction

As this book is on dolphin physiology, much of the focus is on the bottlenose dolphin (*Tursiops truncatus*), which has often been the most convenient species to study, and their numbers in managed care settings have been the highest. However, studies from other species, such as killer whales (*Orcinus orca*) and beluga whales (*Delphinapterus leucas*) have also contributed immensely to our knowledge. Free-ranging populations of small cetaceans have been and continue to be heavily impacted by human activities globally, largely as fisheries by-catch as well as through habitat destruction, thus the knowledge that we have from all small cetacean species is important and largely applicable across genera. While relatively few ex-situ or captive breeding programs exist, their use in sustaining small and dwindling populations may be extremely valuable should these programs expand. Thus, this chapter incorporates additional information related to reproductive biology and physiology of small coastal and freshwater cetaceans.

The benefits of being able to access and obtain biological samples from dolphins in managed care settings has been invaluable in understanding reproductive physiology of dolphins. High-quality long-term programs can sustain generations of dolphins with high survival rates and successful breeding programs. Many of these programs had their origins in entertainment and recreation for humans, as well as military or defense-related programs. And while the purpose of the managed care setting has uniformly shifted toward education and outreach, several programs have maintained research efforts that have been productive and yielded substantial information, particularly on reproductive biology and behavior.

Complementing the knowledge gained from dolphins in managed care settings, a few long-term programs have been successfully maintained on free-ranging dolphins (e.g., Wells et al., 2004). These programs are of immense benefit in that dolphins are exposed to a wider range of ambient conditions, many of which are completely natural and highlight a variety of environmental parameters that may influence reproduction. Together with the studies in managed care settings, knowledge of small cetacean reproduction has grown over the past four decades.

This chapter will focus on some key areas of reproductive physiology and technologies that may be used to assist or limit reproduction in specific settings. It will present the basic anatomy (Fig. 1) and review the basic life history events of sexual maturation, reproductive cycles, gestation, and lactation before turning to recent research on reproductive anomalies and abnormalities. These include events such as pseudopregnancy, dystocia, and other birthing problems, such as spontaneous abortions and failure to thrive. Contraception and assisted reproductive technologies have been

The Physiology of Dolphins
https://doi.org/10.1016/B978-0-323-90516-9.00012-9

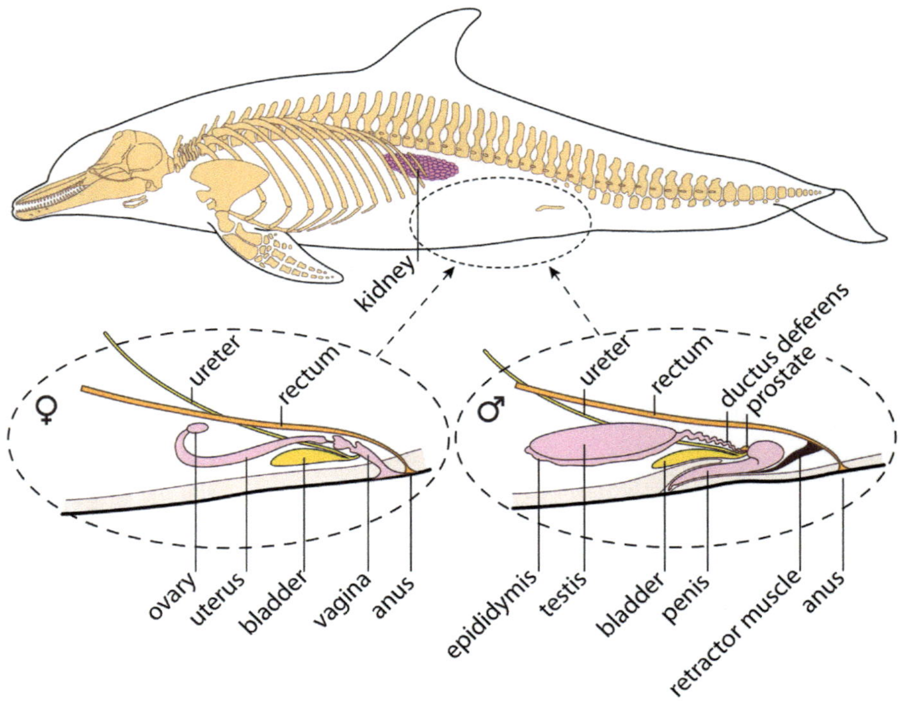

FIG. 1 Reproductive tract of the bottlenose dolphin, with both female and male anatomy shown. *Illustration by Uko Gorter.*

commonly used in managed care settings, and there may be need for additional uses of these technologies that can improve reproductive outcomes for various species (Taylor et al., 2020). Knowledge of impacts on reproduction are increasingly in need as environmental and climate conditions change both regionally and globally, and these are briefly covered. The chapter concludes with the application of ex-situ conservation efforts focused on reproductive programs that are targeted around captive breeding or programs to improve reproductive success. These potential or active programs may need to increase in frequency or duration as many coastal and freshwater small cetacean populations face increased anthropogenic pressures.

Onset of sexual maturation

It is appropriate to start a review of dolphin reproduction with the onset of sexual maturity. This is the period of time that major physiological changes take place in animals, including the development of secondary sexual characteristics, the onset of reproductive cycles, and changes in behavior that are largely driven by hormones related to reproductive processes (Atkinson and Yoshioka, 2007). At the gonadal level, the cyclical changes in cetacean ovaries have been documented for nearly 100 years (Ridgway, 1972; Harrison and Ridgway, 2009). Ovaries lack any signs of ovulation, corpora lutea (CL), or corpora albicantia (CA) in immature dolphins but may possess numerous ovarian primordial and antral follicles that presumably may grow and reach preovulatory size or be resorbed as atretic follicles (Fig. 2). In males, the testes also are generally reduced in size prior to sexual maturity, at which time growth and the cycle of spermatogenesis is initiated. The onset of sexual maturity in male bottlenose dolphin is accompanied by testicular morphological and physiological changes such as testis weight, increased diameter of seminiferous tubules, spermatozoa within the seminiferous tubules, and seminal fluid in the epididymis (Atkinson and Yoshioka, 2007; Atkinson, 2009).

The average age of sexual maturity (ASM) is a life history parameter that reflects the average age of ovulation in the female portion of a population (DeMaster, 1984). It is typically calculated from longitudinal data on the ovulation status of known (e.g., photographically identified) individuals or from data on age-specific ovulation rates (e.g., from directed hunts or from beach cast animals). The ASM is closely related to the average age of first birth, where the difference is a function of the average duration of gestation and the rate at which pregnancies are prematurely terminated (DeMaster, 1984). Some of the earlier studies in dolphin reproduction used body length to assess when sexual maturity had taken place (Perrin and Reilly, 1984). In bottlenose dolphins, gonadal activity was reported when females were over 220 cm in length. For males, 260 cm was the length at which sexual activity was reported. The ratio of the length at sexual maturity to the adult length has been documented for a number of small cetaceans and generally falls in the range of 80–87.6 and varied little between species. Female dolphins held in managed care settings reach sexual

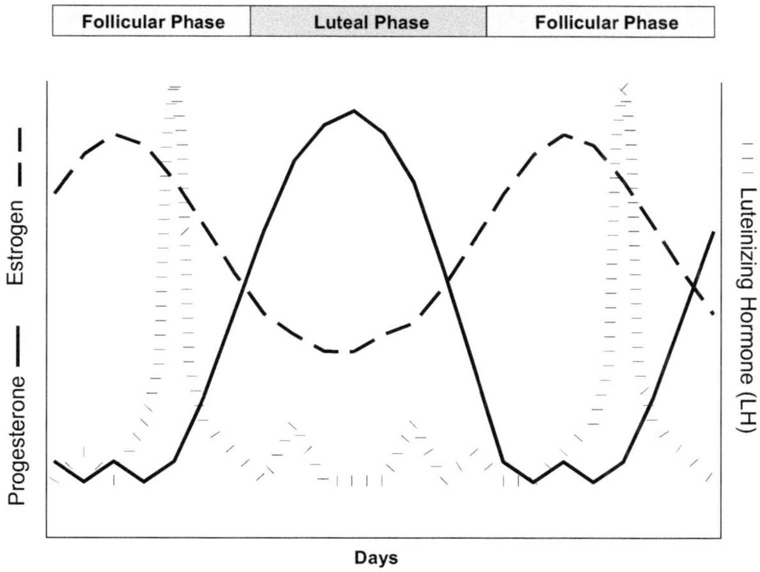

FIG. 2 Progesterone, estrogen, and luteinizing hormone profiles relative to the days of the bottlenose dolphin estrous cycle. Within each estrous cycle, a follicular and luteal phase exists. Ovulation is followed by the formation of a transitory corpus hemorrhagicum and then a corpus luteum.

maturity at 4–10 years of age, and one was reported to give birth at 4 years, but this age is variable and several things may influence it (Table 1). For males, the ASM is greater than for females, and 8–10 years is often used as a benchmark but is also variable (e.g., 7–20 years) and may be a couple of years earlier when the ambient conditions are ideal. For free-ranging bottlenose dolphins, the ASM may occur between 5 and 15 years of age. Similarly, the spread of ASM for small coastal and freshwater cetaceans is 3–16 years (Table 1).

Reproductive cycles

Bottlenose dolphins are known to be spontaneous ovulators, where ovulations occur without external stimulation (Sawyer-Steffan et al., 1983; Kirby and Ridgway, 1984; Robeck et al., 2018). While seasonality is noted in some populations, dolphin pregnancies and births have been recorded in every month of the year, so this seasonality is largely related to the geographic location of the resident population (Urian et al., 1996). For example, female dolphins in Hawai'i are reported to have ovulated one to six times per year, primarily between August and November. Dolphins are also known to be seasonally polyestrous, meaning that they can repetitively ovulate if conception does not occur from a previous ovulation. The cycle length of dolphins is variable with 21–42 days reported based on hormone profiles, although where multiple cycles occur, 30–35 days is considered a normal cycle (Robeck et al., 2018). During this time, the physiology of the ovary goes through typical mammalian phases of follicular development, resulting in ovulation, which is followed by a luteal phase (Fig. 2). During the follicular phase, the ovaries containing numerous antral follicles are active in having small follicles (2–3 mm) grow in diameter, with one becoming a preovulatory follicle (10–28 mm). The process of ovulation results in a corpus hemorrhagicum (CH), which is a transitory phase followed by the formation of the corpus luteum (CL). It is the production of progesterone by the CL that sustains the pregnancy. The cells of the CL continue to produce progesterone until luteolysis, which is a process that breaks down the CL and transforms it to a scar on the ovary, known as a corpus albicantium (CA) (Fig. 3).

Monitoring of estrous cycles is commonly conducted with endocrine profiles (Atkinson and Yoshioka, 2007; Galligan et al., 2018; Robeck et al., 2018). Progesterone is the primary hormone that has historically been monitored, with other hormones such as estradiol, total estrogens, testosterone, luteinizing hormone, corticosteroids, thyroid hormones, relaxin, inhibin, and catecholamines periodically appearing in scientific publications. Progesterone is a key hormone that is primarily produced in the ovaries and is related to ovulation and the ability of the female to sustain a pregnancy. Progesterone is secreted in the granulosa cells that surround the developing ovum in the antral follicle. However, its biosynthesis is triggered by the process of luteinization, which typically occurs with a preovulatory surge of luteinizing hormone (LH). While estrogens can be argued to play the most important role in ovulation, the forms of estrogen vary considerably and the concentrations are relatively low and fleeting around the time of ovulation. Thus, estrogens generally aren't the best indicators of reproductive activity. However, with both estrogen measurement and

TABLE 1 Scientific taxonomy, common name, age of sexual maturity, calving interval, gestation length, lactation duration, known ex-situ studies, notes on reproductive parameters, and sources of information for bottlenose dolphins and other small coastal and freshwater cetaceans.

Family/Genus/ Species	Common name	Age of sexual maturity (years)	Calving interval (years)	Gestation length (months)	Lactation duration	Ex-situ studies	Reproductive notes	Sources
Delphinidae								
Tursiops truncatus	Bottlenose dolphin	5–13 females, 9–14 males		12	4 mos-6 yrs	Yes	Polyestrous, spontaneous ovulators. Diffusely seasonal spring-summer	Wells and Scott (2018)
Tursiops aduncus	Indo-Pacific bottlenose dolphin	12–15 females, 10–15 males	3–6	12	3–5 yrs	Yes	Polyestrous, spontaneous ovulators	Wang (2018)
Sotalia guianensis	Guiana dolphin	8	~1	11–12		Yes[a]	Calving is year-round	Flores et al. (2018)
Sotalia fluviatilis	Tucuxi			11			Calving in fall	Flores et al. (2018)
Sousa chinensis	Indo-Pacific humpback dolphin	9–10 females, >10 males		11			Calving spring/summer; may be location effect	Para and Jefferson (2018), Jefferson et al. (2012), Zeng et al. (2020)
Sousa plumbea	Indian Ocean humpback dolphin							Para and Jefferson (2018)
Sousa teuszii	Atlantic Ocean humpback dolphin	10 females, 12-13 males	3	10–12	Can be >2yrs			Para and Jefferson (2018)
Orcinus orca	Killer whale	11–13 female, 15 males	2–14~5 avg	15–18	1–2yrs	Yes	Polyestrous, calving occurs year-round	Ford (2018)
Orcaella brevirostris	Irrawaddy dolphin			14		Yes	Calving peaks premonsoon season	Smith (2018)
Orcaella heinsohni	Australian snubfin dolphin	4–6		14	2 yrs			Beasley and Brown (2018)
Steno bredanensis	Rough-toothed dolphin	10 females, 14 males				Yes		Jefferson (2018b)
Cephalorhynchus hectori	Hector's dolphin	5–8 females, 5–9 males	2–4	10–11				Dawson (2018)
Iniidae								
Inia geoffrensis	Amazon or Orinoco river dolphin	6–7 females, males later	2–3	11	>1 yr		Mating at low water, calving at river flood levels	da Silva and Martin (2018)
Lipotidae								
Lipotes vexillifer	Baiji or Yangtze River dolphin	6 females, 4 males						Zhou (2018)

Phocoenidae								
Phocoena phocoena	Harbor porpoise	3–7 females, 3–5 males	1–2	10.5	<1 yr		Summer calving, summer to fall mating	Bjorge and Tolley (2018)
Phocoena sinus	Vaquita	<6 females		10–11			Seasonal calving in spring	Rojas-Bracho et al. (2018)
Phocoena spinipinnis	Burmeister's porpoise		1	11–12			Mating in austral summer; lactation may be coincident with pregnancy	Reyes (2018)
Phocoena dioptrica	Spectacled porpoise	3–16				Yes[b]	Calving in austral spring-summer	Natalie and Goodall (2009)
Phocoenoides dalli	Dall's porpoise	4–7 females, 3–8 males		10–12	likely < 1 yr	Yes[b]	Main calving season is summer	Jefferson (2018a)
Neophocaena phocaenoides	Indo-Pacific finless porpoise	4–6	2	11	7 months		Calving season differs with locality	Amano (2018)
Neophocaena asiaeorientalis	Narrow-ridged finless porpoise							Amano (2018)
Platanistidae								
Platanista gangetica	Ganges River dolphin			~12	1–12 mos		Peak calving early winter/early summer	Reeves and Martin (2018)
Platanista minor	Indus River dolphin			~12				Reeves and Martin (2018)
Pontoporidae								
Pontoporia blainvillei	Franciscana dolphin	2.7	2	10.5–11.1	9 mos		Mating austral summer, calving in austral spring	Crespo (2018)

[a] *Susceptible to capture stress.*
[b] *Very few, poor survival.*

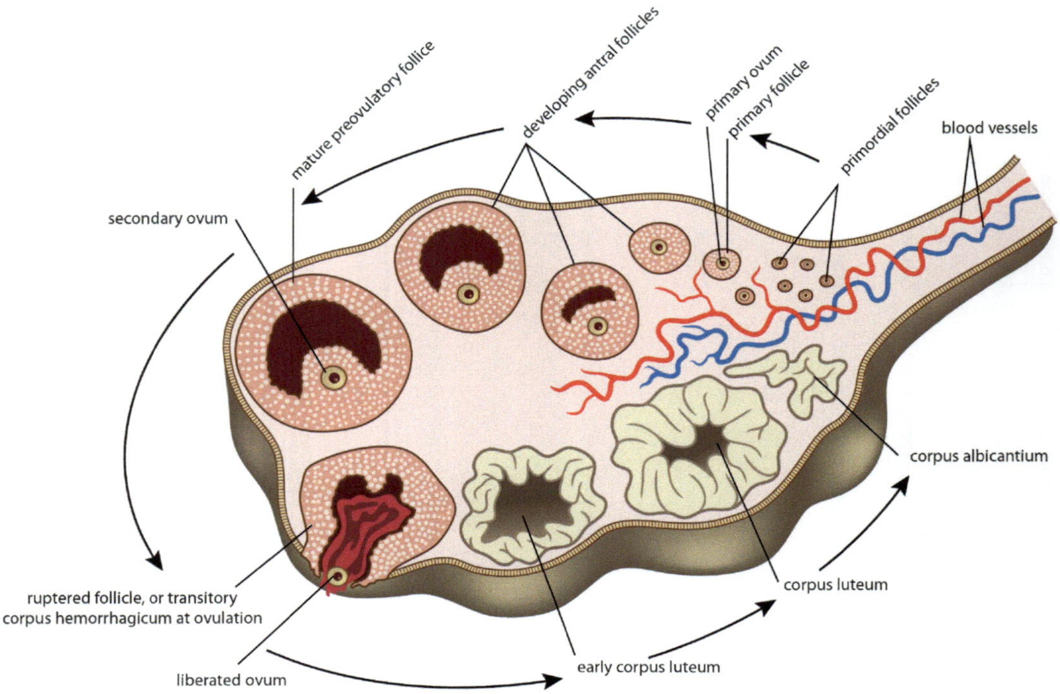

FIG. 3 Schematic detail of the dolphin ovary showing primordial follicles, developing antral follicles, a mature preovulatory follicle, a ruptured follicle or transitory corpus hemorrhagicum at ovulation, a corpus luteum, and a corpus albicantium. *Illustration by Uko Gorter.*

the use of ultrasound, the dominant follicle can be identified up to 10 days before ovulation. A typical hormone profile that depicts the process of ovulation is shown in Fig. 2.

For males, testosterone is a prime hormone that is primarily produced in the gonads and is related to sexual maturity (Atkinson and Yoshioka, 2007). Specifically, testosterone is produced by the Leydig cells in the seminiferous tubules. The biosynthesis is typically stimulated by LH, which is secreted in pulses and has receptors in the Leydig cells. The increasing secretion of testosterone promotes the process of spermatogenesis, particularly the development and maturation of spermatozoa from spermatogonia. Typically, the increase in testosterone precedes the increase in sperm production by 4–8 weeks (Schroeder and Keller, 1989). In the common bottlenose dolphin, reproductive cycles are considered diffuse with seasonal changes in testosterone production, but fertility is maintained throughout the year.

Behavioral and physical signs of estrus are physiologically driven by endocrine changes. For female dolphins, the physical signs of estrus may include subtle changes, including the presence of a pinkish color on the ventral side of the animal, and in some cases, specifically around the genital slit. Behavioral changes in females may also be subtle and include a greater acceptance of the interest of other dolphins in tactile and genital exploration and prodding. These are consistent with estrus being a time of sexual receptiveness, with successful breeding being the ultimate desired behavior. For male dolphins, physical signs often are seen during sex-related behaviors. For example, they may actively pursue females and position their rostrum near the female's genital slit. While this positioning may simply be exploratory, it is also possible that pheromonal communication in dolphins is primarily gustatory, thus this behavioral interaction may be the result of a biochemical role that cues off of the physiology of the female. During or associated with these pursuits, male dolphins will often be seen with penile erections. Assertive or aggressive behaviors between males may occur when females are in estrus, likely reflecting competition between males for breeding access to the target female. Many males in managed care settings are not subtle in their sexual pursuits, and bottlenose dolphins have long had the reputation of being sexually active.

The use of bottlenose dolphins in managed care has been critical for understanding the longitudinal sequences of hormone secretions related to estrous cycles and have been successful in improving the diagnostic abilities of single or snapshot measurements. Longitudinal sampling has occurred in serum, feces, and milk, with additional sampling in blubber, saliva, and respiratory blow in a variety of cetacean species (Richkind and Ridgway, 1975; Yoshioka et al., 1986; Atkinson et al., 1999, 2021; West et al., 2000; Morita et al., 2003; Kellar et al., 2006; Pérez et al., 2011; Steinman et al., 2016; Bergfelt et al., 2017; Richard et al., 2017; Legacki et al., 2020; Atkinson et al., 2022). These matrixes have been used mostly in bottlenose dolphins, false killer whales (*Pseudorca crassidens)*, beluga whales, killer whales, with limited uses in Pacific white-sided (*Lagenorhynchus obliquidens*), common (*Delphinus delphis*), and rough-toothed dolphins (*Steno bredanensis*) in managed care settings. The sampling frequencies can be as often as daily, bi-weekly, or monthly for trained cetaceans. Age, season, time of day, and feeding schedules all have been demonstrated to have significant impacts on serum hormone concentrations in bottlenose dolphins in managed care setting. Time of day has been shown to vary with the majority of hormones measured, with winter being higher than spring and morning being higher than afternoon or evening for cortisol (Houser et al., 2021). The phase of the moon was also related to progesterone peaks. Seasonality of many hormones also exists and is often linked to complex interactions with age and sex (Houser et al., 2021). While many of the studies examining season or time of day were conducted for metabolic hormones (Suzuki et al., 2003), with males increasing with age and females decreasing with age (Houser et al., 2021), the interaction with sex increases the complexity of interpreting all reproductive physiological profiles. Thus having contextual information on individuals is very important to be able to decipher physiological processes that are reflected by measured hormone profiles. Also, consistency in sampling programs is recommended to minimize variations due to collection protocols (Suzuki et al., 2003; Houser et al., 2021).

Gestation and lactation

Several studies have shown that bottlenose dolphins and many small cetaceans have a predominance for ovulating on the left ovary and carrying pregnancies in the left uterine horn (Fig. 4). While this is not true for all cetaceans, it is considered an anatomical peculiarity that is without an obvious explanation. Once conception has occurred, the reproductive phase referred to as gestation occurs. The duration of gestation is one of the least variable reproductive parameters within a given species. For bottlenose dolphins, as well as most of the small coastal and freshwater cetaceans, this period is fairly consistently reported as 10–12 months in duration (Perrin and Reilly, 1984). However, studies that monitored gestation from the known date of conception to parturition provide a range of 355–395 days, with a tight clustering of 365–376 days (Table 1).

The endocrinology of gestation is reasonably well known. The elevation of progesterone occurs post ovulation, where it remains elevated throughout gestation and only declines at luteolysis. Evidence indicates that the corpus luteum is the gland responsible for maintaining the elevated progesterone (Robeck et al., 2012a, b). Typically, a significant elevation of progesterone is anticipated around 3 weeks post-conception, and two or more measurements separated by 3 weeks are used to detect pregnancy. This is a standard procedure and most breeding facilities utilize this, especially for early pregnancy detection. However, as was previously covered, progesterone is also elevated in the

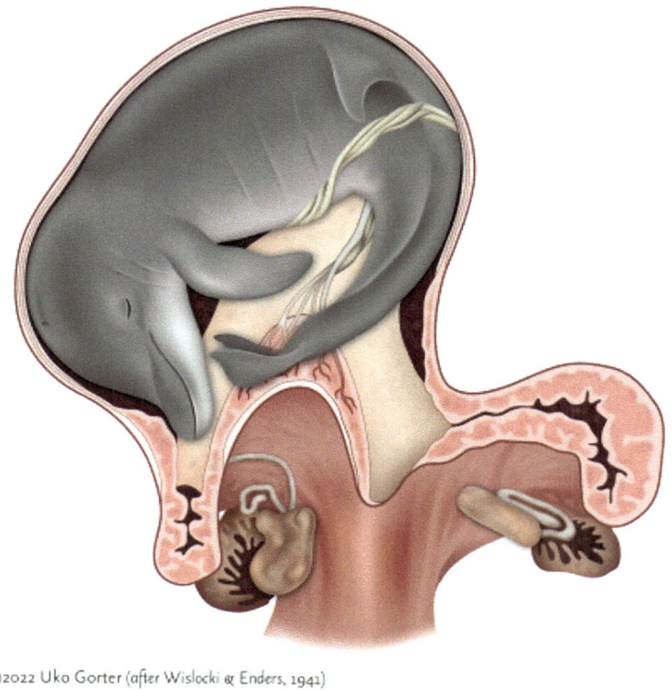

FIG. 4 Anatomy of the pregnant dolphin reproductive tract showing the outline of the uterus with a fetus in the uterine horn. *Illustration by Uko Gorter.*

periovulatory phase as well as during pseudopregnancies (covered below), and those reproductive states may lead to an overestimation of potential or presumed pregnancies. Estrogens and androgens both undergo increases during gestation, estrogens initially and androgens throughout gestation. Relaxin is another hormone, whose role is associated with parturition, as it increases dramatically around that time (Bergfelt et al., 2017). The thyroid hormones are known to be critical for fetal development, and the source is primarily of maternal origin in early gestation. At least in other species, the thyroid hormones promote neuronal development and myelination in the fetal brain, and a deficiency of thyroid hormones is associated with birthing problems in bottlenose dolphins (discussed below).

Ultrasound is also a very good technique for pregnancy detection and provides additional data on the fetal development (Brook, 2001). The detection of a fetus most reliably occurs after day 50 of gestation in bottlenose dolphins, although the presence of an embryonic vesicle in conjunction with the thickening of the endometrium may be visualized prior to that date. Progesterone concentrations are elevated prior to a pregnancy stage that is visible using ultrasound, and therefore it may be considered as a priority for early pregnancy detection. Conversely, because ultrasonic detection of pregnancy is most reliable later in gestation, it may actually underestimate detections as reproductive loss may have already occurred prior to the procedure occurring. Monitoring fetal growth and development can also provide indication of gestational abnormalities and problems, adding benefit to the use of ultrasonic monitoring of gestation. Specifically, perinatal fetal size can be monitored, and based on survivability, a ~15.5 cm axillary width is targeted (Sweeney et al., 2010). Similarly, monitoring the size or echogenicity of the CL can indicate the end of the luteal phase or the oncoming parturition.

The calving interval can vary substantially both within a species and between species, ranging from annually to 4 years (Table 1). Dolphins in managed care settings often have shorter reproductive cycles (intercalving intervals) when compared to their free-ranging counterparts. This is likely due to well-managed care settings, which typically have well-fed animals, housed in stable environments, with routine and appropriate veterinary and husbandry care. The ambient conditions of a high-quality habitat can often dictate the wellbeing of animals, such that reproductive success measured in births can be used as an indicator of the health of a population or group of cetaceans. This is true for small cetaceans in both free-ranging and managed care settings. Births often occur at night, and the female may display a lack of interest in food for a day or two prior to parturition. A normal parturition has the calf presenting tail first.

Development of the mammary glands occurs rapidly within 2–3 days of parturition. Nursing will typically begin within 8 h of birth, but some females will not begin nursing until the placenta is delivered which may be around 6 h postpartum. Unsuccessful nursing by 12 h postpartum is cause for concern, and intervention is recommended by 24 h postpartum if consistent nursing has not been observed (Sweeney et al., 2010). The duration of lactation is variable in bottlenose dolphins and can range from 4 months to 6 years (Table 1). While neonates are not routinely weighed in managed care settings, they are relatively thin at birth and appear to lose body mass the first few days. This pattern turns around during the first week and weight gain becomes the norm. In calves being fed formula in a managed care setting, 200 kcal/kg/day was used as the caloric requirement of a bottlenose neonate (Sweeney et al., 2010). Many calves will show an interest in fish any time from 4 months onward but may still nurse after they start eating fish. Lactation is energetically costly and can be influenced by body condition, prey availability, and other environmental factors.

During lactation, the behavior of the mother is generally characterized as protective of her offspring but also conservative of energy expenditure. Wild dolphin neonates are maintained very close in proximity to their mothers. Nonetheless, bottlenose dolphins are somewhat unique in that they do not experience an obligate lactational anestrous period, meaning that subsequent ovulations are not blocked by lactation. The result of this is that lactating dolphins can become pregnant, resulting in approximately annual births for some individuals. While this is not the norm, it does occur with some frequency, particularly in dolphin mothers that experience a peri-natal mortality of their calf.

Reproductive anomalies and abnormalities

Several anomalies and abnormalities can occur related to dolphin reproductive physiology. In some of these cases, the cause can simply be related to normal maturation processes being compromised in young females. Nonetheless, there is scientific and medical interest in understanding the physiology of several of these conditions, including pseudopregnancy, dystocia, and other birthing problems, including spontaneous abortions and failure to thrive syndromes. These are presented and discussed below:

Pseudopregnancy

Pseudopregnancies have long been recognized in a multitude of species, both terrestrial and marine species. Pseudopregnancies occur when an adult female displays physiological and sometimes behavioral signs of being pregnant but physiologically is not carrying an embryo or fetus. The term translates to a false pregnancy, because without the presence of the embryo or fetus, the animal is not pregnant, and in many cases, the female has not been around or housed with a male. In cetaceans, pseudopregnancies have been reported in some, but not all, species (Atkinson et al., 1999; Atkinson and Yoshioka, 2007; Robeck et al., 2018). The rate of occurrence of pseudopregnancies is 6%–12% of detected pregnancies and is based solely on bottlenose dolphins in managed care facilities. The endocrinology of pseudopregnancy is not completely known, but progesterone is typically elevated post ovulation, past a normal luteal phase. One study reported that the mean length of pseudopregnancy in bottlenose dolphins was 178 days or approximately 6 months. While progesterone concentrations declined during the pseudopregnancy, they remained elevated above those of a nonpregnant dolphin. In addition, androgen and estrogen concentrations were significantly lower in the middle and late stages of the pseudopregnancy compared to a normal pregnancy. This suite of bioindicators may become the diagnostic means of detecting pseudopregnancies.

Spontaneous abortions, dystocia, perinatal loss, and failure to thrive

The loss of a pregnancy can occur at several stages after the time of conception. Early pregnancy losses are very difficult to detect as the presence of a conceptus may not even be recognized. This is generally true up until about 50 days into gestation when ultrasound imaging can reliably detect pregnancy. Recent studies have presented the findings of a suite of hormones, including progesterone, androgens, estrogen, and cortisol, at various stages of pregnancy and when reproductive losses occur (Robeck et al., 2012a, b; Steinman et al., 2021). More detail of abnormal conditions such as dystocia, or difficult birth, may be related to a large fetus or on that is not well positioned on the reproductive tract at the initiation of parturition. While dystocias can have negative outcomes for the calves or mothers, good veterinary support will improve the likelihood of a positive outcome. Dystocia may be caused by endocrine imbalances and has been related to hypothyroid mothers. Similarly, relaxin has been shown to be reduced in dolphins delivering stillbirths as opposed to those who experienced live births. While none of these studies has found a single hormone or perfect combination to predict reproductive loss, they report the usefulness of using multiple bioindicators to profile the physiological state of pregnant dolphins with different outcomes. In addition, ultrasonic monitoring has informed researchers and veterinarians that prenatal calves with an axillary width >15.5cm are known to be prone to dystocia (Sweeney et al., 2010).

Live births occur approximately 81% of the time, and the rate of spontaneous abortions and prenatal loss is reported to be about 11.5%–14.9% within the first 30 days postpartum, for dolphins in a managed care setting (Sweeney et al., 2010). For free-ranging Sarasota dolphins, 83% of detected pregnancies are documented as resulting in live births (Wells et al., 2014). In the managed care populations, calf survival >30 days is 85%–88% and to a full year is 80.7%–85.2%. The rates of perinatal loss can be broken down to stillbirths, reported to be 5.2%, and live calves that do not live beyond 24h postpartum are 6.3%. These rates are similar to those reported for reproductive loss in a well-studied wild population that had 17% as the estimated rate of reproductive loss (Wells et al., 2014). The causes for the mortalities in the first 24h are reportedly due to trauma by conspecifics or failure to thrive, which may be related to poor birth condition (Sweeney et al., 2010). Failure to thrive in neonates that are older than 24h but still in their first week postpartum results primarily from nutritional insufficiency and the resulting poor body condition and poorly developing immune function. Infection that is due largely to bacterial pneumonia was the primary cause of mortalities that occur between one week and one month of age (Sweeney et al., 2010). Some stillbirths have been reported to occur more than once from the same females in a managed care setting. The reasons for this are obscure, and recognizing the physiological and environmental conditions surrounding individual females that have had stillbirths is important to determining the cause.

Multiple factors may influence the reproductive success of bottlenose dolphins. One study reported that the age of the mother was significantly younger for pregnancies that went full term than for pregnancies that resulted in stillbirths or spontaneous abortions (O'Brien and Robeck, 2012). Still births in managed care facilities have been reduced as management and knowledge of nutrition has improved (Sweeney et al., 2010). Calves born and experiencing perinatal mortality often are smaller in mass and may reflect inadequate nutritional status of their mother.

Contraception and assisted reproductive technologies

Among the reasons that bottlenose dolphins were commonly kept in managed care settings is that they are fairly robust as a species, easily trained, and reproductively active. The reproductive activity easily resulted in pregnancies in many facilities, such that management of genetics and avoidance of inbreeding became concerning. In addition, many facilities had space limitations that could create a less-than-optimal social situation or render a facility noncompliant with animal welfare regulations, an example of which is the Animal Welfare Act in the United States. Where lack of reproductive success was not a problem, the need for contraception arose. Following the success of contraceptive pharmaceuticals, the reduction of numbers of dolphins in former military programs, and reduction of the public acceptance of using wild caught dolphins to stock managed care programs in some parts of the world, the outlook and planning for the numbers of dolphins desired in human managed programs was revised. During these years, the recognition arose that dolphins in captive care settings may be best managed if they were self-sufficient, reducing the need to obtain dolphins from the wild. The result was better management of reproductive processes that were undertaken at the same time that assisted reproductive technologies were being developed and refined for bottlenose dolphins. Thus this section will cover both contraception and assisted reproductive technologies.

Contraception

The ideal contraceptive is easily administered, results in minimal or no side effects, is reliable and effective, and is completely reversible. Contraception is only needed in managed care settings and is generally never considered in free-ranging settings. While physical separation can accomplish most of the desired goals of an effective contraceptive, many facilities do not have the space available to keep adult males and females separated. In addition, separation of individual dolphins can result in behavior problems ranging from anxiety, irritability, and aggression to eating disorders (Robeck et al., 2018). Dolphins are social animals and if keeping them in solitary confinement is needed for managing reproductive processes, and may be accompanied by behavior issues, then other forms of contraception should be sought. Several pharmaceutical agents have been used to avoid pregnancies (Robeck et al., 2018). The oldest and perhaps the most commonly used are the progestogens, in the forms of altrenogest (trademarked as Regu-Mate) and medroxyprogesterone acetate (trademarked as Provera). These synthetic progestogens act by suppressing the growth of antral follicles as well as by suppressing the preovulatory LH surge. While the progestogens are not 100% effective and pregnancies have occurred when animals were believed to be contracepted, they are the most commonly used form of hormonal contraceptive. GnRH agonists (leuprolide acetate) and GnRH vaccines have also been attempted as contraceptives, particularly in male bottlenose dolphins and other marine mammals (Atkinson et al., 1998; O'Brien and Robeck, 2012). These were mainly experimental approaches and further studies are warranted. While GnRH agonists may render males to be oligospermic or even aspermic, the reversibility and use in females needs additional study.

Assisted reproductive technologies

A major consideration of most assisted reproductive technologies is to increase genetic diversity in species. These technologies have included artificial insemination (AI), gamete recovery, cryopreservation of the conceptus, and embryo transfer. However, in cetaceans, the majority of assisted reproductive technologies have been AI, which has been successfully conducted in several species (Schroeder et al., 1990; Robeck et al., 1994, 2013). There are four major areas associated with AI, including (1) semen collection, handling, and storage; (2) synchronization of estrus and ovulation; (3) insemination timing and techniques; and (4) post insemination monitoring. Extensive research and practical applications have been conducted on semen collection, handling, and storage for dolphins in managed care settings (Robeck and O'Brien, 2004; Yuen et al., 2009). Semen collection in the majority of mammals, including bottlenose dolphins and other marine mammals, has been accomplished with electroejaculation, through the use of an artificial vagina, or through manual manipulation. Live sperm are extremely sensitive to temperature changes, and the handling of sperm and all seminal products requires great care and exacting protocols. While insemination techniques using fresh or frozen sperm have been used, cryopreservation requires an additional layer of complexity. The second area that requires technique perfection is that of estrous and ovulation synchronization. The need to conduct inseminations when the ovum is mature and available is critical and time sensitive. Hence, synchronizing estrus and ovulation has been accomplished using the synthetic progestogen, altrenogest (Robeck et al., 2018). Insemination timing and technique has been closely tied to the development of semen collection protocols, especially when the insemination is being conducted with fresh semen. Delivery of the semen is typically within the cervix or the uterine body. The timing of insemination in relation to the timing of ovulation is reported as being 6–12h before ovulation. This work has successfully resulted in pregnancies, births, and healthy offspring. A more sophisticated technology of sex selection based on separating sperm carrying the Y chromosome has also been conducted, as the sperm carrying the Y chromosome tends to be smaller than the X chromosome (O'Brien and Robeck, 2006). Once the inseminations have taken place, monitoring for the early detection of pregnancy begins.

External impacts on reproduction

Nutritional, social, and environmental factors are also known to influence the onset of sexual maturity, reproductive cycles, and gestation (Moore and Ridgway, 1996; Atkinson and Yoshioka, 2007). As with all mammals, sexual maturity is not achieved until the animal body is physically capable of performing the functions that can lead to successful reproduction. Therefore, underfed or malnourished animals are often noted as not being able to successfully reproduce, reach sexual maturity, or may take a reproductive "rest" to restore energy or nutritional status in order to resume being reproductively active or successful.

Similarly, the social setting of bottlenose dolphins can be a major influence on reproduction, especially with young or maturing male dolphins. Several anecdotes exist about male dolphins not reaching sexual maturity in the presence of a dominant male or in a tightly controlled social setting, where unlike in the wild, young males may be unable to escape from the presence of dominant males in captivity. Environmental influences on sexual maturity also exist. Extreme examples are reported during unusual phenomenon such as marine heatwaves, but any time a population is impacted (e.g., disease outbreak), changes in the abundance of a population can trigger density-dependent operatives that influence physiological mechanisms of the reproductive system. For example, during the *Deep Water Horizon* oil spill in the Gulf of Mexico, the major influence limiting reproduction was likely caused by the oil and dispersants used to manage a prolonged release of oil, and included both direct mortality, grossly underweight dolphins, and high rates of perinatal strandings (Kellar et al., 2017). The reproductive success rates during and after the spill were less than one third of those reported for unimpacted areas. In addition, there is a strong inverse relationship between reproductive failure and total leukocyte counts, which indicates that the high reproductive failure rates were consistent with published studies that report links between petroleum exposure and reproductive abnormalities and failures (Kellar et al., 2017).

In general, long-term reproductive history helps to better understand the fate of a population, especially in the case of small or declining populations. The emphasis of life history traits and environmental influences is of prime importance for the health and sustainability of any population (Wells et al., 2004). If animals are young and just entering their prime reproductive years, the future will be brighter than in populations with adults that are aging or not reproducing. A good example of this exists with the AT1 killer whale population in Prince William Sound, Alaska, which has failed to successfully reproduce for decades, or the southern resident killer whales in the Pacific Northwest with mostly older females, whose fate seems more tied to anthropogenic disturbances. Cook Inlet beluga whales also fall into the category of a small population that has experienced a major decline, and their population numbers are concerning because a single stochastic event could have severe and long-term impacts. In all of those situations, it becomes imperative for managers to deal with urgent threats to the population.

ASM is one of four life history parameters that can be used to estimate a population's rate of change in abundance over time (i.e., ASM, along with the recruitment rate to ASM, birth rate, and adult rate of survival). It is generally assumed that as a population of cetaceans approaches their carrying capacity, ASM, along with the recruitment rate, will be one of the first life history parameters to change. That is, generally a population with a higher density of animals relative to their carrying capacity will have a greater ASM than a population at a lower relative density (Eberhardt and Siniff, 1977). Conversely, dolphins from exploited populations had declines in ASM (Kasuya et al., 1984). Thus, the impacts of numerous external factors on reproduction renders many small cetaceans as bellwethers of the health of the environment that they inhabit.

Ex-situ conservation applications

Field-based programs have benefited from the numerous studies conducted on small cetaceans in managed care settings, and routine single or snapshot samples are often collected from free-ranging small cetaceans. Care must be taken with biopsy sampling of small cetaceans as the sex ratio of samples has been shown to be skewed in favor of males for bottlenose, Atlantic spotted (*Stenella frontalis*), and common dolphins (*Delphinus delphis*) (Quérouil et al., 2010). The skew was likely influenced by selectivity of biopsy collectors, predominance of males in seasonal or year-round settings, as well as sampling duration and intensity, which led to a tendency for females to avoid the sampling vessel. Single samples are often used for detecting pregnancy, and this technique has been applied to bottlenose and

Indo-Pacific bottlenose dolphins, long-finned pilot whales (*Globicephala melas*), killer whales, Dall's porpoise (*Phocoenoides dalli*), striped dolphins (*Stenella coeruleoalba*), harbor porpoises (*Phocoena phocoena*), and beluga whales. While several short-term studies have been conducted on bottlenose dolphins, long-term studies undertaken off Sarasota and Indian River Lagoon, Florida, and on Indo-Pacific bottlenose dolphins in Shark Bay, Western Australia, have been notable. While long-term studies are logistically difficult to initiate and financially costly to sustain over time, the benefits to knowledge of the health and physiology of the dolphins are immeasurable. This is particularly true as it relates to understanding the physiological health and reproductive success rates for these populations, as well as using these populations as sentinels of ecosystem health (Wells et al., 2004, 2014). An example in bottlenose dolphins is an 83% survival rate of fetuses from females that were detected as being pregnant (Wells et al., 2014). While endocrine profiles are routinely used in monitoring free-ranging dolphins, ultrasonic assessment of pregnancy and evaluations of body condition are also used to complement endocrine profiles and provide more holistic health monitoring (Wells et al., 2014).

Many species of small coastal and freshwater odontocetes are impacted by human activities, and the need to support these populations has become imperative. The species included in this category are listed in Table 2. Bottlenose dolphins as well as healthy populations of coastal and freshwater cetaceans are generally top predators and long-lived.

TABLE 2 Common name, scientific name (including a few subpopulations), International Union for Conservation of Nature (IUCN) species status, and total abundance of multiple small coastal and freshwater cetacean species and dates of publication.

Common name	Scientific name/subpopulation location	IUCN species status	IUCN total abundance	Assessment year
Delphinidae				
Bottlenose dolphin	*Tursiops truncatus*	LC	~750,00	2019
Indo-Pacific bottlenose dolphin	*Tursiops aduncus*	NT	U	2019
Guiana dolphin	*Sotalia guianensis*	NT	U	2018
Tucuxi	*Sotalia fluviatilis*	DD	U	2010
Indo-Pacific humpback dolphin	*Sousa chinensis*	VU	U	2017
Indian Ocean humpback dolphin	*Sousa plumbea*	EN	U	2017
Atlantic Ocean humpback dolphin	*Sousa tueszii*	CR	<3000	2017
Killer whale	*Orcina orca*	DD	U	2017
Irrawaddy dolphin	*Orcaella brevirostris*	EN	<8000	2017
	Ayeyarwady River subpopulation, Myanmar	CR	~60	2004
	Mahakam River subpopulation, Indonesia	CR	~70	2008
	Malampaya Sound subpopulation, Philippines	CR	~75	2004
	Mekong River subpopulation, Cambodia/Laos	CR	~70	2004
	Songkhla Lake subpopulation, Thailand	CR	<50	2004
	Iloilo-Guimaras subpopulation	CR	<25	2019
Australian snubfin dolphin	*Orcaella heinsohni*	VU	9000–10,000	2017
Rough-toothed dolphin	*Steno bredanensis*	LC	U	2019
Hector's dolphin	*Cephalorhynchus hectori*	EN	7381	2013
Iniidae				
Amazon River dolphin	*Inia geoffrensis*	EN	U	2018
Yangtze finless porpoise	*Neophocaena asiaeorientalis asiaeorientalis*	CR	~1800	2012

TABLE 2 Common name, scientific name (including a few subpopulations), International Union for Conservation of Nature (IUCN) species status, and total abundance of multiple small coastal and freshwater cetacean species and dates of publication—cont'd

Common name	Scientific name/subpopulation location	IUCN species status	IUCN total abundance	Assessment year
Lipotidae				
Baiji or Yangtze River dolphin	*Lipotes vexillifer*	CR	Possibly extinct	2017
Phocoenidae				
Harbor porpoise	*Phocoena phocoena*	LC	U	2020
Vaquita	*Phocoena sinus*	CR	18	2017
Burmeister's porpoise	*Phocoena spinipinnis*	NT	U	2018
Spectacled porpoise	*Phocoena dioptrica*	LC	U	2018
Dall's porpoise	*Phocoena dalli*	LC	U	2018
Indo-Pacific finless porpoise	*Neophocaena phocaenoides*	VU	U	2017
Narrow-ridged finless porpoise	*Neophocaena asiaeorientalis*	EN	U	2017
Platanistidae				
South Asian river dolphin	*Platanista gangetica*	EN	~5000	2017
Indus River dolphin	*Platanista gangetica minor*	EN	~1500	2004
Ganges River dolphin	*Platanista gangetica gangetica*	EN	~3500	2004
Pontoporidae				
Franciscana	*Pontoporia blainvillei*	VU	~35,000	2017
	Rio Grande do Sul/Uruguay subpopulation	VU	42,000	2003

Designations: NE, not evaluated; DD, data deficient; LC, least concern; NT, near threatened; VU, vulnerable; EN, endangered; CR, critically endangered; EW, extinct in the wild; EX, extinct; from the IUCN Red List of Threatened Species Designations.

These characteristics make them excellent sentinels of the health of their habitats. As was previously discussed, a key parameter for determining the health of a population is the ASM, as well as reproductive rates and success. An example is the analysis of age composition and reproductive status from two species of dolphins taken in the Japanese drive fishery for pantropical spotted (*Stenella attenuata*) and striped dolphins, where the decline in ASM and a shortening of the reproductive cycle (i.e., inter-calving interval) were measured and used to evaluate the populations' response to exploitation (Kasuya et al., 1984).

Increasingly, international groups of scientists are coming together to help establish ex-situ conservation applications for individual species (Taylor et al., 2020). Some of these efforts are built on studies that have focused on reproduction in species such as the Amazon river dolphin or boto (*Inia geoffrensis*) or the Yangtze river dolphin or baiji (*Lipotes vexillifer*) (Caldwell et al., 1989; Bonar et al., 2007; Hao et al., 2007; Boede et al., 2018; da Silva et al., 2018; Martin and da Silva, 2018). Other efforts were more focused on the immediate need to save individual vaquita (*Phocoena sinus*), with an ultimate goal of attempting captive breeding (Rojas-Bracho et al., 2019). With river dolphins, the severe degradation of terrestrial habitats may render greater impact than those in marine environments. Thus all river dolphins fall into a category of potentially needing ex-situ conservation efforts. Several coastal marine species of small cetaceans also are prime candidates for ex-situ efforts, and these primarily occur where fishing efforts have large by-catch of the dolphins. While some of these efforts have focused on the ASM or the physiology and endocrinology of reproductive cycles and gestation, they all serve to improve our understanding of general reproductive biology in lesser-known species, some of which critically need attention lest they succumb to extinction. All of these efforts will benefit now or in the future from base knowledge that is obtained from bottlenose dolphins that are either studied in the wild or in managed care settings.

References

Amano, M., 2018. Finless Porpoises. In: Würsig, B., Thewissen, J.B.M., Kovacs, K. (Eds.), Encyclopedia of Marine Mammals. Academic Press, pp. 372–375.

Atkinson, S. 2009. Male reproductive systems. *In* Encyclopedia of Marine Mammals. Academic Press. pp. 667–671.

Atkinson, S., Combelles, C., Vincent, D., Nachtigall, P., Pawloski, J., Breese, M., 1999. Monitoring of progesterone in captive female false killer whales, (*Pseudorca crassidens*). Gen. Comp. Endocrinol. 115, 323–332.

Atkinson, S., Mashburn, K., Vos, D., Romano, T., Mahoney, B. 2022. Hormone profiles from Cook Inlet, Bristol Bay, and aquarium beluga whales. Polar Res. 41, 5525.

Atkinson, S., Ragen, T.J., Gilmartin, W.G., Becker, B.L., Johanos, T.C., 1998. Use of a GnRH agonist to suppress testosterone in wild male Hawaiian monk seals (*Monachus schauinslandi*). Gen. Comp. Endocrinol. 112 (2), 178–182.

Atkinson, S., Rogan, A., Baker, C.S., Dagdag, R., Redlinger, M., Polinski, J., Urban, J., Sremba, A., Branson, M., Mashburn, K., Pallin, L., Klink, A., Steel, D., Bortz, E., Kerr, I., 2021. Genetic, endocrine, and microbiological assessments of blue, humpback, and killer whale health using unoccupied aerial systems. Wildlife Society Bull. 1–16.

Atkinson S., Yoshioka, M. 2007. Endocrinology of reproduction. In: Reproductive Biology and Phylogeny of Cetacea, Vol. 7, (Miller, D.L., ed). Enfield: Science Publishers, pp. 171–192.

Beasley, I., Brown, A.M., 2018. Australian Snubfin Dolphin. In: Würsig, B., Thewissen, J.B.M., Kovacs, K. (Eds.), Encyclopedia of Marine Mammals. Academic Press, pp. 47–49.

Bergfelt, D.R., Blum, J.L., Steinetz, B.G., Steinman, K.J., O'Brien, J.K., Robeck, T.R., 2017. Relaxin as a hormonal aid to evaluate pregnancy and pregnancy loss in bottlenose dolphins (*Tursiops truncatus*). Gen. Comp. Endocrinol. 242, 24–29.

Bjorge, A., Tolley, K.A., 2018. Harbor porpoise. In: Würsig, B., Thewissen, J.B.M., Kovacs, K. (Eds.), Encyclopedia of Marine Mammals. Academic Press, pp. 448–451.

Boede, E.O., Mujica-Jorquera, E., Boede, F., Varela, C., 2018. Reproductive management of the Orinoco river dolphin *Inia geoffrensis humboliana* in Venezuela. Int. Zoo Yearbook 52, 1–13.

Bonar, C.J., Boede, E.O., Hartmann, M.G., Lowenstein-Whaley, J., Mujica-Jorquera, E., Parish, S.V., Parish, J.V., Garner, M.M., Stadler, C.K., 2007. A retrospective study of the pathologic findings in the Amazon and Orinoco River dolphin (*Inia geoffrensis*) in captivity. J. Zoo Wildlife Med. 38 (2), 177–191.

Brook, F.M., 2001. Ultrasonographic imaging of the reproductive organs of the female bottlenose dolphin, *Tursiops truncatus aduncas*. Reproduction 121 (3), 419–428.

Caldwell, M.C., Caldwell, D.K., Brill, R.L., 1989. *Inia geoffrensis* in captivity in the United States. In: Biology and Conservation of the River Dolphins. Perrin, W.F., Brownell Jr, R.L., Kaiya, Z., Jiankang, L., (Eds.) IUCN, Species Survival Commission Occasional Paper No. 5. pp 35–41.

Crespo, E.A., 2018. Franciscana dolphin. In: Würsig, B., Thewissen, J.B.M., Kovacs, K. (Eds.), Encyclopedia of Marine Mammals. Academic Press, pp. 388–391.

Dawson, S.M., 2018. *Cephalorhynchus* dolphins. In: Würsig, B., Thewissen, J.B.M., Kovacs, K. (Eds.), Encyclopedia of Marine Mammals. Academic Press, pp. 166–172.

DeMaster, D. 1984. Review of techniques used to estimate the average age at attainment of sexual maturity in marine mammals. Report of the International Whaling Commission. 6: 175–179.

Eberhardt, L.L., Siniff, D.B., 1977. Population dynamics and marine mammal management policies. J. Fish. Res. Board Can. 34 (2), 183–190.

Flores, P.A.C., da Silva, V.M.F., de Fettuccia, D.C., 2018. Tucuxi and Guiana dolphins. In: Würsig, B., Thewissen, J.B.M., Kovacs, K. (Eds.), Encyclopedia of Marine Mammals. Academic Press, pp. 1024–1027.

Ford, J.K.B., 2018. Killer whale. In: Würsig, B., Thewissen, J.B.M., Kovacs, K. (Eds.), Encyclopedia of Marine Mammals. Academic Press, pp. 531–537.

Galligan, T.M., Schwacke, L.H., Houser, D.S., Wells, R.S., Rowles, T., Boggs, A.S.P., 2018. Characterization of circulation steroid hormone profiles in the bottlenose dolphin (*Tursiops truncatus*) by liquid chromatography-tandem mass spectrometry (LC-MS/MS). Gen. Comp. Endocrinol. 263, 80–91.

Hao, Y.J., Chen, D.Q., Zhao, Q.Z., Wang, D., 2007. Serum concentrations of gonadotropins and steroid hormones of *Neophocaena phocaenoides asiaorientalis* in middle and lower regions of the Yangtze River. Theriogenology 67, 673–680.

Harrison, R.J., Ridgway, S.H., 2009. Gonadal activity in some Bottlenose dolphins (*Tursiops truncatus*). J. Zool. 165(3):355–366.

Houser, D.S., Champagne, C.D., Wasser, S.K., Booth, R.K., Romano, T., Crocker, D.E., 2021. Influence of season, age, sex, and time of day on the endocrine profile of the common bottlenose dolphin (*Tursiops truncatus*). Gen. Comp. Endocrinol. (313)113889 ISSN 0016-6480.

Jefferson, T.A., 2018a. Dall's porpoise. In: Würsig, B., Thewissen, J.B.M., Kovacs, K. (Eds.), Encyclopedia of Marine Mammals. Academic Press, pp. 239–242.

Jefferson, T.A., 2018b. Rough-toothed dolphin. In: Würsig, B., Thewissen, J.B.M., Kovacs, K. (Eds.), Encyclopedia of Marine Mammals. Academic Press, pp. 838–840.

Jefferson, T.A., Hung, S.K., Robertson, K.M., Archer, F.I., 2012. Life history of the Indo-Pacific humpback dolphin in the Pearl River Estuary, southern China. Mar. Mamm. Sci. 28 (1), 84–104.

Kasuya, T., Tobayama, T., Matsui, S., 1984. Review of the live-capture of small cetaceans in Japan. Rep. int. Whal. Commn. 34, 597–602.

Kellar, N.M., Speakman, T.R., Smith, C.R., Lane, S.M., Balmer, B.C., Trego, M.L., Catelani, K.N., Robbins, M.N., Allen, C.D., Wells, R.S., Zolman, E.S., Rowles, T.K., Schwacke, L.H., 2017. Low reproductive success rates of common bottlenose dolphins *Tursiops truncatus* in the northern Gulf of Mexico following the *Deepwater Horizon* disaster (2010-2015). Endanger. Species Res. 33, 143–158.

Kellar, N.M., Trego, M.L., Marks, C.I., Dizon, A.E., 2006. Determining pregnancy from blubber in three species of delphinids. Mar. Mamm. Sci. 22 (1), 1–16.

Kirby, V.L., Ridgway, S.H., 1984. Hormonal evidence of spontaneous ovulation in captive dolphins (*Tursiops truncatus* and *Delphinus delphis*). Report of the International Whaling Commission. Special Issue 6:459–464.

Legacki, E.L., Robeck, T.R., Steinman, K.J., Conley, A.J., 2020. Comparative analysis of steroids in cyclic and pregnant killer whales, beluga whales, and bottlenose dolphins by liquid chromatography tandem mass spectrometry. Gen. Comp. Endocrinol. 285, 1–8.

Martin, A.R., da Silva, V.M.F., 2018. Reproductive parameters of the Amazon river dolphin or boto, *Inia geoffrensis* (Cetacea: Iniidae); an evolutionary outlier bucks no trends. Biol. J. Linn. Soc. 123, 666–676.

Moore, S.E., Ridgway, S.H., 1996. Patterns of dolphin sound production and ovulation. Aquat. Mamm. 22, 175–184.

Morita, Y., Fujise, Y., Ueda, K., Uchida, S., Yoshioka, M., Kashiwagi, M., 2003. Serum concentrations of estrone sulfate in pregnant females of three species of small odontocetes and the Antarctic Minke whale. In: Proc.The 15th Biennial Conference on the Biology of Marine Mammals. Greensboro, NC. *December 14-19th*, 2003.

Natalie, R., Goodall, P., 2009. Spectacled porpoise: *Phocoena dioptrica*. In: Encyclopedia of Marine Mammals. Academic Press, pp. 1087–1091.

O'Brien, J.K., Robeck, T.R., 2006. Development of sperm sexing and associated assisted reproductive technology for sex preselection of captive bottlenose dolphins (*Tursiops truncatus*). Reprod. Fertil. Dev. 18 (3), 319–329.

O'Brien, J.K., Robeck, T.R., 2012. The relationship of maternal characteristics and circulating progesterone concentrations with reproductive outcome in the bottlenose dolphin (*Tursiops truncatus*) after artificial insemination, with and without ovulation induction, and natural breeding. Theriogenology 78, 469–482.

Para, G.J., Jefferson, T.A., 2018. Humpback dolphins. In: Würsig, B., Thewissen, J.B.M., Kovacs, K. (Eds.), Encyclopedia of Marine Mammals. Academic Press, pp. 483–489.

Pérez, S., García-López, A., De Stephanis, R., Giménez, J., García-Tiscar, S., Verborgh, P., Mancera, J.M., Martínez-Rodriguez, G., 2011. Use of blubber levels of progesterone to determine pregnancy in free-ranging live cetaceans. Mar. Biol. 158 (7), 1677–1680.

Perrin, W.F., Reilly, S.B., 1984. Reproductive parameters if dolphins and small whales of the family Delphinidae. Rep. Int. Whal. Commn. 6 (Special Issue), 97–133.

Quérouil, S., Luís Freitas, L., Dinis, A., Alves, F., Cascão, I., Prieto, R., Silva, M.A., Magalhães, S., Matos, J.A., Santos, R.S., 2010. Sex bias in biopsy samples collected from free-ranging dolphins. Eur. J. Wildlife Res. 56, 151–158.

Reeves, R.R., Martin, A.R., 2018. River Dolphins. In: Würsig, B., Thewissen, J.B.M., Kovacs, K. (Eds.), Encyclopedia of Marine Mammals. Academic Press, pp. 827–829.

Reyes, J.C., 2018. Burmeister's porpoise. In: Würsig, B., Thewissen, J.B.M., Kovacs, K. (Eds.), Encyclopedia of Marine Mammals. Academic Press, pp. 146–148.

Richard, J.T., Robeck, T.R., Osborn, S.D., Naples, L., McDermott, A., LaForge, R., Romano, T.A., Sartini, B.L., 2017. Testosterone and progesterone concentrations in blow samples are biologically relevant in belugas (*Delphinapterus leucas*). Gen. Comp. Endocrinol. 246, 183–193.

Richkind, M., Ridgway, S.H., 1975. Estrogens, corticosteroids and progestagen patterns in the pregnant and non-pregnant bottle-nosed dolphin *Tursiops truncatus* following the intramuscular administration of NIH-FSH-OVINE-S9. J. Steroid Biochem. 6, 15–20.

Ridgway, S., 1972. Mammals of the sea: biology and medicine. In: Mammals of the Sea: Biology and Medicine. Charles C Thomas, Springfield, Illinois, pp. 392–635.

Robeck, T.R., Curry, B.E., McBain, J.F., Kraemer, D.C., 1994. Reproductive biology of the bottlenose dolphin (*Tursiops truncatus*) and the potential application of advanced reproductive technologies. J. Zoo Wildlife Med. 25 (3), 321–336.

Robeck, T.R., Gili, C., Doescher, B.M., Sweeny, J., De Laender, P., Van Elk, C.E., O'Brien, J.K., 2012. Altrenogest and progesterone therapy during pregnancy in bottlenose dolphins (*Tursiops truncatus)* with progesterone insufficiency. J. Zoo Wildlife Med. 43 (2), 296–308.

Robeck, T.R., Montano, G.A., Steinman, K.J., Smolensky, P., Sweeney, J., Osborn, S., O'Brien, J.K., 2013. Development and evaluation of deep intra-uterine artificial insemination using cryopreserved sexed spermatozoa in bottlenose dolphins (*Tursiops truncatus*). Anim. Reprod. Sci. 139 (1–4), 168–181.

Robeck, T.R., O'Brien, J.K., 2004. Effect of cryopreservation methods and precryopreservation storage on bottlenose dolphin (*Tursiops truncatus*) spermatozoa. Biol. Reprod. 70 (5), 1340–1348.

Robeck, T.R., O'Brien, J.K., Atkinson, S., 2018. In: Gulland, F.M.D., Dierauf, L.A., Whitman, K.L. (Eds.), CRC Handbook of Marine Mammal Medicine: Health, Disease and Rehabilitation, third edition. CRC Press, Boca Raton, FL, pp. 204–215.

Robeck, T.R., Steinman, K.J., Parry, C.B., Gomez, F.M., Jensen, E.D., 2012. Comparisons of serum progesterone and progestogen concentrations in normal and abnormal bottlenose dolphin (*Tursiops truncatus*) pregnancies. Front. Mar. Sci. 8, 1–16.

Rojas-Bracho, L., Gulland, F.M.D., Smith, C.R., Barbara Taylor, R.S., Wells, P.O., Thomas, B.B., et al., 2019. A field effort to capture critically endangered vaquitas Phocoena sinus for protection from entanglement in illegal gillnets. Endanger. Species Res. 38, 11–27.

Rojas-Bracho, L., Jaramillo-Legoretta, M., Nieto-Garcia, E., Cardenas-Hinojosa, G., 2018. Vaquita. In: Würsig, B., Thewissen, J.B.M., Kovacs, K. (Eds.), Encyclopedia of Marine Mammals. Academic Press, pp. 1031–1035.

Sawyer-Steffan, J.E., Kirby, V.L., Gilmartin, W.G., 1983. Progesterone and estrogens in the pregnant and nonpregnant dolphin, *Tursiops truncatus*, and the effects of induced ovulation. Biol. Reprod. 28, 897–901.

Schroeder, J.P., Keller, K.V., 1989. Seasonality of serum testosterone levels and sperm density in *Tursiops truncatus*. J. Exp. Zool. 249 (3), 316–321.

Schroeder, J., Keller, P., Keller, K., 1990. 26 – Artificial Insemination of Bottlenose Dolphins. Naval Ocean Systems Center, Hawaii Laboratory, P.O. Box 997, Kailua, HI 96734-0997.

da Silva, V.M.F., Freitas, C.E.C., Dias, R.L., Martin, A.R., 2018. Both cetaceans in the Brazilian Amazon show sustained, profound population declines over two decades. PLoS One 13(5): e0191304.

da Silva, V.M.F., Martin, A.R., 2018. Amazon river dolphin. In: Würsig, B., Thewissen, J.B.M., Kovacs, K. (Eds.), Encyclopedia of Marine Mammals. Academic Press, pp. 21–23.

Smith, B.D., 2018. Irrawaddy dolphin. In: Würsig, B., Thewissen, J.B.M., Kovacs, K. (Eds.), Encyclopedia of Marine Mammals. Academic Press, pp. 525–529.

Steinman, K.J., Robeck, T.R., Montano, G.A., 2021. Characterization of circulating androgens, cortisol and estrogens during normal, abnormal and false pregnancy in bottlenose dolphins (*Tursiops truncatus*) under managed care. Front. Mar. Sci. 1-19.

Steinman, K.J., Robeck, T.R., O'Brien, J.K., 2016. Characterization of estrogens, testosterone, and cortisol in normal bottlenose dolphin (*Tursiops truncatus*) pregnancy. Gen. Comp. Endocrinol. 226:102-112. Epub 2015 Dec 21. PMID: 26718081.

Suzuki, M., Uchida, S., Ueda, K., Tobayama, T., Katsumata, E., Yoshioka, M., Aida, K., 2003. Diurnal and annual changes in serum cortisol concentrations in Indo-Pacific bottlenose dolphins *Tursiops aduncus and* killer whales *Orcinus orca*. Gen. Comp. Endocrinol. 132 (3), 427–433. PMID: 12849966.

Sweeney, J.C., Stone, R., Campbell, M., McBain, J., St. Leger, J., Xitco, M., Jensen, E., Ridgway, S., 2010. Comparative Survivability of *Tursiops* neonates from three U.S. Institutions for the decades 1990-1999 and 2000-2009. Aquat. Mamm. 36, 248–261.

Taylor, B.L., Abel, G., Miller, P., Gomez, F., von Fersen, L., DeMaster, D., Reeves, R.R., Rojas-Bracho, L., Wang, D., Hao, Y. & Cipriano, F. (eds.) (2020). Ex situ options for cetacean conservation. Report of the 2018 workshop, Nuremberg, Germany. Occasional Paper of the IUCN Species Survival Commission No. 66. Gland, Switzerland: IUCN.

Urian, K.W., Duffield, D.A., Read, A.J., Wells, R.S., Shell, D.D., 1996. Seasonality of reproduction in bottlenose dolphins, *Tursiops truncatus*. J. Mammal. 77, 394–403.

Wang, J.Y., 2018. Bottlenose dolphin, *Tursiops aduncus*, Indo-Pacific bottlenose dolphin. In: Würsig, B., Thewissen, J.B.M., Kovacs, K. (Eds.), Encyclopedia of Marine Mammals. Academic Press, pp. 125–130.

Wells, R.S., Hanson, L.J., Sweeny, J.C., Townsend, F.I., Stone, R., Casper, D.R., Scott, M.D., Hohn, A.A., Rowles, T.K., 2004. Bottlenose dolphins as marine ecosystem sentinels: developing a health monitoring system. EcoHealth 1, 246–254.

Wells, R.S., Scott, M.D., 2018. Bottlenose dolphin, *Tursiops truncatus*, common bottlenose dolphin. In: Würsig, B., Thewissen, J.B.M., Kovacs, K. (Eds.), Encyclopedia of Marine Mammals. Academic Press, pp. 118–125.

Wells, R.S., Smith, C.R., Sweeney, J.C., Townsend, F.I., Fauquier, D.A., Stone, R., Langan, J., Schwacke, L.H., Rowles, T.K., 2014. Fetal survival of common bottlenose dolphins (*Tursiops truncatus*) in Sarasota Bay. Florida. Aquat. Mamm. 40 (3), 252–259.

West, K.L., Atkinson, A., Carmichael, M.J., Sweeny, J.C., Krames, B., Krames, J., 2000. Concentrations of progesterone in milk from bottlenose dolphins during different reproductive states. Gen. Comp. Endocrinol. 117, 218–224.

Yoshioka, M., Mohri, E., Tobayama, T., Aida, K., Hanyu, I., 1986. Annual changes in serum reproductive hormone levels in the captive female bottle-nosed dolphins. Bull. Japan. Soc. Sci. Fish, 1939–1946.

Yuen, Q.W., Brook, F.M., Kinoshita, R.E., Ying, M.T., 2009. Semen collection and ejaculate characteristics in the Indo-Pacific bottlenose dolphin (*Tursiops aduncus*). J. Androl. 30 (4), 432–439.

Zeng, Q., Lin, W., Dai, Y., Zhong, M., Wang, X., Zhu, Q., 2020. Modeling demographic parameters of an edge-of-range population of Indo-Pacific humpback dolphin in Xiamen Bay, China. Reg. Stud. Mar. Sci. 40, 101462.

Zhou, K., 2018. Baiji. In: Würsig, B., Thewissen, J.B.M., Kovacs, K. (Eds.), Encyclopedia of Marine Mammals. Academic Press, pp. 54–56.

CHAPTER

13

Immunology

Sylvain De Guise[a], *Milton Levin*[a], *Tracy A. Romano*[b,c], *and Laura A. Thompson*[b,c]

[a]Department of Pathobiology and Veterinary Science, University of Connecticut, Storrs, CT, United States [b]Research Department, Mystic Aquarium, Mystic, CT, United States [c]Department of Marine Sciences, University of Connecticut, Groton, CT, United States

Abstract

In this chapter, we will summarize current knowledge and discuss emerging concepts on dolphin immunology. Contrary to many other organ systems, emphasis will be less on anatomy and more on the dynamic and functional aspects of blood immune cells and how they interact to respond to challenges. This chapter will further investigate the relationships between the immune system and other aspects of the physiology of marine mammals, including stress, diving physiology, and response to toxicants.

Introduction to immunology

The immune system functions to protect individuals from infections with viruses, bacteria, fungi, and other potential pathogens, in addition to recognizing and responding to tumors. This is especially important as all creatures live in environments with ever present microorganisms. To perform this significant task, the immune system consists of a complex and intricate network of tissues, cells, and soluble messengers, all coordinated to respond when necessary, without over-responding and causing unnecessary damage. Furthermore, the immune system must function in coordination with and under the influence of the nervous, endocrine, and other systems to maintain health and homeostasis.

The mammalian immune system comprises two relatively distinct portions, the innate and the acquired or adaptive immunity (Fig. 1). Innate immunity is the first line of defense and includes cells such as phagocytic cells (neutrophils and macrophages) and natural killer (NK) cells and soluble molecules such as the complement that immediately recognize and respond to insults. It is fast and nonspecific and is not enhanced upon re-exposure to an antigen, i.e. it will always respond to an antigen as if it saw it for the first time. Inflammatory cells such as neutrophils, eosinophils, and basophils can readily release the content of their granules to trigger and sustain the inflammation and help attract immune cells to the site of an insult. On the other hand, acquired immunity uses antibodies and cells such as lymphocytes that are responsible for long-term protection against non-self particles such as bacteria and viruses (antigens). In other words, antigens (the recognizable portion of foreign particles) can trigger the production of antibodies that help eliminate such foreign particles and offer lasting protection against future encounters of the same antigen. The acquired immune response is slower, antigen specific, and requires complex steps such as antigen presentation and cell activation. A hallmark of acquired immunity is the enhanced and quicker response upon re-exposure to a given antigen, referred to as immunological memory, which is long-lasting. A common practice taking advantage of immunological memory is vaccination, which uses inoculation with a whole or portion of an organism such as a virus, which affords immunological protection upon encountering the same infectious virus later in life. However, vaccination against an antigen (e.g., measles) does not offer protection against a different antigen (e.g., the flu), exemplifying the specificity of the acquired immune system, whereas innate immunity would respond indiscriminately to either virus without memory.

While most systems are well-defined anatomically (respiration occurs in the lungs, digestion in the gastrointestinal tract, etc.), the immune system is much less so to combat infections wherever they occur. Most immune cells, including

The Physiology of Dolphins
https://doi.org/10.1016/B978-0-323-90516-9.00004-X

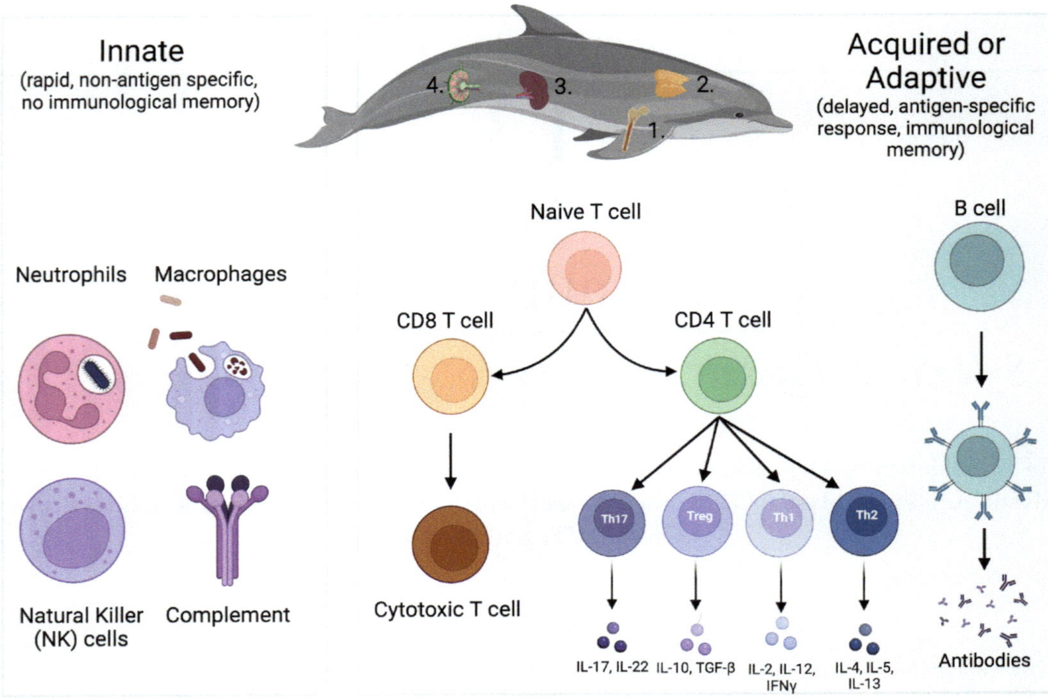

FIG. 1 Immune cells originate from the bone marrow (1) and some T lymphocytes require maturation and undergo a selection process in the thymus (2). The spleen (3) and lymph nodes (4) represent organized cellular architecture that favor the recognition and presentation of antigens. The innate immune system, which provides a rapid, nonspecific response, but is devoid of immunological memory, is comprised of circulating phagocytic cells, such as neutrophils and monocytes, which can engulf and then destroy pathogens through the respiratory burst, as well as natural killer (NK) cells specialize in the rapid killing of virus-infected cells and tumor cells, and the soluble complement that helps recognize and eliminate pathogens. The acquired or adaptive immune system, which provides a delayed, antigen-specific response and generates immunological memory is comprised of T lymphocytes, which can differentiate into CD4 T helper (Th) and CD8 T cytotoxic (Tc) cells and antibody- or immunoglobulin (Ig)-secreting B lymphocytes. Cytokines (e.g., interleukin 2 [IL-2], interferon gamma [IFNγ], transforming growth factor beta [TGFß], and tumor necrosis factor alpha [TNFα]) are produced by immune cells and help in the direction and magnitude of a proper immune response, including through the further differentiation of Th cells into a pro-inflammatory Th17 response, dampening through a regulatory T (Treg) cell response, a cell-mediated Th1 response, or an antibody-mediated Th2 response.

those involved in innate and acquired immunity, originate from the bone marrow, which generates red blood cells (erythrocytes) and white blood cells (leukocytes, which include lymphocytes, monocytes, neutrophils, eosinophils, and basophils) in a phenomenon called hematopoiesis. T lymphocytes require maturation and undergo a selection process in the thymus, an immune organ in the thoracic cavity that physiologically involutes with age without functional consequences. Most immune cells circulate through the blood and lymph, which they can exit to reach the site of insults, where they are needed to mount an immune response. In addition several immune cell aggregations sit at likely encounter sites for invaders, such as the mucosal surfaces of the respiratory and digestive tracts. Such aggregates are unsurprisingly named after such locations, including the mucosa-associated lymphoid tissues (MALTs) and the gut-associated lymphoid tissues. The spleen and lymph nodes represent organized cellular architecture that favor the recognition and presentation of antigens that would have breached natural barriers such as the skin and mucosae. This architecture favors the interactions among complementary cell types to favor the development of the immune response. The immune system comprises different cell types and soluble molecules such as complement and antibodies that coordinate to perform different but complementary functions. While several organs and aggregations of immune cells are recognized, the immune system is "mobile" to mount a response that is effective at the site of an insult rather than being restricted to a single organ. This is especially true for the innate immune response, which is mostly comprised of cells that exit the bone marrow and enter the blood stream, fully ready to exit the vascular system and respond to an insult. For the acquired immune response, while the thymus is essential to the maturation of T lymphocytes, other immune organs such as the spleen or lymph nodes facilitate the interactions among cell types but are not required for an effective immune response.

As part of the innate immune system, phagocytosis is the mechanism by which immune cells engulf pathogens or debris to remove an insult. Phagocytic cells are therefore important in the defense mechanisms against invasion by pathogens such as bacteria. Phagocytic cells are fast acting and require no previous activation; therefore, they are always ready. Some phagocytic cells are circulating, such as neutrophils and monocytes, and are rapidly and strongly attracted to areas of inflammation or tissue damage, whereas others, like macrophages, are resident in tissues such as the lungs and await for potential invaders. Once ingested, microorganisms are destroyed by the generation of oxygen-free radicals by the same phagocytic cells, a process known as respiratory burst. Failure to generate such a respiratory burst would result in the survival of the pathogens that would then be distributed wherever phagocytic cells travel.

NK cells, also part of the innate immune system, specialize in the rapid killing of virus-infected cells and tumor cells, without any requirement for antigen specificity or antigen presentation. In other words, the same NK cells could kill a virus-infected cell and a tumor cell. While performing functions somewhat similar to cytotoxic T cells, NK cells' mechanism of recognition of target cells is independent of the antigen presentation required for cytotoxic T cells, and these cell types have been referred to as mirror images.

Lymphocytes are the major cells of the acquired immune system. While T lymphocytes are the effectors of cell-mediated immunity, resulting in the killing of the target cells, B cells produce soluble antibodies that recognize specific sites on the surface of microorganisms and help eliminate it, known as humoral immunity. T cells are further subdivided into T helper (Th) cells, which dictate the direction and magnitude of the immune response by activating different subsets of effector cells, and cytotoxic T (Tc) cells that kill virus-infected cells after they are activated by Th cells. Furthermore, there are several subsets of Th cells, including Th1 cells that drive a cell-mediated immune response, Th2 cells that drive a humoral immune response, Th17 cells that drive a pro-inflammatory response, and regulatory T cells (Treg) that dampen the immune response to avoid an excessive response and the tissue damage that would result from it. Following encounter with an antigen, a subset of B and T cells develop into memory cells, which require lower thresholds and activation upon re-encounter of the same antigen, allowing for a more rapid response, often resulting in resistance to a challenge. These lymphocyte subsets are morphologically indistinguishable, although their functions differ profoundly, but specific surface molecules allow their identification (see below).

The following sections will focus on the immune system of dolphins and recent physiological advances from other species of cetaceans. It should be noted that several immunological techniques in laboratory animals such as mice require manipulations or organ collection that would not be ethically possible in marine mammals. However, several immune assays and reagents have been validated for dolphins and other marine mammals that only require whole blood, which can be obtained with nonlethal, routine, and minimally invasive techniques, from both managed-care (zoo and aquarium housed) and wild-caught (restrained during capture-release health assessments programs) individuals. It will be important to distinguish in vitro (in controlled laboratory conditions using isolated cells) and in vivo (in whole animals) approaches in the interpretation of immune assays. The following sections will examine advances in understanding the immune system of dolphins and how it interacts with and is affected by other systems such as the stress response, diving physiology, and interactions with pollutants.

Recent progress in cetacean immunology

Anatomy and histology of immune organs

The overall anatomy, morphology, and histology of the dolphin immune system, including the thymus, spleen, and lymph nodes, is not much different from that of other mammals (Cowan and Smith, 1999). A few particularities in dolphins and other cetaceans include the presence of relatively prominent pharyngeal tonsils and the anal tonsil, which are well-organized lymphoid follicles in the mucosa that, along with the thymus, appear to involute with age (Romano et al., 1993; Cowan and Smith, 1999). As part of the MALT, they are likely important at sites susceptible to insults, but the specific significance of the pharyngeal and the anal tonsils has yet to be determined. While involution of the thymus is a common feature in most species, a study of age-related involution in bottlenose dolphins (*Tursiops truncatus*) documented a precipitous drop in volume of the thymus after 5 years of age, along with a dramatic decline of the MALT in the colon after 10 years of age, but high variability and only moderate involution of the pharyngeal and anal tonsils (Clark et al., 2005). While the functional significance of the differential patterns of immune organs involution is not clear, those features should be understood as part of the physiology of dolphins and not misunderstood for pathologies.

Innate immunity

Innate immune cells

Immune cells such as lymphocytes, monocytes, neutrophils, eosinophils, and basophils are generally morphologically and functionally similar to those of other mammals (Table 1). Their differentiation and gene expression in the bone marrow are typical of mammalian hematopoiesis (Segawa et al., 2011). This section will further discuss assays that have been developed or adapted from other species to quantify innate immune functions in dolphins and other cetacean species.

Cell activation

Even though cells of the innate immune system do not require antigen-specific activation to perform their functions, the activation of immune cells can be detected by the expression of surface molecules. Specifically, CD11/CD18 or ß2-integrin is a cell surface molecule important in cell adhesion, phagocytosis, and respiratory burst. An antibody was developed that recognized the activation of bottlenose dolphin neutrophils and monocytes through their expression of CD11/CD18 (De Guise et al., 2004). Prevention of cell activation with this antibody almost fully inhibited in vitro neutrophil and monocyte phagocytosis and respiratory burst (De Guise et al., 2004), demonstrating the functional importance of CD11/CD18 in dolphins, as is the case in other species. This antibody may therefore be useful to monitor the activation status of circulating dolphin leukocytes, providing a potentially useful diagnostic tool to detect an early event in the immune response.

Phagocytosis

Phagocytosis is the ability for innate immune cells such as circulating neutrophils and monocytes or resident tissue macrophages to recognize and engulf microorganisms or foreign bodies. Methods were developed to quantify phagocytosis of cetacean blood leukocytes using either fluorescent latex beads approximately the same size as bacteria (De Guise et al., 1995) or fluorescently labeled bacteria (Keogh et al., 2011) and flow cytometry for rapid and precise quantification. As in other species, cetacean neutrophils and monocytes can quickly recognize and engulf foreign particles, with neutrophils generally more active than monocytes, reflecting their ability to rapidly respond to insults. Noda et al. also assessed bottlenose dolphin neutrophil phagocytosis using polystyrene latex beads but quantified phagocytosis using microscopy (Noda et al., 2003), which is much more time consuming and allows the assessment of much fewer cells.

Respiratory burst

Once engulfed, it is imperative to kill a microorganism. This is accomplished by the respiratory burst, the generation of reactive oxygen species within phagosomes, the membrane-bound outcome of phagocytosis. The respiratory burst of cetacean neutrophils and monocytes was measured using flow cytometry following incubation with a pro-fluorescent probe that gets trapped in cells and stimulation with either phorbol myristate acetate (PMA), a potent cell activator (De Guise et al., 1995), or bacteria, to simultaneously measure phagocytosis and respiratory burst (Keogh et al., 2011). As for phagocytosis, cetacean neutrophils generally show a more intense respiratory burst than monocytes, reflecting their ability to not only engulf but also kill ingested microorganisms. Respiratory burst was also measured in bottlenose dolphins using the nitroblue tetrazolium reduction test (NBT), where superoxide generation transforms the colorless NBT to a dark blue color using spectrophotometry (Noda et al., 2003). This method does not allow the simultaneous identification of the cell types (neutrophils vs monocytes) responsible for the respiratory burst.

TABLE 1 White blood cells (leukocyte) types.

Cell type	Appearance	Main role
Lymphocyte	Small cell, little cytoplasm	Acquired immunity (antibodies and cell-mediated immunity)
Monocyte	Larger cells with large kidney-shaped nucleus	Precursors of tissue macrophages, active in phagocytosis
Neutrophil	Segmented nucleus, pinkish cytoplasmic granules, most abundant in circulation	First line of defense in several types of inflammation, including most bacterial infections
Eosinophil	Segmented nucleus, reddish cytoplasmic granules	Important in allergies and response to parasites
Basophil	Segmented nucleus, cytoplasmic bluish granules, rare in circulation	Important in allergies

Natural killer cell activity

NK cell cytotoxicity plays an important role in innate immunity against viral infections and tumors. NK cell activity was measured in beluga whales (*Delphinapterus leucas*) by incubating isolated lymphocytes with a tumor cell line and quantifying said tumor cell line lysis (De Guise et al., 1997). As in other species, the response varied with different cell lines, and the addition of human recombinant IL-2 enhanced NK cell activity (De Guise et al., 1997). The ability to measure NK cell activity in cetaceans may be especially important given the viral epizootics observed in marine mammals in the last decades, as NK cells are involved in the initial, rapid response to viral infections.

Acquired immunity

Immune cells and their subsets

Modern immunology has recognized the different roles played by morphologically similar subsets of lymphocytes, and efforts have been made to use monoclonal antibodies developed against similar targets in other species and happen to identify similar targets in an unrelated species of interest, in our case in cetaceans (referred to as cross-reactive antibodies) or to generate species-specific monoclonal antibodies to identify cetacean cell subsets. Those antibodies bind specific cell surface markers for cell subsets, and the use of fluorescent or color tags allows the labeling and identification of cell subsets. These efforts used either flow cytometry to rapidly quantify the labeling of single cells in suspension or immunohistochemistry to label cells in a tissue section. It is important to verify the specificity of those antibodies, i.e. assure that they identify the cell subset they are intended for and do not identify cells other than those they are supposed to. This was generally validated using immunohistochemistry (labeling histologically compatible regions of tissue sections), flow cytometry (labeling of the anticipated proportions of cells from a given tissue/organ, and mutually exclusive labeling of different cell types), immunoprecipitation of the target antigen of relevant molecular weight, or sequencing of a compatible molecular sequence. These methods have allowed the identification of cell types designated by their cluster differentiation (CD), a terminology broadly used in immunology. This section describes results to date to identify lymphocyte subsets in cetaceans.

The differentiation of B- and T-lymphocyte subsets in particular has received significant attention, as it allows the identification of morphologically similar cells performing different functions. An antibody to bovine CD2, a molecule present on and common marker of human T lymphocytes, labeled virtually all beluga blood lymphocytes (De Guise et al., 1997). This is compatible with mice where that molecule is present on both B and T cells (Sen et al., 1990). This antibody was therefore not useful in identifying beluga T cells. However, cetacean-specific monoclonal antibodies to dolphin CD2 successfully identified cetacean T lymphocytes, which represented ~45% of blood lymphocytes in bottlenose dolphins and ~50%–90% of blood lymphocytes in all other cetacean species tested, which is within the range in other species (De Guise et al., 2002). Immunohistochemistry using formalin-fixed tissues demonstrated the cross-reactivity of T-cell markers, including CD3 in harbor porpoise (*Phocoena phocoena*) (Beineke et al., 2001), striped dolphin (*Stenella coeruleoalba*) (Jaber et al., 2003a), and Franciscana (*Pontoporia blainvillei*) (Díaz-Delgado et al., 2018), as well as CD5 in striped and bottlenose dolphins (Centelleghe et al., 2019).

The use of various markers to identify the presence of cell surface immunoglobulins that are characteristic of B lymphocytes allowed the identification of B cells in different cetacean species. B cells accounted for 6% of beluga lymphocytes (De Guise et al., 1997) and 10%–15% of bottlenose dolphin lymphocytes (Romano et al., 1992). Cetacean-specific monoclonal antibodies to dolphin CD19 and CD21, well-known markers of B cells, successfully identified B lymphocytes, which represented 14%–23% of blood lymphocytes in bottlenose dolphins and 4%–30% in other cetacean species (De Guise et al., 2002). Markers of surface immunoglobulins or antibodies to other cell surface molecules characteristic of B lymphocytes also allowed the identification of B lymphocytes in tissue sections of bottlenose dolphins (Centelleghe et al., 2019), striped dolphins (Jaber et al., 2003a; Centelleghe et al., 2019), and Franciscana (Díaz-Delgado et al., 2018). Together, these studies confirm our ability to identify and discriminate dolphin T and B lymphocytes in the blood and in tissue sections. As in other species, cetacean T cells are more abundant than B cells in the blood, and tissue sections reflect overall similar distribution of T and B cells, suggesting similarities with humans and other well-studied mammals.

Further efforts were made to distinguish T-lymphocyte subsets with known functional differences in other species. Antibodies to CD4 allowed the identification of T helper cells, which represented 30% of beluga lymphocytes (De Guise et al., 1997) and 30%–32% of bottlenose dolphin lymphocytes (Romano et al., 1999; Reif et al., 2009; De Guise et al., 2019), a proportion that is relatively similar to that in humans. However, the search for antibodies to identify cetacean CD8 cytotoxic T cells has thus far been unsuccessful. This shortcoming does not yet allow the determination of CD4/CD8 ratios that are commonly used in clinical immunology in humans.

Efforts have been made to identify other functionally important subsets of T lymphocytes. Regulatory T lymphocytes (Treg), specialized CD4+ T cells that function to maintain self-tolerance and immune homeostasis by suppressing the activation, proliferation, and effector functions of various immune cells, have been identified and characterized in the blood of bottlenose dolphins (De Guise et al., 2019) and in tissue sections of Franciscana (Díaz-Delgado et al., 2018). γδ T cells are another important subset of T cells at mucosal surfaces. They represented 31% of beluga blood lymphocytes (De Guise et al., 1997) but could not be identified in bottlenose dolphin blood (Elnaggar et al., 2017). The proportion of cells labeled in belugas is in the range observed for this T-cell subset in ruminants, whereas γδ T cells are much less abundant in human and mouse blood (Tizard, 1992). This may relate to the fact that cetaceans are evolutionally relatively closely related to ruminants and/or represent an adaptation to living in the marine environment. Overall, significant progress has been made in identifying functionally different subsets of T lymphocytes in blood and tissue sections, with at least some that may be important in understanding the adaptation of their immune response to the marine environment.

Major histocompatibility complex (MHC) molecules are important in the identification of self and in antigen presentation. Not surprisingly, MHC class I which is usually present on all nucleated cells was found on virtually all blood lymphocytes in belugas (De Guise et al., 1997) and bottlenose dolphins (Shirai et al., 1997; Elnaggar et al., 2017). MHC class II, which is generally expressed on mouse B but not T cells (Klein, 1986) and human B and activated T cells (Hewitt and Feldmann, 1989), was demonstrated on virtually all circulating (B and T) lymphocytes in bottlenose dolphins (Romano et al., 1992; Shirai et al., 1998; Elnaggar et al., 2017; Nouri-Shirazi et al., 2017) and beluga whales (De Guise et al., 1997), as is the case in pigs (Tizard, 1992) and horses (Barbis et al., 1994). Similarly, MHC class II was observed on most lymphocytes in tissue sections from harbor porpoise (Beineke et al., 2001) and Franciscana (Díaz-Delgado et al., 2018) as well as the antigen-presenting Langerhans cells in the skin of Atlantic bottlenose dolphins (Zabka and Romano, 2003). The distribution of MHC class II in most cetacean species appears to differ from that in humans and be more similar to that in pigs and horses, which are more closely related evolutionarily. The exact significance of those differences have not been definitively identified.

Cytokines

Cytokines are small soluble proteins used for cell-to-cell communications. They are important in the initiation, maintenance, amplification, and resolution of an immune response, for which they can modulate the direction and magnitude. For example, Th1 cytokines such as IL-2, IL-12, and interferon gamma (IFNγ) lead to a cell-mediated immune response that is important to address intracellular infections, whereas Th2 cytokines such as IL-4, IL-5, and IL-13 lead to a humoral (antibody) response that is important to address extracellular infections. On the other hand, Treg cytokines such as IL-10 and TGFß dampen the immune response and are important to prevent autoimmunity, whereas the Th17 cytokine IL-17 promotes inflammation. Several approaches have been adopted to study cytokines in cetaceans. While several studies have investigated molecular sequences of genes for diverse cytokines, reviewed in (Beineke et al., 2010), this section will focus on the quantification and functionality of cytokines in cetaceans.

The last decade saw several advances in the amplification of cetacean cytokine mRNA to quantify the expression of the genes for those cytokines. This includes the detection and quantification of the expression of the genes from immune cells for quantifying the cytokines IL-2, IL-4, IL-10, IL-12, IL-13, IL-18, TNFα, TGFß, and IFNγ in bottlenose dolphins, Pacific white-sided dolphins (*Lagenorhynchus obliquidens*), and beluga whales (Sitt et al., 2008), IL-1α, IL-1β, IL-1RA, IL-2, IL-4, IL-6, IL-8, IL-10, IL-12p40, IL-13, IL-18, IFNγ, and TNFα in bottlenose dolphins (Hofstetter et al., 2017), as well as IL-1ß, IL-2, IL-4, IL-6, IL-10, TNFα, and TGFß in harbor porpoises (Beineke et al., 2004; Fonfara et al., 2007). While the quantification of mRNA represents an important, relatively new tool to quantify cytokine production from immune cells, it should be noted that the expression of mRNA does not assess future translation and secretion of those cytokines and does not equate biological activity.

Immunohistochemistry allows the detection and semi-quantification of proteins in tissue sections using antibodies. The cytokines IL-1ß, IL-4, IL-6, IL-8, IL-10, IFNγ, and TGFß were identified in bottlenose dolphin spleen sections (Eberle et al., 2013). The cytokines IL-1α, IL-1ß, IL-8, TNFα, CD25, IL-2, IL-6, IL-10, IL-4, and IFNγ were recognized in snap-frozen, but not formalin-fixed, lymph node sections from Atlantic spotted dolphins (*Stenella frontalis*), striped dolphins, and fin whales (*Balaenoptera physalus*) (Jaber et al., 2010). Similarly, the cytokines IL-4, IL-10, IFNγ, and TGFß were successfully detected in Franciscana tissues sections (Díaz-Delgado et al., 2018).

Measuring cytokine protein concentrations in a biological medium such as blood or tissue culture may best reflect physiological response and immunological relevance. Species-specific enzyme-linked immunosorbent assays (ELISAs) successfully measured the bottlenose dolphin cytokines IFNγ and TNFα in serum or tissue culture supernatant (Eberle et al., 2018) and killer whale (*Orcinus orca*) serum IL-6 (Funke et al., 2003). Alternatively, De Guise et al. validated the use of commercial kits containing antibodies to porcine (IL-1ß, IL-4, IL-8, and TGFß) and human cytokines (IL-2, IL-5, IL-10, IL-13, IFNγ, and GM-CSF), using the Luminex technology to simultaneously quantify serum cytokines in bottlenose dolphins (De Guise et al., 2017, 2019, 2021). The high throughput of the Luminex technology allows the simultaneous detection of a large number of cytokines, with built-in quality control allowing consistency, repeatability, and rapid turnaround time. Overall, it appears that cetacean cytokines function much like in other species, and our ability to measure them in serum and tissues allows a deeper understanding of the cetacean immune response.

Lymphocyte activation and proliferation

Antigen-specific cell activation is a hallmark early event in an acquired immune response and a target of modern immunology. Specifically, efforts have been made to detect markers of activation on the surface of dolphin lymphocytes. CD25 is the cell surface receptor for IL-2, with the secretion of IL-2, and increased expression of CD25 (often in the same cell) representing early events in cell activation. Antibodies to detect the expression of cetacean CD25 were not found, despite multiple attempts (Elnaggar et al., 2017; De Guise et al., 2019). An alternative approach to detecting cell surface expression of CD25 on activated bottlenose dolphin lymphocytes was to use its ligand, IL-2, labeled with a fluorescent tag (Erickson et al., 1995). However, it appears that this reagent is no longer commercially available. Another alternative to detect activated bottlenose dolphin T lymphocytes was the loss of expression of CD45R upon stimulation (De Guise et al., 1998), suggesting that lymphocytes not expressing CD45R on their surface were activated/memory T cells, as is the case in humans (Akbar et al., 1988).

Cell activation, an early event of the acquired immune response, leads to the proliferation of lymphocytes to augment the pool of cells responsive to the specific antigen triggering an immune response. Mimicking this process, lymphocytes can be stimulated in vitro with mitogens, plant extracts that promote the division of specific T or B lymphocyte subsets, to quantify their ability to proliferate through the incorporation of a labeled (radioactive) form of the amino acid thymidine. Early attempts at quantifying cetacean lymphocyte proliferation used whole blood (Mumford et al., 1975; Colgrove, 1978), which provided a relatively crude estimation as it did not take into account the number or types of cells evaluated or serum constituents (such as cortisol) that may have affected the response. Techniques were later refined to include performing the assay with purified lymphocytes (Romano et al., 1992; Lahvis et al., 1995) and confirmation that mitogens indeed triggered the specific proliferation of B and T cell in cetaceans as it does in other species (De Guise et al., 1996). In order to alleviate the logistical laboratory complications associated with handling radioactivity, the assay was modified to use a nonradioactive synthetic thymidine analog (5-bromo-deoxyuridine or BrdU) that can be detected in dividing cells using antibodies (Beineke et al., 2004; Mori et al., 2006). Overall, the activation and proliferation of lymphocytes subsets in cetaceans parallels what we know in other mammals and can provide early diagnostic tools in health assessment.

Humoral immune response

The humoral immune response results in the synthesis and secretion of antibodies or immunoglobulins in the serum or at mucosal surfaces. While some studies have described the different classes of immunoglobulins and their molecular characterization in marine mammals, this section will focus on more recent studies on the quantification of immunoglobulins or sub-classes of immunoglobulins. Species-specific antibodies were produced to killer whale IgG and used to quantify total serum IgG in killer whale using radial immunodiffusion (Taylor et al., 2002), a classic but rather time-consuming methodology. More rapid colorimetric assays (diverse variations of ELISA) were also developed to measure total serum IgG in bottlenose dolphin (Nollens et al., 2007; Ruiz et al., 2009) and killer whale (Taylor et al., 2002). Those reagents were later used to measure antigen-specific antibodies, such as bottlenose dolphin antibodies to the dolphin pathogen *Erysipelothrix rhusiopathiae* (Nollens et al., 2007, 2016; Melero et al., 2016). Overall, those studies have demonstrated the ability to quantify antibodies in the serum of cetaceans, including antibodies to specific antigens, demonstrating previous exposure and response to such antigen/pathogen and suggesting a certain level of protection against it.

Cellular immune response

While the humoral immune response is relatively easily measured through quantification of serum immunoglobulins, the cell-mediated immune response is much harder to quantify, with the best examples using complex mouse models. However, models for correlates of cetacean cell-mediated immune response were developed. Immunocompromised mice deficient in both functional T and B lymphocytes, and therefore unable to produce antibodies or reject grafts, were injected with mononuclear cells from different species of cetaceans, which reconstituted a functional cetacean immune system in those mice, with the detection of those immune cells in the mice up to 63 days after reconstitution (De Guise and Levin, 2004). Furthermore, those mice were vaccinated with a commercial canine distemper vaccine, and a cetacean cell-mediated immune response was detected, as the increased individual cell proliferation of those cells upon in vitro re-exposure to the vaccine demonstrated a secondary or memory response to the vaccine (De Guise and Levin, 2004). This indirect model demonstrates the ability of dolphin cells to respond to a vaccine, without the potential safety concerns associated with injecting the vaccine into an actual dolphin. A similar approach, the detection of increased individual cell proliferation following in vitro re-exposure of mononuclear cells to the vaccine in vitro, is currently being used to demonstrate the efficacy of a canine distemper vaccine in California sea lions (*Zalophus californianus*) (Ross et al., 2021). A similar approach was used by Sitt et al., with in vitro incubation with the vaccine antigen following in vivo vaccination of dolphins with a sub-unit vaccine against the dolphin pathogen *Erysipelothrix rhusiopathiae*; however, rather than measuring the proliferation of individual cells, the authors quantified the expression for genes associated with a Th1, Th2, or Treg response (Sitt et al., 2010). They observed an increased expression of IFNγ following vaccination, suggesting a Th1 response to the vaccine antigen (Sitt et al., 2010).

Inflammation

Inflammation represents the combination of the vascular and immunological processes involved in the response to infectious or noninfectious insults. A number of studies have used immunohistochemistry to label different types of inflammatory cells in tissue sections. Lysozyme, an antimicrobial enzyme produced by macrophages, was used to identify macrophages, and S100 protein was used to identify dendritic cells in striped dolphins (Jaber et al., 2003a). Franciscana tissue macrophages, also known as histiocytic cells, were identified in tissue sections using antibodies to CD68 and lysozyme (Díaz-Delgado et al., 2018), and macrophages were identified with antibodies in tissue sections of short-finned pilot whales (*Globicephala macrorhynchus*) and Risso's dolphins (*Grampus griseus*) (Kawashima et al., 2004). Those antibodies were used to characterize the nature of cells in inflammatory infiltrates in the liver of striped dolphins (Jaber et al., 2003a, b). The identification of different cell types in tissue sections allows a better understanding of the inflammatory response to different insults in different cetacean species.

Acute phase proteins are highly conserved proteins that become elevated during an inflammatory response in a number of species and as such are often used to diagnose the presence of an inflammatory response. Acute phase proteins investigated as markers of inflammation in cetacean serum include C-reactive protein (CRP), serum amyloid-A (SAA), and haptoglobin (Hp), all well characterized markers of inflammation in humans and other species. All three were detected in bottlenose dolphins using automated methods (Cray et al., 2013), and as would be expected in other species, bottlenose dolphin serum SAA and Hp were increased in animals with inflammatory conditions (Segawa et al., 2013a, 2013b). In harbor porpoises, Hp, but surprisingly not CRP, correlated with white blood cell counts, another indicator of an inflammatory response (Fonfara et al., 2007). This suggests differences in acute phase proteins as markers of inflammation among cetacean species. A comprehensive review of acute phase proteins in marine mammals was recently published (Gelain and Bonsembiante, 2019).

Other tools

A microarray was developed and validated to rapidly assess changes in gene expression in bottlenose dolphins, focusing on genes relevant to the immune system and stress response (Mancia et al., 2007). The microarray was later used to characterize the stress response associated with capture-release health assessment (Mancia et al., 2008) and to successfully identify dolphins from a polluted area (Mancia et al., 2015), as discussed in more details below. However, the microarray has not been broadly used, maybe as a result of the increased accessibility and affordability of broad scale sequencing, as demonstrated in a fur seal (*Callorhinus ursinus*) transcriptomic study (Hoffman et al., 2013).

Immunology and stress

The connection between the nervous and immune systems

It was once thought that the immune system was autonomous since cells of the immune system would react when mixed in a test tube. It is now known that there is integration of the immune system with other systems of the body, and these interactions are much more complex than originally thought (Nutma et al., 2019). One of the main systems that can affect the immune system and vice versa is the nervous system. The nervous system (i.e., brain) "perceives" stressors and activates the physiological response to stressors. There is evidence that stressors as perceived by the brain can affect immunocompetence and an organism's ability to fight off invading pathogens, toxins, and contaminants, leading to morbidity and/or mortality (Dhabhar, 2014; Zhang et al., 2020; Ponomarev et al., 2021).

The connections between the nervous and immune systems have been demonstrated in the beluga (Romano et al., 1994, 2002). Nerve fibers from the autonomic nervous system containing neurotransmitters, which are activated during stressor events, have been demonstrated in specific lymphoid compartments in organs of the immune system in the beluga. The organs examined include the primary lymphoid organs, spleen and thymus, and secondary lymphoid organs including assorted lymph nodes, tonsils, and gut-associated lymphoid tissue. In these organs, nerve fibers are not only observed associated with the smooth muscle of blood vessels as expected but also associate in cellular compartments of these organs coming into close contact with cells of the immune system (Romano et al., 1994, 2002). Norepinephrine (NE) contained in these nerves fulfills the criteria for neurotransmission in the beluga based on the presence and availability of NE-containing nerve fibers in specific compartments of lymphoid organs as well as quantitative measures of these neurotransmitters in lymphoid organs using high-performance liquid chromatography. Moreover, adrenergic receptors have been identified on beluga lymphocytes, and in vitro incubation of beluga lymphocytes with neurotransmitters has resulted in changes in cellular responses. Therefore, when a whale encounters a stressor and the sympathetic nervous system is activated, neurotransmitters released from the nerve terminals can potentially affect the immune response (Romano et al., 1993).

Stressors and marine mammals

Stress can be defined as the "biological response elicited when an individual perceives a threat to its homeostasis, with the threat being the stressor" (Moberg, 2000). It is important to recognize that the stress response is adaptive and its purpose is to return the animal to homeostasis. The acute stress response releases hormones into the bloodstream that enable an organism to "fight-or-flight," with the response being short-lived, for example being chased by a predator for a short time. Prolonged or chronic responses to stressors, however, can cause persistent high levels of stress hormones, resulting in impairment of biological functions and leaving an animal vulnerable to a number of pathologies, e.g., the predator never leaving the area. The physiological response to a stressor varies based on type, duration, magnitude, and prior experience. Each individual responds to different stressors differently. What is stressful for one animal may not be stressful or as stressful to another. Although there are many similarities of the physiological response to stressors in marine mammals to land mammals, there are many differences given evolutionary considerations and adaptations to living in the aquatic environment (see Atkinson et al. (2015) for a review). Hormones belonging to the physiological response to stressors that have been or are currently being measured in cetaceans (the majority of which are bottlenose dolphins and belugas) include cortisol, cortisol-binding protein, corticosterone, adrenocorticotropic hormone, aldosterone, epinephrine, norepinephrine, dopamine, and the thyroid hormones (T3 and T4) (St. Aubin et al., 2011; Fair et al., 2014; Thompson et al., 2014; Flower et al., 2015; Delehanty et al., 2019; Houser et al., 2021). These can be combined with the available immune tests described earlier to assess the relationship between stress and immunity in cetaceans.

Multiple stressors can affect cetaceans, including both environmental or natural stressors and unnatural anthropogenic stressors caused by humans. These stressors have been on the rise in our oceans and today more than ever dolphins and whales are subjected to both stressors in a rapidly changing environment. Climate change poses multiple stressors for cetaceans, including increasing water temperatures, ocean acidification, algal blooms, emerging pathogens, and shifts in both predator and prey distributions (Breed et al., 2017; Williams et al., 2021). For cetaceans living in the Arctic, loss of sea ice renders whales and dolphins more vulnerable to shipping traffic, noise from ships and underwater exploration, an increase in fishing pressures, mining, oil exploration, and drilling (Burek et al., 2008; Kovacs and Lydersen, 2008; Moore and Huntington, 2008). Other stressors cetaceans can be subjected to include contaminants (addressed in this chapter and Chapter 14: Human impacts on dolphins: Physiological effects and conservation), incidental capture/entanglement in fishing gear, and oil spills (St. Aubin et al., 2011; De Guise et al., 2021). What is apparent today more than ever is the cumulative effects of multiple stressors on dolphin and whale populations (Maxwell et al., 2013; Moore et al., 2021). However, perhaps one of the major reasons as to why there isn't a lot of information on the cetacean stress response is the difficulty associated with studying stress physiology in cetaceans.

Studies of the nervous and immune systems in cetaceans under professional human care

Aquaria and other facilities that care for cetaceans under professional human care offer opportunities for controlled studies that are not feasible with wild whales (Romano et al., 2021). First, basic information is known (e.g., age, sex, life history, diet, morphometrics [e.g., weight, length, and girth], health history, and status). Most of this information is not consistently available from wild cetaceans. Environmental parameters, e.g., air and water temperature, water chemistries, are also known. Importantly, whales under professional care can be trained for biological sampling and to participate in experimental paradigms (see Romano et al. (2021) for review).

Collection of biological samples from trained whales offers baseline values without the confounding variable of the stress associated with sample collection such as during live capture-release events or strandings. Moreover, serial samples can be obtained from the same individuals over time which is often difficult to do with wild whales. Importantly, appropriate samples are needed for developing and validating assays for investigating both the nervous (stress) and immune system responses and can be acquired from trained whales. In facilities professionally caring for cetaceans, there are also opportunities to measure the physiological (including immune) response during opportunistic events that whales under human care may experience such as transport from one facility to another, out of water examinations, introduction to new social groups and novel environments as well as participation in interactive programs with the public (Schmitt et al., 2010; Spoon and Romano, 2012; Romano et al., 2021).

Capture-release studies of cetaceans

Capture-release studies of cetaceans collect a great deal of information from wild whales or dolphins (Fair et al., 2014; Norman et al., 2012; De Guise et al., 2021) such as stress hormones (e.g., cortisol, epinephrine), immune function (e.g., lymphocyte subsets, lymphocyte proliferation, and gene expression), complete blood cell counts, and serum chemistries. However, it is important to recognize that samples are usually from a single point in time and some of the health indices may reflect physiological states associated with the chase and capture, confounding results. Some of these studies, including taking a blood sample upon capture and then again immediately before release, allow for determining how stress hormones and immune function change within this 30–60 min time period (Unal et al., 2018; Unal and Romano, 2021). Moreover, neural-immune measures from live capture-release cetaceans can be compared with the same species' baseline under human care to give an idea of the magnitude of the response (Fair et al., 2017; Unal and Romano, 2021). Capture-release studies on bottlenose dolphins, belugas, and spotted and spinner dolphins (*Stenella longirostris*) have revealed an incredible amount of health information on these species and their respective geographical locations/habitats (St. Aubin et al., 2011; Norman et al., 2012; Fair et al., 2014; De Guise et al., 2021). Moreover, these studies have been conducted over several years, providing insight into health and physiology changes over time.

Cetacean strandings

Strandings and unusual mortality events have been on the rise and offer another window to investigate stress and the immune system. There are many theories to explain why cetacean strandings occur; however, commonly, there is no definitive answer as to the reason. One common theme in cetacean strandings (whether en masse or individual) is the stress response and a compromised immune system. It is not known what extent the stress of a stranding contributes to suppressed immune responses or if the immune system is compromised well before leading to the stranding event. Stranded cetaceans that are brought into rehabilitation with the intent of eventual release back into the wild offer the opportunity to serve as their own control and allow for monitoring aspects of both the nervous and immune systems upon stranding, throughout rehabilitation and then again when deemed healthy immediately before release.

Monitoring stress and the immune system in free-ranging cetaceans

Over the last decade, there has been a concerted effort to understand how environmental and anthropogenic stressors are impacting the behavior, hearing, communication, diving, and foraging of cetaceans (Halliday et al., 2017; Isojunno et al., 2017; Putland et al., 2018). Physiology and health have also been a more recent focus with particular emphasis on stress physiology and immunocompetence (Schwacke et al., 2012; Hunt et al., 2014; Apprill et al., 2017; Unal et al., 2018). While the "gold standard" has been measurement of stress hormones and immune parameters in blood, blood sampling from free-ranging cetaceans is not feasible presently. However, there are noninvasive biological samples (e.g., breath, feces, remote skin/blubber biopsies) which reveal important health information that can be collected from free-ranging cetaceans that offer promise in determining physiological status including stress and the immune system.

Remote skin/blubber biopsies have been collected from free-ranging cetaceans for decades. In the past, remote biopsies were used primarily for population genetics, but in recent years, they have been used to measure contaminants, hormones, and age determination and to investigate the expression of relevant genes for stress, immune function, and health (Unal et al., 2018; Bors et al., 2020; Galligan et al., 2020). Gene expression is an especially powerful tool when biopsies from the same animal are utilized to determine all or most of the above to create a broader picture of health of the individual and population.

Exhaled condensate or breath from dolphins and whales has proven to also contain valuable health information such as reproductive, metabolic and stress hormones, immunity, sex identification, and gene expression measures of relevant stress, immune, and health markers (Hunt et al., 2014; Thompson et al., 2014; Apprill et al., 2017; Richard et al., 2017; Burgess et al., 2018). Breath has been collected from wild baleen whales through utilization of drones and/or utilizing a collecting device on the end of a pole suspended from a boat. Studies in aquaria are investigating methods for measuring pertinent health information in breath of toothed whales and how to transition collection to free-ranging whales (Romano et al., 2021). It is recognized that the blow of baleen whales is much larger in magnitude and force than that from toothed whales rendering it more feasible to obtain at a higher height with drones than for the smaller toothed whales or dolphins (Raudino et al., 2019).

Feces from dolphins and whales have been collected successfully using dogs to find fecal material through their sense of smell, and hormones have been successfully quantified in right whales (*Eubalaena glacialis*) (Rolland et al., 2006). The feasibility of utilizing fecal samples from wild dolphins and whales will depend on the species, characteristics of fecal material, and ability to visually observe or find fecal material depending on the habitat.

Although studies of stress and the immune system in cetaceans are difficult to carry out, significant strides have been made (Romano et al., 2002; Romano et al., 2004; Spoon and Romano, 2012; St. Aubin et al., 2013; Fair et al., 2017; Romano et al., 2021; Thompson and Romano, 2021; Unal and Romano, 2021). However, there is still so much to learn and a long way to go toward our understanding of stress, immunity, and disease in cetaceans. Given the ever increasing anthropogenic and environmental stressors impacting cetaceans and their cumulative effects, it is even more critical to learn as much as we can on the effects of stress on the immune system and health in hopes of contributing information for management and recovery efforts of cetaceans worldwide.

Immunology and diving physiology

Diving is a multifaceted behavior which requires marine mammals to contend with challenges associated with oxygen availability and changes in temperature and pressure. Dolphins, as other marine mammals, possess a pronounced mammalian dive response (Chapter 5: Cardiovascular physiology in dolphins and other cetaceans and Chapter 7: Diving physiology in dolphins and human) and anatomical and behavioral adaptations to overcome these challenges. However, the effects of diving on cellular function, particularly immune function, has received comparatively little attention. There is likely a complicated relationship between diving, immune function, and maintaining health in diving mammals as increased pressure (or cycles of compression and decompression) and potential changes in thermoregulation (Chapter 3: Thermoregulation) and hypoxia/reoxygenation can all impact cellular processes and immune responses. Furthermore, aspects of the physiological dive response itself may have the potential to influence cellular immune function. This is an important emergent area of research due to the essential role of diving for marine mammals and shifts in dive behavior due to climate change, changing ocean resources, and human activities.

Pressure

Pressure increases 1 atm with every 10 m depth, meaning that for a dive to 100 m, a dolphin would be subject to 10× the pressure experienced at the sea surface. A major concern of this increase in pressure is the effect on air-filled spaces, such as the lung; adaptations to which are discussed elsewhere (Chapter 6: Respiratory physiology in the dolphin and other whales). Cell processes are also susceptible to influence from increased pressure through changes in membrane fluidity and order, ion channel function, membrane-bound protein function, and gene expression (Somero, 1992; Bartlett et al., 1995; Kato and Hayashi, 1999; Pradillon and Gaill, 2007). Mechanoreceptors, such as PIEZO1, which responds to physical stimuli like pressure, have been identified in immune cells such as monocytes and macrophages and have been linked to proinflammatory states following repeated exposure to increased pressure (Solis et al., 2019). While it might be assumed that dive-adapted animals would be able to avoid entering a proinflammatory state during diving, the presence of such mechanoreceptors on immune cells and their potential response to pressure during diving in cetaceans has not been studied.

Temperature

Both hypo- and hyperthermia are known to influence immune responses. In general, hypothermia is linked with a suppression of immune responses (Lee et al., 2001). Similar to pressure, decreased temperatures can affect enzymatic reactions (e.g., Fig. 1 in Daniel et al., (2008)) and can have ordering effects on cell membranes, resulting in decreased membrane fluidity and function (Magee et al., 2005). In contrast, hyperthermia, such as occurs during fever, is commonly known to augment immune and inflammatory responses (Zhang et al., 2008).

Water is an extremely efficient conductor of heat, and diving mammals have developed anatomical, behavioral, and physiological means of thermoregulation: both retaining body heat in cold water and dissipating extra heat in warm water or during intense activity (Favilla and Costa, 2020) (Chapter 3). The relationship between the dive response and thermoregulation is the subject of ongoing research as a potential product of peripheral vasoconstriction associated with the dive response is retention of excess heat. It has been reported that dolphins delay the release of such excess heat during diving, suggesting there could be a potential hyperthermia of core and tissues, the effects of which are unknown (Williams et al., 1999). In contrast, dolphin skin has been reported to remain within 1°C of water temperature even during exercise (Noren et al., 1999), and elephant seals (*Mirounga angustirostris*) failed to display significant decreases in core temperature, though peripheral vasoconstriction due to the dive response may lead to peripheral or localized cooling (Meir and Ponganis, 2010). Thus, there may be a thermal gradient which occurs during a dive (Favilla and Costa, 2020). More research is needed to fully understand thermoregulation during diving, as well as the potential impact of hypo- and hyperthermia on marine mammal immune responses.

Hypoxia/reperfusion

As reviewed elsewhere in this volume, dolphins possess physiological, anatomical, and behavioral adaptations which not only increase body oxygen stores but also conserve the use of oxygen over the course of a dive (Chapter 2: Energetic costs of rest and locomotion in dolphins, Chapter 4: Muscles and movement, Chapter 5, Chapter 6, and Chapter 8: Genetic and molecular adaptations). It is commonly thought that the majority of dives which occur remain aerobic; however, as dives progress, animals are exposed to increasing levels of hypoxemia and hypoxia. Hypoxia and ischemia (lack of blood flow) have been linked to inflammation and changes in both innate and adaptive immune responses (Eltzschig and Carmeliet, 2011) mediated by expression of the hypoxia inducible factors (HIF; (Chen and Gaber, 2021)) which promotes cell survival and glycolysis and alters cytokine expression, cell function, and inflammatory responses (Hellwig-Bürgel et al., 2005; D'ignazio et al., 2016; Huang et al., 2019). These immune-related changes are variable depending on cell type and may include reduced or enhanced activity. For example, respiratory burst, which uses the production and release of reactive oxygen species to destroy bacteria, is reportedly reduced under hypoxia (McGovern et al., 2011). For professional free divers, increasing inflammation and neutrophil migration and decreased cytotoxic T-cell function associated with diving hypoxia (Eftedal et al., 2016) demonstrates cell-type-specific responses to hypoxia and illustrates the complicated relationship between diving, immune function, and health.

Following ischemia, reperfusion of tissue at the cessation of a breath-hold (the end of the dive) has also been linked to the production of reactive oxygen species and tissue damage marked by neutrophil activity and inflammation (Kaminski et al., 2002; Slegtenhorst et al., 2014). While respiratory burst was reduced under hypoxia, production of reactive oxygen species upon reoxygenation is associated with reperfusion injury (Willy et al., 2000; Eisenhardt et al., 2015; Allen and Vazquez-Medina, 2019).

Mammalian dive response

Aspects of the mammalian dive response may also have direct impacts on regulating immune responses and are intertwined with both thermoregulation and oxygen preservation (Chapters 3 and 5). Splenic contraction increases hematocrit and blood oxygen stores by releasing red cells into circulation. Splenic contraction is also linked to increased circulating white blood cells (Baković et al., 2005). For dolphins, as with other cetaceans, the spleen is smaller as compared with pinnipeds and is thought to play less of a red blood cell storage role; however, it still contains active white pulp and likely maintains a hematopoietic role, functioning in the production of blood cells (Cowan and Smith, 1999). Whether there is an associated change in circulating white blood cell number or cell ratios in diving dolphins has not been investigated.

Such splenic contraction, like peripheral vasoconstriction, is thought to be regulated by release of catecholamines such as epinephrine and norepinephrine, following sympathetic nervous activation (Hurford et al., 1996). The relationship between circulating catecholamines and immune function can be complicated based on situation context, cell type, and duration of exposure, with outcomes ranging from immunosuppression to enhancement (Dhabhar, 2014).

Cellular adaptations to diving in marine mammals

It is unlikely that dolphins, or other diving mammals, are in a perpetual state of inflammation or immune dysfunction due to the influence of repetitive dive behavior. While few studies of cellular adaptation to pressure, temperature, or hypoxia have been published for dolphins, some cellular adaptations have been identified among different marine mammal species.

Fatty acids and cholesterol play important roles in determining membrane fluidity (Van Blitterswijk et al., 1987; Maulucci et al., 2016), and membrane fluidity can directly influence cell signaling, receptor binding, and function, including for immune cells (De Pablo and Alvarez De Cienfuegos, 2000). Changes in fatty acid composition of membranes for example have been linked to depressed lymphocyte function (De Pablo and Alvarez De Cienfuegos, 2000; Li et al., 2005; Gorjão et al., 2007) and increased macrophage phagocytosis (Schumann, 2016). Such studies of membrane composition and relation to function are needed for marine mammals; however, other cell types may offer insight into adaptations. The ratio of cholesterol: fatty acid content of erythrocyte (RBC) membranes in deep-diving elephant seals and harbor seals is lower (with decreased cholesterol and higher fatty acids) as compared with shallower diving seals and terrestrial mammals (Williams et al., 2001). Similarly, elephant seal platelets have a higher cholesterol content as compared with humans and show a reversible cold activation and no high pressure induced activation (Field and Tablin, 2012).

Docosahexaenoic acid (DHA) is a membrane fatty acid that is increased in cell membranes of a vertically migrating marine copepod, conferring both pressure and cold tolerance (Pond et al., 2014). DHA content of erythrocyte membranes in bottlenose dolphins under human care is similar to that of humans (Harris and Schmitt, 2014). Similarly, RBC glucose utilization was reported to remain stable in response to increased pressure in both dolphins and humans (Castellini et al., 2001). These similarities may suggest a similar cellular tolerance to pressure changes during diving, but more research is needed and membrane composition alone is likely not the only important factor determining pressure and cold tolerance. Comparable membrane composition studies have not been carried out for leukocytes but could provide insight as to whether dolphin immune cells may be sensitive to changes in pressure and temperature, and whether this sensitivity varies between shallow and deep-diving ecotypes (Klatsky et al., 2007; Fahlman et al., 2018, 2023).

Immune function studies

Consideration of immune adaptation in the context of diving in marine mammals is an emerging area of research interest, with important implications for evaluating individual and population health, understanding disease trends, and evaluating health impacts of the changing ocean environment and resource availability. To date, the few studies directly evaluating immune function in the context of diving have utilized both in vitro experimental manipulation and in vivo approaches with animals under human care.

Investigating the effects of pressure per se, exposure of beluga and phocid immune cells to increased pressure in vitro demonstrated varied responses between the cetaceans and pinnipeds, between exposures (e.g., durations, targeted final pressure), and between cell types and functions. For example, beluga granulocytes (e.g., neutrophils) exposed to either 2000 psi (1360 m) or 1000 psi (640 m) in vitro demonstrated a general decrease in phagocytosis, which appeared to recover to baseline levels following a recovery period at ambient pressure (Thompson and Romano, 2015). Phagocytosis is highly dependent on membrane fluidity to allow for the ingestion of bacterial or cell debris particles. This apparent decrease may be related to the direct effect of the pressure exposures on the physical properties of the cell membrane. The same 2000 psi exposures resulted in decreased phagocytosis in stranded phocids (whose health would have been compromised), while an increase in function was noted once animals were rehabilitated (Thompson and Romano, 2019).

Suppression of lymphocyte proliferation by exposure to pressure per se was also observed in belugas (Thompson and Romano, 2016). Interestingly, in this study, an increase in cell surface interleukin 2 receptors (IL-2R) was measured. IL-2R is important for T-cell activation and differentiation, and its expression is upregulated when cells are activated. This disconnect between increased activation state and depressed activity warrants further investigation.

The complement system is comprised of a series of proteins which, when activated, trigger a cascade of conformational and functional changes to ultimately result in activation of inflammation and immune responses. Activation of the complement system, and evidence of immune system over-activation have been linked to the development of dive-related injury, such as decompression sickness in humans (Magri et al., 2021). While gas dynamics and gas management during diving are beyond the scope of this chapter, if nitrogen supersaturation does occurs and if gas bubbles do form, Thompson et al. (2020) investigated whether marine mammal complement responds in a similar manner as has been reported in humans and rabbits. They found no significant changes in complement protein C5a (an indicator of complement activation) following in vitro exposures to nitrogen gas bubbles at a flow rate that had previously been reported to activate complement in humans and rabbits, suggesting that marine mammals are less sensitive to gas bubble exposure (Thompson et al., 2020). Measures of C5a also show smaller and less variable changes in deep-diving Weddell seals and narwhal (*Monodon monoceros*), suggesting these species are less responsive to the presence of nitrogen gas bubbles as compared to comparatively shallower diving harbor seals and belugas, respectively.

Further complicating the relationship between diving and immune function is the presence of additional stressors and the effect of these stressors on immune function. In each of the above studies, samples obtained from wild belugas, under the additional stress of capture and restraint, displayed different responses to pressure and/or nitrogen gas bubbles as compared with "baseline" samples collected from aquarium-maintained whales with behavioral participation and positive reinforcement (Thompson and Romano, 2015; Thompson et al., 2020). Additionally, phocids displayed different responses to pressure depending on health status, e.g., compromised due to injury, illness, dehydration, or malnutrition vs. healthy (Thompson and Romano, 2019).

Such information leads to the question of whether altered dive behaviors or physiological challenges may increase susceptibility of marine mammals to dive-related injury or disease occurrence. Further studies on the relationship between diving, immune function, and health should not only focus on basic adaptation of immune cells but incorporate consideration for "abnormal" diving in order to evaluate conditions under which diving may result in adverse health conditions.

No studies currently exist that have measured dolphin immune function in the context of diving, yet dolphins provide an excellent model for continuation of these and further comparative studies, not only as a comparatively shallow diving species vs. beluga or narwhal but also within the same species comparing dolphin ecotypes with varying dive behaviors. Bottlenose dolphins are one of the most common cetacean species maintained under human care (Lauderdale et al., 2021), providing an excellent, accessible, source of sampling for in vitro investigations. In addition, the current need to understand changes which occur in vivo that can be related to specific dive activities or behaviors, and the inability to sample wild dolphins without introducing confounding effects of chase, capture, and restraint, precludes a reliance on wild populations, further making managed animals an invaluable resource for in vivo studies.

Such approaches have been used to look at other aspects of the dive response, including thermoregulation (Williams et al., 1999) (Chapter 3) and effects of apnea on heart rate (Fahlman et al., 2020) (Chapter 5), and provide a crucial look at the effects of multiple physiological adjustments on immunity as it is currently impractical to reliably and simultaneously imitate all of the complicated physiological adjustments which occur over the course of a dive in vitro. For example, while C5a, a marker of complement activation, did not increase in baseline beluga samples exposed to nitrogen bubbles in vitro, the concentration of this protein more than doubled in response to nitrogen exposures after performance of 3.5 min stationary or active dives (Thompson and Romano, 2021). A similar study utilized breath-holds in three bottlenose dolphins under human care to investigate molecular markers of adaptation to apnea associated with diving. Two genes, ALOX5 and IL-5RA, were reported to be significantly upregulated in all study animals and have been implicated in both the dive (ALOX5 promotion of vasoconstriction) and immune (ALOX5 and IL-5 promote inflammation) responses (Blawas et al., 2021). The collection of blood samples following specific dives in both belugas and dolphins demonstrates the feasibility of this approach and functional assays should be applied to future studies.

Immunology and environmental contaminants

Marine mammal immunotoxicology aims to study the toxic effects of compounds, both naturally occurring and anthropogenic, on the immune system of marine mammals. Many environmental contaminants are known to be immunotoxic in laboratory animals, and accumulating evidence suggests the same is true for marine mammals, including dolphins. Although cause-and-effect relationships between contaminants and health effects in dolphins have yet to be fully defined, evidence of effects such as reproductive impairment (Lane et al., 2015), endocrine disruption (Trego et al., 2019), and immune modulation (Schwacke et al., 2012) support the broad hypothesis that environmental contaminants are detrimental to the overall health of dolphins.

Methods and materials to assess the impact of contaminants in dolphins

Tools, species-specific reagents, and functional immune assays to assess the impact of contaminants in dolphins have steadily evolved over the last several decades. Approximately 30 years ago, a panel of functional immune assays to assess the immunotoxic potential of chemicals was developed and validated by the US National Institutes of Health and the National Toxicology Program (Luster et al., 1992, 1993; Luster and Gerberick, 2010) with assays shown to be sensitive and reproducible. Combinations of two or three tests predicted the immunotoxicity with a percentage concordance (sum of specificity and sensitivity) of 90% or more (Luster et al., 1992). However, some assays, such as the plaque forming cell test is impractical and unethical in dolphins and other marine mammals, as it involves injecting sheep red blood cells into the animal and surgical removal of the spleen for testing. Immune assays validated for dolphins and other marine mammals that only require whole blood include assays to measure phagocytosis and respiratory burst, NK cell activity, mitogen-induced lymphocyte proliferation, identification and enumeration of immune cell subtypes, and quantification of serum immunoglobulins and cytokines, as described above, which are useful to detect the effects of contaminants on the sensitive immune system.

Study designs and methodologies

The immunotoxicity of many environmental contaminants are well documented in laboratory animals (e.g., rodents), however, similar characterization of risk for dolphins represents a considerable challenge given the many logistical, financial, and ethical constraints. Nonetheless, several approaches to evaluate if environmental contaminants are immunotoxic in dolphins have been developed. These approaches include assessing relationships between immune functions and (1) natural exposure (in vivo), (2) controlled laboratory exposure (in vitro), and (3) effect modeling (in silico) and offer a broad spectrum of specificity and complexity in terms of effects and mechanisms and come with inherent levels of certainty regarding interpretation of results for free-ranging animals.

In vivo field studies involving naturally exposed dolphin populations represent the most realistic and comprehensive approach when evaluating the effects of acute and chronic exposure to environmental contaminants. These types of studies are, however, complicated by confounding factors, including age, sex, diet, habitat, and temporal offsets between potential exposure and measured immunological changes. However, in vivo natural exposure studies capture the toxicokinetics of contaminant exposure, as well as the complex interplay between cells and tissues of the body. These studies entail the live capture of free-ranging dolphins and sampling of tissues and/or blood for both contaminant analysis and immunological assessments, followed by correlative analysis between the two (Schwacke et al., 2012).

In vitro laboratory exposure studies provide rapid detection of direct effects of a contaminant of interest on the immune system at relatively low cost, compared to live capture-release field studies. These experiments typically utilize live immune cells (isolated from whole blood) collected from managed-care or free-ranging individuals to conduct controlled concentration-response studies using a broad range of environmentally relevant contaminant concentrations. Importantly, these types of studies can examine individual contaminants or complex mixtures of contaminants on one or more particular immune functions (Levin et al., 2004). However, these studies cannot account for important toxicokinetic processes, including absorption, distribution, metabolism, and excretion of contaminants.

Interestingly, a meta-analysis of lymphocyte proliferation vs. pollutant concentration data demonstrated similar dose–response relationships between in vitro and in vivo experiments (Desforges et al., 2016). This was particularly important as in vitro studies are conducted not only to pinpoint specific cellular immune effects but also to predict real life in vivo responses. The authors provide support for the use of in vitro dose-response data in wildlife risk assessment, which is especially important when considering the logistical hurdles when conducting live capture-release dolphin studies.

In silico modeling approaches are the farthest removed from live animal studies; however, they may provide useful information on the risks of immunotoxicity for whole populations of marine mammals. For example, one approach involves physiologically based pharmacokinetic (PBPK) modeling to calculate risk quotients based on actual tissue contaminant levels in the species of interest and critical body residues determined from laboratory rodent studies. In this way, risk quotients provide a quantitative measure of immunotoxicity risk. This approach has been used to evaluate and compare the current risk of immunotoxicity in circumpolar populations of polar bears (*Ursus maritimus*) (Dietz et al., 2015). Using a different modeling approach, an individual-based model was used to assess the impact of polychlorinated biphenyls (PCBs) on cetacean populations through impacts on calf survival and immunity (Hall et al., 2018).

Contaminants and their impacts on dolphins

Polychlorinated biphenyls

Environmental organochlorine (OC) contaminants, most notably polychlorinated biphenyls (PCBs), are man-made chemicals found around the globe in almost all ecosystems. Although the production of PCBs was banned in the

United States since the 1970s, these compounds are still found in commercial and industrial products, such as electrical transformers, heat transfer fluids, and paints. These compounds are very stable in the environment and persist long after their production, use, and dumping has ceased. PCBs have made their way into the environment through direct dumping and accidental releases, resulting in global contamination, predominately through atmospheric distribution. Resistance to biodegradation and high lipophilicity cause these compounds to biomagnify in food chains and bioaccumulate in apex species, including dolphins.

An early report demonstrated a decrease in lymphocyte response in dolphins that was associated with increased concentrations of blood PCBs and DDT (Lahvis et al., 1995). A series of in vitro exposure experiments with individual PCBs or simple mixtures documented that it was mostly the noncoplanar PCBs (138, 153, 180), alone or in mixtures (but not the coplanar PCB 169 or TCDD), that contributed to significant reductions in dolphin neutrophil and monocyte phagocytosis (Levin et al., 2004). The noncoplanar PCBs also induced calcium mobilization, an important second messenger involved in numerous leukocyte functions, which may be involved in the reduction of noncoplanar PCB-induced suppression of phagocytosis (Levin et al., 2007). Individual PCBs or simple mixtures also modulated both T and B mitogen-induced lymphocyte proliferation, mostly driven by the noncoplanar PCBs (Mori et al., 2006, 2008). Coplanar PCBs (77, 126, 169) also failed to reduce bottlenose dolphins in vitro T-cell mitogen-induced lymphocyte proliferation (Nakata et al., 2007). Dolphins sampled from an area of the Georgia coast (United States) with known PCB contamination suffered from anemia and hypothyroidism and a significant negative correlation between T-cell proliferation and total blubber PCBs (Schwacke et al., 2012).

Harmful algal blooms biotoxins

Harmful algal blooms (HAB) biotoxins are chemicals that are produced and released by various algae, including phytoplankton and cyanobacteria, and include domoic acid, brevetoxin, microcystin, and saxitoxin. HABs involving the dinoflagellate *Karenia brevis* are responsible for what are commonly referred to as 'red tides,' which produce neurotoxins known as brevetoxins (PbTx). There are increasing concerns that dolphins may be exposed to repeated/chronic, sublethal concentrations of biotoxins, which may impact their overall health (Chapter 14).

Recurrent, unusual mortality events along the Florida panhandle have been tentatively attributed to poisoning from brevetoxin. An eosinophilia syndrome, manifested as an elevated blood eosinophil count without obvious cause, was observed in 23% of sampled dolphins near St. Joseph Bay, Florida. Elevated eosinophil counts were also associated with decreased T-lymphocyte proliferation and increased neutrophil phagocytosis. Given the unusual immunological findings and concurrent evidence of another biotoxin, domoic acid, within the same population, the authors suggested a need for further investigation to elucidate potential links between chronic, low-level exposure to algal toxins and dolphin immune health (Schwacke et al., 2010).

In vitro exposure studies were performed to further understand potential relationships between biotoxin exposure and changes in dolphin immune functions. Concentrations of brevetoxin within the range measured in the blood of naturally exposed dolphins significantly increased proliferation of T and B lymphocytes, as well as neutrophil and monocyte respiratory burst, but had no significant effects on neutrophil or monocyte phagocytosis or NK cell activity (Gebhard et al., 2015). These results suggest that dolphins naturally exposed to brevetoxin may be at risk for immunomodulation.

Per- and polyfluoroalkyl substances (PFAS)

PFAS are a diverse group of man-made aliphatic compounds containing one or more perfluoroalkyl moieties. The two most common PFASs are perfluorooctanoic acid (PFOA) and perfluorooctane sulfonic acid (PFOS), which exhibit both hydrophobic (water repellent) and lipophobic (lipid, grease repellent) properties (Dhore and Murthy, 2021). They therefore have widespread applications as water or grease repellents in water-resistant clothing, stain-resistant coatings used on carpets, and nonstick cookware. PFASs are also found in aqueous film forming foam (AFFF), a fire suppressant used to extinguish flammable liquid fires such as fuel fires, for example, at airports. These chemicals can leach into waterways and make their way to the oceans.

In a population of Atlantic bottlenose dolphins highly exposed to PFAS around Charleston, South Carolina, USA, age-adjusted linear regression models showed statistically significant positive associations between exposure to PFAS and lymphocytes subsets (T cells, helper Th cells, B cells, MHCII+ cells), B-cell proliferation, serum IgG1, and granulocytic/monocytic phagocytosis (Fair et al., 2013). Similarly, dolphin peripheral blood leukocytes exposed in vitro demonstrated a positive relationship between B-cell proliferation and environmentally relevant PFOS concentration, without alterations to NK cell activity and T-cell proliferation (Wirth et al., 2014). In the same study, the effects on mouse cells differed, suggesting that mice could not predict effects in dolphins. In another study, dolphin plasma PFOS concentrations were associated with significantly increased peripheral blood CD4+ and CD8+ T-cell proliferation, and in vitro PFOS exposure increased T-cell proliferation and intracellular IFNγ (but not IL-4) in a concentration-dependent manner (Soloff et al., 2017). Altogether, these suggest that PFOS dysregulates the dolphin cellular immune system and has implications for health hazards.

Nanomaterials

Nanomaterials (NMs) are a diverse class of small-scale substances that have structural components smaller than 1 μm in at least one dimension. Due to their novel nanoscale size, NMs possess unique chemical, biological, and physical properties as compared to larger particles of the same material (USEPA, 2008). While some nanomaterials naturally exist in the environment, they can also be manufactured from a variety of products, such as carbon or minerals like silver and gold. Manufactured NM are used in thousands of commercial products, including medical imaging, antimicrobials, UV blockers, fuel additives, cosmetics, and more (USEPA, 2008).

In vitro exposure of dolphin cells to silver nanoparticles induced cytotoxicity in both polymorphonuclear cells (neutrophils, eosinophils, and basophils) and mononuclear cells (lymphocytes and monocytes), which involved apoptosis, and sublethal concentrations significantly reduced PMNs phagocytosis, significantly increased respiratory burst, and significantly reduced ConA-induced T-cell proliferation (Li et al., 2018).

Brominated flame retardants

Brominated flame retardants (BFRs) are man-made organobromine compounds, which have an inhibitory effect on combustion chemistry and tend to reduce the flammability of products containing them. Although BFRs have helped reduce fire incidence in the United States over the last several decades, there is growing concern being raised because of their persistence, bioaccumulation, and potential for toxicity, both in animals and in humans.

Bottlenose dolphin NK cell activity and lymphocyte proliferation were not altered after in vitro exposure to environmentally relevant concentrations of a penta-polybrominated diphenyl ether (PBDE) mixture (Wirth et al., 2015). Exposure of a fibroblast cell line established from a pantropic spotted dolphin (*Stenella attenuata*) to PBDE resulted in a significant increase in IL-6 upon exposure to BDE-100 and BDE-209, whereas IL-1β and IL-8 were increased with BDE-47 and BDE-100 (Rajput et al., 2018).

Heavy metals

Heavy metals are defined as metallic chemical elements that have a relatively high density compared to water (Tchounwou et al., 2012) and can be toxic at low concentrations. Examples of heavy metals include mercury (Hg), cadmium (Cd), arsenic (As), chromium (Cr), thallium (Tl), and lead (Pb). Sources of heavy metals include industrial, agricultural, pharmaceutical, domestic effluents, and atmospheric sources.

A significant reduction in bottlenose dolphin lymphocyte proliferation and phagocytosis was induced by in vitro exposure to Hg, Cd, Al, and Pb, but not Cr, at concentrations within the environmental ranges reported in the tissues of bottlenose dolphins (Camara Pellisso et al., 2008). These are compatible with the immunotoxic impact of heavy metals previously documented in beluga whales (Bernier et al., 1996; Frouin et al., 2012).

Polycyclic aromatic hydrocarbons/crude oil

Polycyclic aromatic hydrocarbons are an important class of environmental pollutants associated with crude oil spills and have known immunotoxic effects (Davila et al., 1996; Reynaud and Deschaux, 2005; Zaccaria and McClure, 2013). Following the April 2010 Deepwater Horizon (DWH) oil spill, a comprehensive health assessment on bottlenose dolphins in Barataria Bay (BB), Louisiana, was conducted to assess potential health effects from exposure to oil, compared to dolphins that inhabit Sarasota Bay (SB), Florida, which were not exposed to DWH oil. A consistent increase in T and B lymphocyte proliferation was observed in oiled dolphins, along with a cytokine balance tilted toward a T helper 2 (Th2), rather than a T helper 1 (Th1), response (De Guise et al., 2017), which persisted over time (De Guise et al., 2021). In vitro oil exposure of bottlenose dolphin lymphocytes also resulted in an enhanced Th2 responsiveness based on the balance of cytokine expression as measured by RT-PCR (De Guise et al., 2021). An increase in the Th2 response would favor an immune response to extracellular pathogens, while a decrease in the Th1 response may inhibit an immune response to intracellular pathogens, such as viruses and intracellular bacteria. When measured for the first time in 2018, BB dolphins also had significantly more circulating T regulatory (Treg) cells than SB dolphins (De Guise et al., 2021). In a separate study, following in vitro exposure of bottlenose dolphin cells, Louisiana sweet crude oil high-energy media-accommodated fraction significantly increased B- and T-cell proliferation, whereas the chemical dispersant Corexit dispersed oil significantly decreased B- and T-cell proliferation and increased NK cell activity (White et al., 2016).

Future directions

The increasing use of genomics and other advanced molecular tools in ecotoxicology and marine mammal science will surely provide exciting new avenues for marine mammal immunotoxicology for years to come. For example, a study was conducted to investigate the potential use of microarray technology and gene expression profile analysis to screen bottlenose dolphins for exposure to environmental contaminants (pesticides, PCBs, PBDEs) through the immunological and/or endocrine perturbations associated with these agents (Mancia et al., 2014). A dolphin microarray representing 24,418 unigene sequences (including immune response activation) was used to analyze blood samples, and a group of dolphins with the highest and the lowest measured values of PCBs in their blubber was used as strata to determine the differential gene expression of the exposure extremes through machine learning classification algorithms. The authors reported that the algorithms worked very well (100% success rate) at classifying dolphins according to the contaminant load accumulated in their blubber and suggested that gene expression profile analysis may provide a valuable means to screen for indicators of chemical exposure.

Conclusions

The last few decades have seen significant progress in the development of tools and assays to better understand and measure different aspects of the immune system in bottlenose dolphins and other cetaceans. This includes efforts to better understand the physiology of the immune system at the molecular and cellular levels, as well as implications for the health of individuals and populations. It is likely that efforts will continue to refine our understanding of dolphin immunology, which continues to lag way behind the rapidly evolving science in immunology in human and rodent models, given the paucity of commercially available species-specific reagents in marine mammals. Given the central role of the immune system in the defense against infectious diseases, advances in immunology will be important in better understanding, diagnosing, managing, and preventing infectious diseases in dolphin populations under managed-care as well as in wild-ranging populations of dolphins.

Furthermore, significant progress has been made to understand the relationships between the immune system and other aspects of the physiology of dolphins. Although there is much more to learn about the cetacean stress response and impact on the immune system, significant strides have been made. Much has also been learned about the inherent physiological complexity of diving, a common behavior for dolphins. The recent decades benefited from significant technological advances and advances in comparative physiology that helped understand the effects of pressure, temperature, and hypoxia, as well as the interactions between diving and other systems, most recently the immune system. Valuable information has been learned from cetaceans in human care, live capture and release studies of cetaceans, cetacean strandings as well as from free-ranging wild cetaceans, and these practices should continue.

The weight of evidence clearly demonstrates that dolphins are at risk for contaminant-induced immunotoxicity, associated with a broadening range of legacy and emerging contaminants, from anthropogenic or natural origin. Additional work, however, is still required to understand the mechanisms behind the immunotoxic effects, as well as the complex interplay in chemical mixtures and multiple-stressor environments. The increasing use of genomics and other advanced molecular tools in ecotoxicology and marine mammal science will surely provide exciting new avenues for marine mammal immunotoxicology for years to come.

Additional tools, reagents, and databases could greatly advance the field of marine mammal immunology. For example, there could be value in establishing statistically derived reference intervals (i.e., baseline values), as has been done to generate diagnostic reference intervals for adrenal hormones commonly associated with the stress response (i.e., cortisol, aldosterone) (Hart et al., 2015), body condition (Hart et al., 2013), and serum/plasma biochemistry (Schwacke et al., 2009) in bottlenose dolphins. Such reference intervals can provide baseline functional immune assay values, thus allowing the evaluation of individual animal health relative to a 'normal' healthy population. This has been done for T-lymphocyte proliferation in bottlenose dolphins from the reference population in SB, Florida, with deviations from the "healthy" reference intervals observed in BB dolphins exposed to the Deepwater Horizon oil spill (De Guise et al., 2017). These data can be used to help diagnose deviations from normal immune function values that may be associated with contaminants, disease, or other health problems or be included as part of a broader health assessment.

Measures of functions of the immune system along with other relevant physiological parameters (such as nervous system) and health parameters can be included in marine mammal models that will inform health impacts at the individual and population levels. As we move forward, it is critical now more than ever that we understand how several types of environmental and man-made stressors, including chemical pollutants, that cetaceans are exposed to in the wild affect their physiology, immune system, and health. Importantly, this health information can contribute to policy and management decisions toward the conservation of cetaceans worldwide.

References

Akbar, A.N., Terry, L., Timms, A., Beverley, P.C., Janossy, G., 1988. Loss of CD45R and gain of UCHL1 reactivity is a feature of primed T cells. J. Immunol. 140, 2171–2178.

Allen, K.N., Vazquez-Medina, J.P., 2019. Natural tolerance to ischemia and hypoxemia in diving mammals: a review. Front. Physiol. 10, 1199.

Apprill, A., Miller, C., Moore, M., Durban, J., Fearnbach, H., Barrett-Lennard, L., 2017. Extensive core microbiome in drone-captured whale blow supports a framework for health monitoring. mSystems 2, e00119-17.

Atkinson, S., Crocker, D., Houser, D., Mashburn, K., 2015. Stress physiology in marine mammals: how well do they fit the terrestrial model? J. Comp. Physiol. B 185, 463–486.

Baković, D., Eterovic, D., Saratlija Novakovic, Z., Palada, I., Valic, Z., Bilopavlovic, N., Dujić, Z., 2005. Effect of human splenic contraction on variation in circulating blood cell counts. Clin. Exp. Pharmacol. Physiol. 32, 944–951.

Barbis, D.P., Bainbridge, D., Crump, A.L., Zhang, C.H., Antczak, D.F., 1994. Variation in expression of MHC class II antigens on horse lymphocytes determined by MHC haplotype. Vet. Immunol. Immunopathol. 42, 103–114.

Bartlett, D.H., Kato, C., Horikoshi, K., 1995. High pressure influences on gene and protein expression. Res. Microbiol. 146, 697–706.

Beineke, A., Siebert, U., Van Elk, N., Baumgartner, W., 2004. Development of a lymphocyte-transformation-assay for peripheral blood lymphocytes of the harbor porpoise and detection of cytokines using the reverse-transcription polymerase chain reaction. Vet. Immunol. Immunopathol. 98, 59–68.

Beineke, A., Siebert, U., Wohlsein, P., Baumgartner, W., 2010. Immunology of whales and dolphins. Vet. Immunol. Immunopathol. 133, 81–94.

Beineke, A., Siebert, U., Wunschmann, A., Stott, J.L., Prengel, I., Kremmer, E., Baumgartner, W., 2001. Immunohistochemical investigation of the cross-reactivity of selected cell markers from various species for characterization of lymphatic tissues in the harbour porpoise (*Phocoena phocoena*). J. Comp. Pathol. 125, 311–317.

Bernier, J., Fournier, M., De Guise, S., Martineau, D., Beland, P., 1996. Effects of in vitro exposure of beluga whale splenocytes and thymocytes to heavy metals. Environ. Toxicol. Chem. 15, 1357–1364.

Blawas, A.M., Ware, K.E., Schmaltz, E., Zheng, L., Spruance, J., Allen, A.S., West, N., Devos, N., Corcoran, D.L., Nowacek, D.P., Eward, W.C., Fahlman, A., Somarelli, J.A., 2021. An integrated comparative physiology and molecular approach pinpoints mediators of breath-hold capacity in dolphins. Evol. Med. Public Health 9, 420–430.

Bors, E.K., Baker, C.S., Wade, P.R., O'neill, K., Shelden, K.E.W., Thompson, M.J., Fei, Z., Jarman, S., Horvath, S., 2020. An epigenetic clock to estimate the age of living beluga whales. bioRxiv. 2020.09.28.317610.

Breed, G.A., Matthews, C.J., Marcoux, M., Higdon, J.W., Leblanc, B., Petersen, S.D., Orr, J., Reinhart, N.R., Ferguson, S.H., 2017. Sustained disruption of narwhal habitat use and behavior in the presence of Arctic killer whales. Proc. Natl. Acad. Sci. U. S. A. 114, 2628–2633.

Burek, K.A., Gulland, F.M., O'hara, T. M., 2008. Effects of climate change on Arctic marine mammal health. Ecol. Appl. 18, S126–S134.

Burgess, E.A., Hunt, K.E., Kraus, S.D., Rolland, R.M., 2018. Quantifying hormones in exhaled breath for physiological assessment of large whales at sea. Sci. Rep. 8, 10031.

Camara Pellisso, S., Munoz, M.J., Carballo, M., Sanchez-Vizcaino, J.M., 2008. Determination of the immunotoxic potential of heavy metals on the functional activity of bottlenose dolphin leukocytes in vitro. Vet. Immunol. Immunopathol. 121, 189–198.

Castellini, M.A., Castellini, J.M., Rivera, P.M., 2001. Adaptations to pressure in the RBC metabolism of diving mammals. Comp. Biochem. Physiol. A: Physiol. 129, 751–757.

Centelleghe, C., Da Dalt, L., Marsili, L., Zanetti, R., Fernandez, A., Arbelo, M., Sierra, E., Castagnaro, M., Di Guardo, G., Mazzariol, S., 2019. Insights Into dolphins' immunology: immuno-phenotypic study on mediterranean and atlantic stranded cetaceans. Front. Immunol. 10.

Chen, Y., Gaber, T., 2021. Hypoxia/HIF modulates immune responses. Biomedicine 9.

Clark, L.S., Turner, J.P., Cowan, D.F., 2005. Involution of lymphoid organs in bottlenose dolphins (*Tursiops truncatus*) from the western Gulf of Mexico: implications for life in an aquatic environment. Anat. Rec. A Discov. Mol. Cell. Evol. Biol. 282A, 67–73.

Colgrove, G.S., 1978. Stimulation of lymphocytes from a dolphin (*Tursiops truncatus*) by phytomitogens. Am. J. Vet. Res. 39, 141–144.

Cowan, D.F., Smith, T.L., 1999. Morphology of the lymphoid organs of the bottlenose dolphin, *Tursiops truncatus*. J. Anat. 194 (Pt 4), 505–517.

Cray, C., Arheart, K.L., Hunt, M., Clauss, T., Leppert, L.L., Roberts, K., McCulloch, S.D., Goldstein, J.D., Gonzalez, C., Sweeney, J., Stone, R., Fair, P.A., Bossart, G.D., 2013. Acute phase protein quantitation in serum samples from healthy Atlantic bottlenose dolphins (*Tursiops truncatus*). J. Vet. Diagn. Invest. 25, 107–111.

Daniel, R.M., Danson, M.J., Eisenthal, R., Lee, C.K., Peterson, M.E., 2008. The effect of temperature on enzyme activity: new insights and their implications. Extremophiles 12, 51–59.

Davila, D.R., Romero, D.L., Burchiel, S.W., 1996. Human T cells are highly sensitive to suppression of mitogenesis by polycyclic aromatic hydrocarbons and this effect is differentially reversed by alpha-naphthoflavone. Toxicol. Appl. Pharmacol. 139, 333–341.

De Guise, S., Bernier, J., Dufresne, M., Martineau, D., Beland, P., Fournier, M., 1996. Immune functions in beluga whales (*Delphinapterus leucas*): evaluation of mitogen-induced blastic transformation of lymphocytes from peripheral blood, spleen and thymus. Vet. Immunol. Immunopathol. 50, 117–126.

De Guise, S., Bernier, J., Martineau, D., Beland, P., Fournier, M., 1997. Phenotyping of beluga whale blood lymphocytes using monoclonal antibodies. Dev. Comp. Immunol. 21, 425–433.

De Guise, S., Erickson, K., Blanchard, M., Dimolfetto, L., Lepper, H.D., Stott, J.L., Ferrick, D.A., 2004. Characterization of F21.A, a monoclonal antibody that recognize a leucocyte surface antigens for killer whale homologue to ß-2 integrin. Vet. Immunol. Immunopathol. 97, 195–206.

De Guise, S., Erickson, K., Blanchard, M., Dimolfetto, L., Lepper, H., Wang, J., Stott, J.L., Ferrick, D.A., 1998. Characterization of a monoclonal antibody that recognizes a lymphocyte surface antigen for the cetacean homologue to CD45R. Immunology 94, 207–212.

De Guise, S., Erickson, K., Blanchard, M., Dimolfetto, L., Lepper, H.D., Wang, J., Stott, J.L., Ferrick, D.A., 2002. Monoclonal antibodies to lymphocyte surface antigens for cetacean homologues to CD2, CD19 and CD21. Vet. Immunol. Immunopathol. 84, 209–221.

De Guise, S., Flipo, D., Boehm, J.R., Martineau, D., Beland, P., Fournier, M., 1995. Immune functions in beluga whales (*Delphinapterus leucas*): evaluation of phagocytosis and respiratory burst with peripheral blood leukocytes using flow cytometry. Vet. Immunol. Immunopathol. 47, 351–362.

De Guise, S., Levin, M.J., 2004. Cetacean-reconstituted severe combined immunodeficient (SCID) mice respond to vaccination with canine distemper vaccine. Vet. Immunol. Immunopathol. 97, 177–186.

De Guise, S., Levin, M., Gebhard, E., Jasperse, L., Burdett Hart, L., Smith, C.R., Venn-Watson, S., Townsend, F., Wells, R., Balmer, B., Zolman, E., Rowles, T., Schwacke, L., 2017. Changes in immune functions in bottlenose dolphins in the northern Gulf of Mexico associated with the deepwater Horizon oil spill. Endanger. Species Res. 33, 291–303.

De Guise, S., Levin, M., Jasperse, L., Herrman, J., Wells, R.S., Rowles, T., Schwacke, L., 2021. Long-term immunological alterations in bottlenose dolphin a decade after the Deepwater Horizon oil spill in the Northern Gulf of Mexico: potential for multigenerational effects. Environ. Toxicol. Chem. 40, 1308–1321.

De Guise, S., Levin, M., Jasperse, L., Risatti, G., Wells, R.S., 2019. T helper cell subsets and their functions in common bottlenose dolphins (*Tursiops truncatus*). Front. Immunol. 10, 1578.

De Guise, S., Ross, P.S., Osterhaus, A.D., Martineau, D., Beland, P., Fournier, M., 1997. Immune functions in beluga whales (*Delphinapterus leucas*): evaluation of natural killer cell activity. Vet. Immunol. Immunopathol. 58, 345–354.

De Pablo, M.A., Alvarez De Cienfuegos, G., 2000. Modulatory effects of dietary lipids on immune system functions. Immunol. Cell Biol. 78, 31–39.

Delehanty, B., Bossart, G., Champagne, C., Crocker, D., Fair, P., Haulena, M., Houser, D., Richardson, E., Lunn, N., Romano, T., Boonstra, R., 2019. Methods in the study of marine mammal stress: measuring binding affinity of corticosteroid binding globulin. Mar. Mamm. Sci. 35.

Desforges, J.P., Sonne, C., Levin, M., Siebert, U., De Guise, S., Dietz, R., 2016. Immunotoxic effects of environmental pollutants in marine mammals. Environ. Int. 86, 126–139.

Dhabhar, F.S., 2014. Effects of stress on immune function: the good, the bad, and the beautiful. Immunol. Res. 58, 193–210.

Dhore, R., Murthy, G.S., 2021. Per/polyfluoroalkyl substances production, applications and environmental impacts. Bioresour. Technol. 341, 125808.

Díaz-Delgado, J., Ressio, R., Groch, K.R., Catão-Dias, J.L., 2018. Immunohistochemical investigation of the cross-reactivity of selected cell markers in formalin-fixed, paraffin-embedded lymphoid tissues of Franciscana (*Pontoporia blainvillei*). Vet. Immunol. Immunopathol. 200, 52–58.

Dietz, R., Gustavson, K., Sonne, C., Desforges, J.P., Riget, F.F., Pavlova, V., McKinney, M.A., Letcher, R.J., 2015. Physiologically-based pharmacokinetic modelling of immune, reproductive and carcinogenic effects from contaminant exposure in polar bears (*Ursus maritimus*) across the Arctic. Environ. Res. 140, 45–55.

D'ignazio, L., Bandarra, D., Rocha, S., 2016. NF-κB and HIF crosstalk in immune responses. FEBS J. 283, 413–424.

Eberle, K.C., Venn-Watson, S.K., Jensen, E.D., Labresh, J., Sullivan, Y., Kakach, L., Sacco, R.E., 2018. Development and testing of species-specific ELISA assays to measure IFN-gamma and TNF-alpha in bottlenose dolphins (*Tursiops truncatus*). PLoS One 13, e0190786.

Eberle, K.C., Waters, T.E., Jensen, E.D., Venn-Watson, S.K., Sacco, R.E., 2013. Development and application of specific cytokine assays in tissue samples from a bottlenose dolphin with hyperinsulinemia. Front. Endocrinol. 4, 134.

Eftedal, I., Flatberg, A., Drvis, I., Dujic, Z., 2016. Immune and inflammatory responses to freediving calculated from leukocyte gene expression profiles. Physiol. Genomics 48, 795–802.

Eisenhardt, S.U., Weiss, J.B., Smolka, C., Maxeiner, J., Pankratz, F., Bemtgen, X., Kustermann, M., Thiele, J.R., Schmidt, Y., Bjoern Stark, G., Moser, M., Bode, C., Grundmann, S., 2015. MicroRNA-155 aggravates ischemia-reperfusion injury by modulation of inflammatory cell recruitment and the respiratory oxidative burst. Basic Res. Cardiol. 110, 32.

Elnaggar, M.M., Abdellrazeq, G.S., Venn-Watson, S.K., Jensen, E.D., Hulubei, V., Fry, L.M., Sacco, R.E., Davis, W.C., 2017. Identification of monoclonal antibodies cross-reactive with bottlenose dolphin orthologues of the major histocompatibility complex and leukocyte differentiation molecules. Vet. Immunol. Immunopathol. 192, 54–59.

Eltzschig, H.K., Carmeliet, P., 2011. Hypoxia and inflammation. N. Engl. J. Med. 364, 656–665.

Erickson, K.L., Dimolfetto-Landon, L., Wells, R.S., Reidarson, T., Stott, J.L., Ferrick, D.A., 1995. Development of an interleukin-2 receptor expression assay and its use in evaluation of cellular immune responses in bottlenose dolphin (*Tursiops truncatus*). J. Wildl. Dis. 31, 142–149.

Fahlman, A., Cozzi, B., Manley, M., Jabas, S., Malik, M., Blawas, A., Janik, V.M., 2020. Conditioned variation in heart rate during static breath-holds in the bottlenose dolphin (*Tursiops truncatus*). Front. Physiol. 11, 604018.

Fahlman, A., Jensen, F.H., Tyack, P.L., Wells, R.S., 2018. Modeling tissue and blood gas kinetics in coastal and offshore common bottlenose dolphins, *Tursiops truncatus*. Front. Physiol. 9, 838.

Fahlman, A., Tyson Moore, R.B., Stone, R., Sweeney, J., Faulkner Trainor, R., Barleycorn, A., McHugh, K., Allen, J., Wells, R.S., 2023. Deep diving by offshore bottlenose dolphins (*Tursiops spp.*). Mar. Mamm. Sci. 39, 1251–1266.

Fair, P.A., Romano, T., Schaefer, A.M., Reif, J.S., Bossart, G.D., Houde, M., Muir, D., Adams, J., Rice, C., Hulsey, T.C., Peden-Adams, M., 2013. Associations between perfluoroalkyl compounds and immune and clinical chemistry parameters in highly exposed bottlenose dolphins (*Tursiops truncatus*). Environ. Toxicol. Chem. 32, 736–746.

Fair, P.A., Schaefer, A.M., Houser, D.S., Bossart, G.D., Romano, T.A., Champagne, C.D., Stott, J.L., Rice, C.D., White, N., Reif, J.S., 2017. The environment as a driver of immune and endocrine responses in dolphins (*Tursiops truncatus*). PLoS One 12, e0176202.

Fair, P.A., Schaefer, A.M., Romano, T.A., Bossart, G.D., Lamb, S.V., Reif, J.S., 2014. Stress response of wild bottlenose dolphins (*Tursiops truncatus*) during capture-release health assessment studies. Gen. Comp. Endocrinol. 206, 203–212.

Favilla, A.B., Costa, D.P., 2020. Thermoregulatory strategies of diving air-breathing marine vertebrates: a review. Front. Ecol. Evol. 8, 555509.

Field, C.L., Tablin, F., 2012. Response of Northern Elephant Seal platelets to pressure and temperature changes: a comparison with human platelets. Comp. Biochem. Physiol. A: Physiol. 162, 289–295.

Flower, J.E., Allender, M.C., Giovanelli, R.P., Summers, S.D., Spoon, T.R., St Leger, J.A., Goertz, C.E., Dunn, J.L., Romano, T.A., Hobbs, R.C., Tuttle, A.D., 2015. Circulating concentrations of thyroid hormone in Beluga whales (*Delphinapterus leucas*): influence of age, sex, and season. J. Zoo Wildl. Med. 46, 456–467.

Fonfara, S., Siebert, U., Prange, A., 2007. Cytokine and acute phase proteins as markers for infection in harbour porpoises (*Phocoena phocoena*). Mar. Mamm. Sci. 23, 931–942.

Frouin, H., Loseto, L.L., Stern, G.A., Haulena, M., Ross, P.S., 2012. Mercury toxicity in beluga whale lymphocytes: limited effects of selenium protection. Aquat. Toxicol. 109, 185–193.

Funke, C., King, D.P., Mcbain, J.F., Adelung, D., Stott, J.L., 2003. Expression and functional characterization of killer whale (*Orcinus orca*) interleukin-6 (IL-6) and development of a competitive immunoassay. Vet. Immunol. Immunopathol. 93, 69–79.

Galligan, T.M., Boggs, A.S.P., Balmer, B.C., Rowles, T., Smith, C.R., Townsend, F., Wells, R.S., Kellar, N.M., Zolman, E.S., Schwacke, L.H., 2020. Blubber steroid hormone profiles as indicators of physiological state in free-ranging common bottlenose dolphins (*Tursiops truncatus*). Comp. Biochem. Physiol. A: Physiol. 239, 110583.

Gebhard, E., Levin, M., Bogomolni, A., De Guise, S., 2015. Immunomodulatory effects of brevetoxin (PbTx-3) upon in vitro exposure in bottlenose dolphins (*Tursiops truncatus*). Harmful Algae 44, 54–62.

Gelain, M.E., Bonsembiante, F., 2019. Acute phase proteins in marine mammals: state of art, perspectives and challenges. Front. Immunol. 10, 1220.

Gorjão, R., Cury-Boaventura, M.F., De Lima, T.M., Curi, R., 2007. Regulation of human lymphocyte proliferation by fatty acids. Cell Biochem. Funct. 25, 305–315.

Hall, A.J., McConnell, B.J., Schwacke, L.H., Ylitalo, G.M., Williams, R., Rowles, T.K., 2018. Predicting the effects of polychlorinated biphenyls on cetacean populations through impacts on immunity and calf survival. Environ. Pollut. 233, 407–418.

Halliday, W.D., Insley, S.J., Hilliard, R.C., De Jong, T., Pine, M.K., 2017. Potential impacts of shipping noise on marine mammals in the western Canadian Arctic. Mar. Pollut. Bull. 123, 73–82.

Harris, W.S., Schmitt, T.L., 2014. Unexpected similarity in RBC DHA and AA levels between bottlenose dolphins and humans. Prostaglandins Leukot. Essent. Fatty Acids 90, 55–59.

Hart, L.B., Wells, R.S., Kellar, N., Balmer, B.C., Hohn, A.A., Lamb, S.V., Rowles, T., Zolman, E.S., Schwacke, L.H., 2015. Adrenal hormones in common bottlenose dolphins (*Tursiops truncatus*): influential factors and reference intervals. PLoS One 10, e0127432.

Hart, L.B., Wells, R.S., Schwacke, L.H., 2013. Reference ranges for body condition in wild bottlenose dolphins Tursiops truncatus. Aquat. Biol. 18, 63–68.

Hellwig-Bürgel, T., Stiehl, D.P., Wagner, A.E., Metzen, E., Jelkmann, W., 2005. Review: hypoxia-inducible factor-1 (HIF-1): a novel transcription factor in immune reactions. J. Interferon Cytokine Res. 25, 297–310.

Hewitt, C.R., Feldmann, M., 1989. Human T cell clones present antigen. J. Immunol. 143, 762–769.

Hoffman, J.I., Thorne, M.A., Trathan, P.N., Forcada, J., 2013. Transcriptome of the dead: characterisation of immune genes and marker development from necropsy samples in a free-ranging marine mammal. BMC Genomics 14, 52.

Hofstetter, A.R., Eberle, K.C., Venn-Watson, S.K., Jensen, E.D., Porter, T.J., Waters, T.E., Sacco, R.E., 2017. Monitoring bottlenose dolphin leukocyte cytokine mRNA responsiveness by qPCR. PLoS One 12, e0189437.

Houser, D.S., Champagne, C.D., Wasser, S.K., Booth, R.K., Romano, T., Crocker, D.E., 2021. Influence of season, age, sex, and time of day on the endocrine profile of the common bottlenose dolphin (*Tursiops truncatus*). Gen. Comp. Endocrinol. 313, 113889.

Huang, H., Fan, Y., Gao, Z., Wang, W., Shao, N., Zhang, L., Yang, Y., Zhu, W., Chen, Z., Hu, J., Ding, G., 2019. HIF-1α contributes to Ang II-induced inflammatory cytokine production in podocytes. BMC Pharmacol. Toxicol. 20, 59.

Hunt, K., Rolland, R., Kraus, S., 2014. Detection of steroid and thyroid hormones via immunoassay of North Atlantic right whale (*Eubalaena glacialis*) respiratory vapor. Mar. Mamm. Sci. 30.

Hurford, W.E., Hochachka, P.W., Schneider, R.C., Guyton, G.P., Stanek, K.S., Zapol, D.G., Liggins, G.C., Zapol, W.M., 1996. Splenic contraction, catecholamine release, and blood volume redistribution during diving in the Weddell seal. J. Appl. Physiol. 80, 298–306.

Isojunno, S., Sadykova, D., Deruiter, S., Curé, C., Visser, F., Thomas, L., Miller, P., Harris, C., 2017. Individual, ecological, and anthropogenic influences on activity budgets of long-finned pilot whales. Ecosphere 8.

Jaber, J.R., Fernandez, A., Herraez, P., De Los, E., Monteros, A., Ramirez, G.A., Garcia, P.M., Fernandez, T., Arbelo, M., Perez, J., 2003a. Cross-reactivity of human and bovine antibodies in striped dolphin paraffin wax-embedded tissues. Vet. Immunol. Immunopathol. 96, 65–72.

Jaber, J.R., Perez, J., Arbelo, M., Herraez, P., De Los, E., Monteros, A., Rodnguez, F., Fernandez, T., Fernandez, A., 2003b. Immunophenotypic characterization of hepatic inflammatory cell infiltrates in common dolphins (*Delphinus delphis*). J. Comp. Pathol. 129, 226–230.

Jaber, J.R., Perez, J., Zafra, R., Herraez, P., Rodriguez, F., Arbelo, M., De Los Monteros, A.E., Fernandez, A., 2010. Cross-reactivity of anti-human, anti-porcine and anti-bovine cytokine antibodies with cetacean tissues. J. Comp. Pathol. 143, 45–51.

Kaminski, K.A., Bonda, T.A., Korecki, J., Musial, W.J., 2002. Oxidative stress and neutrophil activation—the two keystones of ischemia/reperfusion injury. Int. J. Cardiol. 86, 41–59.

Kato, M., Hayashi, R., 1999. Effects of high pressure on lipids and biomembranes for understanding high-pressure-induced biological phenomena. Biosci. Biotechnol. Biochem. 63, 1321–1328.

Kawashima, M., Nakanishi, M., Kuwamura, M., Takeya, M., Yamate, J., 2004. Immunohistochemical detection of macrophages in the short-finned pilot whale (*Globicephala macrorhynchus*) and Risso's dolphin (*Grampus griseus*). J. Comp. Pathol. 130, 32–40.

Keogh, M.J., Spoon, T., Ridgway, S.H., Jensen, E., Van Bonn, W., Romano, T.A., 2011. Simultaneous measurement of phagocytosis and respiratory burst of leukocytes in whole blood from bottlenose dolphins (*Tursiops truncatus*) utilizing flow cytometry. Vet. Immunol. Immunopathol. 144, 468–475.

Klatsky, L.J., Wells, R.S., Sweeney, J.C., 2007. Offshore bottlenose dolphins (*Tursiops truncatus*): movement and dive behavior near the Bermuda pedestal. J. Mammal. 88, 59–66.

Klein, J., 1986. Natural History of the Major Histocompatibility Complex. Wiley, New York.

Kovacs, K.M., Lydersen, C., 2008. Climate change impacts on seals and whales in the North Atlantic Arctic and adjacent shelf seas. Sci. Prog. 91, 117–150.

Lahvis, G.P., Wells, R.S., Kuehl, D.W., Stewart, J.L., Rhinehart, H.L., Via, C.S., 1995. Decreased lymphocyte responses in free-ranging bottlenose dolphins (*Tursiops truncatus*) are associated with increased concentrations of PCBs and DDT in peripheral blood. Environ. Health Perspect. 103 (Suppl 4), 67–72.

Lane, S.M., Smith, C.R., Mitchell, J., Balmer, B.C., Barry, K.P., McDonald, T., Mori, C.S., Rosel, P.E., Rowles, T.K., Speakman, T.R., Townsend, F.I., Tumlin, M.C., Wells, R.S., Zolman, E.S., Schwacke, L.H., 2015. Reproductive outcome and survival of common bottlenose dolphins sampled in Barataria Bay, Louisiana, USA, following the Deepwater Horizon oil spill. Proc. R. Soc. Lond. B 282, 20151944.

Lauderdale, L.K., Mellen, J.D., Walsh, M.T., Granger, D.A., Miller, L.J., 2021. Towards understanding the welfare of cetaceans in accredited zoos and aquariums. PLoS One 16, e0255506.

Lee, S.L., Battistella, F.D., Go, K., 2001. Hypothermia induces T-cell production of immunosuppressive cytokines. J. Surg. Res. 100, 150–153.

Levin, M., Morsey, B., De Guise, S., 2007. Non-coplanar PCBs induce calcium mobilization in bottlenose dolphin and beluga whale, but not in mouse leukocytes. J. Toxicol. Environ. Health A 70, 1220–1231.

Levin, M., Morsey, B., Mori, C., Guise, S., 2004. Specific non-coplanar PCB-mediated modulation of bottlenose dolphin and beluga whale phagocytosis upon in vitro exposure. J. Toxicol. Environ. Health A 67, 1517–1535.

Li, W.T., Chang, H.W., Yang, W.C., Lo, C., Wang, L.Y., Pang, V.F., Chen, M.H., Jeng, C.R., 2018. Immunotoxicity of silver nanoparticles (AgNPs) on the leukocytes of common bottlenose dolphins (*Tursiops truncatus*). Sci. Rep. 8, 5593.

Li, Q., Wang, M., Tan, L., Wang, C., Ma, J., Li, N., Li, Y., Xu, G., Li, J., 2005. Docosahexaenoic acid changes lipid composition and interleukin-2 receptor signaling in membrane rafts. J. Lipid Res. 46, 1904–1913.

Luster, M.I., Gerberick, G.F., 2010. Immunotoxicology testing: past and future. Methods Mol. Biol. 598, 3–13.

Luster, M.I., Portier, C., Pait, D.G., Rosenthal, G.J., Germolec, D.R., Corsini, E., Blaylock, B.L., Pollock, P., Kouchi, Y., Craig, W., et al., 1993. Risk assessment in immunotoxicology. II. Relationships between immune and host resistance tests. Fundam. Appl. Toxicol. 21, 71–82.

Luster, M.I., Portier, C., Pait, D.G., White Jr., K.L., Gennings, C., Munson, A.E., Rosenthal, G.J., 1992. Risk assessment in immunotoxicology. I. Sensitivity and predictability of immune tests. Fundam. Appl. Toxicol. 18, 200–210.

Magee, A.I., Adler, J., Parmryd, I., 2005. Cold-induced coalescence of T-cell plasma membrane microdomains activates signalling pathways. J. Cell Sci. 118, 3141–3151.

Magri, K., Eftedal, I., Petroni Magri, V., Matity, L., Azzopardi, C.P., Muscat, S., Pace, N.P., 2021. Acute effects on the human peripheral blood transcriptome of decompression sickness secondary to scuba diving. Front. Physiol. 12, 660402.

Mancia, A., Abelli, L., Kucklick, J.R., Rowles, T.K., Wells, R.S., Balmer, B.C., Hohn, A.A., Baatz, J.E., Ryan, J.C., 2015. Microarray applications to understand the impact of exposure to environmental contaminants in wild dolphins (*Tursiops truncatus*). Mar. Genomics 19, 47–57.

Mancia, A., Lundqvist, M.L., Romano, T.A., Peden-Adams, M.M., Fair, P.A., Kindy, M.S., Ellis, B.C., Gattoni-Celli, S., Mckillen, D.J., Trent, H.F., Chen, Y.A., Almeida, J.S., Gross, P.S., Chapman, R.W., Warr, G.W., 2007. A dolphin peripheral blood leukocyte cDNA microarray for studies of immune function and stress reactions. Dev. Comp. Immunol. 31, 520–529.

Mancia, A., Ryan, J.C., Van Dolah, F.M., Kucklick, J.R., Rowles, T.K., Wells, R.S., Rosel, P.E., Hohn, A.A., Schwacke, L.H., 2014. Machine learning approaches to investigate the impact of PCBs on the transcriptome of the common bottlenose dolphin (*Tursiops truncatus*). Mar. Environ. Res. 100, 57–67.

Mancia, A., Warr, G.W., Chapman, R.W., 2008. A transcriptomic analysis of the stress induced by capture-release health assessment studies in wild dolphins (*Tursiops truncatus*). Mol. Ecol. 17, 2581–2589.

Maulucci, G., Cohen, O., Daniel, B., Sansone, A., Petropoulou, P.I., Filou, S., Spyridonidis, A., Pani, G., De Spirito, M., Chatgilialoglu, C., Ferreri, C., Kypreos, K.E., Sasson, S., 2016. Fatty acid-related modulations of membrane fluidity in cells: detection and implications. Free Radic. Res. 50, S40–S50.

Maxwell, S.M., Hazen, E.L., Bograd, S.J., Halpern, B.S., Breed, G.A., Nickel, B., Teutschel, N.M., Crowder, L.B., Benson, S., Dutton, P.H., Bailey, H., Kappes, M.A., Kuhn, C.E., Weise, M.J., Mate, B., Shaffer, S.A., Hassrick, J.L., Henry, R.W., Irvine, L., McDonald, B.I., Robinson, P.W., Block, B.A., Costa, D.P., 2013. Cumulative human impacts on marine predators. Nat. Commun. 4, 2688.

McGovern, N.N., Cowburn, A.S., Porter, L., Walmsley, S.R., Summers, C., Thompson, A.A.R., Anwar, S., Willcocks, L.C., Whyte, M.K.B., Condliffe, A.M., Chilvers, E.R., 2011. Hypoxia selectively inhibits respiratory burst activity and killing of Staphylococcus aureus in human neutrophils. J. Immunol. 186, 453–463.

Meir, J.U., Ponganis, P.J., 2010. Blood temperature profiles of diving elephant seals. Physiol. Biochem. Zool. 83, 531–540.

Melero, M., Giménez-Lirola, L.G., Rubio-Guerri, C., Crespo-Picazo, J.L., Sierra, E., García-Párraga, D., García-Peña, F.J., Arbelo, M., Álvaro, T., Valls, M., Sánchez-Vizcaíno, J.M., 2016. Fluorescent microbead-based immunoassay for anti-Erysipelothrix rhusiopathiae antibody detection in cetaceans. Dis. Aquat. Org. 117, 237–243.

Moberg, G.P., 2000. Biological response to stress: Implications for animal welfare. In: Moberg, G.P., Mench, J.A. (Eds.), The Biology of Animal Stress: Basic Principles and Implications for Animal Welfare. CAB eBooks, pp. 1–21.

Moore, S.E., Huntington, H.P., 2008. Arctic marine mammals and climate change: impacts and resilience. Ecol. Appl. 18, s157–s165.

Moore, M.J., Rowles, T.K., Fauquier, D.A., Baker, J.D., Biedron, I., Durban, J.W., Hamilton, P.K., Henry, A.G., Knowlton, A.R., McLellan, W.A., Miller, C.A., Pace III, R.M., Pettis, H.M., Raverty, S., Rolland, R.M., Schick, R.S., Sharp, S.M., Smith, C.R., Thomas, L., Van Der Hoop, J.M., Ziccardi, M.H., 2021. REVIEW Assessing North Atlantic right whale health: threats, and development of tools critical for conservation of the species. Dis. Aquat. Org. 143, 205–226.

Mori, C., Morsey, B., Levin, M., Gorton, T.S., De Guise, S., 2008. Effects of organochlorines, individually and in mixtures, on B-cell proliferation in marine mammals and mice. J. Toxicol. Environ. Health A 71, 266–275.

Mori, C., Morsey, B., Levin, M., Nambiar, P.R., De Guise, S., 2006. Immunomodulatory effects of in vitro exposure to organochlorines on T-cell proliferation in marine mammals and mice. J. Toxicol. Environ. Health A 69, 283–302.

Mumford, D.M., Stockman, G.D., Barsales, P.B., Whitman, T., Wilbur, J.R., 1975. Lymphocyte transformation studies of sea mammal blood. Experientia 31, 498–500.

Nakata, H., Sasaki, H., Takemura, A., Yoshioka, M., Tanabe, S., Kannan, K., 2007. Bioaccumulation, temporal trend, and geographical distribution of synthetic musks in the marine environment. Environ. Sci. Technol. 41, 2216–2222.

Noda, K., Aoki, M., Akiyoshi, H., Asaki, H., Shimada, T., Ohashi, F., 2003. Evaluation of the polymorphonuclear cell functions of bottlenose dolphins. J. Vet. Med. Sci. 65, 727–729.

Nollens, H.H., Giménez-Lirola, L.G., Robeck, T.R., Schmitt, T.L., Dirocco, S., Opriessnig, T., 2016. Evaluation of anti-Erysipelothrix rhusiopathiae IgG response in bottlenose dolphins *Tursiops truncatus* to a commercial pig vaccine. Dis. Aquat. Org. 121, 249–256.

Nollens, H.H., Green, L.G., Duke, D., Walsh, M.T., Chittick, B., Gearhart, S., Klein, P.A., Jacobson, E.R., 2007. Development and validation of monoclonal and polyclonal antibodies for the detection of immunoglobulin G of bottlenose dolphins (*Tursiops truncatus*). J. Vet. Diagn. Invest. 19, 465–470.

Noren, D.P., Williams, T.M., Berry, P., Butler, E., 1999. Thermoregulation during swimming and diving in bottlenose dolphins, *Tursiops truncatus*. J. Comp. Physiol. B 169, 93–99.

Norman, S.A., Goertz, C.E., Burek, K.A., Quakenbush, L.T., Cornick, L.A., Romano, T.A., Spoon, T., Miller, W., Beckett, L.A., Hobbs, R.C., 2012. Seasonal hematology and serum chemistry of wild beluga whales (*Delphinapterus leucas*) in Bristol Bay, Alaska, USA. J. Wildl. Dis. 48, 21–32.

Nouri-Shirazi, M., Bible, B.F., Zeng, M., Tamjidi, S., Bossart, G.D., 2017. Phenotyping and comparing the immune cell populations of free-ranging Atlantic bottlenose dolphins (*Tursiops truncatus*) and dolphins under human care. BMC Vet. Res. 13, 78.

Nutma, E., Willison, H., Martino, G., Amor, S., 2019. Neuroimmunology—the past, present and future. Clin. Exp. Immunol. 197, 278–293.

Pond, D.W., Tarling, G.A., Mayor, D.J., 2014. Hydrostatic pressure and temperature effects on the membranes of a seasonally migrating marine copepod. PLoS One 9, e111043.

Ponomarev, S., Kalinin, S., Sadova, A., Rykova, M., Orlova, K., Crucian, B., 2021. Immunological aspects of isolation and confinement. Front. Immunol. 12, 697435.

Pradillon, F., Gaill, F., 2007. Pressure and life: some biological strategies. Rev. Environ. Sci. Biotechnol. 6, 181–195.

Putland, R.L., Merchant, N.D., Farcas, A., Radford, C.A., 2018. Vessel noise cuts down communication space for vocalizing fish and marine mammals. Glob. Chang. Biol. 24, 1708–1721.

Rajput, I.R., Xiao, Z., Yajing, S., Yaqoob, S., Sanganyado, E., Ying, H., Fei, Y., Liu, W., 2018. Establishment of pantropic spotted dolphin (*Stenella attenuata*) fibroblast cell line and potential influence of polybrominated diphenyl ethers (PBDEs) on cytokines response. Aquat. Toxicol. 203, 1–9.

Raudino, H., Tyne, J., Smith, A., Ottewell, K., McArthur, S., Kopps, A., Chabanne, D., Harcourt, R., Pirotta, V., Waples, K., 2019. Challenges of collecting blow from small cetaceans. Ecosphere 10.

Reif, J.S., Peden-Adams, M.M., Romano, T.A., Rice, C.D., Fair, P.A., Bossart, G.D., 2009. Immune dysfunction in Atlantic bottlenose dolphins (*Tursiops truncatus*) with lobomycosis. Med. Mycol. 47, 125–135.

Reynaud, S., Deschaux, P., 2005. The effects of 3-methylcholanthrene on lymphocyte proliferation in the common carp (*Cyprinus carpio* L.). Toxicology 211, 156–164.

Richard, J.T., Robeck, T.R., Osborn, S.D., Naples, L., McDermott, A., Laforge, R., Romano, T.A., Sartini, B.L., 2017. Testosterone and progesterone concentrations in blow samples are biologically relevant in belugas (*Delphinapterus leucas*). Gen. Comp. Endocrinol. 246, 183–193.

Rolland, R., Hamilton, P., Kraus, S., Davenport, B., Gillett, R., Wasser, S., 2006. Faecal sampling using detection dogs to study reproduction and health in North Atlantic right whales (*Eubalaena glacialis*). J. Cetacean Res. Manag. 8, 121–125.

Romano, T.A., Felten, S.Y., Olschowka, J.A., Felten, D.L., 1993. A microscopic investigation of the lymphoid organs of the beluga, *Delphinapterus leucas*. J. Morphol. 215, 261–287.

Romano, T.A., Felten, S.Y., Olschowka, J.A., Felten, D.L., 1994. Noradrenergic and peptidergic innervation of lymphoid organs in the beluga, *Delphinapterus leucas*: an anatomical link between the nervous and immune systems. J. Morphol. 221, 243–259.

Romano, T.A., Keogh, M.J., Kelly, C., Feng, P., Berk, L., Schlundt, C.E., Carder, D.A., Finneran, J.J., 2004. Anthropogenic sound and marine mammal health: measures of the nervous and immune systems before and after intense sound exposure. Can. J. Fish. Aquat. Sci. 61, 1124–1134.

Romano, T.A., Olschowka, J.A., Felten, S.Y., Quaranta, V., Ridgway, S.H., Felten, D.L., 2002. Immune response, stress, and environment: Implications for cetaceans. In: Pfeiffer, C.J. (Ed.), Molecular and Cell Biology of Marine Mammals. Krieger Publishing Co, pp. 253–279.

Romano, T.A., Ridgway, S.H., Felten, D.L., Quaranta, V., 1999. Molecular cloning and characterization of CD4 in an aquatic mammal, the white whale *Delphinapterus leucas*. Immunogenetics 49, 376–383.

Romano, T.A., Ridgway, S.H., Quaranta, V., 1992. MHC class II molecules and immunoglobulins on peripheral blood lymphocytes of the bottlenosed dolphin, *Tursiops truncatus*. J. Exp. Zool. 263, 96–104.

Romano, T.A., Thompson, L.A., Driscoll, M.V., Unal, E., Tuttle, A.D., Sirpenski, G., Mateleska, M.E., Wolbrink, D., 2021. The role of aquaria in beluga research and conservation. Polar Res. 40, 5567.

Ross, K., Le-Bert, C., De Guise, S., Levin, M., Saliki, J., Sanchez, S., Linnehan, B., Meegan, J., Lutmerding, B., Bgomex, F., Jensen, E., 2021. Cellular and Humoral Immune Responses of the California Sea Lion (*Zalophus californianus*) to Canarypox Vectored Canine Distemper Vaccine. AAZV. Virtual.

Ruiz, C.L., Nollens, H.H., Venn-Watson, S., Green, L.G., Wells, R.S., Walsh, M.T., Nolan, E.C., Mcbain, J.F., Jacobson, E.R., 2009. Baseline circulating immunoglobulin G levels in managed collection and free-ranging bottlenose dolphins (*Tursiops truncatus*). Dev. Comp. Immunol. 33, 449–455.

Schmitt, T., Aubin, D., Schaefer, A., Dunn, J.L., 2010. Baseline, diurnal variations, and stress-induced changes of stress hormones in three captive beluga, *Delphinapterus leucas*. Mar. Mamm. Sci. 26, 635–647.

Schumann, J., 2016. It is all about fluidity: fatty acids and macrophage phagocytosis. Eur. J. Pharmacol. 785, 18–23.

Schwacke, L.H., Hall, A.J., Townsend, F.I., Wells, R.S., Hansen, L.J., Hohn, A.A., Bossart, G.D., Fair, P.A., Rowles, T.K., 2009. Hematologic and serum biochemical reference intervals for free-ranging common bottlenose dolphins (*Tursiops truncatus*) and variation in the distributions of clinicopathologic values related to geographic sampling site. Am. J. Vet. Res. 70, 973–985.

Schwacke, L.H., Twiner, M.J., De Guise, S., Balmer, B.C., Wells, R.S., Townsend, F.I., Rotstein, D.C., Varela, R.A., Hansen, L.J., Zolman, E.S., Spradlin, T.R., Levin, M., Leibrecht, H., Wang, Z., Rowles, T.K., 2010. Eosinophilia and biotoxin exposure in bottlenose dolphins (*Tursiops truncatus*) from a coastal area impacted by repeated mortality events. Environ. Res. 110, 548–555.

Schwacke, L.H., Zolman, E.S., Balmer, B.C., De Guise, S., George, R.C., Hoguet, J., Hohn, A.A., Kucklick, J.R., Lamb, S., Levin, M., Litz, J.A., Mcfee, W.E., Place, N.J., Townsend, F.I., Wells, R.S., Rowles, T.K., 2012. Anaemia, hypothyroidism and immune suppression associated with polychlorinated biphenyl exposure in bottlenose dolphins (*Tursiops truncatus*). Proc. R. Soc. Lond. B 279, 48–57.

Segawa, T., Amatsuji, H., Suzuki, K., Suzuki, M., Yanagisawa, M., Itou, T., Sakai, T., Nakanishi, T., 2013a. Molecular characterization and validation of commercially available methods for haptoglobin measurement in bottlenose dolphin. Results Immunol. 3, 57–63.

Segawa, T., Itou, T., Suzuki, M., Moritomo, T., Nakanishi, T., Sakai, T., 2011. Hematopoietic cell populations in dolphin bone marrow: analysis of colony formation and differentiation. Results Immunol. 1, 1–5.

Segawa, T., Otsuka, T., Itou, T., Suzuki, M., Karatani, N., Sakai, T., 2013b. Characterization of the circulating serum amyloid A in bottlenose dolphins. Vet. Immunol. Immunopathol. 152, 218–224.

Sen, J., Rosenberg, N., Burakoff, S.J., 1990. Expression and ontogeny of CD2 on murine B cells. J. Immunol. 144, 2925–2930.

Shirai, K., Sakai, T., Fukuda, M., Oike, T., 1998. A monoclonal antibody against dolphin lymphocytes (6E9) which recognizes bovine MHC class II antigens. J. Vet. Med. Sci. 60, 291–293.

Shirai, K., Watanabe, H., Weerasinghe, A., Sakai, T., Sekikawa, H., Abo, T., 1997. A monoclonal antibody, DL10, which recognizes a sugar moiety of MHC class I antigens expressed on NK cells, NK+ T cells, and granulocytes in humans. J. Clin. Immunol. 17, 510–523.

Sitt, T., Bowen, L., Blanchard, M.T., Gershwin, L.J., Byrne, B.A., Dold, C., Mcbain, J., Stott, J.L., 2010. Cellular immune responses in cetaceans immunized with a porcine erysipelas vaccine. Vet. Immunol. Immunopathol. 137, 181–189.

Sitt, T., Bowen, L., Blanchard, M.T., Smith, B.R., Gershwin, L.J., Byrne, B.A., Stott, J.L., 2008. Quantitation of leukocyte gene expression in cetaceans. Dev. Comp. Immunol. 32, 1253–1259.

Slegtenhorst, B.R., Dor, F.J., Rodriguez, H., Voskuil, F.J., Tullius, S.G., 2014. Ischemia/reperfusion injury and its consequences on immunity and inflammation. Curr. Transplant. Rep. 1, 147–154.

Solis, A.G., Bielecki, P., Steach, H.R., Sharma, L., Harman, C.C.D., Yun, S., De Zoete, M.R., Warnock, J.N., To, S.D.F., York, A.G., Mack, M., Schwartz, M.A., Dela Cruz, C.S., Palm, N.W., Jackson, R., Flavell, R.A., 2019. Mechanosensation of cyclical force by PIEZO1 is essential for innate immunity. Nature 573, 69–74.

Soloff, A.C., Wolf, B.J., White, N.D., Muir, D., Courtney, S., Hardiman, G., Bossart, G.D., Fair, P.A., 2017. Environmental perfluorooctane sulfonate exposure drives T cell activation in bottlenose dolphins. J. Appl. Toxicol. 37, 1108–1116.

Somero, G.N., 1992. Adaptations to high hydrostatic pressure. Annu. Rev. Physiol. 54, 557–577.

Spoon, T.R., Romano, T.A., 2012. Neuroimmunological response of beluga whales (*Delphinapterus leucas*) to translocation and a novel social environment. Brain Behav. Immun. 26, 122–131.

St. Aubin, D., Forney, K., Chivers, S., Scott, M., Danil, K., Romano, T., 2011. Hematological, serum, and plasma chemical constituents in pantropical spotted dolphins (*Stenella attenuata*) following chase, encirclement, and tagging. Mar. Mamm. Sci. 29.

St. Aubin, D.J., Forney, K.A., Chivers, S.J., Scott, M.D., Danil, K., Romano, T.A., Wells, R.S., Gulland, F.M.D., 2013. Hematological, serum, and plasma chemical constituents in pantropical spotted dolphins (*Stenella attenuata*) following chase, encirclement, and tagging. Mar. Mamm. Sci. 29, 14–35.

Taylor, B.C., Brotheridge, R.M., Jessup, D.A., Stott, J.L., 2002. Measurement of serum immunoglobulin concentration in killer whales and sea otters by radial immunodiffusion. Vet. Immunol. Immunopathol. 89, 187–195.

Tchounwou, P.B., Yedjou, C.G., Patlolla, A.K., Sutton, D.J., 2012. Heavy metal toxicity and the environment. Exp. Suppl. 101, 133–164 (2012).

Thompson, L.A., Hindle, A.G., Black, S.R., Romano, T.A., 2020. Variation in the hemostatic complement (C5a) responses to in vitro nitrogen bubbles in monodontids and phocids. J. Comp. Physiol. B 190, 811–822.

Thompson, L.A., Romano, T.A., 2015. Beluga (*Delphinapterus leucas*) granulocytes and monocytes display variable responses to in vitro pressure exposures. Front. Physiol. 6, 128.

Thompson, L.A., Romano, T.A., 2016. Pressure induced changes in adaptive immune function in belugas (*Delphinapterus leucas*); implications for dive physiology and health. Front. Physiol. 7, 442.

Thompson, L.A., Romano, T.A., 2019. Effects of health status on pressure-induced changes in phocid immune function and implications for dive ability. J. Comp. Physiol. B 189, 637–657.

Thompson, L.A., Romano, T.A., 2021. The immune response and diving: conservation considerations for belugas (*Delphinapterus leucas*) in a changing Arctic environment. Polar Res. 40, 5499.

Thompson, L.A., Spoon, T.R., Goertz, C.E., Hobbs, R.C., Romano, T.A., 2014. Blow collection as a non-invasive method for measuring cortisol in the beluga (*Delphinapterus leucas*). PLoS One 9, e114062.

Tizard, I., 1992. Veterinary Immunology, An Introduction. WB Saunders Company, Montreal.

Trego, M.L., Hoh, E., Whitehead, A., Kellar, N.M., Lauf, M., Datuin, D.O., Lewison, R.L., 2019. Contaminant exposure linked to cellular and endocrine biomarkers in Southern California bottlenose dolphins. Environ. Sci. Technol. 53, 3811–3822.

Unal, E., Goertz, C.E.C., Hobbs, R.C., Suydam, R., Romano, T., 2018. Investigation of molecular biomarkers as potential indicators of health in wild belugas (*Delphinapterus leucas*). Mar. Biol. 165, 182.

Unal, E., Romano, T.A., 2021. Of whales and genes: unraveling the physiological response to stressors in belugas (*Delphinapterus leucas*) at the molecular level. J. Zool. Bot. Gard. 2, 559–575.

USEPA, 2008. Nanotechnology for Site Remediation Fact Sheet. Solid Waste and Emergency Response.

Van Blitterswijk, W.J., Van Der Meer, B.W., Hilkmann, H., 1987. Quantitative contributions of cholesterol and the individual classes of phospholipids and their degree of fatty acyl (un)saturation to membrane fluidity measured by fluorescence polarization. Biochemistry 26, 1746–1756.

White, N.D., Godard-Codding, C., Webb, S.J., Bossart, G.D., Fair, P.A., 2016. Immunotoxic effects of in vitro exposure of dolphin lymphocytes to Louisiana sweet crude oil and Corexit. J. Appl. Toxicol. 37, 676–682.

Williams, R., Lacy, R.C., Ashe, E., Hall, A., Plourde, S., Mcquinn, I.H., Lesage, V., 2021. Climate change complicates efforts to ensure survival and recovery of St. Lawrence Estuary beluga. Mar. Pollut. Bull. 173, 113096.

Williams, T.M., Noren, D., Berry, P., Estes, J.A., Allison, C., Kirtland, J., 1999. The diving physiology of bottlenose dolphins (*Tursiops truncatus*). III. Thermoregulation at depth. J. Exp. Biol. 202, 2763–2769.

Williams, E.E., Stewart, B.S., Beuchat, C.A., Somero, G.N., Hazel, J.R., 2001. Hydrostatic-pressure and temperature effects on the molecular order of erythrocyte membranes from deep-, shallow-, and non-diving mammals. Can. J. Zool. 79, 888–894.

Willy, C., Dahouk, S., Starck, C., Kaffenberger, W., Gerngross, H., Plappert, U.G., 2000. DNA damage in human leukocytes after ischemia/reperfusion injury. Free Radic. Biol. Med. 28, 1–12.

Wirth, J.R., Peden-Adams, M.M., White, N.D., Bossart, G.D., Fair, P.A., 2014. In vitro PFOS exposure on immune endpoints in bottlenose dolphins (*Tursiops truncatus*) and mice. J. Appl. Toxicol. 34, 658–666.

Wirth, J.R., Peden-Adams, M.M., White, N.D., Bossart, G.D., Fair, P.A., 2015. In vitro exposure of DE-71, a penta-PBDE mixture, on immune endpoints in bottlenose dolphins (*Tursiops truncatus*) and B6C3F1 mice. J. Appl. Toxicol. 35, 191–198.

Zabka, T.S., Romano, T.A., 2003. Distribution of MHC II (+) cells in skin of the Atlantic bottlenose dolphin (*Tursiops truncatus*): an initial investigation of dolphin dendritic cells. Anat. Rec. A Discov. Mol. Cell. Evol. Biol. 273, 636–647.

Zaccaria, K.J., McClure, P.R., 2013. Using immunotoxicity information to improve cancer risk assessment for polycyclic aromatic hydrocarbon mixtures. Int. J. Toxicol. 32, 236–250.

Zhang, H.G., Mehta, K., Cohen, P., Guha, C., 2008. Hyperthermia on immune regulation: a temperature's story. Cancer Lett. 271, 191–204.

Zhang, A., Zou, T., Guo, D., Wang, Q., Shen, Y., Hu, H., Ye, B., Xiang, M., 2020. The immune system can hear noise. Front. Immunol. 11, 619189.

C H A P T E R

14

Human impacts on dolphins: Physiological effects and conservation

Randall S. Wells[a] *and Andreas Fahlman*[b,c,d]

[a]Chicago Zoological Society's Sarasota Dolphin Research Program, c/o Mote Marine Laboratory, Sarasota, FL, United States
[b]Global Diving Research SL., Valencia, Spain [c]Fundación Oceanogràfic de la Comunitat Valenciana, Valencia, Spain
[d]Kolmården Wildlife Park, Kolmården, Sweden

Abstract

In this chapter, we describe aspects of the growing pressure on coastal and inshore common bottlenose dolphins (*Tursiops truncatus*, hereafter referred to as dolphins) as human use of these marine environments increases. While the ultimate outcomes of many threats from humans have been well-documented, less attention has been given to the physiological mechanisms underlying these impacts. We address these physiological mechanisms by using bottlenose dolphins as the basis of discussion, as much is known about this species, and it is an important apex predator that serves as a sentinel species for coastal ecosystem health. For much of this chapter, we will draw upon information available from the dolphin community in Sarasota Bay, Florida, USA, that has been studied for more than 50 years. In some cases, the bottlenose dolphin may also be useful as a model for understanding or predicting impacts to less-studied cetacean species. We make comparisons as appropriate with other cetacean species to look at how anthropogenic disturbances, such as entanglement in fishing gear, boat strikes and disturbance, pollution, and climate disruption may affect the health, survival, and reproductive success.

Introduction

Around the world, dolphins face increasing threats from human activities, especially in coastal and inshore waters where the overlap of people and dolphins is greatest. Dolphins must contend with threats from fishing activities, vessel traffic, disturbance and harassment, pollution, habitat alteration, and climate disruption. In many cases, dolphins are exposed to multiple concurrent threats. Some of these threats result in acute impacts, leading quickly to death, e.g., asphyxiation, while others may be chronic or sublethal, affecting the ability of the animal to function normally or reproduce successfully. In nearly all cases, physiological considerations play a documented or hypothetical role and may ultimately affect survival. In this chapter, we therefore focus how human activities and alterations to the environment result in disruption of the basic mechanisms of life. In this context, physiology is a broad concept where secondary trauma, for example, an oil spill, results in physiological changes that result in morbidity or direct impacts which are beyond the physiological capacity of the individual and result in death.

Due to the species' accessibility to researchers, extensive work with bottlenose dolphins (*Tursiops truncatus*) has identified many examples of impacts to dolphins from human activities in coastal and inshore waters (Wells and Scott, 2018; Wells et al., 2019). Threats and responses have been studied in considerable detail in the southeastern United States (SEUS), so this chapter will focus primarily on this region, with recognition of the fact that findings and conclusions from the SEUS can be extrapolated to many parts of the world (Vollmer and Rosel, 2013). Patterns of long-term, year-round residency in many of the bays, sounds, and estuaries of the SEUS facilitate understanding of exposures of dolphins to threats and assessment of impacts (Wells and Scott, 2018).

To understand human impacts on dolphins, long-term studies, like those conducted by the Sarasota Dolphin Research Program, working in conjunction with Mote Marine Laboratory's Stranding Investigations Program to document the lives of dolphins, can be very illuminating. The studies performed in Sarasota Bay, Florida, have provided long-term data on physiology, metabolism, behavior, and overall health, along with survival and reproductive success of dolphins. During the period 1985–2020, 26% of identifiable causes of death for Sarasota Bay dolphins have

The Physiology of Dolphins
https://doi.org/10.1016/B978-0-323-90516-9.00007-5

resulted from human interactions (G.N. Lovewell, pers. comm.). Research on Sarasota Bay dolphins began in 1970 and became the world's longest-running dolphin conservation research program (Wells, 2020). During the first 50 years of this study, the research team has seen tremendous increases in human utilization of, and impacts on, Sarasota Bay and surrounding areas. This long-term resident dolphin community serves as an important reference group against which more at-risk populations can be compared. For example, health indices for the Sarasota Bay dolphins have been compared with those from populations in areas where environmental pollution, such as the *Deepwater Horizon* oil spill, has occurred (e.g., Schwacke et al., 2014; Smith et al., 2017). Differences between sites can build a case for identifying likely impacts from the pollution. For many cetacean species and populations, health and population status information is limited. For this reason, the data provided by health assessments and population and ecosystem monitoring, along the lines of those performed by the Sarasota Dolphin Research Program, are and will be vital to help determine how human use of marine ecosystems affects these species and their survival.

Below we present examples of documented threats to bottlenose dolphins in the SEUS, focusing on what has been learned from Sarasota Bay and vicinity and then discuss physiological considerations associated with each threat.

Fishing activities

Issues

Accidental bycatch or entanglement of marine species is unfortunately quite frequent, and hundreds of thousands of whales and dolphins die each year as bycatch from fisheries interactions (Read et al., 2006; Hamer et al., 2012). Even more individuals are hurt, wounded, or mutilated by fishing hooks, lines, and nets, which likely result in intense suffering and reduced welfare, survival, and/or reproductive success for the individual animal (e.g., Wells et al., 2008). In the shallow inshore waters along the west coast of Florida, three kinds of fishing activities impact dolphins: recreational hook and line angling, crab pots, and to a much lesser extent, gillnets.

Fishing gear is the leading cause of human-related mortality for Sarasota Bay dolphins, accounting for 20% of stranding cases for which a cause of death could be determined and 79% of all human interaction cases (G.N. Lovewell, pers. comm.). Dolphins die or are seriously injured from ingestion of fishing gear, entanglement in fishing line, terminal tackle, nets or crab trap float lines, or from hooking (Wells et al., 1998, 2008). Following a severe red tide (*Karenia brevis*) harmful algal bloom event in 2005–2006, 2% of the Sarasota Bay resident dolphins died from ingestion of fishing gear as dolphins and anglers were vying for the few remaining fish (Powell and Wells, 2011). In some cases, interactions with anglers and associated gear is a learned behavior from mothers who frequent angling situations and place their naïve calves at risk, often reducing their own reproductive success (Wells, 2019). Dolphins can also come to recognize anglers as a food source, through direct provisioning, from discarded catch or bait, or by depredating fishing gear (Powell and Wells, 2011). Dolphins that engage in unnatural foraging behaviors through interacting with humans are at increased risk of future injuries or mortality from such interactions (Christiansen et al., 2016).

Physiological impacts

Dolphins become entangled in fishing nets or crab trap float lines as they swim into undetected nets, attempt to depredate fish from nets, or try to access crab trap bait or catch fish around the traps. On occasion, they become entangled at such depths that they are unable to reach the surface to breathe, and if they are unable to escape within a few minutes of entanglement, they die from asphyxiation. As a variant of this injury, fishing line attached to catch or bait can become tightly wrapped, multiple times, around the epiglottal beak ("goosebeak") that connects the blowhole to the lungs, through repeated swallowing and regurgitation. As the fish or gear is expelled orally or passes through the gastrointestinal tract, the line can tighten to the point of strangling a dolphin (Wells et al., 2008). Other ingestion-related injuries, including fish hooks embedded in the throat, goosebeak, or esophagus, typically lead to death.

Dolphins with gear involvement that don't die from drowning in gear or strangulation may suffer in other ways from fishing gear-related injury and trauma. Fishing line or fishing nets can easily cut into the soft skin of a dolphin and may lead to severe infections or necrosis or amputation of fins. Multiple, constrictive wraps of line around fin insertions can lead to amputation, blood loss, impaired mobility, or infection (Fig. 1). Dolphins with such severe entanglements may swim away with the gear, but likely die later. Some dolphins survive amputations of the distal ends of their fins (Wells et al., 2008). Amputations, while not necessarily fatal, may limit power production and movement and/or maneuverability, or may reduce a dolphin's ability to use its tail as a weapon against predators, prey, or conspecifics (Tolley et al., 1995; see Chapter 4: Muscles and movement). Entangled fishing line, crab traps, or parts of fishing nets

FIG. 1 Emaciated yearling bottlenose dolphin calf in 2011 in Sarasota Bay, Florida, with gear from five different entanglement events. Although the gear was removed by rescuers, the calf was no longer with its mother several days later and was presumed to have died. *Photo by the Chicago Zoological Society's Sarasota Dolphin Research Program, taken under National Marine Fisheries Service Scientific Research Permit No. 15543.*

also cause increased drag, which increases the energetic cost of swimming and increases the need to obtain more food (van der Hoop et al., 2014; van der Hoop et al., 2018). This gear often collects algae, which causes significant additional drag and can deepen cuts and prevent healing. If the increased drag is substantial, the reduced power output or maneuverability may affect the foraging efficiency, resulting in starvation, or may interfere with predator avoidance or social interactions. Similarly, increased drag from entangling gear is also a major issue for large whales, which result in severe suffering and pain and is a major animal welfare issue for wild populations of cetaceans (Moore, 2021).

Mitigation possibilities

Mitigation through responding in the moment to entanglements is difficult, as they often occur in the absence of trained personnel who can disentangle live dolphins or they are not noted until after the animal has died. Details of entanglements and bycatch situations are rarely obtained as they are often not seen, or only seen by fishermen, and unless there is an independent observer onboard, these mortalities may not be reported. Dedicated fisheries observers are deployed on a limited number of boats for certain commercial fisheries, which helps improve the accuracy of the estimated number of fisheries interactions, and in some cases, autonomous video systems can also provide a level of monitoring, but the extent of loss of cetaceans as bycatch is largely unknown (Kindt-Larsen et al., 2012; Scheidat et al., 2018). This monitoring helps to provide a better understanding of the reasons for the interactions particularly if necropsies have been possible on the cadavers. Dolphins depredate fish from gillnets and mobile trawls and asphyxiate if they become entangled in the nets. The necropsies show full stomachs, net trauma, and evidence of terminal decompression-like symptoms (Moore et al., 2009, 2013).

Some of the tools that can be employed to reduce net fishery bycatch include collaboration with stakeholders, such as local communities and fishermen, international organizations, and governments to help reduce impact. For example, with the help of fishermen, improved knowledge of the driving forces that cause dolphins to be attracted to fish in nets may yield new tools and/or methods to mitigate and reduce bycatch. Approaches may include avoiding certain fishing modalities in some areas during part of the year when migrating animals are present. Another option may be to develop new fishing gear that helps prevent nontarget species from entanglement or capture. This may include nets or lines that larger species are able to break free from and the use of reflective gear, acoustically enhanced gillnets, or acoustic sources (Larsen et al., 2007). For some species, such as harbor porpoises (*Phocoena phocoena*) (Dähne et al., 2017) and common dolphins (*Delphinus delphis*) (Carretta and Barlow, 2011; Larsen and Eigaard, 2014), sound sources referred to as "pingers" are attached to nets to make them aware of the presence of nets and to help them avoid them. However, for some other species, such as bottlenose dolphins, pingers appear to be less effective as a deterrent, and may even serve as "dinner bells" attracting the dolphins to nets, where depredation may occur (Dawson et al., 2013).

Along the west coast of Florida, where recreational angling and crab traps are the primary fisheries of concern for dolphins, education and outreach, including stakeholder meetings, are used extensively to increase public awareness of the issues. In recent years, the Sarasota Dolphin Research Program has adopted an approach of helping the public to appreciate the dolphins as individually identifiable long-term residents—their mammalian neighbors who live in the bay year-round, across multiple decades and generations—rather than anonymous gray creatures of unknown backgrounds. The hope is that if people can better relate to the animals as individuals, they will care more about their potential impacts on them.

Fishery interactions are, at times, exacerbated by dolphins learning to approach humans, especially anglers, because they have obtained food from humans previously (Cunningham-Smith et al., 2006). Each time a dolphin obtains a fish from a human, it reinforces the behavior of approaching humans, making the risky behavior more likely to occur again. Feeding wild dolphins is illegal in the United States under the federal Marine Mammal Protection Act. To help get the word out about why dolphins should not be fed, a 30-s public service announcement has been developed, available here: https://sarasotadolphin.org/help-dolphins/dont-feed-wild-dolphins/.

While prevention is the most desirable approach for minimizing fishery interactions, entanglements continue to occur, leading to a need for mitigation through direct interventions. Where suitable facilities exist nearby, entangled live-stranded dolphins with sufficiently serious injuries can be taken to rehabilitation facilities for treatment, ideally leading to reintroduction to their original range once they have recovered. Such interventions, when successful, can not only provide population-level conservation benefits in terms of increasing dolphin abundance (McHugh et al., 2021), but can also provide opportunities to increase awareness to the general public and local communities. Research has shown that interventions prior to stranding have a greater probability for at least short-term success, as measured by survival beyond 6 weeks postrelease, presumably because the animal is in better condition (Wells et al., 2013). In some cases, it is possible to remove fishing gear by means of long-handled tools, without having to restrain the dolphin (e.g., Wells et al., 1998). In many parts of the SEUS, trained teams of veterinarians, biologists, and handlers work together to catch entangled individuals via long seine nets in shallow water, remove gear, evaluate the injury and condition of the animal, and either release it on-site or bring it into a rehabilitation facility. Current research is exploring the development of sedation via darting to be able to access entangled dolphins in waters where traditional capture techniques are not feasible (e.g., due to water depth). Recent research has documented long-term success of interventions, with 92% of rescued animals surviving longer than 6 weeks, no decline in survivorship over the first 5 years postrelease, and successful reproduction by rescued females (McHugh et al., 2021).

Although individuals may survive fisheries-related injuries, the physiological impact of these injuries are difficult to assess. However, injuries and amputations may, in addition to severe pain and suffering, result in elevated metabolic cost of travel and diving, reduced foraging efficiency and effectiveness, and reduced ability to avoid boats (see Section "Boat strikes and disturbance"). Depending on other environmental changes or disturbances, this may push injured animals closer to the physiological limitations for survival. Thus, understanding the impact of fisheries interaction injuries on physiology may not only be a welfare issue but may also be important for dolphin conservation.

Boat strikes and disturbance

Issues

Threats to dolphins from vessels are manifested as collisions or disturbances. Direct boat strikes are becoming a major issue of mortality and morbidity for many cetaceans that either live in populated areas or migrate close to shore or through busy shipping lanes. For inshore dolphins, increasing vessel use of shallow waters and seagrass meadows where dolphins cannot dive beneath the vessel makes it more difficult for dolphins to avoid direct strikes (Fig. 2). Along the west coast of Florida, about 4% of mortalities for which cause of death could be determined are ascribed to vessel collisions (G.N. Lovewell, pers. comm.), and about 4% of resident dolphins bear scars from collisions with boats (Wells and Scott, 1997). Dolphins in Sarasota Bay experience a powerboat passing within 100 m once every 6 min during daylight hours (Nowacek et al., 2001). These dolphins often change their behavior as boats approach, including changing whistling patterns, forming a tighter group, diving longer, swimming faster, and changing heading as the vessel gets closer (Nowacek et al., 2001; Buckstaff, 2004).

FIG. 2 Vessels pass within 100 m of each dolphin in Sarasota Bay once every 6 min during daylight hours (Nowacek et al., 2001). *Photo Credit: Photo by Sarasota Dolphin Research Program, taken under NMFS MMPA Scientific Research Permit.*

FIG. 3 Adult male bottlenose dolphin in Sarasota Bay, Florida, in 2013 with wounds from a boat propeller. The dolphin did not survive these injuries. *Photo by the Chicago Zoological Society's Sarasota Dolphin Research Program, taken under National Marine Fisheries Service Scientific Research Permit No. 15543.*

Physiological impacts

Boat strikes include direct collision with the hull or injury from the propeller, which may result in immediate death. Less severe boat strikes may result in broken bones and concussion. The former may result in pain which may affect the ability to swim. Broken ribs may cause puncturing of the pulmonary pleura, causing pneumothorax and collapse of the lung. This results in respiratory complications, which prevent gas exchange. Propeller injury results in open wounds, which may result in inflammation and infection. When a propeller damages the fins or flukes, the injury and resulting fin mutilation may affect the ability to move and result in similar problems as entanglement (Fig. 3).

Most boat injuries near Sarasota, Florida, involve propellers slicing into dorsal fins and are not lethal (Wells et al., 2008). Potential concerns about damage to the dolphin's dorsal fin cooling system (Rommel et al., 1998; see Chapter 3: Thermoregulation) adversely affecting a mother's ability to successfully produce offspring appear to not be well supported, as mothers with mutilated or missing dorsal fins have been observed to successfully give birth and rear calves (Wells et al., 2008).

Observations of dolphin responses to disturbance by vessels led to concerns about the potential energetic impacts of the frequently repeated need to alter their behavior (Nowacek et al., 2001; Buckstaff, 2004). However, recent telemetry studies to assess the energetic impact of boat avoidance have found that the energetic costs of individual occasions of avoiding boats in Sarasota Bay appear to be minimal (Allen, 2021). Although the direct energetic costs of each avoidance event may be low (Allen, 2021; up to 3.7% of total daily metabolic cost), the additional costs to the animals of frequent and repeated disruption of important activities and behaviors such as feeding, mating, or nursing in order to avoid being struck by boats, and associated stress, should not be underestimated. Thus, whether this increase in metabolic cost has a significant impact on survival depends on the metabolic and physiological limits for this species and population. This highlights the need to better understand the physiological limits of dolphins, which will help target conservation efforts. Similarly, vessel noise can interfere with acoustic communication with conspecifics (Buckstaff, 2004), which may affect collaborative foraging and also the ability for finding soniferous prey fish, with concomitant impacts on dolphin energetics (Berens McCabe et al., 2010). In an area such as Sarasota Bay, where there are more than 40,000 vessels registered within the two counties that encompass the home range of the resident dolphin community, the cumulative costs of frequent behavioral changes can add up.

Mitigation possibilities

Since it is unlikely that boat traffic will decline as more people move to coastal areas, what approach is best to reduce risks from vessels to dolphins? In Florida, boat speed zones are an effective means of reducing vessel collisions with manatees (*Trichechus manatus latirostris*) in shallow areas they frequent (e.g., Nowacek et al., 2001), and these benefits likely transfer to dolphins using these waters as well. Slow speed zones may also help reduce boat noise and thereby reduce noise pollution. Expanding slow speed zones to encompass all of the waters frequented by dolphins would be impractical and would be met with much resistance by the large proportion of the boating public that is reluctant to accept existing limited regulations.

Technological solutions have been suggested, such as requiring the installation of noise-making devices on vessels, but it is already well documented that dolphins (and manatees) are able to hear approaching vessels without the need for adding more noise to the environment. What is needed is sufficient time for animals to get out of the way of the approaching vessels. The installation of propeller guards can reduce damage from propellers to dolphins, manatees, and sea turtles, but this comes with a cost of decreased engine efficiency and handling and does little to reduce impact injuries.

Along the west coast of Florida, outreach and education efforts are underway to increase awareness among members of the boating public about the risks of boats to dolphins. Of particular importance is increasing vigilance by boat operators, especially in shallow waters where there is insufficient depth to allow a boat to pass over a dolphin without striking it, and educating boaters about the misconception that dolphins are too nimble to be struck by boats. Among the outreach tools in use are downloadable pocket-sized cards that describe federal guidelines about keeping boats at least 50 yards from dolphins to reduce disturbance, as well as other actions that can improve a boater's experience on the water while minimizing risks to dolphins. For examples and videos see: https://sarasotadolphin.org/wp-content/uploads/2020/11/Dolphin-Friendly-Tips-Card-English-16Oct08.pdf; https://sarasotadolphin.org/videos-and-downloads/. With regards to other boat-based human activities that can cause disturbance of dolphins, in the United States, swimming with wild dolphins is prohibited under the federal Marine Mammal Protection Act.

Pollution

Increasing urbanization and use of the ocean and coastal waters also result in increasing pollution. Pollution covers a broad spectrum of adverse anthropogenic alterations to marine ecosystems. These may come from point sources or nonpoint sources and cause direct or indirect effects for dolphins. For the purposes of this chapter, we will focus on five pollution threats to dolphins: (1) persistent organochlorine pollutants (POPs), (2) oil spills, (3) nutrient loading and harmful algal blooms, (4) freshwater diversions, and (5) noise. Other kinds of pollution, including, for example, heavy metals (Bryan et al., 2007; Woshner et al., 2008), perfluorinated alkyl compounds (Houde et al., 2006), and microplastics and plasticizers (e.g., Hart et al., 2020, 2022) are of additional concern, but concerns and mitigation measures largely mirror those of POPs as presented in this chapter.

Issues

POPs

POPs include a large variety of chemicals, such as the well-known PCBs (polychlorinated biphenyls) used in industrial and commercial applications, and pesticides such as DDTs (dichlorodiphenyltrichloroethane) and their metabolites. Some of these ecotoxicants have demonstrated impacts on health and/or reproduction of humans and other animals such as dolphins, leading to the prohibition of their manufacture and use in most industrialized countries around the world (Tornero et al., 2014). In spite of these prohibitions decades ago, some of these chemicals can still be found in many ecosystems and dolphin tissues, earning them their characterization as "persistent" (Kucklick et al., 2011). PCBs can originate from local industrial sites, such as the Environmental Protection Agency Superfund Site off Brunswick, Georgia (Balmer et al., 2011), or may be transported from long distances through airborne deposition, likely from incineration of materials containing PCBs (Wells et al., 2005). DDT and its metabolites typically originate from local urban or agricultural sources (Tornero et al., 2014). These compounds bioaccumulate in fatty tissues and are biomagnified through food webs, resulting in very high concentrations in species at the highest trophic levels, such as dolphins (Tanabe, 2002).

Oil spills

Oil spill impacts on dolphins vary with the size of the spill, the nature of the oil and dispersant chemicals, and the duration of exposure (Helm et al., 2015). Dolphins can be exposed to crude and weathered oils from spills and seeps, as well as associated dispersant chemicals through direct contact with the skin, eyes, mouth, and blowhole, and they can also inhale volatile petroleum fractions at the water's surface, ingest oil directly, and consume oil components in food. These exposures can initiate a suite of physiological responses that lead to health and/or reproductive impacts. In addition, oil spills can have indirect effects on short- and long-term prey availability. Resident dolphins maintaining limited ranges in bays, sounds, or estuaries can be at particular risk because of their inability to avoid oil (Tornero et al., 2014; Helm et al., 2015; Wells et al., 2017).

Nutrient loading and harmful algal blooms

Even human behavior on land can have direct or indirect impacts on dolphins. One of these is run-off from land that results in high nutrient loads in nearshore waters. Such run-off may result from land clearing, or urbanization where the buried soil gets exposed and nutrients are carried to the ocean by rain. Other kinds of run-off come from golf courses, residential yards, agriculture, or farming, where fertilizer or animal excrement washes into nearby waterways. Similarly, human wastes from sewage system failures or failing septic tanks can increase nutrient loading. This results in increasing nutrient concentrations in the water, such as nitrogen and phosphorous. Changes in nutrient loads are measurable in top-level predators such as bottlenose dolphins. Measurements of nitrogen stable isotope in dolphin teeth tracked changes in nitrogen loads in Sarasota Bay, mirroring improvements in seagrass coverage with implementation of advanced wastewater treatment (Tomasko et al., 2005; Rossman et al., 2013).

Run-off from farm land results in elevated levels of nutrients, called eutrophication. Aquatic organisms, like plankton and algae, can use these nutrients as food to grow and multiply which can lead to algal blooms or dense growth of algae and plankton. As the plankton and algae reproduce, they can block sunlight which causes underlying algae such as seagrass, which are important for filtering the water, to die (Tomasko et al., 2005). The bacteria that help decompose the decaying algae and microorganisms deplete oxygen in the water and thereby cause areas of water with low oxygen levels. These hypoxic or anoxic regions cause fish to move or die, which reduces food resources for predators at higher trophic levels like dolphins.

The excess nutrients can also exacerbate naturally occurring harmful algal blooms, such as *K. brevis* red tides, that occur along the west coast of Florida. When physical and chemical conditions are appropriate, these dinoflagellates are transported from offshore into nearshore waters, and they reproduce rapidly, turning waters reddish-brown. The organisms contain neurotoxins known as brevetoxins (Pierce and Henry, 2008). When the cells lyse, the toxins are released, and in sufficient concentrations, the toxins can kill marine life (fish, sea turtles, seabirds, manatees, dolphins) in large numbers (Fig. 4) (Steidinger et al., 1973; Rycyk et al., 2020).

Freshwater diversions

The bottlenose dolphin is considered a marine species, primarily found in salinities of >8 ppt, and generally >30 ppt (Hornsby et al., 2017). Potentially adverse anthropogenic alterations to marine ecosystems can include waterway management that changes salinity regimes in areas frequented by dolphins, reducing salinity to below physiologically tolerable limits, leading to potential health and ecological impacts (McClain et al., 2020). Such projects have been undertaken, and are planned on a large scale, in the estuarine systems of Louisiana, to try to mitigate long-term wetland loss (Coastal Protection and Restoration Authority of Louisiana, 2017). For example, the diversion of massive quantities of freshwater and associated sediment from the Mississippi River into Barataria Bay as part of the planned Mid-Barataria Sediment Diversion project, a restoration project resulting from the *Deepwater Horizon* oil spill settlement, could greatly reduce salinity in a region inhabited by thousands of bottlenose dolphins (Coastal Protection and Restoration Authority of Louisiana, 2017; Takeshita et al., 2021).

FIG. 4 Fish kill caused by red tide harmful algal bloom, off Sarasota, Florida, in 2021. Fish typically begin to die when *Karenia brevis* cell counts exceed about 100,000 cells/L.

Noise pollution

Noise pollution and acoustic disturbance are of increasing concern in the marine environment, and in particular, for cetaceans due to their reliance on sound for communication, navigation, and finding food. Increasing ocean noise, through increased boat traffic, sonar use or even echosounders, may make it more difficult for cetaceans to hear one another and may make it more difficult to use echolocation to navigate and find and catch prey. It has been reported that in stranded dolphins, as many as 57% of all individuals had significant hearing deficits (Mann et al., 2010), which may play a significant role in strandings. Sound may also cause acute stress, and in deep-diving species, exposure to sonar has been linked to mass-strandings (Frantzis, 1998; Jepson et al., 2003). While most of these reports have been associated with naval sonar and deep-diving beaked whales, there are reports that link sonar to strandings in the common dolphin, a species usually found offshore (Jepson et al., 2005; Bernaldo de Quirós et al., 2019; Fahlman et al., 2021). Closer to shore, noise from boat traffic, industrial activities, and marine construction and demolition may be of greater concern (e.g., Buckstaff, 2004; Buckstaff et al., 2013).

Physiological impacts

POPs

Lipophilic POPs can have a variety of health and reproductive effects on dolphins, especially through immune system suppression (see Chapter 13: Immunology) or through reduced calf survival from transfer via milk (Jepson et al., 2016). Blubber, the primary site of metabolic lipid storage, has been shown to also be the primary POP storage site, containing more than 90% of the whole-body burden (Wells and Scott, 2018). Bottlenose dolphins in Sarasota Bay exhibit seasonal variation in blubber dynamics (Wells, 1993), with blubber thickness increasing by more than 30% and lipid content increasing by more than 50% during fall/winter (Wells, 2010). As lipids mobilize from blubber during spring/summer, contaminants may redistribute, leading to elevated tissue concentrations and toxicity (Yordy et al., 2010a). Health effects in bottlenose dolphins include anemia and endocrine system impacts, including reduced thyroid hormone levels and total thyroxine, free thyroxine, and triiodothyronine negatively correlated with blubber PCB concentrations in bottlenose dolphins in the SEUS (Schwacke et al., 2012). Immune system effects suggesting increased susceptibility to infectious disease have also been noted (see Chapter 13) (Lahvis et al., 1995; Schwacke et al., 2012). Individuals with reduced blubber lipid may be at increased risk for exposure-related health effects (Yordy et al., 2010a). It is interesting to note that Yordy et al. (2010a) provided evidence that the melon, a metabolically inert lipid-rich structure, may serve as an alternate storage site for POPs, preventing some blubber contaminants from being directly available to other tissues.

Reproductive success can also be impacted by POPs through calf mortality (Schwacke et al., 2002). Circulating POPs are incorporated into lipid-rich milk and transferred to calves via nursing (Cockcroft et al., 1989; Wells et al., 2005; Yordy et al., 2010b). The process of depuration involves a female transferring nearly 80% of her residue load to her calf during the first few months of nursing, a period that overlaps the period when lipids and their associated contaminants are being mobilized from blubber to accommodate summer water temperatures (Cockcroft et al., 1989; Wells et al., 2005). First-born calves receive higher concentrations than subsequent calves, as they are receiving the contaminants accumulated over the first 10 years of a mother's life, on average, which may be a toxic load. Subsequent calves, born within just 1–2 years of the initial flushing of mother's POPs into the first-born calf, receive lower concentrations and have higher survival rates (Wells et al., 2005).

Oil spills

Oil pollution may have a number of physiological effects (Helm et al., 2015). In recent years, much has been learned about these effects through studies of dolphins presumed to have been exposed to oil and associated chemicals from the *Deepwater Horizon* spill in 2010 off Louisiana, the worst environmental disaster in history in the United States (NOAA, 2015) (Fig. 5). Schwacke et al. (2017) estimated 30,347 lost cetacean years (95% CI: 11,511–89,746) for dolphins inhabiting Barataria Bay, Louisiana, which was heavily impacted by the spill. Estimated time to recovery for Barataria Bay dolphins was 39 years (95% CI: 24–80). Data from catch-and-release health assessments and population monitoring through photographic identification surveys of oiled sites such as Barataria Bay compared to similar kinds of data from unoiled reference sites have provided a great deal of insight about acute and chronic health and reproductive impacts (Schwacke et al., 2014; Lane et al., 2015; De Guise et al., 2017; Smith et al., 2017).

Oil tends to float at the surface, which may cause direct effects on the skin or respiratory problems (Schwacke et al., 2014; Venn-Watson et al., 2015). Immersion in oil has been shown to cause adrenal lesions and hypoadrenocorticism, which may result in death (Venn-Watson et al., 2015). Dolphins in Barataria Bay, Louisiana, following the *Deepwater*

FIG. 5 Bottlenose dolphins in Chandeleur Sound, Louisiana, swim beneath oil from the *Deepwater Horizon* in 2010. *AP Photo, Alex Brandon.*

Horizon oil spill had a five times higher incidence of moderate-severe lung disease, including alveolar interstitial syndrome, lung masses, and pulmonary consolidation as compared to the reference population in Sarasota Bay, Florida (Schwacke et al., 2009, 2014; Smith et al., 2017). Health assessments done in 2011, 2013, and 2014 on dolphins in Barataria Bay and Mississippi Sound showed that although dolphins in the affected population had improved their overall health in years since, there were still a considerable number of dolphins that suffered from moderate to severe respiratory issues as long as 4 years following the disaster (Smith et al., 2017). Subsequent Barataria Bay health assessments during 2016–2018 found that the prevalence of moderate to severe lung disease did not decrease in the years following the oil spill and in fact may have worsened (Smith et al., 2022).

Impoverished respiratory function affects the ability to dive and forage efficiently. Reduced ability to exchange gases while at the surface may increase the recovery duration at the surface, which in turn reduces the foraging efficiency. Consolidated lungs result in a decreased total lung capacity (see Chapter 6: Respiratory physiology in the dolphin and other whales), which for breath-hold diving vertebrates may reduce the total oxygen store. As the total oxygen store affects the aerobic dive limit, respiratory problems may also affect the dive duration and thereby the foraging efficiency. Reduced respiratory efficiency may also affect the ability for high speed or sustained travel, which in turn may make dolphins less able to escape predators or avoid boat collisions. While the effect of inhalation of aerosolized hydrocarbons is known, the effect of elevated pressure, as happens during a dive, is not known. It is likely that the higher pressure inside the lung that develops during the breath-hold may exacerbate the damage.

Lane et al. (2015) reported on survival and reproductive impacts of the *Deepwater Horizon* oil spill. They found that only 20% of the pregnant dolphins in Barataria Bay produced viable calves, compared to a pregnancy success rate of 83% in the Sarasota Bay, Florida reference population. In Barataria Bay, 57% of unsuccessful pregnant females had been previously diagnosed with moderate-severe lung disease. Overall estimated annual survival rate of the Barataria Bay dolphins was low (86.8%, 95% CI: 80.0%–92.7%) as compared with rates of 95.1% and 96.2% from two unoiled reference populations. In addition, long-term presumed oil-related impacts on immune system function have been found for Barataria Bay dolphins, with trends appearing exaggerated in dolphins born after the spill, suggesting continued exposure to oil or possible multigenerational health impacts (see Chapter 13, De Guise et al., 2021).

Nutrient loading

Nutrient loading can have direct and indirect effects on dolphins. Wastewater associated with some forms of nutrient loading can contain pathogens that adversely impact dolphin health. Indirect, ecological impacts are of great importance to dolphins. Habitat degradation through loss of seagrasses or eutrophication can reduce prey availability. The exacerbation of natural phenomena such as red tides can have dramatic direct and indirect effects on dolphins. Most directly, the brevetoxins can block sodium channels and cause paralysis and death of fish, invertebrates, seabirds, marine turtles, manatees, and dolphins (Twiner et al., 2012). Dolphins can inhale aerosolized brevetoxins or ingest them in prey fish (Fire et al., 2008). Indirect effects of red tides can reduce dolphin prey fish by more than half (Gannon et al., 2009; Berens McCabe et al., 2010; Rycyk et al., 2020). Loss of prey can lead to behavioral changes, such

FIG. 6 Resident bottlenose dolphin of Sarasota Bay, Florida, with fishing gear ingested during a red tide in 2006.

as shifts in habitat use and social patterns (McHugh et al., 2011). Such shifts, along with the need to shift to different, small prey such as clupeids that remain abundant during red tides could have energetic consequences, when additional foraging, capture, and prey handling effort is required to obtain sufficient numbers of smaller fish to obtain the requisite calories (Bejarano et al., 2017). Thus, nutrient loading can have both direct effects on respiratory function from brevetoxins (see Chapter 6) or cause increased metabolic demand to find sufficient food due to shift in prey availability (see Chapter 2: Energetic costs of rest and locomotion in dolphins). In addition, as mentioned above, when dolphins interact and compete with anglers for the few remaining desirable fish during or after a red tide, dolphins can be seriously injured or die from ingestion of, hooking by, or entanglement in fishing gear (Fig. 6).

Freshwater diversions

Prolonged (days to weeks) immersion in very low salinity waters (~0 ppt) can lead to health impacts through direct contact with the skin and through ingestion. The ability of skin to act as an effective barrier will decline as skin lesions develop, followed by algal, fungal, and/or bacterial overgrowth that may sometimes penetrate the skin, followed in turn by secondary infections (McClain et al., 2020; Takeshita et al., 2021). Consumption of too much freshwater can affect absorption and increase the potential for infections (see Chapter 11: Kidneys and osmoregulation). As suggested by Takeshita et al. (2021), these two routes of exposure may lead to systemic physiological changes, including osmotic imbalance, biochemical aberrations, cellular damage, and the potential for localized or systemic secondary infections or septicemia. In 2019, an Unusual Mortality Event was declared by NOAA in the northern Gulf of Mexico in response to dramatically increased strandings of dolphins with a high prevalence of freshwater-like lesions, following record-breaking precipitation (NOAA, 2019). On occasion, it has been possible to document resolution of health issues when an individual is returned to waters of normal salinity (e.g., Deming et al., 2020). Much remains to be learned about the conditions leading to effects from freshwater exposure. For example, following historic flooding in the Florida panhandle in 2014 that led to lowered salinities over several months, no significant increase in mortality rates was noted, and an anticipated widespread skin lesion outbreak did not occur (Toms et al., 2021). In addition to the potential health consequences of reduced salinity, prey availability is likely to change as well as fish move or die in response to the changing environmental conditions.

Noise pollution

In both water and air, sound is produced by propagation of acoustic waves. Sound waves can travel through any substance, but they travel faster in dense media, e.g., five times faster in water as compared with air. Sound also travels farther in water as compared with air, although the distance traveled is dependent on the frequency of the sound, water depth, salinity, presence of submerged vegetation, physiography, and other features. These physical characteristics of sound affect both dolphins and other cetaceans that use sound to communicate over long distances and echolocation to find and catch prey. Masking from increasing noise is likely to make it more difficult for cetaceans to hear each other or may interfere with the ability to find prey (see Chapter 10: Sensory physiology in delphinids, Buckstaff, 2004). Thus, hearing deficiency may reduce the ability to capture food, interact with conspecifics, avoid predators or boat traffic, or the ability to navigate and may in part explain why a high proportion of stranded dolphins had hearing deficiency (Mann et al., 2010). The only Sarasota Bay resident subject of Mann et al. (2010) was a young female that had separated prematurely from her mother. She had deep cuts in her peduncle from a boat propeller strike, fishing line embedded in the cuts in the peduncle, and high frequency hearing loss; this individual is not believed to have survived postrelease following rehabilitation (Wells et al., 2013). In part to address the question of whether repeated, frequent exposure to noise from intensive small boat traffic has affected the hearing abilities of inshore bottlenose dolphins, auditory evoked potential tests were conducted with Sarasota Bay resident dolphins during health assessments (Wells et al., 2004). No significant hearing loss was found for any of the long-term resident dolphins (Cook, 2006), but it should be noted that these measures were from living, relatively healthy dolphins inhabiting shallow waters, where they are not exposed to noise from large vessels, military activities, or large-scale industrial activities; hearing loss may be an issue in areas with greater exposure to loud noise.

The effect of man-made sonar has received considerable attention and been a matter of intense debate since initial reports that linked naval sonar to mass-strandings of deep-diving beaked whales (Jepson et al., 2003; Fernandez et al., 2005; Bernaldo de Quirós et al., 2019). A considerable international effort followed these initial reports, and as of 2021, it is generally accepted that exposure to powerful sonar results in behavioral changes (e.g., DeRuiter et al., 2013). It has been proposed that both behavioral and physiological changes occur following exposure to powerful sonar that may result in the formation of symptomatic gas emboli (Fahlman et al., 2021). Recent work in the bottlenose dolphin, which has been used as an animal model to understand the physiology of deep-diving cetaceans, suggests that these species have a mechanism that helps regulate gas exchange (see Chapter 6 for the *Selective Gas Exchange* hypothesis) and prevents uptake of N_2. When this mechanism fails, as during potential stress following sonar exposure, more N_2 is taken up, which leads to development of gas emboli that results in severe symptoms (see Figs. 2 and 4 in Fahlman et al., 2021). This hypothesis, if correct, provides a simple means for mitigation under some circumstances as the stress may be minimized during training or testing exercises by desensitizing cetaceans to the sonar. One method could be to ramp up the sonar source level giving the whales time to leave the area without stress and/or making sure that the exposure duration of each event is short enough for the whales to recover in between exposures. More work is needed on the potential effects of noise on stress in dolphins.

Mitigation possibilities

Pollution mitigation can involve cessation of polluting activities, clean-up, and/or restoration actions; however, in many cases, this may not be possible or practical. For example, production and use of some of the most toxic POPs were halted decades ago in the United States, but concentrations of concern for dolphin health and reproduction are still found at many sites.

POPs remain highly concentrated in some areas where they were produced or released, such as Brunswick, Georgia, leading to concentrations in dolphin tissues that vary inversely with distance of the dolphin ranges from the point source (Balmer et al., 2011). Clean-up of the Environmental Protection Agency Superfund Site near Brunswick has been proposed, with a goal of reducing contaminant concentrations in dolphins. Widespread nonpoint-source deposition of toxic POPs in many other parts of the world precludes the possibility of such clean-up. In these cases, only the passage of time can reduce availability of these contaminants. For example, in Sarasota Bay, Florida, the concentrations of legacy POPs in bottlenose dolphin blubber declined from 2000 through 2009 but leveled off through 2016 at concentrations that may continue to suppress population growth (Kucklick et al., 2022). Elimination of legacy POPs as environmental threats will not happen soon. Hopefully, lessons learned from these legacy pollutants will reduce future threats from development and application of toxic chemicals.

Restoration following oil spills can be very challenging, depending on the size of the area impacted and the severity of the spill. Efforts to restore marine mammals impacted by the *Deepwater Horizon* oil spill are currently in early stages (Deepwater Horizon Natural Resource Damage Assessment Trustees, 2016). While marine mammal populations cannot be restored directly, efforts to provide optimal habitat for population recovery and reduction of other human impacts can contribute to restoration, but much time will be required given the life history of these animals.

Mitigation of nutrient pollution involves preventing nutrients from entering the waterways in the first place. Mechanisms exist that can help control these sources of pollution, including reduced application of fertilizers, advanced wastewater treatment, elimination of septic tanks, and increased use of retention areas that trap nutrients and prevent them from reaching the waterways. As excess nutrients are removed, natural processes such as filtration by seagrass meadows can expedite the process (Tomasko et al., 2005; Rossman et al., 2013). As water clarity improves and sunlight can penetrate deeper, seagrass meadows can expand into deeper waters, providing a positive feedback loop for improving water quality. However, the implementation and effectiveness of these approaches depend on public support and political will, which can sometimes be difficult to mobilize in coastal communities facing a multitude of needs and priorities.

As is the case with nutrient pollution, the best method for mitigating the problems for dolphins associated with exposure to too much freshwater is to not engage in diversions of such magnitude as to make dolphin habitats uninhabitable. However, in Louisiana, where such diversions are occurring and planned, these competing interests appear to have taken precedence over dolphin restoration. Efforts are moving forward to re-establish wetlands lost to previous engineering projects, storms, and sea-level rise, so that they can provide a protective buffer for future storms and sea-level changes, as well as habitat for other forms of wildlife.

In some cases, it may be possible to reduce anthropogenic noise in the coastal marine environment. Underwater construction projects can sometimes use noise barriers, such as steel coffer cells, or bubble curtains to reduce noise (e.g., Buckstaff et al., 2013; Dähne et al., 2017). In some cases, dolphins have demonstrated the ability to adapt their behavior to at least some acoustic changes in their environment. For example, in Sarasota Bay, dolphins did not entirely avoid an area of bridge construction, but dolphin density increased significantly following completion (Buckstaff et al., 2013). Near John's Pass, Florida, female presence declined during a 5-yr bridge construction project, while males showed no response (Weaver, 2015). Although these reports suggest behavioral changes, it is unclear whether there were physiological effects for those individuals that remained in the area.

What may happen if the noise level in the ocean is increased permanently (e.g., the EU Marine Strategy Framework Directive; https://ec.europa.eu/environment/marine/good-environmental-status/descriptor-11/index_en.htm)? One option may be for the animals to increase their level of sound emission. This is similar to what happens in a noisy restaurant where it becomes increasingly difficult to have a conversation as the restaurant becomes increasingly noisy as the evening is progressing. The solution is to increase the source level of your voice so your friends can hear you, but such a compensation only works for a limited noise range and may alter the metabolic cost of communication. However, recent work has shown that sound production is metabolically cheap and communicating louder may not result in a significant metabolic burden (Pedersen et al., 2020). In a confined terrestrial context, disturbed animals can relocate to quieter areas, but because sound transmits well in the ocean, there are fewer options for dolphins to hide from increasing noise levels. Every effort should be made to reduce noise in the marine environment, to not let it get to the point where it results in dolphin hearing loss or behavioral disturbance or forces potentially detrimental changes in habitat use (Fig. 7).

FIG. 7 Demolition of previous bridge supports during construction of new bridge in Sarasota Bay in 2003 (Buckstaff et al., 2013).

Climate change

Climate change is likely to have multiple effects on dolphin populations. As greenhouse gases trap more heat, the average temperature of the earth is increasing. Since water has more than four times higher specific heat capacity than air, the ocean acts as a heat sink and moderates the increase in air temperatures. Thus, with time, the water temperature of the surface layers will begin to increase, and this increase will penetrate deeper into the water column. This will affect physiology and could possibly affect thermoregulation (see Chapter 3). Increased water temperatures, in conjunction with increased availability of nutrients from human activities, are likely to increase the occurrence and strength of harmful algal blooms, including red tides, which may have direct health effects and also indirectly reduce food availability, as described above. Increasing water temperatures are also likely to have an impact both directly and indirectly on dolphins and other cetaceans (see Chapter 3). As the temperature of the surface layers increase, currents change, water level rises, and pH decreases, the distribution of prey is likely to change. Temperature-sensitive prey species can either move further down into the water column (not possible in shallow inshore habitats) or move to higher latitudes.

Physiological impacts

Dolphins, like all mammals, maintain a stable core body temperature. Within a given temperature range, called the thermoneutral zone, there is no increase in the metabolic rate. However, temperatures below or above the thermoneutral zone cause an increase in metabolic rate (see Chapters 2 and 3). The increased metabolic rate is caused by physiological processes that help produce additional heat at lower temperatures or increase blood flow to help dump or remove heat at higher temperatures. Both responses help to maintain a stable core temperature. The thermoneutral zone of inactive dolphins accustomed to being housed in human care in ocean pens ranges from 5°C to 30°C (Williams et al., 2001; Yeates and Houser, 2008). As dolphins inhabit warm and temperate regions worldwide, it is likely that the thermoneutral range may vary among populations and also among nearshore, shallow-diving or offshore, deep-diving populations. Still, increased activity during active swimming and prey capture events will increase heat production. In such a case, dolphins may have lower exercise tolerance before overheating. Bottlenose dolphins in Sarasota Bay, Florida, have been found to have significantly higher field metabolic rates in summer than in winter (Costa et al., 2013). During summer, inshore water temperatures can exceed 30°C and approach body temperature, thus limiting the ability of the animals to cool through heat conduction to surrounding waters. Already, dolphin mortality rates are significantly higher during warm seasons than cool seasons, when water-borne pathogens are more prevalent (Buck et al., 2006), but the potential role of water temperature in these mortalities has not been defined (Wells, 2010).

Increasing water temperatures may result in dolphins having thinner blubber layers. As adipose tissue is also a store of energy and water, dolphins in warmer waters may be less capable of handling food shortages and may be unable to fast for long durations. Thinner blubber may also lead to higher circulating concentrations of environmental contaminants that might otherwise be sequestered in blubber (Wells, 2010). Thinner blubber also means a thinner protective layer to be penetrated by predator bites before reaching more vital tissues (Wells, 1993).

Changes in water temperature may not only have direct thermoregulatory effects on dolphins but could affect their prey species. The vertebrate and invertebrate prey of dolphins are ectotherms (they do not regulate their internal body temperature) and so are directly affected by temperature. Most of these are able to survive in a large temperature range but often have a preferred temperature where they live. As water temperatures increase, it is likely that at least some prey species will move, and resident dolphins will need to change their diet. It is possible that some new prey may benefit the dolphins, but changes in diet will likely come with a cost. Bottlenose dolphins eat a wide variety of fish, but they disproportionately select some species because of their sound production characteristics, and they develop specific foraging techniques for different species (Wells, 2019). If new prey have a lower energetic yield, then maintaining an energetic balance may be challenging. The overall effect of changing prey availability is that the combination of energetic cost and the potential need to learn and perfect new foraging and prey capture and/or handling techniques is likely to reduce the foraging efficiency. Thus, to catch the same amount of food, or more as the cost of catching food is increasing, the dolphins may have to spend more time searching and hunting for prey. While the proportion of time spent foraging is likely flexible and can vary seasonally, annually and with prey species, an increasing proportion of time spent in this activity will reduce the time available for other activities that may be important for learning and creating and maintaining social bonds. Eventually, if the metabolic cost of foraging exceeds the energy gain, this might lead to a negative energy balance with the possibility that the dolphin will lose body mass and eventually starve to death if conditions do not change.

Alternatively, if prey species migrate to higher latitudes, the dolphins may follow their familiar prey, which increases the cost of transport and introduces a variety of stressors from changes in habitat, predators, and other ecological parameters. Such movements at the extreme of the species' range have been demonstrated, as identifiable individual bottlenose dolphins in southern California were documented to shift their range more than 600 km to the north during an El Niño warm water event (Wells et al., 1990). However, many inshore bottlenose dolphins away from the extremes of the species range live in year-round, multidecadal, multigenerational community ranges, and they may not leave these ranges easily (Wells and Scott, 2018). Members of some of these communities have not left their long-term ranges in spite of catastrophic environmental changes, from oil spills, major hurricanes, or severe harmful algal blooms that reduced prey by more than 75%—these animals are considered to live in "ecological cul-de-sacs" and whether they will be able to adapt to changing conditions if they refuse to leave their ranges is of concern (Wells, 2010; Tornero et al., 2014). In addition, the impacts from their moving into established ranges of other communities must also be considered.

Whether offshore bottlenose dolphins are strongly bound to a specific home range is not well documented. In Bermuda, for example, re-sightings of individual animals have been reported for several years and it is likely that these populations show site fidelity. For these deep-diving populations, climate change may result in deeper dives to access prey, which result in reduced foraging efficiency as a greater proportion of the dive duration is part of the transport. The greater effort may reduce the aerobic dive limit as increased cost of swimming also elevates the metabolic rate. It is also likely that the elevated metabolic rate increases hypoxia, which may result in molecular changes (see Chapter 8: Genetic and molecular adaptations) and have consequences for immune function (see Chapter 13). Regardless, the metabolic consequences of moving or not moving may result in lower foraging efficiency. Therefore, understanding the metabolic requirements of dolphins is important, especially with regards to how variation in prey alters the metabolic cost for different activities (see Chapter 2). In addition, greater time spent at depth may also increase uptake of N_2 and cause greater risk of formation of gas emboli (Fahlman et al., 2021, Fig. 8).

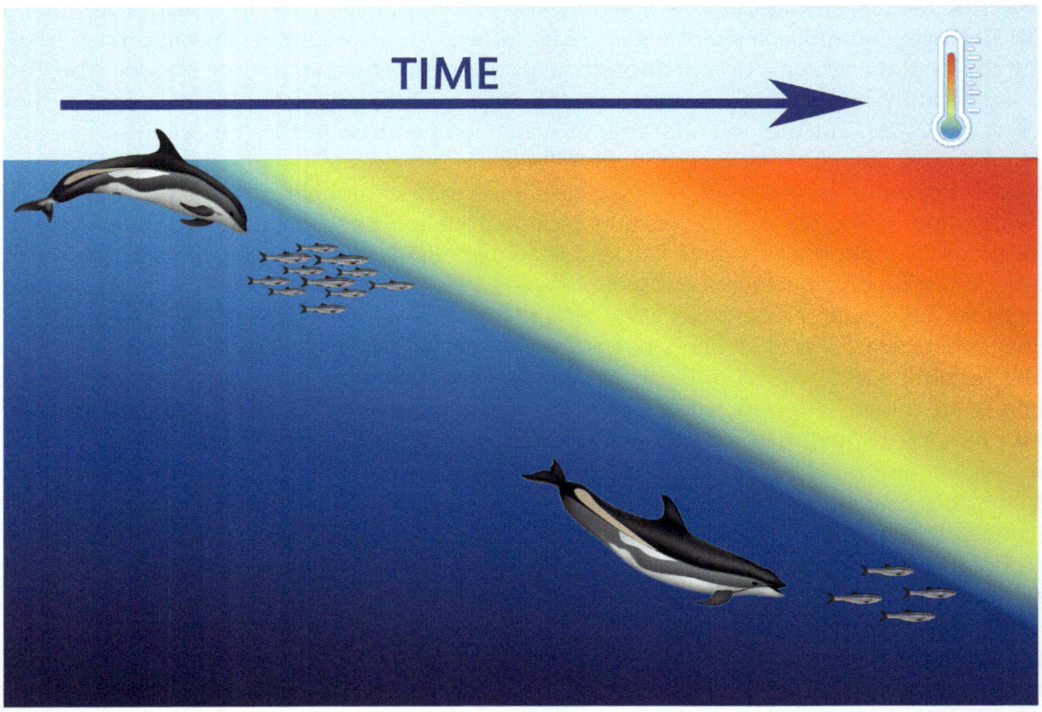

FIG. 8 Potential effect of climate change where water temperature increases with time (x-axis from left to right), pushing prey deeper in the water column (y-axis increasing depth from top to bottom) with increasing pressure. Dolphins then have to spend more time and energy to get to the prey, which will reduce the total foraging time and efficiency. Colors indicate changes in water temperature, see thermometer. *Illustration by Uko Gorter.*

Mitigation possibilities

The only effective mitigation option for climate disruption is the cessation of global warming from anthropogenic sources. However, this will require international political (e.g., https://www.ipcc.ch) and economic will that is well beyond what has occurred so far. Increased levels of education and outreach and acceptance of the now-well-established facts surrounding climate change will be necessary to bring about the large-scale changes in attitudes and actions that are required.

Concurrent and multiple threats

Although the community of resident dolphins in Sarasota Bay has been increasing at a modest average annual rate of 2.1% since 1993 (Lacy et al., 2021), cumulative human impacts may reach a point when compensatory mechanisms may not suffice, which may have detrimental impact on the population. Dolphins in inshore waters are usually not given the option of dealing with one threat at a time. They face multiple concurrent and cumulative threats from natural and anthropogenic sources, and many of these threats have physiological consequences (Tyack et al., 2022). It is up to humans to reduce the threats that are within their power to control, to give the dolphins more degrees of freedom to cope with and adapt to the remaining threats. Inshore bottlenose dolphin populations are bellwethers of coastal ecosystem health. The animals swim through the same waters, breathe the same air, and catch and eat the same fish as their human neighbors in these coastal areas. It is in both the humans' and dolphins' best interest to care for this shared environment.

References

Allen, A.S., 2021. Estimating the Cost of Locomotion in Common Bottlenose Dolphins: Calibration, Validation, and Application to Study the Impacts of Disturbance, PhD thesis. Duke University Department of Marine Science and Conservation, 143 pp.

Balmer, B.C., Schwacke, L.H., Wells, R.S., George, R.C., Hoguet, J., Kucklick, J.R., Lane, S.M., Martinez, A., McLellan, W.A., Rosel, P.E., Rowles, T.K., Sparks, K., Speakman, T., Zolman, E.S., Pabst, D.A., 2011. Relationship between persistent organic pollutants (POPS) and ranging patterns in common bottlenose dolphins (*Tursiops truncatus*) from coastal Georgia, USA. Sci. Total Environ. 409, 2094–2101.

Bejarano, A.C., Wells, R.S., Costa, D.P., 2017. Development of a bioenergetic model for estimating energy requirements and prey biomass consumption of the bottlenose dolphin *Tursiops truncatus*. Ecol. Model. 356, 162–172.

Berens McCabe, E.J., Gannon, D.P., Barros, N.B., Wells, R.S., 2010. Prey selection by resident common bottlenose dolphins (*Tursiops truncatus*) in Sarasota Bay, Florida. Mar. Biol. 157, 931–942.

Bernaldo De Quirós, Y., Fernandez, A., Baird, R.W., Brownell, R.L., Aguilar De Soto, N., Allen, D., Arbelo, M., Arregui, M., Costidis, C., Fahlman, A., Frantzis, A., Gulland, F.M.D., Iñíguez, M., Johnson, M., Komnenou, A., Koopman, H., Pabst, D.A., Roe, W.D., Sierra, E., Tejedor, M., Schorr, G., 2019. Advances in research on the impacts of anti-submarine sonar on beaked whales. Proc. R. Soc. Lond. B 286, 20182533.

Bryan, C.E., Christopher, S.J., Balmer, B.C., Wells, R.S., 2007. Establishing baseline levels of trace elements in blood and skin of bottlenose dolphins in Sarasota Bay, Florida: implications for non-invasive monitoring. Sci. Total Environ. 388, 325–342.

Buck, J.D., Wells, R.S., Rhinehart, H.L., Hansen, L.J., 2006. Aerobic microorganisms associated with free-ranging bottlenose dolphins in coastal Gulf of Mexico and Atlantic Ocean waters. J. Wildl. Dis. 42, 536–544.

Buckstaff, K.C., 2004. Effects of watercraft noise on the acoustic behavior of bottlenose dolphins, *Tursiops truncatus*, in Sarasota Bay, Florida. Mar. Mamm. Sci. 20, 709–725.

Buckstaff, K.C., Wells, R.S., Gannon, J.G., Nowacek, D.P., 2013. Responses of bottlenose dolphins (*Tursiops truncatus*) to construction and demolition of coastal marine structures. Aquat. Mamm. 39, 174–186.

Carretta, J.V., Barlow, J., 2011. Long-term effectiveness, failure rates, and "dinner bell" properties of acoustic pingers in a gillnet fishery. Mar. Technol. Soc. J. 13, 7–19.

Christiansen, F., McHugh, K.A., Bejder, L., Siegal, E.M., Lusseau, D., Berens McCabe, E., Lovewell, G., Wells, R.S., 2016. Food provisioning increases the risk of injury in a long-lived marine top predator. Royal Soc. Open Sci. 3, 160560.

Coastal Protection and Restoration Authority of Louisiana, 2017. Louisiana's 2017 Comprehensive Master Plan for a Sustainable Coast. State of Louisiana.

Cockcroft, V.G., De Kock, A.C., Lord, D.A., Ross, G.J.B., 1989. Organochlorines in bottlenose dolphins *Tursiops truncatus* from the east coast of South Africa. S. Afr. J. Mar. Sci. 8, 207–217.

Cook, M.L.H., 2006. Behavioral and Auditory Evoked Potential (AEP) Hearing Measurements in Odontocete Cetaceans. PhD thesis, University of South Florida, 124 pp.

Costa, D.P., Worthy, G.A., Wells, R.S., Read, A.J., Waples, D., Scott, M.D., 2013. Patterns of seasonal metabolic rate variation for bottlenose dolphins in Sarasota Bay, Florida. In: 20th Biennial Conference on the Biology of Marine Mammals, 9–13 December 2013 Dunedin, New Zealand.

Cunningham-Smith, P., Colbert, D.E., Wells, R.S., Speakman, T., 2006. Evaluation of human interactions with a provisioned wild bottlenose dolphin (*Tursiops truncatus*) near Sarasota Bay, Florida, and efforts to curtail the interactions. Aquat. Mamm. 32, 346–356.

Dähne, M., Tougaard, J., Carstensen, J., Rose, A., Nabe-Nielsen, J., 2017. Bubble curtains attenuate noise from offshore wind farm construction and reduce temporary habitat loss for harbour porpoises. Mar. Ecol. Prog. Ser. 580, 221–237.

Dawson, S.M., Northridge, S., Waples, D., Read, A.J., 2013. To ping or not to ping: the use of active acoustic devices in mitigating interactions between small cetaceans and gillnet fisheries. Endanger. Species Res. 19, 201–221.

De Guise, S., Levin, M., Gebhard, E., Jasperse, L., Burdett Hart, L., Smith, C.R., Venn-Watson, S., Townsend, F., Wells, R., Balmer, B., Zolman, E., Rowles, T., Schwacke, L., 2017. Changes in immune functions in bottlenose dolphins in the northern Gulf of Mexico associated with the Deepwater Horizon oil spill. Endanger. Species Res. 33, 291–303.

De Guise, S., Levin, M., Jasperse, L., Herrman, J., Wells, R.S., Rowles, T., Schwacke, L., 2021. Long-term immunological alterations in bottlenose dolphin a decade after the Deepwater Horizon oil spill in the northern Gulf of Mexico: potential for multigenerational effects. Environ. Toxicol. Chem. 40, 1308–1321.

Deepwater Horizon Natural Resource Damage Assessment Trustees, 2016. Deepwater Horizon oil spill: Final Programmatic Damage Assessment and Restoration Plan and Final Programmatic Environmental Impact Statement. http://www.gulfspillrestoration.noaa.gov/restoration-planning/gulf-plan.

Deming, A.C., Wingers, N.L., Moore, D.P., Rotstein, D., Wells, R.S., Ewing, R., Hodanbosi, M.R., Carmichael, R.H., 2020. Health impacts and recovery from prolonged freshwater exposure in a common bottlenose dolphin (*Tursiops truncatus*). Front. Vet. Sci. 7, 235.

Deruiter, S.L., Southall, B.L., Calambokidis, J., Zimmer, W.M.X., Sadykova, D., Falcone, E.A., Friedlaender, A.S., Joseph, J.E., Moretti, D., Schorr, G.S., Thomas, L., Tyack, P.L., 2013. First direct measurements of behavioural responses by Cuvier's beaked whales to mid-frequency active sonar. Biol. Lett. 9, 1–5.

Fahlman, A., Moore, M.J., Wells, R.S., 2021. How do marine mammals manage and usually avoid gas emboli formation and gas embolic pathology? Critical clues from studies of wild dolphins. Front. Mar. Sci. 8, 598633.

Fernandez, A., Edwards, J.F., Rodruiquez, F., De Los, E., Monteros, A., Herraez, M.P., Castro, P., Jaber, J.R., Martin, V., Arbelo, M., 2005. "Gas and fat embolic syndrome" involving a mass stranding of beaked whales (family *Ziphiidae*) exposed to anthropogenic sonar signals. Vet. Pathol. 42, 446–457.

Fire, S.E., Flewelling, L.J., Naar, J., Twiner, M.J., Henry, M.S., Pierce, R.H., Gannon, D.P., Wang, Z., Davidson, L., Wells, R.S., 2008. Prevalence of brevetoxins in prey fish of bottlenose dolphins in Sarasota Bay, Florida. Mar. Ecol. Prog. Ser. 368, 283–294.

Frantzis, A., 1998. Does acoustic testing strand whales? Nature 392, 29.

Gannon, D.P., Berens, E.J., Camilleri, S.A., Gannon, J.G., Brueggen, M.K., Barleycorn, A., Palubok, V., Kirkpatrick, G.J., Wells, R.S., 2009. Effects of *Karenia brevis* harmful algal blooms on nearshore fish communities in southwest Florida. Mar. Ecol. Prog. Ser. 378, 171–186.

Hamer, D., Childerhouse, S., Gales, N., 2012. Odontocete bycatch and depredation in longline fisheries: a review of available literature and of potential solutions. Mar. Mamm. Sci. 28, E345–E374.

Hart, L.B., Dziobak, M.K., Pisarski, E.C., Wirth, E.F., Wells, R.S., 2020. Sentinels of synthetics—a comparison of phthalate exposure between common bottlenose dolphins (*Tursiops truncatus*) and human reference populations. PLoS One 15, e0240506.

Hart, L.B., Dziobak, M., Wells, R.S., Ertel, B., Weinstein, J., 2022. Microplastics in gastric samples from common bottlenose dolphins (*Tursiops truncatus*) residing in Sarasota Bay, FL (USA). Front. Mar. Sci. 9, 947124.

Helm, R.C., Costa, D.P., Debruyn, T.D., O'Shea, T.J., Wells, R.S., Williams, T.M., 2015. Overview of effects of oil spills on marine mammals. In: Fingas, M.F. (Ed.), Handbook of Oil Spill Science and Technology, first ed. John Wiley & Sons, Inc, Hoboken, NJ, pp. 455–475.

Hornsby, F.E., McDonald, T.L., Balmer, B.C., Speakman, T.R., Mullin, K.D., Rosel, P.E., Wells, R.S., Telander, A.C., Marcy, P.W., Klaphake, K.C., Schwacke, L.H., 2017. Using salinity to identify common bottlenose dolphin habitat in Barataria Bay, Louisiana, USA. Endanger. Species Res. 33, 181–192.

Houde, M., Balmer, B.C., Brandsma, S., Wells, R.S., Rowles, T.K., Solomon, K.R., Muir, D.C., 2006. Perfluoroalkyl compounds in relation to life-history and reproductive parameters in bottlenose dolphins (*Tursiops truncatus*) from Sarasota Bay, Florida, U.S.A. Environ. Toxicol. Chem. 25, 2405–2412.

Jepson, P.D., Arbelo, M., Deaville, R., Patterson, I.A.P., Castro, P., Baker, J.R., Pocknell, A.M., Rodruiquez, F., Howie, F.E., Espinosa, A., Reid, R.J., Jaber, J.R., Martin, V., Cunningham, A.A., Fernandez, A., 2003. Gas-bubble lesions in stranded cetaceans. Nature 425, 575–576.

Jepson, P.D., Deaville, R., Patterson, I.A.P., Pocknell, A.M., Ross, H.M., Baker, J.R., Howie, F.E., Reid, R.J., Colloff, A., Cunningham, A.A., 2005. Acute and chronic gas bubble lesions in cetaceans stranded in the United Kingdom. Vet. Pathol. 42, 291–305.

Jepson, P.D., Deaville, R., Barber, J.L., Aguilar, À., Borrell, A., Murphy, S., Barry, J., Brownlow, A., Barnett, J., Berrow, S., Cunningham, A.A., Davison, N.J., Ten Doeschate, M., Esteban, R., Ferreira, M., Foote, A.D., Genov, T., Giménez, J., Loveridge, J., Llavona, Á., Martin, V., Maxwell, D.-L., Papachlimitzou, A., Penrose, R., Perkins, M.W., Smith, B., De Stephanis, R., Tregenza, N., Verborgh, P., Fernandez, A., Law, R.J., 2016. PCB pollution continues to impact populations of orcas and other dolphins in European waters. Sci. Rep. 6, 18573.

Kindt-Larsen, L., Dalskov, J., Stage, B., Larsen, F., 2012. Observing incidental harbour porpoise *Phocoena phocoena* bycatch by remote electronic monitoring. Endanger. Species Res. 19, 75–83.

Kucklick, J., Schwacke, L., Wells, R., Hohn, A., Guichard, A., Yordy, J., Hansen, L., Zolman, E., Wilson, R., Litz, J., Nowacek, D., Rowles, T., Pugh, R., Balmer, B., Sinclair, C., Rosel, P., 2011. Bottlenose dolphins as indicators of persistent organic pollutants in the Western North Atlantic Ocean and Northern Gulf of Mexico. Environ. Sci. Technol. 45, 4270–4277.

Kucklick, J., Boggs, A., Huncik, K., Moors, A., Davis, E., Ylitalo, G., McConnell, M., Makris, C., Wells, R.S., 2022. Temporal trends of persistent organic pollutants in Sarasota Bay common bottlenose dolphins (*Tursiops truncatus*). Front. Mar. Sci. 9 (763918).

Lacy, R.C., Wells, R.S., Scott, M.D., Allen, J.B., Barleycorn, A.A., Urian, K.W., Hofmann, S., 2021. Assessing the viability of the Sarasota Bay community of bottlenose dolphins. Front. Mar. Sci. 8 (788086).

Lahvis, G.P., Wells, R.S., Kuehl, D.W., Stewart, J.L., Rhinehart, H.L., Via, C.S., 1995. Decreased lymphocyte responses in free-ranging bottlenose dolphins (*Tursiops truncatus*) are associated with increased concentrations of PCBs and DDT in peripheral blood. Environ. Health Perspect. 103 (Suppl 4), 67–72.

Lane, S.M., Smith, C.R., Mitchell, J., Balmer, B.C., Barry, K.P., McDonald, T., Mori, C.S., Rosel, P.E., Rowles, T.K., Speakman, T.R., Townsend, F.I., Tumlin, M.C., Wells, R.S., Zolman, E.S., Schwacke, L.H., 2015. Reproductive outcome and survival of common bottlenose dolphins sampled in Barataria Bay, Louisiana, USA, following the Deepwater Horizon oil spill. Proc. R. Soc. Lond. B 282, 20151944.

Larsen, F., Eigaard, O.R., 2014. Acoustic alarms reduce bycatch of harbour porpoises in Danish North Sea gillnet fisheries. Fish. Res. 153, 108–112.

Larsen, F., Eigaard, O.R., Tougaard, J., 2007. Reduction of harbour porpoise (*Phocoena phocoena*) bycatch by iron-oxide gillnets. Fish. Res. 85, 270–278.

Mann, D., Hill-Cook, M., Manire, C., Greenhow, D., Montie, E., Powell, J., Wells, R., Bauer, G., Cunningham-Smith, P., Lingenfelser, R., DiGiovanni Jr., R., Stone, A., Brodsky, M., Stevens, R., Kieffer, G., Hoetjes, P., 2010. Hearing loss in stranded odontocete dolphins and whales. PLoS One 5, e13824.

McClain, A.M., Daniels, R., Gomez, F.M., Ridgway, S.H., Takeshita, R., Jensen, E.D., Smith, C.R., 2020. Physiological effects of low salinity exposure on bottlenose dolphins (*Tursiops truncatus*). J. Zool. Bot. Gard. 1, 61–75.

McHugh, K.A., Allen, J.B., Barleycorn, A.A., Wells, R.S., 2011. Severe harmful algal bloom events influence juvenile common bottlenose dolphin behavior and sociality in Sarasota Bay, Florida. Mar. Mamm. Sci. 27, 622–643.

McHugh, K.A., Barleycorn, A.A., Allen, J.B., Bassos-Hull, K., Lovewell, G., Boyd, D., Panike, A., Cush, C., Fauquier, D., Mase, B., Lacy, R.C., Greenfield, M.R., Rubenstein, D.I., Weaver, A., Stone, A., Oliver, L., Morse, K., Wells, R.S., 2021. Staying alive: long-term success of bottlenose dolphin interventions in Southwest Florida. Front. Mar. Sci. 7, 624729.

Moore, M.J., 2021. We Are All Whalers: The Plight of Whales and Our Responsibility. The University of Chicago Press, Chicago.

Moore, M.J., Bogomolni, A.L., Dennison, S.E., Early, G., Garner, M.M., Hayward, B.A., Lentell, B.J., Rotstein, D.S., 2009. Gas bubbles in seals, dolphins, and porpoises entangled and drowned at depth in gillnets. Vet. Pathol. 46, 536–547.

Moore, M.J., Van Der Hoop, J., Barco, S.G., Costidis, A.M., Gulland, F.M., Jepson, P.D., Moore, K.T., Raverty, S., McLellan, W.A., 2013. Criteria and case definitions for serious injury and death of pinnipeds and cetaceans caused by anthropogenic trauma. Dis. Aquat. Org. 103, 229–264.

NOAA, 2015. Injury to Natural Resources. Deepwater Horizon Oil Spill: Draft Programmatic Damage Assessment and Restoration Plan and Draft Programmatic Environmental Impact Statement. National Oceanic and Atmospheric Administration.

NOAA, 2019. Bottlenose Dolphin Unusual Mortality Event along the Northern Gulf of Mexico. National Oceanic and Atmospheric Administration.

Nowacek, S.M., Wells, R.S., Solow, A.R., 2001. Short-term effects of boat traffic on bottlenose dolphins, *Tursiops truncatus*, in Sarasota Bay, Florida. Mar. Mamm. Sci. 17, 673–688.

Pedersen, M.B., Fahlman, A., Borque-Espinosa, A., Madsen, P.T., Jensen, F.H., 2020. Whistling is metabolically cheap for communicating bottlenose dolphins (*Tursiops truncatus*). J. Exp. Biol. 223, jeb.212498.

Pierce, R.H., Henry, M.S., 2008. Harmful algal toxins of the Florida red tide (*Karenia brevis*): natural chemical stressors in South Florida coastal ecosystems. Ecotoxicology 17, 623–631.

Powell, J.R., Wells, R.S., 2011. Recreational fishing depredation and associated behaviors involving common bottlenose dolphins (*Tursiops truncatus*) in Sarasota Bay, Florida. Mar. Mamm. Sci. 27, 111–129.

Read, A.J., Drinker, P., Northridge, S., 2006. Bycatch of marine mammals in U.S. and global fisheries. Conserv. Biol. 20, 163–169.

Rommel, S.A., Pabst, D.A., McLellan, W.A., 1998. Reproductive thermoregulation in marine mammals: how do male cetaceans and seals keep their testes cool without a scrotum? It turns out to be the same mechanism that keeps the fetus cool in a pregnant female. Am. Sci. 86, 440–448.

Rossman, S., Barros, N.B., Ostrom, P.H., Stricker, C.A., Hohn, A.A., Gandhi, H., Wells, R.S., 2013. Retrospective analysis of bottlenose dolphin foraging: a legacy of anthropogenic ecosystem disturbance. Mar. Mamm. Sci. 29, 705–718.

Rycyk, A.M., Tyson Moore, R.B., Wells, R.S., McHugh, K.A., Berens McCabe, E.J., Mann, D.A., 2020. Passive acoustic listening stations (PALS) show rapid onset of ecological effects of harmful algal blooms in real time. Sci. Rep. 10, 17863.

Scheidat, M., Couperus, B., Siemensma, M., 2018. Electronic Monitoring of Incidental Bycatch of Harbour Porpoise (*Phocoena phocoena*) in the Dutch Bottom Set Gillnet Fishery (September 2013 to March 2017). Wagening University and Research, Wageninge, NL.

Schwacke, L.H., Voit, E.O., Hansen, L.J., Wells, R.S., Mitchum, G.B., Hohn, A.A., Fair, P.A., 2002. Probabilistic risk assessment of reproductive effects of polychlorinated biphenyls on bottlenose dolphins (*Tursiops truncatus*) from the Southeast United States Coast. Environ. Toxicol. Chem. 21, 2752–2764.

Schwacke, L.H., Hall, A.J., Townsend, F.I., Wells, R.S., Hansen, L.J., Hohn, A.A., Bossart, G.D., Fair, P.A., Rowles, T.K., 2009. Hematologic and serum biochemical reference intervals for free-ranging common bottlenose dolphins (*Tursiops truncatus*) and variation in the distributions of clinicopathologic values related to geographic sampling site. Am. J. Vet. Res. 70, 973–985.

Schwacke, L.H., Zolman, E.S., Balmer, B.C., De Guise, S., George, R.C., Hoguet, J., Hohn, A.A., Kucklick, J.R., Lamb, S., Levin, M., Litz, J.A., McFee, W.E., Place, N.J., Townsend, F.I., Wells, R.S., Rowles, T.K., 2012. Anaemia, hypothyroidism and immune suppression associated with polychlorinated biphenyl exposure in bottlenose dolphins (*Tursiops truncatus*). Proc. R. Soc. Lond. B 279, 48–57.

Schwacke, L.H., Smith, C.R., Townsend, F.I., Wells, R.S., Hart, L.B., Balmer, B.C., Collier, T.K., De Guise, S.D., Fry, M.M., Guillette Jr., L.J., Lamb, S.V., Lane, S.M., McFee, W.E., Place, N.J., Tumlin, M.C., Ylitalo, G.M., Zolman, E.S., Rowles, T.K., 2014. Health of common bottlenose dolphins (*Tursiops truncatus*) in the Gulf of Mexico following the Deepwater Horizon oil spill. Environ. Sci. Technol. 48, 93–103.

Schwacke, L.H., Thomas, L., Wells, R.S., McFee, W.E., Hohn, A.A., Mullin, K.D., Zolman, E.S., Quigley, B.M., Rowles, T.K., Schwacke, J.H., 2017. Quantifying injury to common bottlenose dolphins from the Deepwater Horizon oil spill using an age-, sex- and class-structured population model. Endanger. Species Res. 33, 265–279.

Smith, C.R., Rowles, T.K., Hart, L.B., Townsend, F.I., Wells, R.S., Zolman, E.S., Balmer, B.C., Quigley, B., Ivancic, M., Mckercher, W., Tumlin, M.C., Mullin, K.D., Adams, J.D., Wu, Q., McFee, W., Collier, T.K., Schwacke, L.H., 2017. Slow recovery of Barataria Bay dolphin health following the Deepwater Horizon oil spill (2013–2014), with evidence of persistent lung disease and impaired stress response. Endanger. Species Res. 33, 127–142.

Smith, C.R., Rowles, T.K., Gomez, F.M., Ivančić, M., Colegrove, K.M., Takeshita, R., Townsend, F.I., Zolman, E.S., Morey, J.S., Cendejas, V., Meegan, J.M., Musser, W., Speakman, T.R., Barratclough, A., Wells, R.S., Schwacke, L.H., 2022. Poor pulmonary health in Barataria Bay dolphins in the eight years following the Deepwater Horizon oil spill. Front. Mar. Sci. 9, 975006.

Steidinger, K.A., Burklew, M., Ingle, R., 1973. The effects of *Gymnodinium breve* toxin on estuarine animals. In: Marine Pharmacognosy: Action of Marine Toxins at the Cellular Level. Academic Press, New York, pp. 179–202.

Takeshita, R., Balmer, B.C., Messina, F., Zolman, E.S., Thomas, L., Wells, R.S., Smith, C.R., Rowles, T.K., Schwacke, L.H., 2021. High site-fidelity in common bottlenose dolphins despite low salinity exposure and associated indicators of compromised health. PLoS One 16, e0258031.

Tanabe, S., 2002. Contamination and toxic effects of persistent endocrine disrupters in marine mammals and birds. Mar. Pollut. Bull. 45, 69–77.

Tolley, K.A., Read, A.J., Wells, R.S., Urian, K.W., Scott, M.D., Irvine, A.B., Hohn, A.A., 1995. Sexual dimorphism in wild bottlenose dolphins (*Tursiops truncatus*) from Sarasota, Florida. J. Mammal. 76, 1190–1198.

Tomasko, D.A., Corbett, C.A., Greening, H.S., Raulerson, G.E., 2005. Spatial and temporal variation in seagrass coverage in Southwest Florida: assessing the relative effects of anthropogenic nutrient load reductions and rainfall in four contiguous estuaries. Mar. Pollut. Bull. 50, 797–805.

Toms, C.N., Stone, T., Och, T., 2021. Skin lesion and mortality rate estimates for common bottlenose dolphins (*Tursiops truncatus*) in the Florida Panhandle following a historic flood. PLoS One 16 (10), e0257526.

Tornero, V., Sylvina, T.J., Wells, R.S., Singh, J., 2014. Eco-toxicants: a growing global threat. In: Yamagiwa, J., Karczmarski, L. (Eds.), Primates and Cetaceans: Field Research and Conservation of Complex Mammalian Societies, Primatology Monographs. Springer, Tokyo, pp. 309–332.

Twiner, M.J., Flewelling, L.J., Fire, S.E., Bowen-Stevens, S.R., Gaydos, J.K., Johnson, C.K., Landsberg, J.H., Leighfield, T.A., Mase-Guthrie, B., Schwacke, L., Van Dolah, F.M., Wang, Z., Rowles, T.K., 2012. Comparative analysis of three brevetoxin-associated bottlenose dolphin (*Tursiops truncatus*) mortality events in the Florida Panhandle region (USA). PLoS One 7, e42974.

Tyack, P.L., Thomas, L., Costa, D.P., Hall, A.J., Harris, C.M., Harwood, J., Kraus, S.D., Miller, P.J.O., Moore, M., Photopoulou, T., Pirotta, E., Rolland, R.M., Schwacke, L.H., Simmons, S.E., Southall, B.L., 2022. Managing the effects of multiple stressors on wildlife populations in their ecosystems: developing a cumulative risk approach. Proc. R. Soc. Lond. B 289, 20222058.

Van Der Hoop, J.M., Fahlman, A., Hurst, T., Rocho-Levine, J., Shorter, A.K., Petrov, V., Moore, M.J., 2014. Bottlenose dolphins modify behavior to reduce metabolic effect of tag attachment. J. Exp. Biol. 217, 4229–4236.

Van Der Hoop, J.M., Fahlman, A., Shorter, K.A., Gabaldon, J., Rocho-Levine, J., Petrov, V., Moore, M.J., 2018. Swimming energy economy in bottlenose dolphins under variable drag loading. Front. Mar. Sci. 5, 00465.

Venn-Watson, S., Colegrove, K.M., Litz, J., Kinsel, M., Terio, K., Saliki, J., Fire, S., Carmichael, R., Chevis, C., Hatchett, W., Pitchford, J., Tumlin, M., Field, C., Smith, S., Ewing, R., Fauquier, D., Lovewell, G., Whitehead, H., Rotstein, D., McFee, W., Fougeres, E., Rowles, T., 2015. Adrenal gland and lung lesions in Gulf of Mexico common bottlenose dolphins (*Tursiops truncatus*) found dead following the Deepwater Horizon oil spill. PLoS One 10, e0126538.

Vollmer, N.L., Rosel, P.E., 2013. A review of common bottlenose dolphins (*Tursiops truncatus truncatus*) in the Northern Gulf of Mexico: population biology, potential threats, and management. Southeast. Nat. 12, 1–43.

Weaver, A., 2015. Sex differences in bottlenose dolphin sightings during a long-term bridge construction project. Anim. Behav. Cogn. 2 (1), 1–13.

Wells, R.S., 1993. Why all the blubbering? Bison 7, 12–17.

Wells, R.S., 2010. Feeling the heat—potential climate change impacts on bottlenose dolphins. Whalewatcher J. Am. Cetacean Soc. 39, 12–17.

Wells, R.S., 2019. Common bottlenose dolphin foraging: behavioral solutions that incorporate habitat features and social associates. In: Würsig, B. (Ed.), Ethology and Behavioral Ecology of Odontocetes, Ethology and Behavioral Ecology of Marine Mammals. Springer Nature, Switzerland, pp. 331–344.

Wells, R.S., 2020. The Sarasota Dolphin Research Program in 2020: celebrating 50 years of research, conservation, and education. Aquat. Mamm. 25, 502–503.

Wells, R.S., Scott, M.D., 1997. Seasonal incidence of boat strikes on bottlenose dolphins near Sarasota, Florida. Mar. Mamm. Sci. 13, 475–480.

Wells, R.S., Scott, M.D., 2018. Bottlenose dolphin: common bottlenose dolphin (*Tursiops truncatus*). In: Würsig, B., Thewissen, J.G.M., Kovacs, K. (Eds.), Encyclopedia of Marine Mammals, third ed. Academic Press/Elsevier, San Diego, CA, pp. 118–125.

Wells, R.S., Hansen, L.J., Baldridge, A., Dohl, T.P., Kelly, D.L., Defran, R.H., 1990. Northward extension of the range of bottlenose dolphins along the California coast. In: Leatherwood, S., Reeves, R.R. (Eds.), The Bottlenose Dolphin. Academic Press, San Diego.

Wells, R.S., Hofmann, S., Moors, T.L., 1998. Entanglement and mortality of bottlenose dolphins, *Tursiops truncatus*, in recreational fising gear in Florida. Fish. Bull. 96, 647–650.

Wells, R.S., Rhinehart, H.L., Hansen, L.J., Sweeney, J.C., Townsend, F.I., Stone, R., Casper, D.R., Scott, M.D., Hohn, A.A., Rowles, T.K., 2004. Bottlenose dolphins as marine ecosystem sentinels: developing a health monitoring system. EcoHealth 1, 246–254.

Wells, R.S., Tornero, V., Borrell, A., Aguilar, A., Rowles, T.K., Rhinehart, H.L., Hofmann, S., Jarman, W.M., Hohn, A.A., Sweeney, J.C., 2005. Integrating life-history and reproductive success data to examine potential relationships with organochlorine compounds for bottlenose dolphins (*Tursiops truncatus*) in Sarasota Bay, Florida. Sci. Total Environ. 349, 106–119.

Wells, R.S., Allen, J.B., Hofmann, S., Bassos-Hull, K., Fauquier, D.A., Barros, N.B., Sutton, G., Socha, V., Scott, M., 2008. Consequences of injuries on survival and reproduction of common bottlenose dolphins (*Tursiops truncatus*) along the west coast of Florida. Mar. Mamm. Sci. 24, 774–794.

Wells, R.S., Fauquier, D.A., Gulland, F.M.D., Townsend, F.I., DiGiovanni Jr., R., 2013. Evaluating post-intervention survival of free-ranging odontocete cetaceans. Mar. Mamm. Sci. 29, E463–E483.

Wells, R.S., Schwacke, L.H., Rowles, T.K., Balmer, B.C., Zolman, E., Speakman, T., Townsend, F.I., Tumlin, M.C., Barleycorn, A., Wilkinson, K.A., 2017. Ranging patterns of common bottlenose dolphins *Tursiops truncatus* in Barataria Bay, Louisiana, following the Deepwater Horizon oil spill. Endanger. Species Res. 33, 159–180.

Wells, R.S., Natoli, A., Braulik, G., 2019. *Tursiops truncatus*. In: The IUCN Red List of Threatened Species. International Union for Conservation of Nature and Natural Resources. e.T22563A50377908.

Williams, T.M., Haun, J., Davis, R.W., Fuiman, L.A., Kohin, S., 2001. A killer appetite: metabolic consequences of carnivory in marine mammals. Comp. Biochem. Physiol. A: Physiol. 129, 785–796.

Woshner, V., Knott, K., Wells, R., Willetto, C., Swor, R., O'Hara, T., 2008. Mercury and selenium in blood and epidermis of bottlenose dolphins (*Tursiops truncatus*) from Sarasota Bay, FL: interaction and relevance to life history and hematologic parameters. EcoHealth 5, 360–370.

Yeates, L.C., Houser, D.S., 2008. Thermal tolerance in bottlenose dolphins (*Tursiops truncatus*). J. Exp. Biol. 211, 3249–3257.

Yordy, J.E., Pabst, D.A., McLellan, W.A., Wells, R.S., Rowles, T.K., Kucklick, J.R., 2010a. Tissue-specific distribution and whole-body burden estimates of persistent organic pollutants in the bottlenose dolphin (*Tursiops truncatus*). Environ. Toxicol. Chem. 29, 1263–1273.

Yordy, J.E., Wells, R.S., Balmer, B.C., Schwacke, L.H., Rowles, T.K., Kucklick, J.R., 2010b. Life history as a source of variation for persistent organic pollutant (POP) patterns in a community of common bottlenose dolphins (*Tursiops truncatus*) resident to Sarasota Bay, FL. Sci. Total Environ. 408, 2163–2172.

Index

Note: Page numbers followed by *f* indicate figures, and *t* indicate tables.